Proceedings of the 2nd Annual Gas Processing Symposium

11–14 January, 2010, Doha, Qatar

Advances in Gas Processing

Volume 2: Proceedings of the 2nd Annual Gas Processing Symposium
(Farid Benyahia, Fadwa T. Eljack, Editors)

Proceedings of the 2nd Annual Gas Processing Symposium

11–14 January, 2010, Doha, Qatar

Edited by

Farid Benyahia
Department of Chemical Engineering, Qatar University, Doha, Qatar

Fadwa T. Eljack
Department of Chemical Engineering, Qatar University, Doha, Qatar

ELSEVIER

Amsterdam • Boston • Heidelberg • London • New York • Oxford
Paris • San Diego • San Francisco • Singapore • Sydney • Tokyo

Elsevier
Radarweg 29, PO Box 211, 1000 AE Amsterdam, The Netherlands
The Boulevard, Langford Lane, Kidlington, Oxford OX5 1GB, UK

First edition 2010

British Library Cataloguing in Publication Data
A catalogue record for this book is available from the British Library

Library of Congress Cataloging-in-Publication Data
A catalog record for this book is available from the Library of Congress

ISBN: 978-0-444-53588-7
ISSN: 1876-0147

Transferred to Digital Printing in 2010

List of Contents

Preface ix

International Technical Committee xi

Environmental Sustainability

Biogasification of Waste Monoethanolamine Generated in Post Combustion CO_2 Capture
D. Botheju, Y. Li, J. Hovland, T. Risberg, H. A. Haugen, C. Dinamarca and R. Bakke 1

Application of Post Combustion CO_2 Capture to Natural Gas Liquefaction Plants
Babatunde Oyenekan, Stanley Huang and Ian Lindsay 11

Activated DEEA Process for CO_2 Capture
Prashanti B. Konduru, Prakash D. Vaidya, Eugeny Y. Kenig 21

CO_2 capture into aqueous solutions of the mixed solvent Cesar 1
Peter Brúder, Andreas Grimstvedt, Thor Mejdell, Eirik F. da Silva, Hallvard F. Svendsen 31

Q-Chem Steam Boilers NOx Emissions Reduction
Hesham ELZeini 41

The experience in using LDAR for monitoring fugitive emissions of Volatile Organic Compounds with special reference to RasGas' experience
Chris Horne, Roger Medalla, Julius Bacani 51

Development of Al Dahkhirah Sewage Treatment Works to Reuse Treated Water for Construction Works, Qatar
Khaled El Emadi , Ghazy Abdel Kerim , Moaaz Hashim 59

Bio-ethanol from Municipal Solid Waste (MSW): the Environmental Impact Assessment
Majeda Khraisheh, Aiduan Li 69

Monoethanolamine biodegradation processes
Deshai Botheju, Jon Hovland, Hans Aksel Haugen, and Rune Bakke 77

Environmental Sustainability and the Role of LNG in a Carbon Constrained World
Susan L. Sakmar 87

Industrial Water Discharge and Biocide Fate Simulations with Nonlinear Conversion
Johannes Lawen, Yu Huaming, Patrick Linke, Ahmed Abdel-Wahab 99

Sustainable Water Management: A Systems Integration Approach
S. Atilhan, P. Linke, A. Abdel-Wahab, A. Bin Mahfouz, and M. El-Halwagi 107

Overview of Conversion of Greenhouse Gas Carbon dioxide to Hydrocarbons
S. M. Javaid Zaidi 115

Sulphur Sustainable Applications: Initial Field Monitoring and Performance of Shell Thiopave Trial Road in Qatar
Marwa Al-Ansary, Eyad Masad, and David Strickland 121

Utilization of Byproduct Sulfur for Chromium Waste Treatment
Ahmed Abdel-Wahab, Ali Dashti, Ali Anany, Khaled Elsaid 131

Natural Gas Processing Technologies

Removal of Acid gases from Natural Gas Streams by Membrane Technology
S. M. Javaid Zaidi 139

Catalytic Dry Reforming of Methane Using Ni/MgO-ZrO_2 Catalyst
Teddy G. Monroy, Leonila C. Abella, Susan M. Gallardo, Hirofumi Hinode 145

Smart Leak Detection and Repair at Q-Chem
Ahmad M Homssi 153

Gas processing and Integrated Environmental Management
Per Gerhard Grini, Jostein Pettersen, Henrik Solgaard Andersen 163

Dehydration of Acid Gas Prior to Injection
Eugene Grynia , John J. Carroll , Peter J. Griffin 177

Adsorption Thermodynamics of Natural Gas Storage onto Pitch-Based Activated Carbons
Kazi A. Rahman, Wai S. Loh, Anutosh Chakraborty, Bidyut B. Saha, and Kim C. Ng 187

Opportunities for Selective Control of Fischer-Tropsch Synthesis Hydrocarbons Product Distribution
Nimir O. Elbashir 197

Infrared (IR) Thermography for Condition Monitoring at Q-Chem
Salahuddin Ghori 209

Industrial Operation of HySWEET®, a New Hybrid Solvent for Improved Mercaptan Removal
Renaud Cadours, Virenkumar Shah, Claire Weiss, Damien Roquet and François Lallemand ... 221

FLEXSORB® SE A Proven Reliable Acid Gas Enrichment Solvent
L. E. Parks, D. Perry , R. Fedich 229

Energy Efficiency in Operations

Computational Study of a Lifted Turbulent Jet Flame in a Cross-flow: Flame Length and Emissions
M.S. Lawal, M. Fairweather, D.B. Ingham, L. Ma, M. Pourkashanian, and A. Williams 237

Towards Novel Hybrid Biomass and Coal Processes for Satisfying Transportation Fuel Demands
Richard C. Baliban, Josephine A. Elia, & Christodoulos A. Floudas 247

The Impact of PHEV Adoption on Natural Gas Demand in Electricity Generation
Shisheng Huang, Bri-Mathias S. Hodge, Joseph F. Pekny, Gintaras V. Reklaitis 257

Design & Safety

Design of Sustainable Processes: Systematic Generation & Evaluation of Alternatives
Ana Carvalho, Rafiqul Gani , Henrique A. Matos 267

Steady State Simulation for Optimal Design and Operation of a GTL Process
Mehdi Panahi, Sigurd Skogestad, Ramprasad Yelchuru 275

Techno-Economic Analysis of Gas-To-Liquid process
Buping Bao and Mahmoud M. El-Halwagi, Nimir O. Elbashir 287

The Role of Molecular Thermodynamics and Simulation in Natural Gas Sustainable Processes
Ioannis G. Economou, and Zoi A. Makrodimitri 299

Efficient Design, Operating and Control Strategies for LNG Plants
Francois Chantant, Matthew Swatton 311

Thermodynamic Analysis on Post Combustion CO_2 Capture of Natural Gas Fired Power Plant
Zeinab Amrollahi, Ivar S. Ertesvåg, Olav Bolland 321

Self-Optimizing and Control Structure Design for a CO_2 Capturing Plant
Mehdi Panahi, Mehdi Karimi, Sigurd Skogestad, Magne Hillestad, Hallvard F. Svendsen 331

Pilot Plant Study of 3-(methylamino)Propylamine Sarcosine for Post-combustion CO_2 Capture
Ugochukwu E. Aronu, Hallvard F. Svendsen, Karl Anders Hoff, Hanna Knuutila 339

Qatar, LNG, Spill Experiments and Process Safety
R. Qi, M. AlBaker, O. Basha, R. Hassiba, S. Mannan, T. Olewski, and S. Waldram 349

Optimization of Flare Header Platform Design in a Liquefied Natural Gas Plant
E. Mahdi, K. Nasser and M. Gharbia 359

A Method to Design an Advanced Gas-to-Liquid Technology Reactor for Fischer-Tropsch Synthesis Nimir O. Elbashir, Fadwa T. Eljack 369

Recent Developments in Identification from Step Response
Salim Ahmed 379

Multi-Objective Optimization for Operational Excellence
Hasan Y. Alhammadi 389

On the PρT, PηT and Phase Envelope Behavior Characterization of Qatari Type Natural Gas Mixtures
M. Atilhan, Aparicio-Martinez, K.R. Hall, M. El-Halwagi, G.A. Iglesias-Silva 399

Operational Excellence

Studying Influence of Changing Fuel on the Operation and Total Annual Cost of the Total Site Using R-curve
Shervin Karimkashi, Majid Amidpour 409

Chemicals in Gas Processing (CHIGP): An industrial project for the thermodynamics of complex petroleum fluids
G. M. Kontogeorgis, I. Tsivintzelis, M. L. Michelsen, M. P. Breil and E. H. Stenby 419

Hybrid Models for Monitoring & Optimization of Hydrocarbon Separation Equipment
Vladimir Mahalec, Asaad Hashim, Yoel Sanchez 427

Optimal Unloading Procedure for a Mixed Operation of Above-ground and In-ground LNG Storage Tank using Dynamic Simulation
Chul-Jin Lee, Youngsub Lim, Chansaem Park, Sangho Lee, Chonghun Han 437

Inhibition of Gas Hydrate Formation by Low-dosage, Environmentally Benign Inhibitors
Lars Jensen, Hans Ramløv, Kaj Thomsen, Nicolas von Solms 445

Strategies for Mitigating Impacts on LNG Plant Capacity by Variations in Ambient Temperature
Stanley Huang, Meredith Chapeaux and Bobby Martinez 455

Environmental Sustainability: Industry – Academia Perspective
Farid Benyahia and Fadwa T. Eljack 465

Index **473**

Preface

Natural gas is becoming the fuel of choice for power generation and a multitude of thermo-mechanical applications in the oil and gas industries. It is also becoming a key feedstock in a wide range of petrochemical processes. This trend is driven by environmental, economic and supply considerations with a balance clearly tilting in favor of natural gas.

Despite the recent global economic hitch affecting mainly the finance sector, the oil and gas industry is expected to continue its growth globally. The expansion in LNG capacity beyond 2009 and 2010 coupled with planned and on-stream GTL plants poses real technological and environmental challenges. These important developments coupled with a global concern on green house gas emissions that were linked to global warming provided a fresh impetus to engage in new and more focused research activities aimed at mitigating or resolving challenges that ensue.

For these very reasons, the second annual gas processing symposium covered themes closely related to sustainability in gas processing. The main themes covered at the symposium constitute the parallel sessions and chapters of this book:

1. Natural Gas Processing Technologies
2. Environmental Sustainability
3. Energy Efficiency in Operations
4. Design & Safety
5. Operational Excellence

Given the importance of environmental sustainability in the gas processing sector and beyond, two tracks were scheduled in the technical program, making this chapter a particularly significant one in this book. The rest of the chapters provide an excellent insight into the state of the art in the areas listed above.

The organizing committee of the second annual gas processing symposium wishes to express its deepest gratitude to a number of people who made this event possible and enjoyable. Grateful acknowledgement is given to His Highness Sheikh Tamim bin Hamad Al-Thani, the Heir Apparent of the State of Qatar under whose patronage this symposium was held. The support of her Excellency, Professor Sheikha Al-Misnad, President of Qatar University, is gratefully acknowledged. The strong support from the College of Engineering and the Gas processing Center staff at Qatar University is also gratefully acknowledged.

The organizing committee also wishes to extend their gratitude to the symposium sponsors and co-sponsors: Qatar Petroleum, Dolphin Energy, Shell, ExxonMobil, and co-sponsors: Qatari Ministry of Environment, American Institute of Chemical Engineers (AIChE), Gas Processing Association (GPA) – GCC Chapter, and the European Federation of Chemical Engineers (EFCE).

The organizing committee would like to thank members of the international technical committee for their expert review and advice.

Finally, the editors would like to thank the authors who shared their knowledge and expertise in the various papers constituting the core content of this book.

Farid Benyahia, Qatar University, Qatar
Fadwa T. Eljack, Qatar University, Qatar
Second Annual Gas Processing Symposium book editors

International Technical Committee Members (Reviewers)

Name	Affiliation
Abdelmajeed Hamouda	Qatar University
Ber Gerhard Grini	Statoil Hydro
Dinesh Seth	Qatar University
Elsadiq Mahdi	Qatar University
Farayi Musharafati	Qatar University
Hassan Al-Hammadi	University of Bahrain
Hazim Qiblawi	Qatar University
Ioannis Economou	National Center for Scientific Research "Demokritos"
James Rigby	ExxonMobil, Qatar
Jong Woon	Qatar University
Kenneth Hall	Texas A&M University
Mahmoud El-Halwagi	Texas A&M University
Mert Atilhan	Qatar University
Nimir El-Bashir	Texas A&M University, Qatar
Prof. Faisal Khan	Qatar University
Prof.Farid Benyahia	Qatar University
Ramazan Kahraman	Qatar University
Rex Reklaitis	Purdue University
Saud Ghani	Qatar University
Tarek Elmekkawy	Qatar University
Vladimir Mahalec	McMaster University

Proceedings of the 2nd Annual Gas Processing Symposium
Farid Benyahia and Fadwa T. Eljack (Editors)

Biogasification of Waste Monoethanolamine Generated in Post Combustion CO_2 Capture

Deshai Botheju[a], Yuan Li[a], Jon Hovland[b], Trond Risberg[b], Hans Aksel Haugen[b], Carlos Dinamarca[a] and Rune Bakke[a]

[a]*Telemark University College, Faculty of Technology, Porsgrunn, Norway*
[b]*Telemark Technological Research and Development Centre (Tel-Tek), Porsgrunn, Norway*

Abstract

Monoethanolamine (MEA) contaminated liquid waste generated in post combustion CO_2 capture poses a disposal issue. Anaerobic biogasification potential of waste MEA is evaluated in a series of laboratory experiments conducted under different operating conditions. Provision of a limited amount of oxygen may enhance the methane potential of waste MEA. Co-digestion of MEA together with a readily biodegradable substrate is found to be a solution to overcome ammonia and pH inhibition caused by low C/N ratio and high alkalinity. Substrate inhibition caused by waste MEA can be overcome by acclimatization. MEA degradation pathways via acetic and ammonium, with and without oxygen, are included in an anaerobic digestion model (ADM 1-Ox). Model simulations predicted the experimentally observed digestion characteristics with a fair accuracy.

Keywords: ADM 1, amine wastes, biogasification, biodegradation, monoethanolamine

1. Introduction

Aerobic and anoxic biodegradability of monoethanolamine has previously been demonstrated while anaerobic biodegradation has only been suggested before, but not demonstrated. The scope of this study is to investigate the suitable operating conditions for converting waste MEA into a CH_4 rich biogas through anaerobic biogasification. A batch test assay was used to study the effects of using different amounts of oxygen on the biogas generation potential of waste MEA. Effects of supplying a limited amount of oxygen in anaerobic digestion of organic wastes has been discussed elsewhere (Botheju *et al.,* 2009). A continuous bioreactor was operated to study the feasibility of adopting a co-substrate feed approach for waste amine treatment. A series of aerobic biodegradation tests was also carried out initially in order to recognize an operable range of feed MEA concentration with minimal biomass inhibition. Another aim of this article is to simulate and validate a model developed to predict the biogasification potential of waste MEA under strict anaerobic and micro-aerobic conditions.

2. Experimental Methodology

2.1. Feed substrates

The amine waste used for the batch tests was collected while emptying the Aker Kvaerner pilot facility at Kårstø, Norway. This plant has been used to capture CO_2 (0.2 ton CO_2/h) from natural gas combustion. The short listed waste composition is shown in

Table 1. MEA constitutes a major portion of the waste and the ammonia level indicates the extent of amine degradation.

Table 1: Characteristics of the amine wastes used in the batch tests

Component	Concentration	Component	Concentration
Monoethanolamine	210 g/l	COD	232 g/l
N-Kjeldahl	40 g/l	pH	10.9
Nitrate (NO_3^-)	186 mg/l	Phosphorus	1280 μg/l
Ammonium (NH_4^+)	14.3 g/l	Sulphur	3.5 mg/l
Sulfate (SO_4^{--})	< 144 mg/l		

The feed used for the continuous bioreactor was a mixture of pure MEA (~30 % COD) and apple juice (drinking grade, 100 %) constituting a total COD load of about 13.7 g/L.

2.2. Aerobic degradation Tests

Two series of aerobic biodegradation batch tests were carried out at 35 °C temperature, in order to explore the biodegradation potential of waste MEA at different initial concentrations, while assessing the possibility of microbial inhibition at these concentrations.

2.3. Micro-aerobic batch tests

This experimental series was conducted to test the anaerobic biodegradability of the amine waste under different initial oxygen loading conditions. Miniature anaerobic reactors of 60 ml total volume were used, operated for 522 hrs at 35 °C. Reactor liquid volume of 20 ml was used and the rest was a variable head space volume. Different initial air head spaces were used to introduce different initial oxygen loads. Two amine waste concentrations were tested; 125 mg /L and 500 mg/L (as MEA equivalent).

2.4. Continuous flow experiment

A continuously fed bioreactor of 0.5 L effective volume was operated for 240 days at 35 °C. During the first 100 days, the reactor was fed a diluted pure MEA solution (analytical grade). Unstable reactor performance was observed due to extreme pH conditions together with NH_4^+ and VFA accumulation. A co-substrate was then introduced to stabilise the process; pure MEA mixed with apple juice (drinking grade) at an approximate COD ratio of 1:4 (MEA:apple juice). A vitamin and a mineral solution was also fed together with a pH buffer. The reactor hydraulic retention time was reduced by steps from 40 days on day 101 to 20 days on day 217.

2.5. Analytical methods

Gas generation, gas composition, soluble and total COD (chemical oxygen demand), pH, VFA (volatile fatty acids - acetic, propionic, butyric, iso-butyric, valeric and iso-valeric acids), ethanol, NH_4^+-N, MEA, TSS/VSS (total/volatile suspended solids), TS/VS (total/volatile solids) and alkalinity were determined by standard methods [3].

3. Model simulation

As described in a separate paper (Botheju *et al.*, 2010b), simplified MEA biodegradation mechanisms are incorporated into a previously developed oxygen included ADM 1-Ox model (Botheju *et al.*, 2009), which is based on the generally accepted Anaerobic Digestion Model no. 1 structure (ADM1- Batstone *et al.*, 2002). The model simulations are carried out in AQUASIM 2.1 (Reichert, 1998). ADM1 kinetic and stoichiometric constants are used as suggested in Batstone *et al.* (2002). No parameter optimization is attempted as the simulations are done to compare the dynamics of the experimental conditions.

3.1. Simulation of Micro-aerobic Batch tests

Micro-aerobic condition is simulated by introducing initial air head spaces above the liquid reactor zone. A gas transfer diffusive link controls the exchange of gaseous species between the liquid and the gas phase. The same gas exchange coefficient (K_La) is used for all five gaseous species (CH_4, CO_2, O_2, N_2, H_2), while their corresponding non dimensional Henry's Law coefficients (K_H) (Sander, 1999) are used as the conversion factors (Eq. 1).

$$\rho_{g1} = k_L a(S_{liq,g1} - K_{H,g1} p_{g1}) \qquad (1)$$

Three sets of simulations, for air head spaces ranging 0 – 37 ml (equivalent to 0 – 0.39 ml O_2/ml liquid volume), were carried out, *i.e:* gas generation of inoculum alone (gas generation due to the degradation of composite particulate matter (Xc) present in the inoculum), and for the 125 and 500 mg/L initial MEA concentrations.

3.2. Continuous Bioreactor

The bioreactor is approximated by a CSTR configuration with a mixture of MEA and sugar as feed. The apple juice COD is (quite accurately) simulated as sugar. Simulated hydraulic and mass loadings were as in the experiments (Table 2).

Table 2: Different phases of reactor operation, imitated in simulations

Duration (days)	HRT (d)	Flow rate (ml/d)	Feed load (kg COD/m^3)	
			MEA	sugar
100-127 (28)	40	12.5	3.43	10.28
128-144 (17)	30	16.7	3.48	8.12
145-151 (7)	25	20	3.48	8.12
152-184 (33)	20	25	3.48	8.12
185-217 (32)	20	25	4	9.32

4. Results and Discussion

4.1. Aerobic Batch Tests

MEA containing reactors consume more oxygen than the control, demonstrating MEA degradation (Fig 1). 500 mg/L MEA initial concentration show less ultimate BOD and slower reaction compared to 125 mg/L MEA (Fig. 1(a)), explained by MEA induced inhibition. Such inhibition was avoided in the second test by using the culture from the first test after longer adaptation and pH adjusted to neutral (Fig. 1 (b)). MEA toxicity and inhibition has also been noticed by others (Lai and Shieh, 1996).

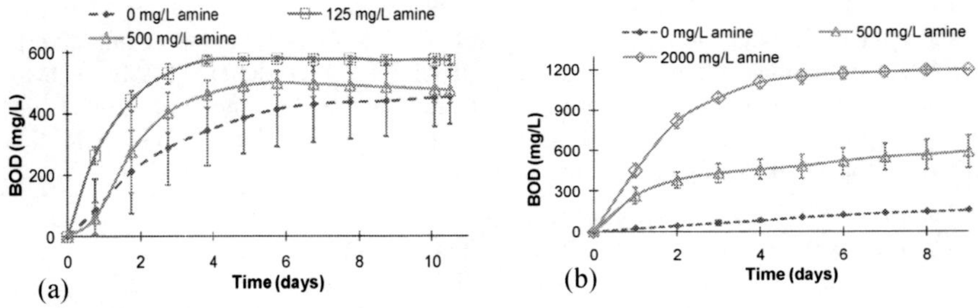

Figure 1: BOD profiles for the aerobic degradation test series; (a) series 1-non-adopted, non diluted inoculum (b) series 2- adopted and diluted (4x) inoculum

First order rate constants for aerobic MEA biodegradation are calculated according to Botheju *et al.* (2010b) using the experimental BOD data (Table 3).

Table 3: Degradation rate constant (K_1) values (d^{-1}) for different MEA concentrations and test series

MEA con. (mg/L)	*Rate constant* (d^{-1})
125	1.08 (by series 1)
500	0.66 (average from series 1 and 2)
2000	0.47 (by series 2)

4.2. Micro-aerobic Batch Tests

4.2.1. Cumulative biogas generation

Biogasification of waste MEA under strict anaerobic and micro-aerobic conditions is observed (Fig. 2). Measured and simulated biogas generation are similar for both waste MEA input levels tested (125 and 500 mg /L of equivalent MEA concentrations). The observed over prediction by the simulations can be due to the un-optimized kinetic constants and dissimilar initial conditions used, and partial inhibition caused by some toxic components in the feed (*e.g.* some MEA degradation compounds or additives, including metallic elements).

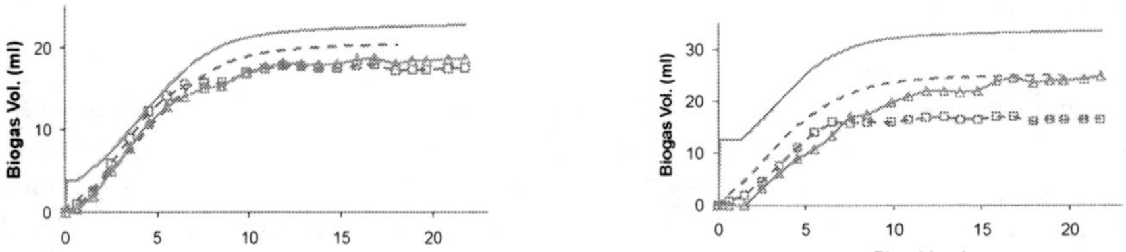

(a) : 125 mg/L MEA input and 0.03 ml O_2/ml liq. micro-aeration

(b): 500 mg/L MEA input and 0.12 ml O_2/ml liq. micro-aeration

Figure 2: Measured (*with marker*) and simulated (*line only*) cumulative biogas generation in micro-aerobic batch tests for anaerobic (square; *dash line*) and micro-aerobic (triangle; *continuous line*) conditions.

4.2.2. Oxygen Effects

Measurements and simulations both suggest that the methane generation potential is positively influenced by the micro-aerobic load within the range of 0 - 0.39 ml O_2/ ml liq. (Fig. 3). This effect is observed also for the control reactors containing only the inoculum (data not shown). Such can be explained by enhanced hydrolysis (Johansen and Bakke, 2006). Simulated CH_4 generation vs. oxygen load has less agreement with experimental observations at higher oxygen load.

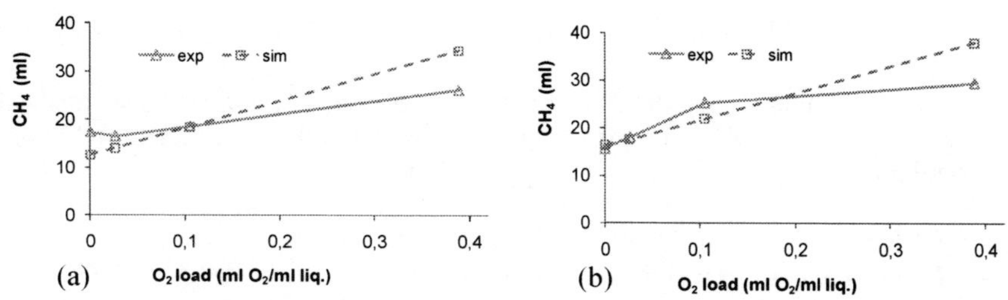

Figure 3: Simulated and experimentally observed methane generation potential of 2 different initial MEA concentrations of (a) 125 mg/L and (b) 500 mg/L, at varying micro-aeration levels.

Simulation shows that the head space supplied oxygen was rather quickly transferred and used up in the liquid phase (Fig.4 (a)) eliminating oxygen inhibition of methanogenic biomass (Fig. 5). The remaining inhibition is due to the presence of ammonia (Fig. 5). An inhibition factor of about 0.6 is predicted for an approximate NH_4^+-N concentration of 560 mg/L and a pH of 7.4, at 35 °C. The simulation indicates a positive effect of oxygen on acetic acid generation, as the peak level of acetic acid is obtained when oxygen depletion is complete (Fig. 4(b)). This acetate is converted to methane subsequently.

The net methane generation from MEA (Fig. 6) is small but a trend of increased yield with increased O_2 loads in the range tested is observed. This effect is more pronounced for the higher MEA concentration.

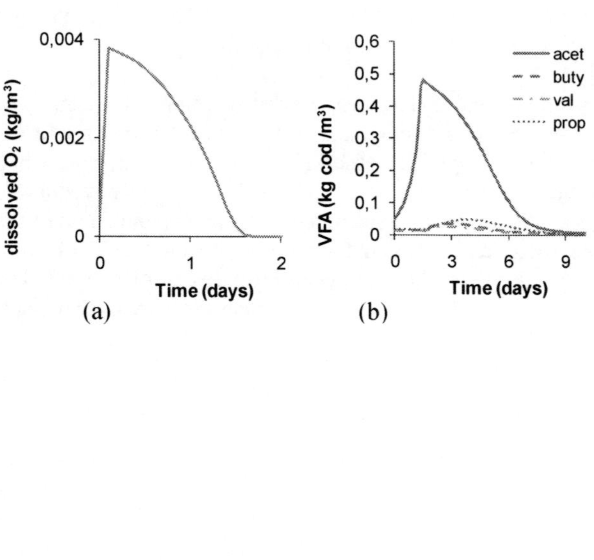

Figure 4: Simulated dissolved oxygen profile (a) and VFA profiles (b) for the initial period of 500 mg/L MEA fed reactor at the oxygen load of 0.39 ml O_2/ml

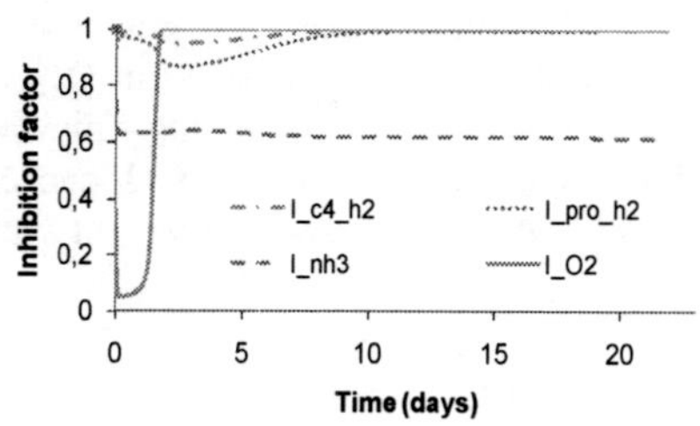

Figure 5: Simulated inhibition functions for 500 mg/L MEA, 0.39 ml O_2 /ml condition

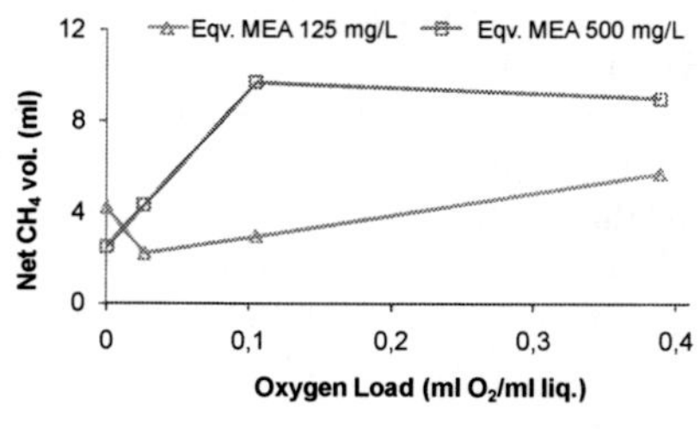

Figure 6: Experimental net methane generation by MEA feeds of 125 and 500 mg/L

4.3. Anaerobic Continuous Flow Bioreactor

4.3.1. Simulated vs. Experimental observations

Experimental and model simulated biogas generation of the continuous bioreactor is presented in Fig. 7. The model predicts the reactor behaviour well during a 40 day time period of stable operation (day 10 –day 50). The simulations are much less satisfactory during the first ten days and during the second half of the test. The initial deviation can be explained as a numerical overshoot problem. The latter deviations appear to be due to unintentional reactor disturbances that the model is not capable of simulating. Steady state gas generation simulations (days 10 to 50) closely resemble the actual data even if they are different from the theoretical maximum gas generation estimated based on the input COD load only. Theoretical maximum figures do not account for the restrictions imposed by reactor operating conditions like process dynamics, biomass activity, temperature etc. Both the simulation and the experiment suggest that it is quite possible to operate the bioreactor with this feed mixture close to the theoretical maximum gas generation if process disturbances can be avoided. A detailed mass balance analysis carried out on 97th day discloses that 86 % of COD intake is converted to biogas. Total COD recovery was 93 %. The remaining 7 % COD might have been accumulated in the reactor, probably embedded in biomass. A chemical analysis carried out for MEA in the effluent stream further confirmed the complete MEA degradation in the bioreactor. High variation in experimental data observed after day 60 (Fig. 7) is considered to be due to a biomass flush out from the reactor caused by the decreasing hydraulic retention time (some unintended changes occurred in the feed source also).

The measured biogas compositions and pH values are also comparable with the simulation (Fig. 8 and Fig.9). Measured and simulated NH_4^+ concentrations are in the same range, but measurements show a slight decrease in NH_4^+ content with time while the simulation suggests a marginal increase (Fig. 10). High NH_4^+ content present in the reactor prior to the 116 days operational period considered here can explain some of the observed NH_4^+ reduction. Measured and simulated total volatile fatty acid (TVFA) concentration in the reactor (Fig. 11) differ, but the general trend is similar. Observed TVFA has a higher maximum and stays high much longer than the simulated values.

This behavioural deviation can be attributed to differences in biomass characteristics and some inhibition not accounted for in the model.

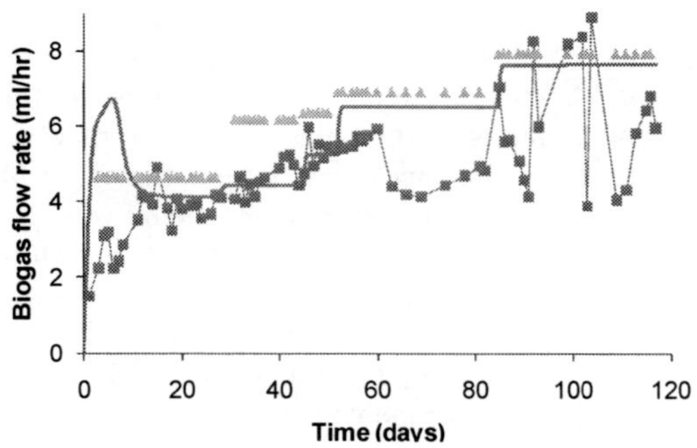

Fig. 7: Biogas flow rates observed *(line with marker)*, simulation predicted *(line without marker)*, and maximum expected *(marker only)* for the continuous flow reactor during the 116 days fed with MEA and apple juice

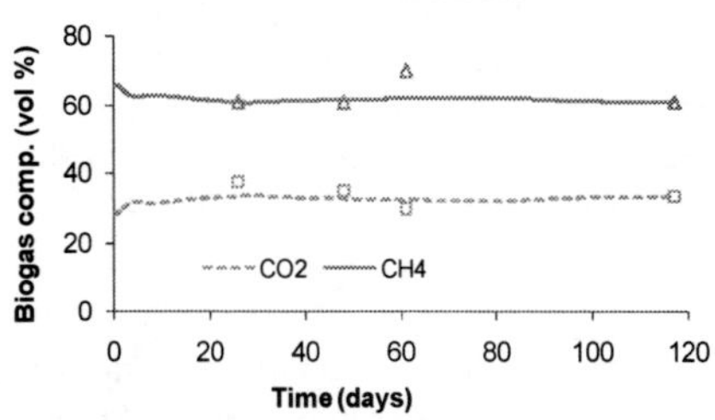

Figure 8: Measured (*markers*) and simulated (*lines*) biogas composition for the continuous flow reactor

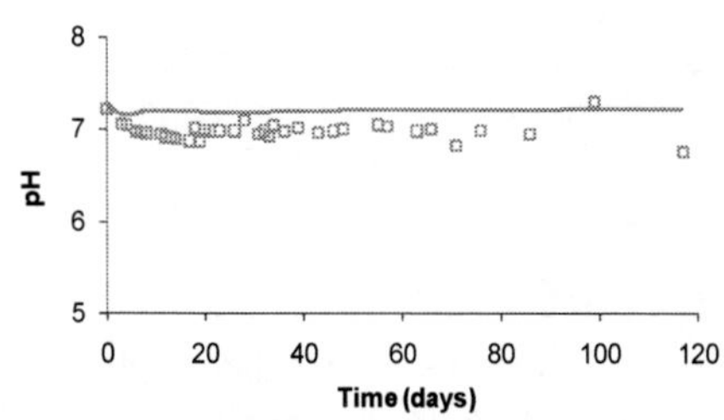

Figure 9: Observed (*marker*) and the predicted (*line*) pH for the continuous flow reactor

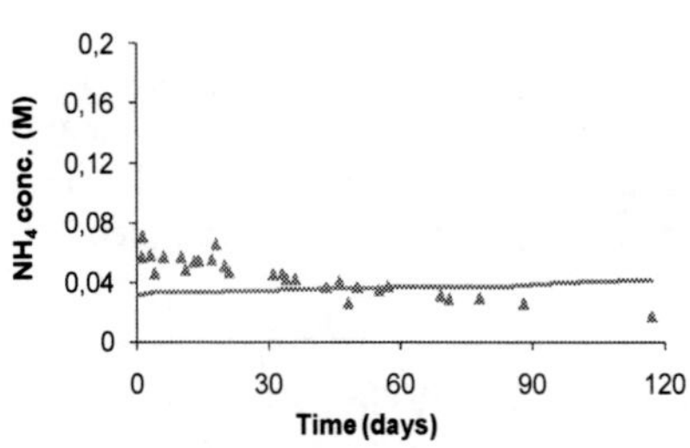

Figure 10: Measured *(marker)* and the predicted *(line)* NH_4^+-N for the continuous flow reactor

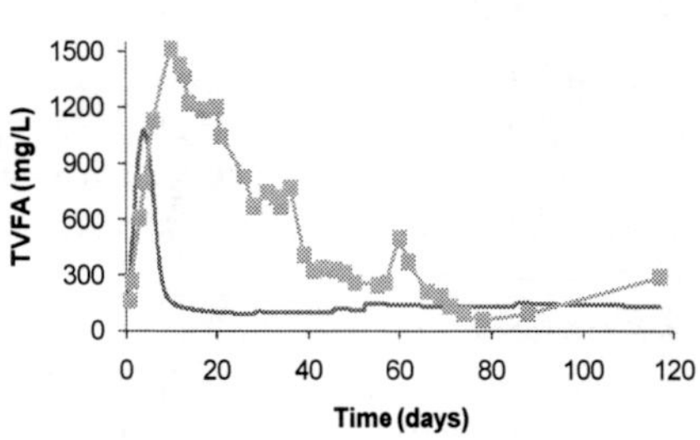

Figure 11: Measured (*with marker*) and the simulated (*line only*) TVFA for the continuous flow reactor

4.3.2. Co-substrate / co-digestion approach

Experimental operation of the MEA/apple juice fed continuous bioreactor confirmed the effectiveness of adopting a co-digestion approach for the successful methanogenic treatment of waste MEA. Isik and Sponza (2008) pointed out the usefulness of adding an easily degradable secondary substrate like glucose, starch or acetic acid in the anaerobic treatment of recalcitrant wastes like textile dyes. Dyes are of interest as they often contain amine- or azo- groups. Georgiou and Aivasidis (2006) also used acetic acid as a co-substrate and a pH regulator when experimenting on the anaerobic de-colorization potential of textile effluents. Acidic substrates like acetic acid or fruit wastes can make preferred co-substrates in amine waste treatment due to the necessity of neutralizing high pH conditions. This co-metabolic approach is essential in anaerobic treatment of amine wastes due to the very low C/N ratios typically found in amine wastes. Commonly used solvent amines of MEA, piperazine AMP, MDEA and DIPA have C/N ratios of 1.71, 1.71, 2.42, 4.28 and 5.13, respectively. The recommended range of C/N ratios for optimum digester performance is 20 - 30 (Verma, 2002). Low C/N ratios can lead to process inhibition by high ammonia and high pH. Thus, co-digesting amine wastes together with other industrial or municipal wastes in a common digester facility is recommended. Domestic wastewater is a good alternative due to its general availability and high buffer capacity, while certain industrial wastes have the advantage of low pH and ammonia content. Special attention must be paid to the possibility of contamination (e.g. heavy metal), which may hinder the possibility of using anaerobic digestate as a soil amendment /fertilizer (Botheju *et al.*, 2007).

5. Conclusions

MEA degradation to methane by anaerobic digestion with no or small quantities of free oxygen is adequately predicted by model simulations. Micro-aeration has a favourable effect on the methane generation potential. A co-substrate approach is required to obtain stable biogasification of waste MEA in a continuous flow reactor, in order to increase the low C/N ratios typical for amine wastes and hence to avoid the ammonia and pH inhibition. Culture acclimatization to MEA is also advantageous.

Acknowledgements

The authors would like to thank Aker Clean Carbon AS, BioTek AS, Gassnova SF, NOAH AS, Norcem AS, Skagerak Kraft AS, StatoilHydro ASA and Østfold Energi AS for financing the study. The technical support provided by the Laboratory Engineer at Telemark University College, Ms. Hildegunn H. Haugen deserves a special mention.

References

Batstone, D.J., J. Keller, I. Angelidaki, S. Kalyuzhnyi, S.G. Pavlostathis, A. Rozzi, W. Sanders, H. Siegrist and V. Vavilin, 2002, Anaerobic digestion Model No.1, IWA publishing, London.

Botheju, D., B. Lie and R. Bakke, 2009, Modeling free oxygen effects in anaerobic digestion, Proceedings of the MATHMOD 2009 - 6th Vienna International Conference on Mathematical Modeling, Vienna.

Botheju, D., J. Hovland, H.A. Haugen and R. Bakke, 2010b, Monoethanolamine biodegradation processes, Proceedings of the 2nd Annual Gas Processing Symposium, Doha, Qatar. ; Submitted manuscript

Botheju, D., Ø. Svalheim, B. Rydtun, J. Johansen, T.K. Haraldsen, and R. Bakke, 2007, Fertilizer production by digestate nitrification, Proceedings of the 12th European Biosolids and Organic

Resources Conference, Manchester, UK.
Georgiou, D. and A. Aivasidis, 2006, Decolouration of textile wastewater by means of a fluidized bed loop reactor and immobilized anaerobic bacteria, J. Hazardous Materials, B135, pp. 372-377.
Isik, M. and D.T. Sponza, 2008, Anaerobic/aerobic treatment of a simulated textile wastewater, Separation and Purification Technology, vol. 60, pp. 64-72.
Johansen, J.E. and R. Bakke, 2006, Enhancing hydrolysis with micro-aeration, Water science and Technology, 2006; 53(8), pp. 43-50.
Lai, B. and W.K. Shieh, 1996, Technical note; Batch monoethanolamine degradation via nitrate respiration, Water Research, vol. 30, No. 10, pp. 2530-2534.
Reichert P., 1998, AQUASIM 2.0-User manual; Computer program for the identification and simulation of Aquatic Systems, Swiss Federal Institute for Environmental Science and Technology (EAWAG).
Sander, S., 1999, Compilation of Henry's Law constants for inorganic and organic species of potential importance in environmental chemistry, Available online on http://www.mpch-mainz.mpg.de/~sander/res/henry.html) ; accessed on 04.08.2009.
Verma, S., 2002, Anaerobic digestion of biodegradable organics in municipal solid wastes, Dept. of Earth & Environmental Eng., Columbia University, Master of Science, Thesis.

Proceedings of the 2nd Annual Gas Processing Symposium
Farid Benyahia and Fadwa T. Eljack (Editors)

Application of Post Combustion CO_2 Capture to Natural Gas Liquefaction Plants

Babatunde Oyenekan, Stanley Huang and Ian Lindsay
Chevron Energy Technology Company, 1400 Smith Street, Houston, TX 77002, U.S.A.

Abstract

Realistic estimates for meeting future energy demand suggest that fossil fuels will play a significant role in the global energy mix at least through 2050. Increased fossil fuel use means increased greenhouse gas emissions. On the other hand, there is increasing concern over greenhouse gases (GHG) and their impact on the global climate. The increased public concern and governments' regulatory action around climate change pose a challenge to the energy industry regarding how to meet the growing world energy demand while reducing GHG emissions. Proposals to curb or reduce industrial GHG emissions include measures such as energy conservation, increased energy efficiency, the use of renewable energy, fuel switching (e.g. coal to natural gas), and carbon capture and storage (CCS).

The liquefied natural gas (LNG) industry is monitoring developments in CCS technologies and is willing to adopt them to reduce its carbon footprint. This paper gives a brief overview of current carbon capture technologies. The focus is on the pros and cons of each technology and its applicability in the LNG industry. Estimates of the GHG emissions reduction and cost of applying post combustion CO_2 capture in a 10 million tonne per annum (10 mtpa) generic LNG plant are presented.

Keywords: liquefied natural gas (LNG), greenhouse gas emissions (GHG), carbon capture and storage (CCS), post combustion capture (PCC), amine solvent scrubbing

1. Introduction

Estimates of future global energy demand suggest that the world's demand for energy in 2030 will be 40% to 70% higher than it was in 2005 (NPC 2007, EIA 2007) due to population growth and increased standard of living. A significant portion of the global energy demand in 2050 is expected to be met by natural gas. However, the global distribution of natural gas reserves does not coincide with major consuming centers. Liquefying natural gas and transporting it by ship is the preferred method for moving large volumes of natural gas over long distances from remote locations without a gas market to major consuming centers.

As greenhouse gas (GHG) regulatory action accelerates around the world, it is prudent for the energy industry to quantify the impact of carbon constraints on both existing and future resource development. This paper focuses on greenhouse gas mitigation at natural gas liquefaction plants. It briefly describes the liquefied natural gas (LNG) value chain and identifies the major sources of greenhouse gas emissions along the chain. The GHG emissions reduction, additional capital and operating costs of applying post combustion CO_2 capture to the gas turbine emissions of a generic 10 million tonne per annum (mtpa) LNG plant and compressing all the CO_2 removed from the acid gas removal unit (AGRU) to 104 barg and 50°C in preparation for transportation and storage, are presented.

2. LNG Chain Emission Sources

An LNG chain is created to transport natural gas from a remote location without a market for the gas to major consuming centers. Oftentimes, such transportation refers to transoceanic shipment for which pipeline connection are either technically challenging or economically unjustifiable. The main components of a typical LNG chain and the greenhouse gas emissions along the chain are described below.

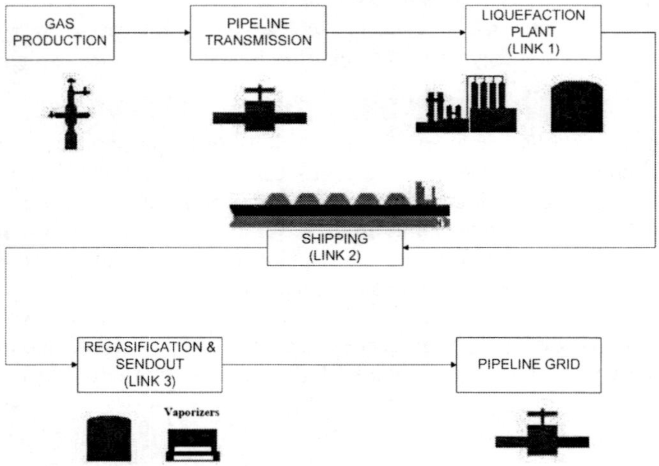

Figure 1: Functional units in an LNG liquefaction plant

2.1. Components of a typical LNG project

A baseload LNG project consists of the following components: gas production, pipeline transmission, liquefaction, shipping, regasification, and send-out to local pipeline grid (Huang et al., 2007). The three components of liquefaction, shipping, and regasification are referred to as the links in the LNG gas monetization chain, because the full functioning of the chain is tightly linked to individual components in a sequential order.

2.2. Carbon Dioxide Emissions from the LNG Chain

The discussion in this section is limited to the LNG chain, which starts from the inlet boundary to an LNG plant and ends at the outlet boundary of a regasification terminal.

2.2.1. Emissions from the liquefaction plant

The liquefaction plant is the most capital-intensive link in the entire LNG chain. Figure 2 shows the functional units of the liquefaction plant. The core of the LNG plant is the cryogenic section, which consists of the refrigeration systems and various chilling units. Other supporting units in an LNG plant are the gas sweetening, dehydration, natural gas liquids (NGL) removal, and N_2 rejection units. A power generation plant supplies the electrical power needs of the entire facility.

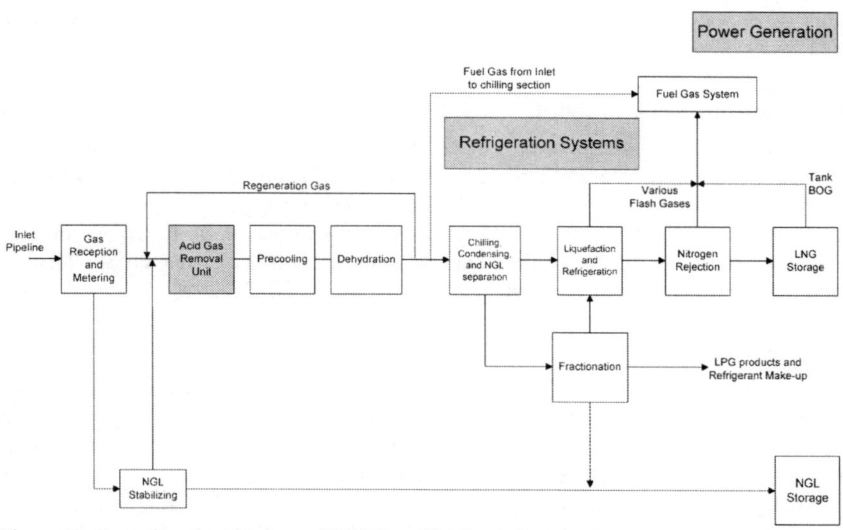

Figure 2: Functional units in an LNG liquefaction plant

The three main sources of routine CO_2 emissions in an LNG liquefaction plant, the power generation facility, the refrigeration system, and the acid gas removal unit (AGRU) are highlighted in Figure 2. Gas turbines, because of their efficiency and versatility, are used in the power generation and refrigeration systems. The CO_2 emissions from these turbines are a direct consequence of fuel gas combustion. In contrast, the emission from the acid gas removal unit is proportional to the amount of CO_2 removed from the feed natural gas prior to liquefaction. Due to its relatively high demand for horsepower in refrigeration loops, the liquefaction plant has received intense scrutiny recently (Pace, 2009). As a point of reference, the liquefaction plant is estimated to account for 60-80% of CO_2 emissions in the LNG chain (Jaramillo, et al., 2007; Ditzel and Hartman, 2008).

2.2.2. Emissions from Shipping

There are two main emission sources from LNG shipping operations: Boil-Off-Gas (BOG) from LNG tanks and fuel used to power the LNG ships. BOG, mostly methane and inert (such as N_2), can be used as ship fuel or can be re-liquefied. In newer ship fleet, re-liquefaction packages have been included.

2.2.3. Emissions at the regasification terminal

There are two emission sources at the regasification terminal: combustion heat for LNG vaporization and power generation for plant operations. In recent constructions, heat from seawater or ambient air has replaced fuel as heat sources. This has dramatically reduced the CO_2 emissions at the regasification terminal. In addition, many receiving terminals are wired to get power from local grids. This further reduces the emission directly attributable to the regasification plant though it does not reduce emissions overall.

3. CO_2 Capture Technologies

CO_2 capture refers to the separation of CO_2 from a flue gas or process stream followed by subsequent compression and drying in preparation for transportation and geological storage. There are three main CO_2 capture technology approaches: post combustion capture, pre-combustion capture and oxy-firing. This section briefly describes the three approaches, their areas of application, advantages, and disadvantages. Detailed descriptions of these technologies can be found elsewhere (IPCC 2005).

3.1. Post Combustion Capture

This involves the separation of CO_2 from the flue gas generated from the combustion of a fossil fuel in air. The captured CO_2 is then compressed in preparation for transportation and geological storage. Post combustion capture is applicable to new and retrofit boilers, heaters and turbines. The main advantages of post combustion capture are: existing combustion technology can be used, it is suitable for existing plant retrofits, and has been demonstrated in natural gas-fired and coal-fired power plants on a small scale. The current technology for post combustion capture is amine solvent scrubbing. The practical challenges with this technology include its very large equipment (and plot space) requirements, solvent oxidative and thermal degradation concerns, and health and environment issues associated with solvent losses from the absorber.

3.2. Pre- Combustion Capture

This involves the capture of CO_2 from syngas (mainly H_2 and CO) generated when a fossil fuel reacts with air/oxygen. Syngas is sent to water shift reactors where steam reacts with the CO to form CO_2 and more H_2. The product from the shift reactor is cooled and sent to a CO_2 removal unit. The product streams after CO_2 capture are a hydrogen rich stream and a CO_2 stream. The hydrogen can be used in a wide range of applications (e.g. turbine firing). Pre-combustion capture is applicable to new plants and existing heaters and boilers. The advantage of pre-combustion capture is that the CO_2 is at a higher concentration and pressure than post combustion capture resulting in smaller equipment (and therefore footprint) for CO_2 removal. The practical challenges with this technology include the additional syngas generation infrastructure required and the high investment costs.

3.3. Oxy-firing

This involves the capture of CO_2 from flue gas generated because of the combustion of a fossil fuel in almost pure oxygen. Recycle of exhaust gas is used to moderate the combustor temperature. The flue gas stream consists mainly of CO_2 and H_2O vapor. The water vapor is condensed while the CO_2 product stream is sent to a CO_2 purification unit. Oxy-firing is most applicable to new facilities e.g. heaters, boilers and possibly turbines. The advantages of oxy-firing are: bulk nitrogen is absent so the flue gas volume and the capture equipment is significantly reduced, the high CO_2 concentration (75-90%) after dehydration results in low capture costs, co-sequestration of CO_2, SO_x, NOx, O_2 and traces of water may be possible. The practical challenges to be addressed with oxy-firing are: reducing the O_2 production cost and energy consumption, re-design of boilers for oxy-firing mode, and cryogenic distillation (a costly operation) may be required to purify CO_2 in flue gas.

4. Application of Post Combustion Capture to Liquefaction Plants

This section describes the application of post combustion capture to liquefaction plants. In the short term, if carbon capture is to be applied as retrofits to existing liquefaction plants, the most likely capture technique to be applied will be post combustion capture. This is because the flue gases from the gas turbines in liquefaction plants are dilute in CO_2 and at low pressure. Post combustion capture with solvent scrubbing is the only commercially demonstrated carbon capture method, albeit on small scales.

4.1.1. Study Basis

This study is based on a generic LNG plant with a nominal capacity of 10 million tonne per annum (mtpa).

4.1.2. How CO_2 emissions from the facility were addressed

The CO_2 emissions from the LNG plant were addressed as follows:

- CO_2 removed from the acid gas removal unit (AGRU) was compressed to 104 barg and 50°C. Since the AGRU is an integral part of LNG production, the additional costs included in this study are those for dehydration and compression of the CO_2 exiting the AGRU stripper.
- The CO_2 containing flue gases from the gas turbines were sent to the CO_2 capture plant where 85% of the CO_2 was captured and compressed to 104 barg and 50°C.
- The CO_2 from the flares and fugitives were left unabated.

4.1.3. CO_2 Capture Plant Description

The capture plant was designed to be stand-alone. The post combustion CO_2 capture plant was an amine-based process with a circulating solution of monoethanolamine with corrosion inhibitors capable of handling oxygen containing streams. A detailed description of the process can be found elsewhere (Heimel and Lowe, 2009, Lindsay et al., 2009). Since the capture plant was designed as a retrofit with no integration between the LNG plant and the capture plant, all utilities (power, steam, and cooling water) were independently generated.

4.1.4. CO_2 Capture Plant and Compression Plant Cost Estimate Basis

The following assumptions were made in estimating the additional costs of the CO_2 capture plant and the compression of the CO_2 from the AGRU:

- For CAPEX calculations, the location was the US Gulf Coast (USGC) with costs based on 4Q 2007.
- For OPEX calculations, a low energy cost location with a fuel gas cost of \$1/MMBtu was assumed.

The CAPEX consists of the major equipment cost, bulks and labor, indirect field costs and EPC contingency. Owner's costs are not included in the estimate. The OPEX consists of fuel gas cost, maintenance costs, operator costs, and cost of chemicals.

5. Results

5.1. Uncontrolled Emissions

The uncontrolled CO_2 emissions of the generic 10 mtpa LNG plant consist of CO_2 removed from natural gas feed, the CO_2 in the exhaust of the gas turbines, and the CO_2 from the flares and fugitives. Figure shows the annual uncontrolled CO_2 emissions range from 3 to 7 $MMTCO_2e$. The annual CO_2 emissions from the gas turbine exhaust, flares and fugitives are ~3 $MMTCO_2e$. Figure 3 shows a breakdown of the CO2 emissions in a generic 10 mtpa LNG plant.

Figure 3: Sources of CO_2 in a 10 mtpa uncontrolled LNG plant. Left: no CO_2 in feed gas, Right: 12% CO_2 in feed

5.2. LNG Plant Emissions with CO_2 Capture

A summary of the total CO_2 generated from the 10 mtpa LNG plant and the CO_2 recovered when the feed natural gas contains 0-12% CO_2 is shown in Table 1.

Table 1: Summary of CO_2 emissions from a 10 mtpa LNG plant with CO_2 capture

CO_2 in natural gas feed on a dry basis (%)	0	12
CO_2 Source	MM tonnes/yr	
Total CO_2 generated without CO_2 capture	3.0	7.0
CO_2 generated in LNG Plant + capture plant + CO_2 from AGRU CO_2 compressor drivers	3.5	7.9
AGRU CO_2 compressed	0	4.0
CO_2 fed to capture plant	2.7	2.7
CO_2 captured	2.5	2.5
Controlled emissions (CO_2 emitted after CO_2 capture of gas turbines' exhaust and compression of AGRU CO_2)	1.0	1.4
CO_2 avoided (Uncontrolled emissions – Controlled emissions)	2.0	5.6

5.3. Cost Estimation Results

The additional costs of CO_2 capture and compression of the CO_2 from the gas turbines' exhaust and compression of all the CO_2 from the stripper overhead in the AGRU to 50°C and 104 barg are presented in this section. The costs of transportation and storage are not included. These costs are highly site specific and thus they were excluded from this study.

5.3.1. Economic Evaluation

For a given carbon capture technology, the CO_2 capture costs are usually expressed in two forms - the "CO_2 capture cost" and "CO_2 avoided cost." In this work, the capital cost for the CO_2 capture and compression plants were annualized using a 10% discount factor and a 20-year plant life. The annualized cost consists of the CAPEX and OPEX. The summary of the CO_2 recovery costs is shown in Table 2.

Table 2: Summary of CO_2 recovery costs (4Q 2007 US$)

CO_2 in natural gas feed on a dry basis (%)	0	12
Additional CAPEX of capture of gas turbine CO_2 and compression of CO_2 from AGRU (MM$)	858	996
OPEX of capture of gas turbine CO_2 and compression of CO_2 from AGRU (MM$)	23	37
CO_2 capture cost (US$/tonne CO_2) = Annualized cost/Annual CO_2 captured	50	24
CO_2 avoided cost (US$/tonne CO_2) = Annualized cost/Annual CO_2 avoided	62	28

* Based on a 10% discount factor and a 20-year life.

** These costs were derived from studies with a -15%/+30% accuracy conducted by a third party engineering contractor.

***The costs presented do not include transportation and storage of the captured CO_2

6. Conclusions

The conclusions from the greenhouse gas quantification and mitigation for a generic 10 mtpa liquefaction plant are:

- The two main sources of CO_2 emissions are the CO_2 removed from the acid gas removal unit (AGRU) and the CO_2 from the gas turbine exhaust.
- The gas turbines from a generic 10mtpa LNG plant emit ~ 3 MMTCO$_2$e/yr. The amount of CO_2 removed from the natural gas feed is an additional emission source and depends on the CO_2 content of the feed gas. This means that an uncontrolled LNG plant is a significant CO_2 emissions source even without adding the CO_2 removed from the natural gas feed.
- By applying amine-based post combustion capture to the gas turbine exhaust and compressing all the CO_2 removed in the AGRU, up to 5.6 MMTCO$_2$e/yr can be avoided from a gas stream containing 12% CO_2 in the feed gas. This represents a net CO_2 avoided rate of 80%.
- For natural gas feed with 0 to 12% CO_2, economic analysis showed that the CO_2 avoided costs vary by about a factor of 2.2 from US$62/tonne to US$28/tonne. These exclude costs of CO_2 transportation and storage.
- This avoided cost for the 12% CO_2 case is lower than that of the 0% CO_2 case because the CO_2 in the natural gas feed is removed by the AGRU as part of the base LNG plant. To prepare this CO_2 for transportation and storage only requires a small investment (CAPEX and OPEX) for compression and dehydration.

References

Ditzel, K. H. and N.A. Hartman, 2008, Rebuttal to the Carnegie Mellon University Researchers' Report - Comparative Life-Cycle Air Emissions of Coal, Domestic Natural Gas, LNG, and SNG for Electricity Generation, Report by CRA International for The Gasification Technologies Council.

Energy Information Administration, http://www.eia.doe.gov/ data downloaded in September, 2007

Heimel, S. and C. Lowe, 2009, Technology Comparison of CO2 Capture for a Gas-to-Liquids Plant, Energy Procedia, Volume 1, Issue 1, 4039-4046.

Huang, S., C.-H. Chiu, D. Elliot, 2007, LNG: Basics of Liquefied Natural Gas", book published by The University of Texas, Continuing Education, Petroleum Extension Services (PETEX), Austin, Texas.

Intergovernmental Panel on Climate Change, 2005, Carbon Dioxide Capture and Storage, Eds B. Metz et al. Cambridge University press, U.K.

Jaramillo, P., W.M. Griffin, and H.S. Matthews, 2007, Comparative Life-Cycle Air Emissions of Coal, Domestic Natural Gas, LNG, and SNG for Electricity Generation, Environ. Sci. Technol. 41, 6290-6296.

Lindsay, I. et al., 2009, Designing a Climate Friendly Hydrogen Plant, Energy Procedia, Volume 1, Issue 1, 4095-4102.

National Petroleum Council, 2007, Hard Truths: Facing the Hard Truths about Energy

Pace (Pace Global Energy Services, LLC), 2009, Life Cycle Assessment of GHG Emissions from LNG and Coal Fired Generation Scenarios: Assumptions and Results, Report prepared for Center for Liquefied Natural Gas (CLNG).

Proceedings of the 2nd Annual Gas Processing Symposium
Farid Benyahia and Fadwa T. Eljack (Editors)

Activated DEEA Process for CO_2 Capture

Prashanti B. Konduru,[a] Prakash D. Vaidya,[a] Eugeny Y. Kenig[b]

[a]*Department of Chemical Engineering, Institute of Chemical Technology, Mumbai-400019, India*

[b]*Faculty of Mechanical Engineering, Chair of Fluid Process Engineering, University of Paderborn, D-33098, Paderborn, Germany*

Abstract

N,N-Diethylethanolamine (DEEA), which can be prepared from renewable resources, represents a candidate alkanolamine having good potential for the bulk removal of CO_2 from gaseous streams. A blend comprising DEEA, piperazine (PZ), and H_2O, which combines the high CO_2 loading capacity of DEEA with the high reactivity of PZ, is attractive for the enhancement of CO_2 capture. In the present work, the chemical kinetics of the CO_2 reaction with this novel amine blend was studied using a stirred cell reactor. The investigated reaction was considered as a reaction between CO_2 and DEEA in parallel with the reaction of CO_2 with PZ. The CO_2 – DEEA – PZ system belongs to the fast pseudo-first-order reaction systems. The observed reaction rate constants for this system were measured. The physical properties and the liquid-side mass transfer coefficient were evaluated, too. The second order rate constant for the CO_2 reaction with PZ was determined from the absorption rate measurements in the activated DEEA solutions, and its value at 303 K was found to be 24450 $m^3/(kmol\ s)$.

Keywords: Carbon Dioxide, N,N-Diethylethanolamine, Piperazine, Kinetics.

1. Introduction

CO_2 represents an undesirable component in a variety of industrial gases (e.g., natural/synthesis gas, flue gas, coke oven gas and refinery off-gas) and it has to be removed down to very low concentrations to meet the required specifications. Although a number of different CO_2 separation technologies are available by now, chemisorption with alkanolamines represents the most feasible option, and it is commonly used in the gas processing industry. Industrially important alkanolamines for CO_2 removal are the primary amine, monoethanolamine (MEA), the secondary amine, diethanolamine (DEA), and the tertiary amine, methyldiethanolamine (MDEA) (Kohl and Nielsen, 1997).

N,N-Diethylethanolamine (DEEA) represents a further candidate alkanolamine for CO_2 removal from gaseous streams. This tertiary amine comprises two ethyl groups replacing the hydrogen atoms of the amino group in MEA. Diethylamine (obtained from ethanol) and ethylene oxide (obtained from the oxidation of ethylene which, in turn, can be prepared by dehydration of ethanol) react to form DEEA. Ethanol, which is the major raw material for the manufacture of this alkanolamine, can be prepared from agricultural products and/or residues. Thus, DEEA is a potentially attractive absorbent for sour gas purification, as it can be prepared from renewable resources. Since there is no hydrogen atom attached to the nitrogen atom, as in case of MEA and DEA, the carbamation reaction cannot take place. Instead, DEEA promotes the CO_2 hydrolysis reaction, which leads to bicarbonate formation. The reaction heat released in bicarbonate formation is lower than that of carbamate formation, thus resulting in lower solvent regeneration costs. Furthermore, DEEA has a CO_2 loading capacity higher than MEA (1 mole of CO_2/ mole of amine).

In our recent paper, we studied the kinetics of the CO_2 reaction with DEEA in aqueous solutions (Vaidya and Kenig, 2008). Due to its tertiary amine characteristics, DEEA has low reactivity with respect to CO_2. We estimated the second order reaction rate constant at 303 K and studied the acceleration of the CO_2 reaction with DEEA by piperazine (PZ) which is applied as an efficient activator within the activated MDEA technology used by BASF SE. We discovered that PZ also enhances CO_2 absorption rates in aqueous DEEA solutions. In this work, the kinetics of the CO_2 reaction with activated DEEA solutions was investigated. Besides, the reaction rate constant for the reaction between CO_2 and PZ was determined from the absorption rate measurements of CO_2 in the blend comprising DEEA, PZ and H_2O.

2. Experimental

A batch-operated stirred cell reactor with a plane, horizontal gas-liquid interface was used to study reaction kinetics (Vaidya and Mahajani, 2005). In each experiment, a certain amount of CO_2 was rapidly introduced into the reactor and the pressure decrease caused by absorption and reaction was recorded. From this record, the absorption rate at a particular temperature and liquid-phase composition was determined. This measurement method based on the fall-in-pressure technique is very simple and straightforward; besides, no analysis of the liquid phase is required for the evaluation of the kinetic parameters. Since CO_2 reacts with amines, its solubility and diffusivity in amine solutions cannot be directly determined. N_2O is a non-reacting gas, whose molecular structure, molecular weight and electronic configuration are similar to those of CO_2. It is used to represent CO_2 behavior in reacting systems. In the present work, the N_2O solubility in the activated DEEA solutions was measured using the same gas-liquid contactor. Some experiments were repeated and the error in the rate measurements was found to be less than 3%.

3. Reaction Mechanism

Donaldson and Nguyen (1980) proposed that tertiary alkanolamines cannot react directly with CO_2. Such amines have a base catalytic effect on the hydration of CO_2. When CO_2 is absorbed into an aqueous DEEA solution, base catalysis could be represented as follows (here DEEA is denoted as $R_2R'N$, where $R = -C_2H_5$ and $R' = -CH_2CH_2OH$):

$$R_2R'N \ + \ H_2O \ + \ CO_2 \ \xrightarrow{k'} \ R_2R'N^+H \ + \ HCO_3^- \tag{1}$$

where k' denotes the reaction rate constant in eq. 1. The base catalysis reaction could also be explained by a zwitterion mechanism earlier proposed by Yu et al. (1985):

$$R_2R'N \ + \ CO_2 \ \longleftrightarrow \ R_2R'NCOO^- \tag{2}$$

$$R_2R'NCOO^- \ + \ H_2O \ \longrightarrow \ R_2R'N^+H \ + \ HCO_3^- \tag{3}$$

Eq. 2 represents a reaction of DEEA with CO_2 to form an unstable complex. Eq. 3 describes the homogeneous hydrolysis reaction. Water reacts with the zwitterion-type complex resulting in a bicarbonate formation. The following reactions may also take place simultaneously in an aqueous DEEA solution:

$$R_2R'N \ + \ H_2O \ \longleftrightarrow \ R_2R'N^+H \ + \ OH^- \tag{4}$$

$$H_2O \ \longleftrightarrow \ H^+ \ + \ OH^- \tag{5}$$

$$CO_2 \ + \ OH^- \ \xleftrightarrow{k_{OH^-}} \ HCO_3^- \tag{6}$$

$$CO_2 \ + \ H_2O \ \xleftrightarrow{k_{H_2O}} \ HCO_3^- \ + \ H^+ \tag{7}$$

Here, k_{OH^-} and k_{H_2O} denote the forward reaction rate constants in eqs. 6 and 7. The amine dissociation reaction (eq. 4) is almost instantaneous. The reaction between CO_2 and OH^- (eq. 6) is also fast. The total rate of all CO_2 reactions in an aqueous solution containing DEEA is denoted by the sum of the reaction rates given by eqs. 1, 6 and 7:

$$r_{DEEA} = \left\{ k_{H_2O}(H_2O) + k_{OH^-}(OH^-) + k'(R_2R'N) \right\}(CO_2) \ = \ k_{obs}(CO_2) \tag{8}$$

where k_{obs} denotes the observed reaction rate constant and (CO_2), (H_2O), (OH^-) and $(R_2R'N)$ denote the concentrations of CO_2, H_2O, hydroxyl ions and DEEA, respectively.

PZ reacts with CO_2 rapidly, which results in carbamate formation. Besides, the dissociation reaction for PZ may occur (Liu et al., 1999; Zhang et al., 2001). The reactions that may take place in presence of PZ (here denoted by $R''(NH)_2$) can be explained by the homogeneous activation mechanism (Xu et al., 1992):

$$R''(NH)_2 \ + \ 2\,CO_2 \ \longleftrightarrow \ R''(NHCOO)_2 \tag{9}$$

$$R''(NHCOO)_2 \ + \ 2\,H_2O \ \longrightarrow \ R''(NH_2)_2^+ \ + \ 2\,HCO_3^- \tag{10}$$

Reaction described by eq. 9 occurs simultaneously with that described by eq. 2, and, CO_2 could be transferred by $R''(NHCOO)_2$ to DEEA. This is reflected by the following reaction:

$$R''(NHCOO)_2 \ + \ 2\,R_2R'N \ \longleftrightarrow \ R''(NH)_2 \ + \ 2\,R_2R'NCOO^- \tag{11}$$

Thus, PZ contributes to absorption of CO_2 and enhances the overall rate of absorption.

4. Results and Discussion

4.1. Estimation of physical properties

The physical properties (e.g., density, viscosity, N_2O diffusivity and solubility) of the blend comprising DEEA, PZ and H_2O were measured, and some values at 303 K are represented in Table 1. The diffusion coefficient of N_2O in the aqueous amine solution was found by using the modified Stokes-Einstein relation (Versteeg and van Swaaij, 1988). The plots of N_2O diffusivity (D) and solubility (H) vs. initial PZ concentration at 303 K are shown in Fig. 1. The values of D_{CO_2} and H_{CO_2} were found using N_2O analogy, which states that the ratios of the CO_2/N_2O diffusivities and CO_2/N_2O solubilities in water are equal to those in the amine solutions (Versteeg and van Swaaij, 1988).

4.2. Estimation of k_L

In order to determine the liquid-side mass transfer coefficient k_L, the physical absorption of CO_2 in water at 303 K was studied and the fall-in-pressure due to absorption was recorded. Using a technique earlier adopted by Littel et al. (1991), the value of k_L was found to be 0.001 cm/s and it is in line with those typical for stirred cell reactors.

4.3. Study of chemical kinetics

The overall reaction of CO_2 with aqueous mixtures of DEEA and PZ can be regarded as a reaction between CO_2 and DEEA in parallel with the reaction of CO_2 with free PZ. This suggestion is based on the fact that free PZ transfers CO_2 to DEEA thereby regenerating itself (cf. eq. 11). If PZ could not transfer CO_2 to DEEA, the free PZ concentration during reaction would quickly decrease and the absorption rate enhancement would not be observed.

To study the reaction kinetics, it is essential that the system belongs to the fast reaction regime, without depletion of the amine at the gas-liquid interface (Danckwerts, 1970). The necessary conditions for the fast pseudo-first-order reaction regime are:

$$10<\sqrt{M}<(E_i-1) \tag{12}$$

where $\sqrt{M}$ is given by

$$\sqrt{M} = \frac{\sqrt{D_{CO_2}\, k_{obs}}}{k_L} \tag{13}$$

and E_i is the enhancement factor for an instantaneous reaction:

$$E_i = 1 + \left[\frac{(\text{amine})_0}{z(CO_2)} \frac{D_{amine}}{D_{CO_2}} \right] \quad (14)$$

Table 1: Physical properties of aqueous mixtures of DEEA and PZ at 303 K

$(DEEA)_0 + (PZ)_0$ ($kmol/m^3$)	Density (kg/m^3)	Viscosity (mPa s)	N_2O solubility x 10^4 ($kmol/(m^3\ kPa)$)	N_2O diffusivity x 10^{10} (m^2/s)
2.0 + 0.1	989	1.98	1.880	10.1
2.0 + 0.2	989	1.98	1.690	10.1
2.0 + 0.3	988	2.16	0.988	9.3
2.0 + 0.4	990	2.15	0.905	9.3
2.0 + 0.5	990	2.15	0.890	9.3

where z, $(\text{amine})_0$ and D_{amine} denote the stoichiometric coefficient, initial amine concentration ($kmol/m^3$) and diffusivity of amine in liquid phase (m^2/s), respectively. In the fast pseudo-first-order reaction regime, the rate of absorption is independent of the liquid-side mass transfer coefficient k_L and hence it should not depend on the agitation speed. We studied this effect experimentally and found practically no change in the absorption rate, while varying the stirring speed in the range 60-120 rpm at 308 K. Hence, it can be concluded that the CO_2 – DEEA – PZ system belongs to the fast pseudo-first-order reaction regime systems. All further experiments were conducted at a speed of 100 rpm. Since pure CO_2 was used in all experiments, the resistance to mass transfer was entirely in the liquid phase. When CO_2 concentration in the bulk liquid is negligible, it can be shown, based on the two-film theory of mass transfer (Danckwerts, 1970), that the following relation holds:

$$\frac{P_{CO_2}}{R_{CO_2}\, a} = \frac{1}{k_L\, a\, H_{CO_2}\, E} \quad (15)$$

where P_{CO_2} and a denote the CO_2 partial pressure in the bulk gas phase (kPa) and gas-liquid interfacial area (m^2/m^3), respectively. For the present case, eq. 15 is rearranged to

$$\frac{P_{CO_2}}{R_{CO_2}} = \frac{1}{H_{CO_2}\sqrt{\left\{D_{CO_2}\left[k_1 (PZ)_0 + k'(DEEA)_0\right]\right\}}} \quad (16)$$

where k_1 denotes the CO_2-PZ reaction rate constant (m^3/(kmol s), while $(DEEA)_o$ and $(PZ)_0$ are the initial DEEA and PZ concentrations (kmol/m^3) respectively. Eq. 16 is further transformed to

$$\frac{P_{CO_2}}{R_{CO_2}}=\frac{1}{H_{CO_2}\sqrt{\left\{D_{CO_2}k_2\left(DEEA\right)_0\right\}}} \tag{17}$$

with k_2 (m^3/(kmol s)) given by

$$k_2=k'+\frac{k_1\left(PZ\right)_0}{\left(DEEA\right)_0} \tag{18}$$

The rate of absorption in the formulated amine solution can be expressed as

$$R_{CO_2}=P_{CO_2}H_{CO_2}\sqrt{\left\{D_{CO_2}k_2\left(DEEA\right)_0\right\}} \tag{19}$$

The CO_2 absorption rates in aqueous solutions containing PZ and DEEA at different temperatures and values of the parameter are represented in Table 2. Using the values of D_{CO_2} and H_{CO_2} at 303 K (see Table 3), the value of k_2 was estimated. Thereafter, the conditions given by eq. 12 were checked and found to be satisfied. The values of $\sqrt{M}$ and E_i, represented in Table 3, provide a further check of our preliminary assumption of a fast pseudo-first order reaction system. The variation in CO_2 absorption rates at 303 K with the initial PZ concentration is shown in Fig. 2.

To quantify the acceleration of the CO_2 reaction with DEEA by PZ, the enhancement factor due to chemical reaction E was chosen (see Fig. 3). We measured the CO_2 absorption rate in aqueous DEEA solution. The enhancement factor for the CO_2 – DEEA system at 303 K and $(DEEA)_0$ = 2 kmol/m^3 was estimated and its value (53) was less than those presented in Fig. 3. The observed reaction rate constants for this system were measured (see Fig. 4). The value of k_{obs} rises with PZ concentration. From eq. 18 and from the knowledge on k' at 303 K (viz., 173 m^3/(kmol s), cf. (Vaidya and Kenig, 2008)), the value of the rate constant for the CO_2 – PZ system, k_1 , was evaluated and it was found to be 24450 m^3/(kmol s).

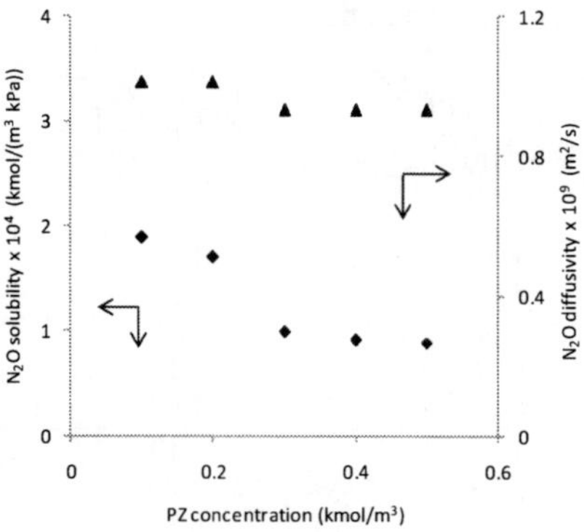

Figure 1: The plots of N_2O diffusivity and solubility vs. $(PZ)_0$ at 303 K

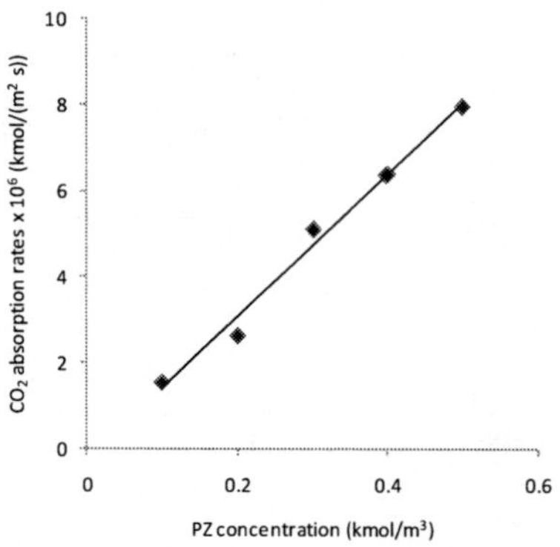

Figure 2: A plot of absorption rates vs. $(PZ)_0$ at 303 K

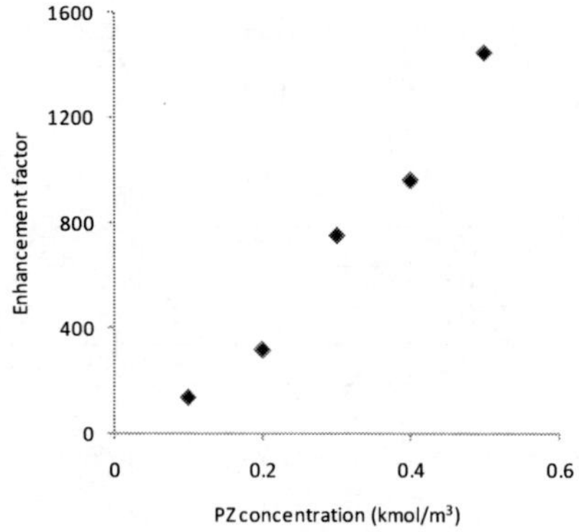

Figure 3: Enhancement factors for CO_2 absorption at 303 K

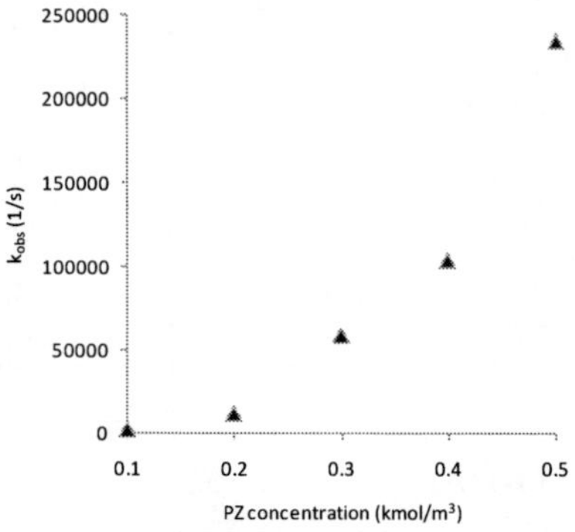

Figure 4: A plot of k_{obs} vs. $(PZ)_0$ at 308 K

Table 2: CO_2 absorption rates in aqueous mixtures of PZ and DEEA

Temperature K	$(DEEA)_o + (PZ)_o$ kmol/m^3	P_{CO_2} kPa	$R_{CO_2} \times 10^6$ kmol/(m^2 s)	$H_{CO_2}\sqrt{(D_{CO_2}\, k_2)} \times 10^7$ kmol$^{1/2}$/(m$^{1/2}$ s kPa)
298	2.0 + 0.1	5.8	1.51	1.84
	2.0 + 0.2	5.6	1.82	2.29
	2.0 + 0.3	6.7	4.52	4.77
	2.0 + 0.4	5.6	6.46	8.15
	2.0 + 0.5	6.1	6.84	7.92
303	2.0 + 0.1	5.0	1.55	2.19
	2.0 + 0.2	5.0	2.64	3.73
	2.0 + 0.3	4.3	5.09	8.37
	2.0 + 0.4	5.0	6.36	8.99
	2.0 + 0.5	6.0	7.95	9.37
308	2.0 + 0.1	5.0	1.73	2.44
	2.0 + 0.2	4.8	3.13	4.61
	2.0 + 0.3	5.0	5.55	7.84
	2.0 + 0.4	5.7	7.36	9.13
	2.0 + 0.5	5.3	8.66	11.55

Table 3: Equilibrium and kinetic characteristics of CO_2 – DEEA – PZ system at 303 K

$(PZ)_0$ kmol/m^3	$H_{CO_2}\sqrt{(D_{CO_2}\, k_2)} \times 10^7$ kmol$^{1/2}$/(m$^{1/2}$ s kPa)	$H_{CO_2} \times 10^4$ kmol/(m^3 kPa)	$D_{CO_2} \times 10^{10}$ m^2/s	k_2 m^3/(kmol s)	$\sqrt{M}$	E_i
0.1	2.19	2.49	10.8	718	106	1687
0.2	3.73	2.24	10.8	2570	202	1965
0.3	8.37	1.31	9.9	41200	807	4083
0.4	8.99	1.20	9.9	56700	947	4000
0.5	9.37	1.18	9.9	63700	1000	3531

Conclusions

The kinetics of the reaction between CO_2 and aqueous mixtures of DEEA and PZ were studied over a temperature range of 298 – 308 K and overall amine concentration range of 2.1 – 2.5 kmol/m^3 using a stirred cell reactor. The experimental data were interpreted using a homogeneous activation mechanism, by which the investigated reaction was considered as a reaction between CO_2 and DEEA in parallel with the reaction of CO_2 with PZ. The solubility and diffusivity of CO_2 in the aqueous amine solution, which are needed to interpret the data, were estimated using the N_2O analogy method. Besides, solution densities and viscosities were measured. The liquid-side mass transfer coefficient was estimated, and its value (0.001 cm/s) is in line with those typical for stirred cell reactors. The observed reaction rate constants for this system, which belongs to the fast pseudo-first order reaction regime, were measured. The second order rate constant for the CO_2 reaction with PZ was determined from the absorption rate measurements in the activated DEEA solutions, and its value at 303 K was found to be 24450 $m^3/(kmol\ s)$. This study proves the importance of the aqueous blend consisting of PZ and DEEA in gas purification applications.

References

Danckwerts, P.V., 1970, Gas-Liquid Reactions, McGraw-Hill, New York.
Donaldson, T. L. and Y. N. Nguyen, 1980, Ind. Eng. Chem. Fundam. 19, 260.
Kohl, A. and R. Nielsen, 1997, Gas Purification, 5th ed., Gulf Publishing Company, Houston, Texas.
Littel, R. J., G. F. Versteeg and W. P. M. van Swaaij, 1991, Chem. Eng. Sci. 46, 3308.
Liu, H. B., C. F. Zhang and G. W. Xu, 1999, Ind. Eng. Chem. Res. 38, 4032.
Vaidya, P. D. and E. Y. Kenig, 2008, Ind. Eng. Chem. Res. 47, 34.
Vaidya, P. D. and V. V. Mahajani, 2005, Ind. Eng. Chem. Res. 44, 1868.
Versteeg, G. F. and W. P. M. van Swaaij, 1988, J. Chem. Eng. Data 33, 29.
Xu, G. W., C. F. Zhang, S. J. Qin and Y. W. Wang, 1992, Ind. Eng. Chem. Res. 31, 921.
Yu, W. C., G. Astarita and D. W. Savage, 1985, Chem. Eng. Sci. 40, 1585.
Zhang, X., C. F. Zhang, S. J. Qin and Z. S. Zheng, 2001, Ind. Eng. Chem. Res. 40, 3785.

Proceedings of the 2nd Annual Gas Processing Symposium
Farid Benyahia and Fadwa T. Eljack (Editors)

CO_2 capture into aqueous solutions of the mixed solvent Cesar 1

Peter Brúder,[a] Andreas Grimstvedt,[b] Thor Mejdell,[b] Eirik F. da Silva,[b]
Hallvard F. Svendsen[a]

[a]Norwegian University of Science and Technology (NTNU) Department of Chemical Engineering, Sem Sælands vei 4, Trondheim NO-7491, Norway
[b]SINTEF Materials and Chemistry, Sem Sælands vei 2A, Trondheim NO-7465, Norway

Abstract

In this work, experimental data and a simplified vapor-liquid equilibrium (VLE) model for the absorption of CO_2 into aqueous solutions of the mixed solvent Cesar 1 are reported. The effect of the base amine/promoter ratio and the total concentration level of amine were studied. The highest possible ratio of base amine/promoter which does not form solid precipitates during the absorption of CO_2 at 40 ^{0}C and high loadings (40 w% amine), was identified. Considering the maximum loading found in the screening tests for Cesar 1 and for 30 w% MEA systems, the Cesar 1 system has about 128 % higher specific cyclic capacity if operating between 40 and 80 ^{0}C, and almost twice the CO_2 partial pressure at 120 ^{0}C compared to MEA.

Keywords: Cesar 1, vapor-liquid equilibrium, CO_2, absorption, capture

1. Introduction

Global warming is one of the most serious problems facing the earth at present. Anthropogenic CO_2 is one of the gases causing this problem, and its main sources stem from fossil fuel based power production (Desideri and Paolucci, 1999), traffic, iron and steel, cement production, and the metal and metal alloy production based on reduction with carbon. CO_2 absorption in aqueous amine solutions is the most widely used process in the industry, and the most viable one at present. However, capital and operating costs are high, and in particular the heat required for absorbent regeneration is high, causing large drops in efficiency for the industry. A better solvent may have higher CO_2 loading capacity, lower heat of absorption, faster absorption kinetics and favorable environmental properties. A combination of a fast promoter and a tertiary or sterically hindered amine can be expected to have high cyclic loading capacity combined with fast kinetics and possibly low heat consumption.

2. Experimental section

Component A (promoter, ≥ 98%) and B (base amine, ≥ 97%), both supplied by Fluka, were used for the experiments without further purification and dissolved in deionized water. The N_2 and CO_2 were of industrial quality and supplied by AGA Gas GmbH. Densities of the prepared unloaded and CO_2 loaded solutions were measured with glass 25 ml Gay-Lussac pycnometer. The analysis of CO_2 was done with the standard $BaCl_2$ method in every case, see Ma'mun et al (2006). For the comparison of the absorption rate and the loading capacity, a screening apparatus was used (Figure 1). The results were compared to the widely used 30 w% (5 M) monoethanolamine (MEA) solution. The screening tests were performed at 40 ^{0}C, to have some safety margin to 50 ^{0}C due to the solid precipitation formation, and as a standard level. The flows of N_2 and CO_2 were controlled by flow controllers, and these regulate also their mixing ratio. The CO_2 flow was set to 0.5 Nl/min and the N_2 to 4.5 Nl/min. This gave a 10 % CO_2 concentration in the inlet gas flow through the amine solution. The volume of absorbent solution was set to 750 ml and was calculated based on the weight and the density of the solution. After absorption in the solution, the gas was passed through a set of 2 coolers, to get as much as possible of the evaporated solution back to the reactor, and additional 2 aftercoolers to cool down the vapors to 10-15 ^{0}C, and get the remaining solvent out of the gas phase before the detector. The CO_2 content in the gas was measured by an IR detector and continuously logged by LabView software (Ver 6.0). An analyzer calibration run was performed before every measurement, and the 2 points used being pure N_2 (0% CO_2) and 0.5N l/min CO_2 + 4.5 Nl/min N_2 (10% CO_2). The program then registered the measured remaining CO_2 in the gas stream, and the difference between 0.5 Nl/min and the registered value by the IR instrument was saved as a function of time. The absorption rate can be calculated then as a function of specific loading (mol CO_2/ mol amine). The specific loading can be calculated from the prepared solution and the CO_2 absorbed into the solution. The measurements were stopped when the measured outlet gas CO_2 concentration reached 90% of the initial value, which means 0.45 Nl/min CO_2 flow in the gas leaving the solution (or 9% CO_2 measured by the detector). After the end of the absorption, a liquid sample was taken for measuring density and CO_2 concentration.

The desorption was performed at 80 ^{0}C, and the stripping gas was 100 % N_2.

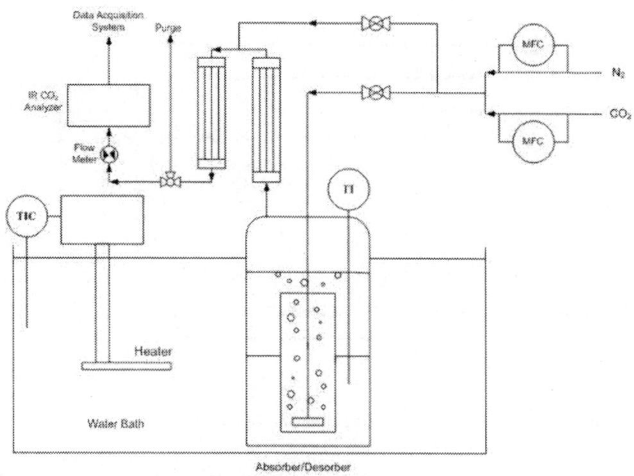

Figure 1: Sketch of the screening equipment

For the equilibrium measurements two equipments were used, an atmospheric unit operating at 40, 60 and 80 ^{0}C, and a metal one for higher pressures (up to 7 bars) and for 100 and 120 ^{0}C.

The glass apparatus consisted of 3 reactors filled with 100 - 150 ml of preloaded amine solution (No. 2, 3 and 4 in Figure 2). For the measurement of the CO_2 concentrations in the vapor phase 4 IR detectors were used (Rosemount Binos 100) with ranges 0 - 2000 ppm, 0 – 1%, 0 – 5%, and up to 20% respectively. The reactors were placed in a waterbath, and heated to the requested temperature. The glass tank No.1 was empty, and served for preventing backflush and possible contamination of the IR detector. The temperature of reactor No 4. and the waterbath were monitored, and the measurement and sampling done when equilibrium was reached. All the glass equipment was placed in a closed thermostated box. The vapor phase is circulated through the amine solution. After the equilibration the vapor phase passes a cooler (10 - 15 ^{0}C) before passing to the IR analyzer. The temperature is monitored and considered as the saturation temperature for water in the gas. Calibrations of all four IR detectors were performed daily. After the CO_2 concentration in the gas phase was stabilized, the values were logged. The CO_2 partial pressure in the equilibrium cell was determined by the procedure described in Ma'mun et al. (2007). A sample of the liquid phase was taken from the reactor No. 4.

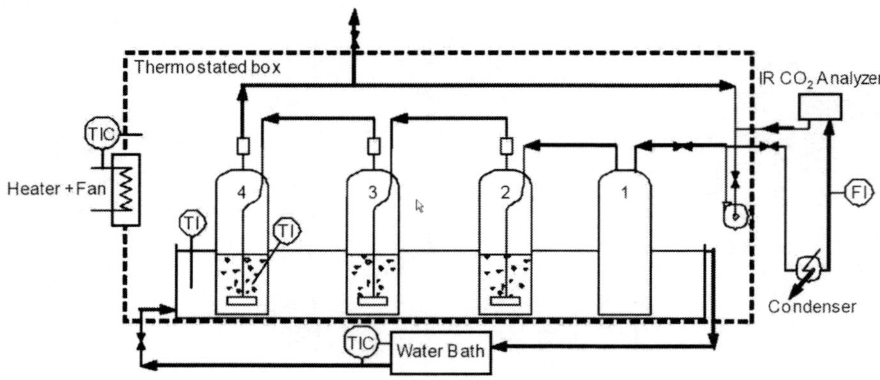

Figure 2: Sketch of the atmospheric vapor-liquid equilibrium apparatus

The apparatus for high temperature equilibrium measurements consisted of two vertically linked jacketed vessels with volumes 1.18 and 0.33 l. The small vessel was filled with ceramic Berl saddles. Thermostated heating oil served as heating medium. The vessels are connected and mounted in a framework, which allows them to be rotated 180 degrees in a vertical plane so the solvent can flow from one vessel to the other. The tilting frequency is governed by an electronically controlled pneumatic piston cylinder. The experiments were performed by pumping 150 ml of unloaded solution into the small vessel. When the set point temperature was reached, the desired pressure was adjusted on the gas flask reduction valve, and the rotation started. After one hour the CO_2 feed was closed and 1-2 hours were used for establishing and stabilizing equilibrium. The time needed was indicated by a stable system pressure and temperature. A liquid sample was then withdrawn to measure the CO_2 loading. For this purpose a closed metal cylinder with known weight of unloaded amine solution was used to avoid any CO_2 loss. From the known dilution, the concentration of CO_2 in the original solution was calculated.

3. Results and discussion

3.1. The screening tests:

The target was to find the amine concentration with the highest possible absorption capacity, absorption rate and cyclic capacity, and with a low heat of reaction. The Cesar 1 syatem forms precipitates with CO_2 at higher amine and CO_2 concentrations, and this was undesired. A list of the tested concentrations and the qualitative observations can be found in Table 1.

Table 1: Visual observations for the tested systems

Concentration [mol/l]		Observations
B	A	
2	3	Around 22 ^{0}C the crystals has been precipitated (formation starts around 33 – 35 ^{0}C)
3	2	All crystals dissolved at 22 ^{0}C if mixed, without mixing very small amount formed during 12 – 16 hours (probably will not make problems in a stable industrial operation). Clogged the apparatus during the measurement.
2.5	2.5	Without heating and mixing the mixture crystallizes at 22 ^{0}C, but the solids are possible to dissolve (formation starts around 33 – 35 ^{0}C). No precipitation by loading with CO_2, but the viscosity seems to be quite high and the solution is sticky.
2	3.5	Full of crystals at 22 ^{0}C and all the solution is quite solid. The crystals are formed at 35 ^{0}C. No crystals were found during CO_2 loading, but the solution appears to be sticky and highly viscous.
4	4	Crystallizes without mixing, and upon loading forms a solid phase
5	2	No crystals at 22 ^{0}C with mixing, but loading with CO_2 causes precipitation.
5	1	Clogged the apparatus during the measurement
2	5	Similar to 4:4 (was heated up to 40 ^{0}C to dissolve the crystals at preparation)
4	1.5	Clogged the apparatus during the measurement
4	1	Successfully measured
3	1.5	Successfully measured

From these observations can be concluded that at high concentration of A, B or a high total amine concentration, the CO_2 loading causes precipitation formation in the solution, and also the viscosity and stickiness is increasing. This may lead to operational problems. The concentration ratios B/A 5/1, 3/2, 4/1.5 clogged the parts of the laboratory glass apparatus with small cross sections, even during the absorption stage. In some cases, the ceramic sinter used in the reactor for bubble distribution could be blocked. The holes are quite small where the CO_2 reacts with amine, and as the CO_2 concentration here is the highest, the probability to form precipitate is higher. Also the viscosity of the solution plays a significant role, mainly for the desorption at 80 ^{0}C and could cause operational problems. The collection of the successful (finished without operational problems) screening tests can be found in the Table 2.

Table 2: The loading and cyclic capacities of the measured amine systems

	Loading capacity (40 ^{0}C) [mol CO_2 /mol amine]	Cyclic capacity (40-80 ^{0}C) [mol CO_2 /mol amine]
5 M MEA	0.46	0.19
4 M B + 1 M A	0.51	0.39
3 M B + 1.5 M A	0.58	0.43

Aronu, et al., (2009) have measured the maximum loading and cyclic capacity for 5 M MEA and found 0.518 mol CO_2/mol amine 0.271 mol CO_2/mol amine respectively in an identical measurement. For 2.5 M B + 0.5 M A system they found 0.671 mol CO_2/mol amine as maximal loading and a cyclic capacity of 0.524 mol CO_2/mol amine. This implies that the B/A 3/1.5 M system should have around 30 % higher loading capacity and 23 % higher cyclic capacity than the 2.5/0.5 system per liter of solution. The comparison of these systems measured by the screening equipment is only semi-quantitative, as the bubble distribution can vary somewhat (Ma'mun, et al., 2007).

3.2. The equilibrium tests:

Based on the results of the screening tests the system with the highest loading capacity and cyclic load a B/A ratio = 3/1.5 was selected as Cesar 1 for equilibrium tests at 5 temperatures resulting in CO_2 partial pressures above the solution as function of temperature and loading. Also the heat of absorption can be estimated. The general Gibbs-Helmholtz equation relates the heat of reaction to the changes of equilibrium constant with temperature:

$$\frac{d \ln K_{eq}}{d(1/T)} = -\frac{\Delta H_{reac}}{R} \qquad (1)$$

The equilibrium reaction for CO_2 absorption on an overall form can be written:

$$CO_2(g) + Am + \sum B_i = \sum B_i COO^- + AmH^+ \qquad (2)$$

Here B_i denotes bases of the system. See Kim et al., (2009) for a more thorough treatment. This reaction contains the following equilibrium constant after making the simplification of setting all activity coefficients γ=1:

$$K_{eq} = \frac{1}{p_{CO_2}} \frac{x_{AmH^+} \prod x_{BCOO^-}}{x_{Am} \prod x_{B_i}} \qquad (3)$$

Differentiation of ln K_{eq} with the inverse of absolute temperature T gives

$$\frac{d \ln\left(\frac{1}{p_{CO_2}}\right)}{d(1/T)} + \frac{d \ln\left(\frac{x_{AmH^+} \prod x_{BCOO^-}}{x_{Am} \prod x_{B_i}}\right)}{d(1/T)} = -\frac{\Delta H_{reac}}{R} \qquad (4)$$

For a specific loading the second term can be assumed negligible compared to the first one, and we get:

$$\frac{d\ln p_{CO_2}}{d(1/T)} = \frac{\Delta H_{reac}}{R} \quad (5)$$

Since reaction (2) includes CO_2 in the gas phase, this describes the total absorption process, the heat of reaction may be denoted the heat of absorption. Equation (5) has been used extensively in the literature to find the heat of absorption from equilibrium data (Lee, et al., 1976; Jou, et al. 1982; Rebolledo-Libreros and Trejo, 2004). It must be emphasised that this is an approximation as changes in the activity coefficient and mol fraction ratios have been disregarded.
The dependence of CO_2 partial pressure from the loading (α) can be fitted with the parameterized function:

$$\ln p_{CO_2} = A\ln\alpha + k_1 + \frac{B}{1 + k_2\exp(-k_3\ln\alpha)} \quad (6)$$

The first term represents the linear part, and the coefficient A is constant for all temperatures, but may be different for different systems. The second term k_1, is a constant that represents the distance between the isotherms. The last term represents a sigmoid function with 3 parameters. The parameter B is constant for each system, while k_2 and k_3 are temperature functions. The parameters of eq. (6) are shown in Table 3.

Table 3: The parameters of the sigmoid function (6) for the Cesar 1 and MEA systems

	CESAR 1	5M MEA
A	2.219363745	1.8
B	10	10
k1	-10364.215 * (1/T) + 33.12	-9155.95 * (1/T) + 28.03
k2	exp (-1275.94 * (1/T) + 4.23)	exp (-6146.18 * (1/T) + 15)
k3	2.567	7527.04 * (1/T) - 16.94

A comparison between the experimental points and model curves is shown in Figure 3. The range of the model, based on the experimental data is up to a loading of about 0.83 mol CO_2 / mol amine for the Cesar 1 system, and 0.57 for the 30 w% MEA systems. The 5 M MEA data and the model in eq. (6) is based on unpublished internal data. It shows good agreement between the experimental points and model over the whole range. It should be emphasised that this model is only a parameterised fit of the data and has no thermodynamic significance.

The partial CO_2 pressure for Cesar 1 at a loading of 0.58 mol CO_2 /mol amine — which is the maximum loading from the screening tests at 40 ^{0}C, see Table 2 — and 120 °C is about 600 kPa. Compared to 30 wt% MEA, where the same screening tests gave a maximum loading of 0.46 mol CO_2 /mol amine, the equilibrium CO_2 pressure is about 350 kPa. This indicates that the Cesar 1 system has a potential for a significant reduction in stripping steam requirement in the desorption. From the screening tests it is also clear that the specific cyclic capacity, measured between 40 and 80 ^{0}C, see Table 2, is also about twice as high, even when considering the 10 % difference in total amine concentration, between the two systems. This indicates that also the sensible heat loss will be significantly lower for the Cesar 1 system

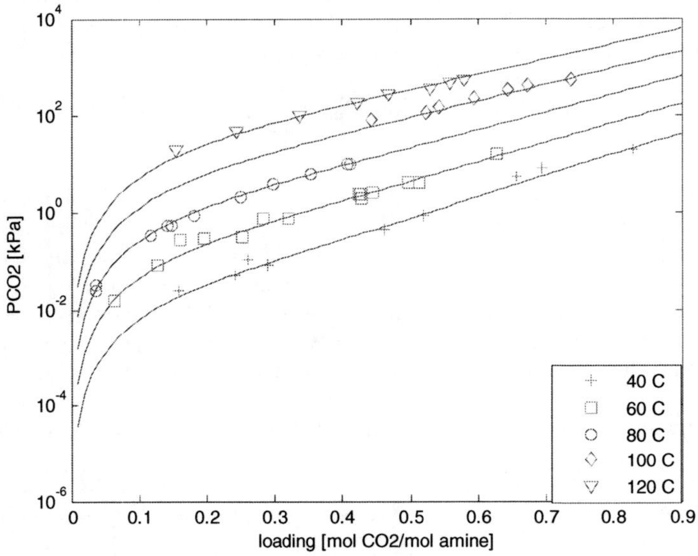

Figure 3: Comparison of experimental equilibrium data to model for the Cesar 1 system

Table 4: Dependence of heat of absorption on loading according to eq. (5)

	Cesar 1		MEA 5 M	
Specific load [mol CO_2/mol amine]	pCO_2 [kPa]	ΔH_{abs} [J/mol]	pCO_2 [kPa]	ΔH_{abs} [J/mol]
0.1	0.006	86002	0.005	77615
0.2	0.033	85199	0.017	82578
0.3	0.103	83519	0.041	88623
0.4	0.285	80937	0.217	84697
0.5	0.778	77627	6.191	58510
0.6	2.147	73914	179.155	31053
0.7	5.946	70188	1157.321	21298
0.8	16.153	66804	2788.016	22518

The heat of absorption as a function of loading, as an average for the temperature interval between 40 and 120 ^{0}C can be calculated according to eq.(5) and is tabulated in Table 4.

It is known that the heat of absorption in 5M MEA is about 83-84000 J/mol CO_2 absorbed, and that is reasonably constant up to a loading of 0.5, whereafter is fall rapidly as the capacity of MEA vanishes (Kim et al., 2007). The heat of reaction for the Cesar 1 system is seen to be about the same as for MEA at the lower loadings but decreases slowly as the base amine takes over the absorption of CO_2, and retains its capacity up to much higher loadings. From the heat of absorption data it is clear that the Cesar 1 system does not offer any advantage over the MEA system in this respect.

4. Conclusion

The Cesar 1 system has about twice the specific cyclic capacity, measured between 40 and 80 ^{0}C, compared to 5M MEA. It also has almost twice the partial CO_2 pressure for a maximum loaded solution at 120 ^{0}C compared to MEA. The heat of absorption of the two systems is comparable. This should imply both significantly reduced stripping steam consumption and sensible heat loss for the Cesar 1 system making this system a very strong contender to 5 M MEA system in exhaust gas absorption plants.

Acknowledgements

The research leading to these results has received funding from the European Community's Seventh Framework Program FP7/2007-2011 under grant agreement n° 213569 (CESAR).

References

Aronu U.E., H.F. Svendsen, K.A. Hoff, O. Juliussen, 2009, Solvent selection for carbon dioxide absorption, Energy Procedia, 1, 1051-1057

Desideri U., A. Paolucci, 1999, Performance modeling of a carbon dioxide removal system for power plants, Energy Conversion and Management, 40, 1899-1915

Jou F.-Y., A.E. Mather, F.D. Otto, 1982, Solubility of acidic gases in aqueous methyldiethanolamine solutions, Industrial and Engineering Chemistry Process Design and Development, 21, 539-544

Kim I., K.A. Hoff, E.T. Hessen, T. Haug-Warberg, and H.F. Svendsen, 2009, Enthalpy of absorption of CO_2 with alkanolamine solutions predicted from reaction equilibrium constants, Chemical Engineering Science, 64, 2027-2038

Kim I. and H.F. Svendsen, 2007, Heat of absorption of carbon dioxide in monoethanolamine and 2-(aminoethyl)ethanolamine solutions, Industrial and Enginnering Chemistry Research, 46, 5803-5809

Lee L.L., 1976, Equilibrium between carbon dioxide and aqueous monoethanolamine solutions, Journal of Applied Chemistry and Biotechnology, 26, 541-549

Ma'mun S., J.P. Jakobsen, H.F. Svendsen, O. Juliussen, 2006, Experimental and modeling study of the solubility of carbon dioxide in aqueous 30 mass % 2-((2-aminoethyl)amino)ethanol solution, Industrial & Engineering Chemistry Research, 45(8), 2505-2512

Ma'mun S., H.F. Svendsen, K.A. Hoff, O. Juliussen, 2007, Selection of new absorbents for carbon dioxide capture, Energy Conversion and Management, 48, 251-258

Rebolledo-Libreros M.E., A. Trejo, 2004, Gas solubility of H_2S in aqueous solutions of N-methyldiethanolamine and diethanolamine with 2-amino-2-methyl-1-propanol at 313, 343 and 393 K in the range 2.5-1036 kPa, Fluid Phase Equilibria, 224, 83-88

Proceedings of the 2nd Annual Gas Processing Symposium
Farid Benyahia and Fadwa T. Eljack (Editors)

Q-Chem Steam Boilers NOx Emissions Reduction

ELZeini,Hesham Y., Sr. Process Engineer, Qatar Chemical Co. ltd., P. O. Box 24646, Doha, Qatar

Abstract

Q-Chem Plant in Mesaieed has three steam boilers. They run on fuel gas that is 90% mol methane and are also designed to co-fire C4-C11 liquid fuel which is a by-product from the De-Propanizer tower. The boilers design and actual efficiency is ~ 90%. Based on the 1999 Qatar state environmental regulations, they were designed for maximum NOx emissions of 250 mg/Nm³. However, the new Ministry of Environment (MOE) regulations limited the NOx emissions to 125 mg/Nm³ maximum from December 2008. Modifying the existing boilers to produce 125 mg/Nm³ max NOx was a big challenge for the technical group especially with a running plant. The steam system is very critical to the plant and its outage leads to a total plant outage. This paper will discuss and reflect the different technical options and how the boilers modification was carried out. Special emphasis will be presented on the required technical modifications, value engineering and operational flexibility to grant the outage on the boilers for a month each.. Some final actual NOx emissions will be presented where the modification resulted in superior NOx reductions up to 90 mg/Nm³ at full firing mode.

Keywords: Liquid Fuel, Dual Firing, NOx emissions

Q-Chem:	Qatar Chemical Company Limited
NOx	Nitrogen Oxides (NO2, NO3 and NO4)
mg/Nm³	unit measurement for the NOx emissions
RX-N	organic compound containing Nitrogen atom
Liquid Fuel	a by-product stream containing olefins and aromatics
Dual Firing	Gas and Liquid fuel firing simultaneously

1. Low NOx Burners industrial Application

The global awareness about the effect of air pollutants is increasing greatly over the past two decades. There is an increasing demand in reducing pollutant emissions of all types from all combustion processes. Continuous Efforts are underway from a broad cross-section of organizations to improve existing techniques and to develop new techniques to minimize combustion related air pollutants emissions.

Many combustion processes are operated at elevated temperatures and high excess air levels. These conditions produce high thermal efficiencies and production throughput rates. On the other hand, such conditions greatly contribute to higher NOx formation rates. NOx emissions are among the primary air pollutants. Nowadays, the combustion technology must adopt the environment NOx emissions regulations while achieving maximum firing efficiency and performance. Q-Chem has done a major upgrade to contain and minimize the NOx emissions from the Boilers by upgrading the burners to Low NOx burners with minimal changes in the fire box and external shell work.

1.1 Low NOx emission control overview:

Emissions of nitrogen oxides, commonly referred collectively as NOx, are regulated because of their adverse effects on health and the environment. They have a major role in acid rain, the formation of harmful ozone and photochemical smog in the lower atmosphere and the depletion of the beneficial ozone in the upper atmosphere. Figure1 shows the exponential increase in these emissions in the last century to date. This big step change was due to the industry evolution and the high energy usage in all industrial and services sectors.

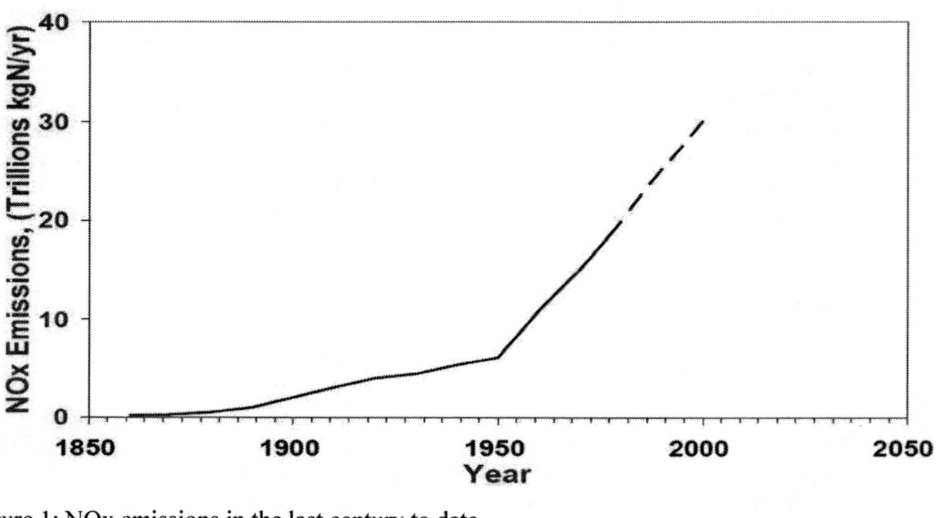

Figure 1: NOx emissions in the last century to date

1.1.1 How NOx pollutants are formed: The Theory:

The most environmentally important oxides of nitrogen are: NO, NO2, & more recently N2O. N2O is a recent concern because it is a "greenhouse" gas which contributes to global warming and because it can aid in destroying the upper atmosphere ozone layer which protects us from Ultraviolet radiation. Fortunately, very little N2O is emitted from the flame of a typical burner. Over 90% of the NOx from a typical flame is in the form of NO and the remainder is NO2. However, since NO is eventually converted to NO2 in the atmosphere, most regulations treat all of the NOx as NO2. NOx emissions from combustion sources are due to the oxidation of atmospheric N_2 and the Oxidation of nitrogen chemically bound in fuel molecules.

There are three major mechanisms for NOx production. Some of these mechanisms are well understood, while others are still under investigation. They are:

1. The Thermal mechanism
2. The Prompt mechanism
3. The Fuel mechanism

1- Thermal NOx is formed by the high – temperature reaction of nitrogen with oxygen, via the well – known Zeldovich mechanism, as given by the simplified reaction:

$$N_2 + O_2 \rightarrow NO, NO_2$$

Thermal NOX increases exponentially with temperature. Above 2000F (1100C), it is generally the predominant mechanism in combustion processes, making it important in most high – temperature heating applications. This means that this mechanism becomes more important when air preheating or oxygen enrichment of the combustion air are used, which normally increases the flame temperature.

2- Prompt NOx is formed by the relatively fast reaction between nitrogen, oxygen, and hydrocarbon radicals. It is given by the overall reaction:

$$CH_4 + O_2 + N_2 \rightarrow NO, NO_2, CO_2, H_2O, \text{trace species}$$

In reality, this very complicated process consists of hundreds of reaction. The hydrocarbon radicals are intermediate species formed during the combustion process. Prompt NOX is generally an important mechanism in lower temperature combustion processes.

3- Fuel NOx is formed by the direct oxidation of organic nitrogen compounds contained in the fuel. It is given by the overall reaction:

$RXN + O_2 \rightarrow NO, NO_2, CO_2, H_2O$, trace species

Figure 2 illustrates the mechanism of NOx formation due to Fuel influence

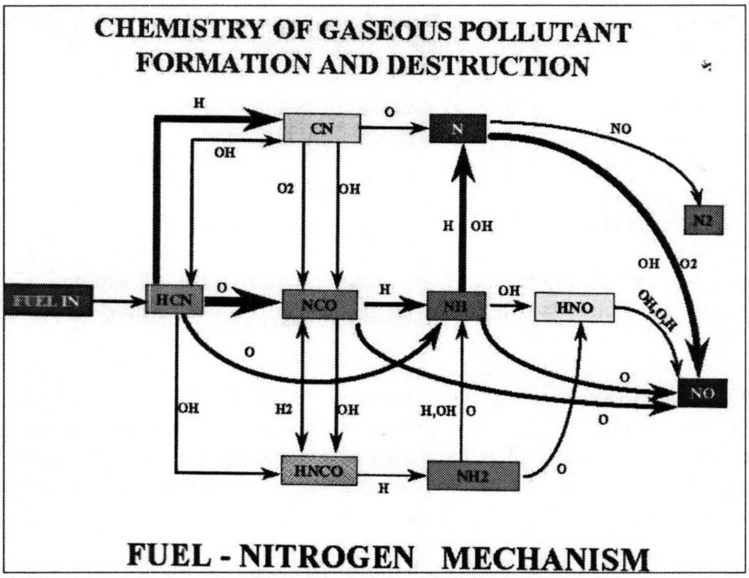

Figure 2: Schematic of fuel NOx formation pathways

1.1.2 Low NOx burner technology application for boilers:

The objective of the modern typical boiler burner retrofit is to reduce NOx. Two NOx techniques for boilers are combustion modification and back-end (post-treatment) cleanup. Combustion modifications include low NOx burners, low excess air, over fired air, flue gas recirculation (FGR), fuel induced recirculation, reburn and water tempering. The Back End cleanup techniques include selective catalytic reduction (SCR) and selective non-catalytic reduction (SNCR). Overhere, the emphasis will be on the combustion modification techniques of low NOx burners, low excess air, over fired air, flue gas recirculation, and fuel induced recirculation.

Today's low NOx burner relies on control of the combustion air in several component streams, as well as the controlled injection of fuel into the air streams at selected point for maintaining stable, attached flames with low NOx generation. Typically, in the venture-style, Primary and secondary air enters the burner radially through the venturi and exits the burner axially with a primary air swirl defined by the fixed blade axial swirler. The swirler determines the size and strength of the recirculation zone. A tertiary stream flows between the venture base and the burner throat quarl. Tertiary air separates some of the combustion air from the main flame, effectively staging combustion and reducing NOx. For natural gas firing, fuel can be introduced through internal or external pokers or gas rings, and can also be injected through a central gas pipe with multiple orifices at the furnace end. A single conventional atomized burner (oil gun), located along the burner centre line, and typically supplies the oil. The oil gun may use dual fluid, mechanical or rotary cup atomization.

2.2 Q-Chem Practical Experience in Low NOx Emissions:

2.2.1 Problem statement:

The NOx emissions have become a great concern lately. Strict regulations worldwide and especially in Qatar are calling to reduce the NOx emissions to lower limits compared to 5 years ago. Q-Chem has reviewed the NOx emissions plant wide and found out that the steam Boilers are exceeding the limits in the NOx emissions (~ > 200 mg/Nm^3). For that, a detailed design was initiated and completed to upgrade the combustion process as a whole in the boilers to meet the new regulations of 125 mg/Nm^3 max. The modification scope was to upgrade the existing burners to low NOx burners along with some changes in the firebox. The modifications on the three boilers were completed and commissioned. The results were outstanding achieving 80-90 mg/Nm^3 maximum NOx emissions at maximum dual firing mode and steam production rate.

2.2.2 Data collection and analysis:

The new Steam quenched burners along with minimum modifications to the boiler structure and firebox were used to control the NOx formation within the allowable limits. Table 1 below shows the major characteristics of both old and new burners. Mainly they differ in the NOx formation and the steam quench additional provision. Else, all the characteristics are the same as the boiler capacity and duty did not change. The new burners have the advantage of higher firing efficiency also.

Table 1: Old and new burner's improvement

	Improvement made	Unit	Old	New
	Design load basis		100%	100%
1	Calculated heat release HHV (Higher Heating Value)	Mcal/h	-	No change
2	Fuel gas flow	kg/h	-	16% more
3	Boiler steam capacity	kg/h	-	No change
4	Air density	kg/Nm3	1.286	1.286
5	Excess air at burners	%	15	15
6	Relative humidity	%	70	60
7	NOx	mg/Nm3	>200	125
8	CO	mg/Nm3	600	500
9	O_2 by vol. in dry flue gas	% vol	-	0.2% lower
10	Boiler efficiency on the LHV (Lower Heating Value)	%	-	1.2 % more

The main changes were limited to the quench steam facilities in the new burners to control the flame tip and prevent the temperature to reach the high NOx formation regime. The design had to consider dual firing option where special liquid fuel is burned jointly with the fuel gas which is mainly Methane (90 wt%).

Table 2 reflects the emissions before and after the burners' replacement in each boiler. Starting with boilers 1 upgrade completion, the NOx emissions dropped to an average of 70-90 mg/Nm^3 compared to the previous full load firing in January 2008 where it exceeded 200 mg/Nm^3 before the upgrade. The same result was achieved for boiler 2 proving the success of the burners modification work. This data included the liquid fuel firing mode and it was the basis for the modification comparison.

Table 2: Q-Chem Boilers NOx emissions comparison before/After modification

	Boiler 1		Boiler 2	
Month	kg/month	mg/Nm³	kg/month	mg/Nm³
Before Modification	10,089.2	210.0	11,530.5	240.0
After Modification	3,766.5	91.4	3,502	85.6

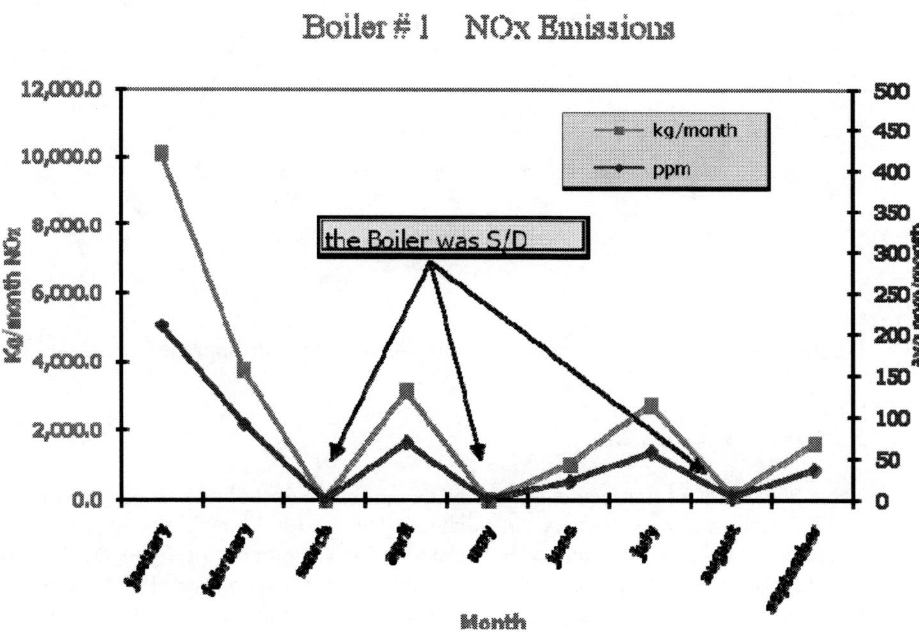

Figure 3: Boiler 1 NOx emissions trends before and after the burners upgrade

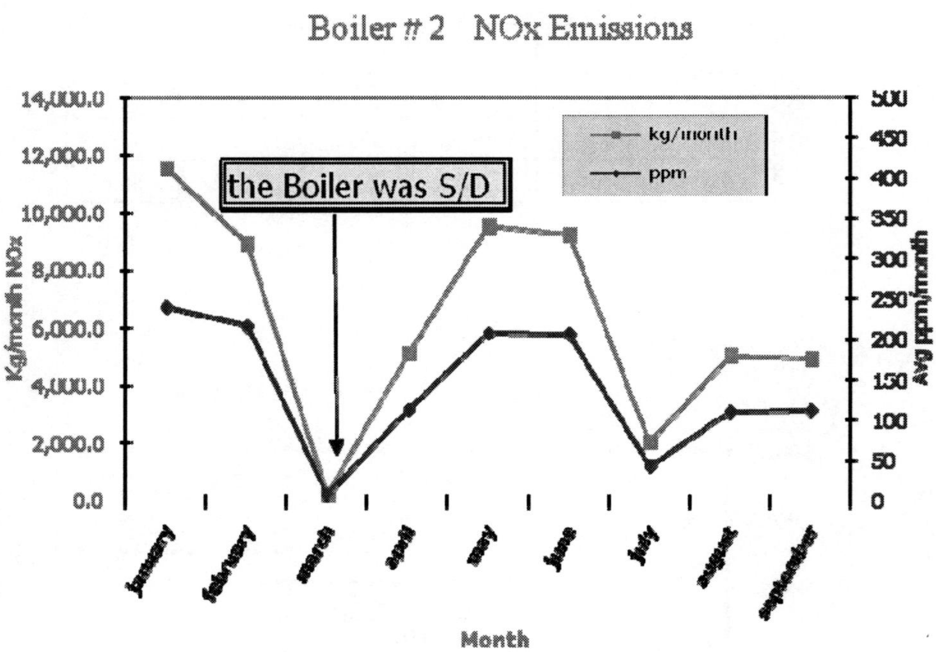

Figure 4: Historical Boiler 2 NOx emissions trends before the upgrade

2.2.3 Analysis Discussion:

Table 2 shows the improvement in the NOX emissions for boiler 1 before and after the modification and new burners' installation. The boiler 1 stack flue gases showed significant NOx emissions reduction by 60 to 70%. The monitoring figures collected on monthly basis reflect this improvement to even lower figures than 125 mg/Nm3, the new set target by the Ministry of Environment (MoE) in Qatar. Considering the liquid fuel firing intermittent need, these new burners were designed to handle maximum liquid fuel firing dual mode also which is a bit a difficult case than the fuel gas firing only. The dual firing burners are common in the industry. However, there are certain quench steam rates that have to be adjusted corresponding to liquid fuel flow to achieve the lox NOx emissions results.

Normally, the operation strategy is to manage the fuel oil to one boiler and maintain constant fuel flow to minimize upsets and necessary steam adjustment. Commonly, the boilers operate on a load basis. Two boilers are in operation and the 3rd one is S/B (STAND BY). In case of higher abnormal loads, the boilers will operate at full load hence generating spikes of NOx until the normal conditions is restored. Such criterion is not prevailing basis, but could happen. The production tends to be proactive and anticipate these load surges ahead and act to minimize the impact. NOx emission in boiler 2 & 3 is showing fluctuation and this was minimized after installing the new burners and carrying out the new modification.

Another potential cause of the NOx fluctuation is the improper firing in the burner. Q-Chem has installed a flame monitoring system that works with both UV and IR signals. This combination grants both proper firing and flame pattern. It also grants safe operation of the boilers to prevent free HC flow inside, hence causing explosion and fire box destruction and damage.

2.2.4 Conclusion:

The boilers low NOx burners are advanced technology nowadays. They require careful flame pattern understanding and combustion mechanism. Q-Chem as with other industries selected the best upgrade option as the option of recirculation the flue gases is not practical and would require consideration in the basic design of the boilers. Nevertheless, the objective was met through minor steam introduction along with burners' replacement.

The process design was changed in a feasible simple modification way serving both: same boiler duty and production and at the same time lower NOx emissions. The current MOE regulations are more stringent toward the control of these air pollutants. The new plants that are under construction must comply with even tighter specs as shown in Table 4.

Table 3- External Combustion Sources (boilers – heaters) emissions regulations

Air pollutants	**Permissible limits**
Particulate matters	5 mg/Nm3
Sulphur dioxide	500 mg/Nm3
Nitrogen Oxides	55 mg/Nm3

References

Charles E. Baukal, Jr. (2001). *The John Zink Combustion Handbook*. CPC Press LLC. Florida, USA.

Q-Chem Air Emissions Monthly Monitoring Reports (Jan to Sept 2008)

Ministry of Environment (MOE, Qatar), the Standards and Limits for Air Quality.

Proceedings of the 2nd Annual Gas Processing Symposium
Farid Benyahia and Fadwa T. Eljack (Editors)

The experience in using LDAR for monitoring fugitive emissions of Volatile Organic Compounds with special reference to RasGas' experience

Chris Horne[a], Roger Medalla[a], Julius Bacani[a]
[a] *RasGas Company Ltd., Ras Laffan Industrial City, Qatar*

Abstract

Controlling fugitive volatile organic compound (VOC) emissions through the implementation of a leak detection and repair (LDAR) program has been in practice at RasGas Company Limited (RasGas), Qatar, since 2007. An LDAR program is designed to identify pieces of equipment that are emitting sufficient amounts of material to warrant reduction of the emissions through repair[1]. The program was initiated by RasGas, on a voluntary basis, on the basis of acceptable environmental practices in other countries susceptible to ozone formation. The United States Environmental Protection Agency (EPA) reports that VOCs contribute to the formation of ground level ozone, which is a major component of smog, and causes or aggravates respiratory disease, particularly in children, asthmatics and healthy adults who participate in moderate exercise[2]. The Ras Laffan Industrial City (RLC) Environmental Regulations (2005) implemented fugitive monitoring and reporting requirements for petroleum plant processing equipment[3], including valves, pump and compressor seals, pressure relief valves, open-ended lines and sampling connection systems. In addition, the State of Qatar Ministry of Environment has included LDAR monitoring requirements in the operating permits for RasGas' facilities.

The United States Environmental Protection Agency (EPA) has determined that leaking equipment, such as valves, pumps and connectors, are the largest source of emissions of VOCs and volatile hazardous air pollutants from petroleum refineries and chemical manufacturing facilities[2]. The LDAR program at RasGas incorporates the requirements of USEPA Method 21[4], while using emerging technologies for leak detection. To measure the success of RasGas' LDAR program, RasGas determines the 'Control Effectiveness', which is based on the relationship between the percentage of equipment pieces that are leaking and the corresponding average leak rate for all of the equipment[1]. The LDAR program at RasGas has to date resulted in the successful detection and repair of leaking equipment, thus reducing product losses and decreasing exposure to hydrocarbon vapors for workers, the facility and surrounding receptors.

Keywords: Fugitives, Emissions, Monitoring

1. Background

Fugitive emissions from equipment leaks are unintentional losses and may arise due to normal wear and tear, improper or incomplete assembly of components, inadequate material specification, manufacturing defects, damage during installation or use, corrosion, fouling and environmental effects. Components also tend to have greater average emissions when subjected to frequent thermal cycling, vibrations or cryogenic service[5].

Volatile organic compounds are defined as organic compounds that participate in photochemical reactions. Any organic compound is presumed to participate in atmospheric reactions, unless the United States EPA determines that it does not. The EPA considers several organic compounds to have negligible photochemical reactivity, including methane, ethane, 1,1,1-trichloroethane, methylene chloride, trichlorofluoromethane, dichlorodifluoromethane, chlorodifluoromethane, trifluoromethane, thrichlorotriflhuoroethane, dichlorotetrafluoroehane, and chloropentafluoroethane[6].

Ozone at ground level is considered a secondary pollutant, and is formed by the reaction of nitrogen oxides (NO_X) and VOCs in the presence of heat and sunlight. NO_X and VOCs primarily result from the combustion of fossil fuels and leakage from piping equipment and storage tanks, respectively. The management of this secondary pollutant is a complex issue and depends upon a considerable number of factors, notably the reduction on reducing ozone precursors, namely VOCs and nitrogen oxides.

In a typical facility, most of the emissions are from valves and connectors because these are the most prevalent components and can number in the thousands. The major cause of emissions from valves and connectors is seal or gasket failure due to normal wear or improper maintenance. Previous EPA studies have estimated that valves and connectors account for more than 90% of emissions from leaking equipment with valves being the most significant source[2].

2. RasGas Tagging, Monitoring and Repair Methods

The onshore RasGas facility has tens of thousands of potential leak sources from equipment in VOC service, which includes all equipment / components handling fluids that contain at least ten percent by weight of VOCs[7]. Considering this number, a significant amount of effort is required to implement and maintain a functioning LDAR program. In order to execute the program effectively, RasGas is completing the program in phases, commencing with the oldest plant equipment (LNG Trains 1 – 3, sales gas Train AKG-1) and progressing to the newer equipment (LNG Trains 4 – 7, sales gas Train AKG-2).

The RasGas LDAR program consists of the following six (6) major elements:

1. Component identification;
2. Component tagging;
3. Monitoring;
4. Data Collection;
5. Repairs; and
6. Database management.

To identify streams in VOC service, the weight percent VOC of each process stream is calculated using the material balance data. The process streams with fluids containing >10% weight VOC are identified on the plant process flow diagrams (PFDs) and piping and instrumentation diagrams (P&IDs). The VOC process streams are color-coded on the diagrams for gas / vapor service and/or light liquid service[8].

Using the color-coded P&IDs, the RasGas LDAR Environmental Technicians locate the components in VOC service in the plant operating units. A metal tag, with a unique identification number, is attached to each piece of equipment. As the tagging is performed, the tag information is entered into an electronic data logger, for uploading into a database management software.

The monitoring of the VOC equipment is conducted using two dedicated instruments;

- Infrared image camera (ThermaCAM GasFindIR); and
- Portable organic / inorganic compound detection instrument (flame ionization and photo ionization detection).

RasGas was the first company in Qatar to utilize an infrared camera as a fugitive emissions detection technology. During monitoring, the infrared image camera is used first to identify leaking equipment. The technology of the camera delivers real-time thermal images of gas leaks, which are viewed through the camera as "black smoke" images. This technique allows the quick and cost effective identification of the largest leakers that contribute the vast majority of the fugitive emissions. The camera lens selects and displays a particular frequency range and filters out the light outside of that frequency range (infrared)[9]. All components that visibly leak, using the infrared camera, are referred to as "potential leakers" and are then subjected to USEPA Method 21 monitoring, which involves placing a portable, properly calibrated, gas-sampling probe near the surface of each piping component seal for a defined period of time, and measuring the corresponding VOC concentration. Leaking equipment with a hydrocarbon concentration equal to, or exceeding 10,000 ppmv, is considered a "leaker". Most EPA New Source Performance Standards (NSPS) have a leak definition of 10,000 ppmv[2]. USEPA Method 21 is intended to locate and classify leaks only, and is not intended to be used as a direct measure of mass emission rates from individual sources[4]. The benefit of using the infrared camera is the ability to assess a significantly higher volume of components than using the procedures of Method 21 alone. Other benefits of using the camera included being able to pin-point a leak, to detect very small leaks, and to identify leaks up to 10 m away, in calm weather conditions.

During windy conditions, or in portions of the plant with limited shelter, the Method 21 procedure (hydrocarbon analyzer) may be used initially, and if a leaking component (hydrocarbon concentration > 10,000 ppmv) is identified, confirmation monitoring is conducted using the infrared camera to pin-point the leak location. This occasional 'reverse' method of monitoring is based on plant conditions and the experience of the RasGas LDAR technicians that some leaks may go undetected if solely relying upon the infrared camera for the initial screening of leaks.

For efficient, reliable and accurate data collection during monitoring, an electronic data logger is used to collect the monitoring data. Prior to monitoring, the data set of the selected group of components is downloaded to the data logger. During the equipment monitoring, the LDAR technicians record the tag number and the measured concentration of each component into the data logger. At the end of each monitoring session, the technician uploads the information collected in the data logger into a database via a USB port connection.

Components with a leak concentration of 100,000 ppmv or greater are considered by RasGas to be of high risk, and repair attempts are made as soon as possible. This concentration (>100,000 ppmv) was selected for this determination, as it is a limitation of the sampling equipment. The actual concentration could be significantly higher than 100,000 ppmv, thus a heightened level of urgency is considered.

Components with a leak concentration between 10,000 ppmv and 100,000 ppmv are repaired and re-monitored within 15 calendar days, unless standards allowing delay are met[3].

The priority and feasibility of repairing a given component will depend on the leak rate, value of the process fluid being lost, cost of repairs, life expectancy of the repair, and the value of various indirect factors such as avoiding safety, health and environmental impacts, avoiding damage to the component, improved process reliability and better performance[5].

Following each monitoring interval, the RasGas Environmental Department will send a "leak report" to the managing team of the monitored area. The managing team has the responsibility for raising the request for repair with the maintenance department. Following the repair attempt, a notification from the managing team to the Environmental Department is made, who will then re-monitor the equipment, using both pieces of equipment. A leak repair will be considered successful if after re-inspection (follow-up monitoring), the VOC instrument reading is below 10,000 ppmv. In most cases, the RasGas maintenance department will inform the Environmental Department of the repair schedule, such that the environmental technicians will accompany the maintenance technicians to the leaking equipment. The advantage of this process allows the maintenance technicians to see the leak using the infrared camera, conduct the repair, then have immediate confirmation (or further repair) of the leaking component.

The field results have determined that gate valves were the pre-dominant type of component that leaks. The gate valves are routinely opened and closed, thus the packing may be disturbed or deteriorates, thus generating leaks. The leaking equipment, however, can generally be repaired through the tightening of the packing bolts. The pre-dominant location of the leaking equipment was on the upper-most working platform of the plant.

The monitoring data is stored and managed through a commercial LDAR database. The database records all monitoring efforts, repair attempts, delay of repair and information regarding the equipment in VOC service. The database also schedules inspections, generates custom reports and provides access to historical data such as leak history, repair records and analyzer calibration information.

RasGas has informed both the Ministry of Environment and RLC of its LDAR program, including presentations and status reports. A summary of the leaking equipment is provided within the quarterly environmental monitoring report, submitted to the Ministry of Environment and RLC.

3. Estimating Equipment Leak Emissions

There are several approaches for estimating equipment leak emissions, including the 'Average Emission Factor' approach, the 'Screening Ranges' approach (formerly known as the leak / no leak approach), the "EPA Correlation' approach and the 'Unit-Specific Correlation' approach. Specific differences between these approaches are described by the USEPA, November (1995[10].

To calculate the leak rate of VOC emissions at RasGas, the 'Screening Ranges' approach has been adopted. The Screening Range approach offers some refinement over the Average Emission Factor approach, thereby allowing some adjustments for individual unit conditions and operation. This approach requires that screening data is collected from the process equipment, and it is assumed that components having screening values greater than 10,000 ppmv have a different average emission rate than components with screening values less than 10,000 ppmv[1].

To determine the 'Control Effectiveness' of the LDAR program, three important factors need to be considered: (1) how a 'leak' is defined; (2) the initial leak frequency before the LDAR program is implemented; and (3) the final leak frequency after the LDAR program is implemented. The leak definition is the screening value at which a 'leak' is indicated if a piece of equipment screens equal to or greater than the leak definition. The leak frequency is the fraction of equipment with screening values equal to or greater than the leak definition[1].

The Control Effectiveness is calculated as follows:

$$Eff = (ILR - FLR) / ILR \times 100$$

Where:

Eff = Control Effectiveness (percent);

ILR = Initial leak rate (kg/hr); and

FLR = Final leak rate (kg/hr).

To calculate VOC emissions using the Screening Ranges approach, the following equation is used:

$$E_{VOC} = (F_G \times N_G) + (F_L \times N_L)$$

Where:

E_{VOC} = VOC Emission rate for an equipment type (kg/hr);

F_G = applicable emission factor for sources with screening values greater than or equal to 10,000 ppmv (kg/hr/source);

N_G = equipment count (specific equipment type) for sources with screening values greater than or equal to 10,000 ppmv;

F_L = applicable emission factor for sources with screening values less than 10,000 ppmv (kg/hr/source); and

N_L = equipment count (specific equipment type) for sources with screening values less than 10,000 ppmv

To date at RasGas, leaking components have only been associated with valves, both in gas and in light liquid service. To establish the Control Effectiveness, the initial and final leak rates were determined for the leaking valves, which are presented in Table 1. Table 1 also includes the RasGas Control Effectiveness, with a leak definition of 10,000 ppmv.

Table 1
RasGas Equipment Leak Rates, Control Effectiveness

Equipment Type and Service	Initial Leak Rate (kg/hr)	Final Leak Rate (kg/hr)	Control Effectiveness (10,000 ppmv leak definition)
Valves – gas	4.46	0.741	83.4
Valves – Light Liquid	1.36	0.318	76.6

The EPA[1] has developed Control Effectiveness percentages for equipment monitored at specific leak definitions and frequencies. Table 2 summarizes the Control Effectiveness for equipment at petroleum refineries for different monitoring intervals, with a leak definition of 10,000 ppmv.

Table 2
Control Effectiveness for LDAR Component Monitoring Frequencies for Petroleum Refineries

	Control Effectiveness	
Equipment Type and Service	**Monthly Monitoring 10,000 ppmv leak definition[A]**	**Quarterly Monitoring 10,000 ppmv leak definition[A]**
Valves - gas	88	70
Valves – light liquid	76	61
Pumps – light liquid	68	45
Compressors - gas	B	33
Connectors – gas and light liquid	C	C
Pressure relief valves - gas	B	44

A – Source: EPA, October 1997
B – monthly monitoring is not required in any control program
C – Information not available

At this stage in the LDAR program, due to the size of the RasGas facility and available resources, annual monitoring of the components is presently taking place. In comparing the RasGas annual Control Effectiveness to the EPA results (Table 2), the RasGas annual Control Effectiveness is higher than the quarterly Control Effectiveness, yet below the monthly Control Effectiveness of the EPA for gas valves. The annual Control Effectiveness for light liquid valves is slightly higher than monthly Control Effectiveness at petroleum refineries. With an increase in the monitoring frequency at RasGas, it is reasonable to expect a correspondingly higher Control Effectiveness.

4. Conclusion

RasGas has successfully launched and is making significant progress with respect to its LDAR program. Successes are being realized through a Control Effectiveness percentage that is consistent with EPA references. These successes are also realized through lower emissions, increased safety for plant personnel and reducing product loss.

As RasGas continues to implement its LDAR program, evaluations will take place periodically, which measure the successes of the program and where focused efforts are necessary. More effort will be directed at components that are more susceptible to leaks, while less effort, based on data, will be directed at components that have a low likelihood of leaking.

References

[1] Preferred and Alternative Methods for Estimating Fugitive Emissions from Equipment Leaks, Eastern Research Group, November 1996.
[2] Leak Detection and Repair, A Best Practices Guide, United States Environmental Protection Agency, October 1997.
[3] Environmental Regulations for Ras Laffan Industrial City, Ras Laffan Industrial City, Revision 1, 2005.
[4] United States Environmental Protection Agency, Method 21, Determination of Volatile Organic Compound Leaks.
[5] Management of Fugitive Emissions at Upstream Oil and Gas Facilities, Canadian Association of Petroleum Producers, 2007.
[6] United States Environmental Protection Agency, Petroleum Fugitive Emissions – Background Information for Promulgated Standards, October 1983.
[7] United States Environmental Protection Agency, Title 40, CFR Part 60, Standards of Performance for New Stationary Sources, Subpart KKK, Equipment Leaks of VOCs from Onshore Natural Gas Processing Plants, July 1991.
[8] URS Consultants – Doha, LDAR Procedures, 2007.
[9] Methods for Quantification of Mass Emissions from Leaking Process Equipment when Using Optical Imaging for Leak Detection, Miriam Lev-On, et al, September 2005.
[10] United States Environmental Protection Agency, Protocol for Equipment Leak Emission Estimates, November 1995.

Proceedings of the 2nd Annual Gas Processing Symposium
Farid Benyahia and Fadwa T. Eljack (Editors)

Development of Al Dahkhirah Sewage Treatment Works to Reuse Treated Water for Construction Works, Qatar

Khaled El Emadi [a], Ghazy Abdel Kerim [b], Moaaz Hashim [c]

[a] Manager of Quality,Safety& Environment Department, PWA*, Doha 22188, Qatar.
[b] Environmental Expert, Prof. of Environmental Engineering, PWA*, Doha 22188, Qatar.
[c] Senior Lab Specialist, Assistant Prof., PWA*, Doha 22188, Qatar.

Abstract

Public Works Authority (PWA) have been established in the state of Qatar in the year 2004, and since then PWA started planning, designing and implementing sanitary projects depending on professional consulting offices as well as international contracting companies. These consultancy and contracting services are to design, build and operate sanitary drainage works that produce treated water suitable for different reuse applications. During these decades, the capacity of STWs will be extended from 140800 m3/d up to 377000, 632500 m3/d, and 713500 m3/d in years 2013, 2020 and 2030 respectively.

PWA plans for developing the existing sewage treatment works to produce TSE suitable for reuse for unrestricted irrigation. In the same time, PWA plans for further upgrade for the STWs projects to extend TSE reuse applications in building works, road works, industrial applications as well as cooling purposes. In 2008, Ashghal initiated demonstration pilot project including upgrading of one sewage treatment work (Al Dhakhirah STW) by ultra filtration UF units & reverse osmosis RO units in order to reuse TSE for construction works in two major projects: Doha North STW Project and North Road Project.

This paper presents the development of one of the existing sewage treatment works (Al Dahkhirah), it describes the additional treatment works, which are implemented. It demonstrates produced treated water quality parameters in terms of Chloride, Sulphate, Alkali Carbonates and bicarbonates, Total Dissolved ions and pH. Moreover, it displays concrete testing regarding comprehensive strength. Furthermore, it expresses future vision of Ashghal with regard to treated wastewater reuse in construction works in the state of Qatar.

Keywords: Domestic, Wastewater, Treatment, Reuse, Construction.

PWA : Public Works Authority

1. Introduction

PWA has taken the responsibility to upgrade the existing sewage treatment works to achieve the settled TSE reuse standards for unrestricted irrigation. In the same time, PWA plan for further upgrade for the STWs projects to extend TSE reuse applications beyond irrigation. The TSE quality will be improved to the higher quality standard to extend the reuse application beyond irrigation. It is estimated that the surpluses TSE would reach 211670 m3/d (77 millions m3/year) up to 2030. Further studies should be carried out to cover other TSE reuse applications in building works, road works, industrial applications as well as cooling purposes.
Accordingly, Quality Safety and Environment Department (QSED) PWA, initiated a pilot project to upgrade one of the existing sewage treatment plant for the purpose of TSE reuse in construction works of one main PWA project. The main purpose of this pilot project is to demonstrate the TSE reuse applications in construction works.

Both Doha North Sewage Treatment Works Project & North Road Project demands a large amount of water for non-potable use during the construction phase. Instead of tapping valuable potable water from the city of Doha, Keppel Seghers is installing a Potabloc System at Al Dahkhirah Sewage Treatment Works to process tertiary treated effluent into high grade reclaimed water to meet the non-potable water requirement of the Doha North Project. Tekfen install additional chlorine injection units to improve TSE quality to meet the non-potable water requirement of the North Road Project

2. Methodology

In year 2008, QSED PWA began to review Qatar Construction Specifications (QCS) 2007, to find that section No. 5 Part 4 is related to water specifications used in concrete works.
In order to modify Qatar Construction Specifications (QCS) 2007, Municipal Treated Water (MTW) should be consider to be used in parallel with potable water in concrete works. Accordingly, QSED PWA, initiated cooperation with both Qatar General Organization for Standards and Metrology (QGOSM) and National Technical Committee for Construction & Building Specifications Sector (NTCCBSS).

First, on February 2008, a small committee was established headed by the manager of QSED and including members from QGOSM staff and QSED PWA staff to set up the recommendations regard (QCS) 2007 modifications. The committee recommendations were based on the following outlines:

- To consider the European Standard EN 1008:2002, superseded British Standards, to be the base for modification of QCS 2007. These standards describe mixing water for concrete specifications regard sampling, testing and assessing the suitability of water, including water recovered from processes in the concrete industry, as mixing water for concrete.

- To consider Municipal Treated Water (MTW) as water recovered from processes in the concrete industry to be used as mixing water for concrete in condition fulfilling the water specifications stated in EN 1008:2002.
- To add health and safety specifications related to applying MTW in concrete works.

Second, the recommendations were sent by the head of this committee to the general director of QGOSM, where in sequence the recommendations were reviewed by Specification & Metrology Department. The reviewed recommendations were passed to NTCCBSS for second review. Finally, the final recommendations were delivered to the general director of QGOSM for preliminary approval in October 2008 to proceed it to the minster level, see table (1).

In order to upgrade one of the existing sewage treatment plant, QSED conducted survey on available small scale plants and selected Al Dahkhirah STWs. Moreover, QSED investigated the contractors who are willing to use TSE in construction works. As a result, cooperation was established with Keppel Seghers, Doha North STWs Project Contractor and Tekfen Construction, North Road Project.

First, QSED and Both Keppel Seghers & Tekfen carried out site visit survey of Al Dahkhirah STWs, where the location of the recommended pilot was identified. The plant was evaluated by studying its design data and samples were taken from its processes and analyzed. Moreover, both Keppel Seghers & Tekfen presented technical offers to QSED, which display the recommended polishing units to be added to the system. The main recommended polishing units are Ultra Filtration, Reverse Osmosis and chlorine injections units.

Second, QSED reviewed the technical offers as well as inspected the pilot influent and effluent connections within the existing facilities. Accordingly, Keppel Seghers considered two main storage tanks: one for TSE taken from the existing final TSE storage tank, the other tank for RO product storage tank. Moreover, Keppel Seghers added a tanker filling station, where RO product is discharged into tankers to transport the RO water into the construction site. In the same time, Tekfen considered chlorine injection units, one main TSE storage tank and TSE tanker filling line.

Third, PWA QSED carried out two agreements with Keppel Seghers, Doha North STWs Project Contractor and Tekfen Construction, North Road Project. The main lines of this agreement are as follow:

- Keppel Seghers & Tekfen supplies install and operate the pilot project during the duration of Doha North STWs project and North Road Project.
- Keppel Seghers and Tekfen carry out sampling and testing as well as water quality weekly reporting and forwarding for QSED for approval.
-

Table 1: Chemical Limitations for Mixing Water
Qatar Construction Specifications QCS 2007 Section 5(4) and Preliminary Approved Modifications.

Parameter	QCS 2007 Test Method	New PRELIMINARY Approved QCS Test Method	QCS Maximum Limit, mg/	New PRELIMINARY Approved QCS Maximum Limit, mg/l
(a) Chloride (as Cl) (a-1) for Pre-stressed concrete. (a-2) for reinforced concrete. (a-3) for concrete without reinforcement.	BS 1377 : Part 3	EN 196-2	500	500 1000 4500
(b) Sulfates (as SO_4^{2-})	BS 1377 : Part 3	EN 196-2	1000	2000
(c) Alkali (c-1) Alkali carbonates and bicarbonates (c-2) Alkali equivalent sodium oxides	BS 2690 : Part 109	BS 6068-2.51 BS EN ISO 9963-1 BS EN ISO 9963-2	500	500 1500
(d) Total dissolved ions, including a, b and c above (d-1) for pre-stressed concrete (d-2) reinforced concrete. (d-3) for concrete without reinforcement.	BS 1377 : Part 3	EN 196-2 BS 1377 : Part 3	2000	1000 2000 5000
(e) pH	BS 2690 : Part 5	BS 6068-2.50	7 - 9	6.5 - 9.0
COD		SM 5220 B		100
Phosphate; expressed as PO_4 - P		SM 4500 P B, C, SM 4500 PD by subtraction		30
Nitrate; expressed as NO_3^- -N		ISO 7890-1		100
Lead; expressed as Pb^{2+}		SM 3125B (ICP/MS)		100
Zinc; expressed as Zn^{2+}		SM 3125B (ICP/MS)		100

- Keppel Seghers obtains 600 m3/d RO product water for use in concrete works during the project period.
- Tekfen obtains 1200 m3/d TSE water for use in earthworks, filling works, and dust control during the project period.

In addition, QSED allocated three Auto samplers for Sampling and Quality control of the pilot connection with the existing facilities. QSED carried out its routine sampling and testing and weekly reporting on the existing plant as well as the RO pilot, as quality control procedures. Finally, all RO pilot and auxiliaries works supply, installation, and commissioning were carried out during the year 2008.

3. Results & Discussion

The following results were focused only on RO water quality produced by Keppel Seghers and the concrete testing produced by RO water.

3.1 The Existing Al Dahkhirah STWs

The existing Al Dahkhirah STWs consists of preliminary treatment, secondary treatment, and sand filtration as well as chlorination. The plant receives around 4000 m3/d, however its design capacity is 1600 m3/d. Its treated sewage effluent TSE is connected to two tankers filling stations, where TSE transported by tanker to be used for trees as well as landscape irrigation in Al Khore Community as well as Al Dahkhirah surroundings. The excess TSE is discharged into natural depression forming lagoon and causing annoyance to the surrounding housing.

3.2 The RO Pilot

The RO Pilot named the Potabloc System is an innovative, self-sustaining and mobile water production unit that is capable of producing high quality water from unconventional sources using technologically advanced membrane separation techniques such as ultra-filtration and reverse osmosis. The Potabloc is designed to receive 1250 m3/d, where Ultra filtration units design capacity is 1100 m3/d with UF recovery 90%. The Reverse Osmosis RO units design capacity is 600 m3/d with RO recovery 56%. Therefore, the RO Product Water capacity of the pilot is 600 m3/d. The reject from the Potabloc system is returned to the TSE storage Tank of Al Dahkhirah STWs.

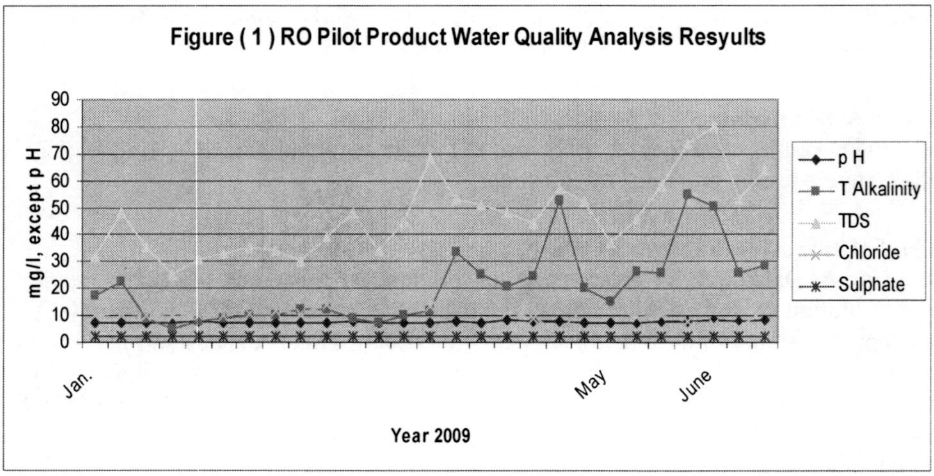

3.3 RO Product Water Quality

The RO Pilot was commissioned in year 2008 and it started regular operation in January 2009, weekly sampling and water quality testing was carried out in external qualified laboratory approved by PWA. Water quality testing were mainly focusing on QCS 2007, section no.5 part 4 parameters such as pH, Total Alkalinity, TDS, Chloride, and Sulfate. Moreover, other water quality parameters were tested to satisfy the requirements of preliminary approved QCS modifications added parameters such as Phosphate, Nitrate, Lead, and Zinc. In addition a comprehensive list of parameters were tested to cover a complete picture of RO Product water quality. These parameters are: Ca, Mg, Na, Fe, Cu, Mn, Al, B, SiO2, K, Hg, as well as EC, CO3, HCO3, F, TKN, TOC, NH3-N, and odor.

The results of RO Product Water quality analysis show that parameters such as pH, Total Alkalinity, TDS, Chloride, and Sulfate are within the acceptable level as indicated in QCS 2007 and preliminary approved QCS modifications, see Figure (1) & Table (1). Moreover, the water quality analysis results demonstrate that Phosphate, Nitrate, Lead, and Zinc are extremely below the acceptable limit as indicated in preliminary approved QCS modifications added parameters, see table (2).

Table 2: RO Water Quality Compared with Modified QCS 2007

Samples No. = 40	Phosphate	Nitrate	Lead	Zinc
	mg/l	mg/l	mg/l	mg/l
Max.	0.1	0.1	0.095	0.15
Min.	0.1	0.1	0.005	0.005
Avg.	0.1	0.1	0.008	0.017
Median	0.1	0.1	0.005	0.005
Standard	30	100	100	100

3.4 Impact of RO Pilot Operation on Existing Al Dahkhirah STWs

QSED was auditing the impact of RO pilot operation on the existing Al Dahkhirah STWs, where the reject from the Potabloc system operation is returned to the TSE storage Tank of Al Dahkhirah STWs. The quality of Secondary Treated Sewage Effluent discharged from the TSE storage tank was recorded before and after RO pilot operation. The TSE water quality data of months January, February, March, April, and May were compared between year 2008 and year 2009. As known that the Potabloc system mainly reduce the TDS level of the RO Product water, and the reject is returned back to Al Dahkhirah STWs, the main parameter which is expected to be the indicator of impact is considered TDS. The comparison results show that there are no significant change of TSE TDS level, with a very slight average increase approximately 1 %. Other TSE water quality parameters were checked such as pH, TSS, and COD as well.

Figure(2) Secondary Treated Sewage Effluent Quality, TDS Level

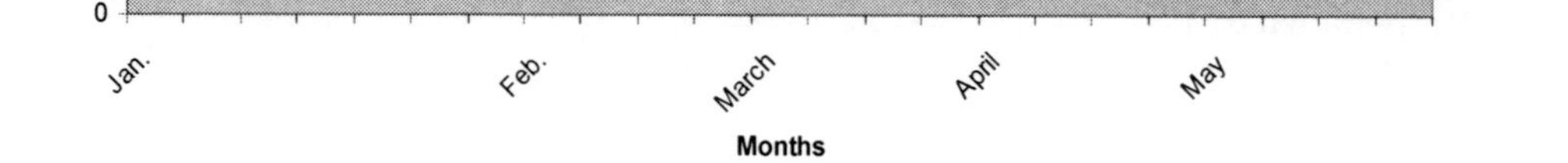

3.5 Quality of concrete produced by using RO water

In order to assess the suitability of RO water for mixing concrete, 519 test cubes were prepared on site during the period from February to June 2009. The test cubes no. 21, 99, 163, 187, and 49 per February, March, April, May and June respectively, were tested for concrete strength and the result was shown in figure (3).

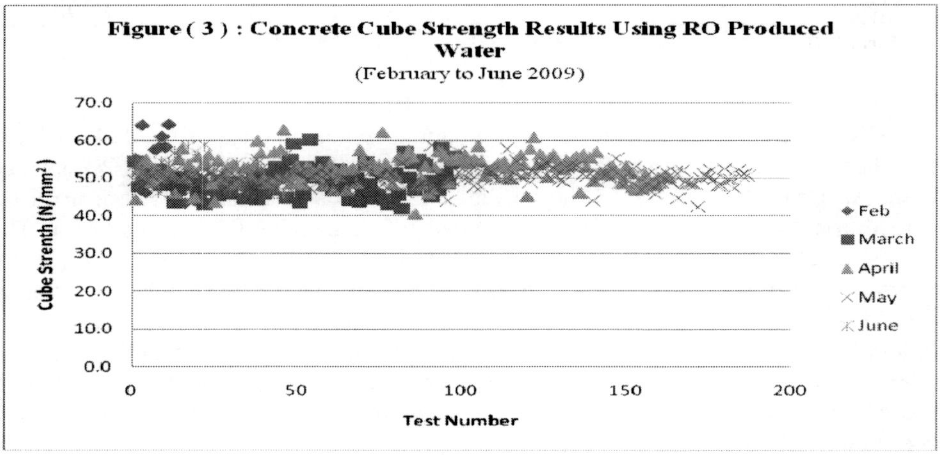

From figure (3), it can be seen that all these cubes revealed a concrete strength of more than a specified strength of **40 KN/mm²** which indicates that the concrete mixed with RO water does not differ from that produced by using potable water with regard to concrete strength.

4. Preliminary Cost Analysis

Initial cost analysis was carried out on treated sewage effluent of STWs. The preliminary data indicates that the cost of treatment of crude sewage to achieve secondary treated water and RO water are around 2 QR/m^3 and 2 QR/m^3 respectively. These preliminary data does not include the cost of crude sewage collection sewerage works and RO water transportation cost.

5. Paper limitation and recommended Further Work

The paper is mainly focusing on the RO water quality and strength of concrete produced by using RO water. Further work is required to test concrete produced by using different types of water (Secondary treated water, Filtration & Ultra filtration water and potable water).

Comparative Cost analysis study should be carried out between concrete productions by using different types of water. Moreover, it is scheduled for testing regime under rigorous conditions to compare concrete mixed with different types of water compared to that with potable water. Further investigation would be carried out to perform comprehensive cost analysis including STWs life span, capacity scale, collection & distribution works fixed and O&M cost, and present values.

6. Conclusion &Recommendations

The RO pilot project results demonstrate the successful improvement of TSE quality achieving high water quality standards by using the polishing treatment units called Potabloc. Moreover, the results show successful application of RO Water in concrete works. It is recommended to continue auditing the Potabloc system during the project period with carrying out recommended further works. Moreover, PWA calls for contractors specifically implementing infrastructures projects to progress their applications to QSED PWA. The purpose of application is: 1) To study of the use of TSE to its undertakings construction activities and to contribute with qualified companies to supply and install polishing treatment units. 2) To improve the TSE quality to match with allowable water quality standards for water use in different construction activities such as concrete, blocks manufacturing, earthworks and filling works as well as dust control. 3) To guarantee successful applications, TSE reuse activities should be controlled and audited by Quality & Safety bodies.

Acknowledgements

The authors acknowledge the General Director of PWA, and all concerned parties including Drainage Project Department, Drainage O&M Department, Legal Affairs Department, Contract Affairs Department, and Public Relation Department as well as Qatar General Organization for Standards and Metrology for their cooperation to bring about this pilot project. Moreover, The authors acknowledge Keppel Seghers Engineering Singapore Pte Ltd, Tekfen Construction and Quality, Safety and Environment Department, PWA for their contribution to accomplish the main job of this pilot project.

References

Qatar Construction Specifications, QCS 2007, Section 5, Part 4, Superseded QCS 2002.

The European Standard EN 1008:2002 superseded British Standards 3148:1980, Mixing Water for Concrete- Specification for sampling, testing and assessing the suitability of water, including water recovered from processes in the concrete industry, as mixing water for concrete.

USA: ASTM 1602 and 1603 superseded ASTM C94.
Qatar Construction Specifications, QCS 2009, Section 5, Part 4, not published yet.
Ghazy Abdel Kerim and Jaber Al Mohanadi, " Ashghal * develop sewage treatment works to reuse treated water for unrestricted irrigation, Qatar" , Euro-Arab Environment Conference & Exhibition, State of Kuwait, 27-29th November 2006.
Ghazy Abdel Kerim, and Al Sharafi S., 14-15 April 2005, "Comparison Study Between Different Tertiary Treatment Processes To Achieve Microbiological Standards for Wastewater Reuse, Gulf Region" , Minia International CONFERENCE " Towards a Safe and Clean Environment " , TSCE 05.
Metcalf & Eddy, 1991, " Wastewater Engineering, Treatment, Disposal, and reuse", Third edition, McGraw-Hill.
WRC, January 1999, Alexandria Effluent and Sludge Reuse Study, WRC Ref: AESRS02.
WHO, 1989, Health Guidelines for the use of Wastewater in Agriculture and Aquaculture, Technical Report Series 778, Geneva.

- م. عصام الربيعان، 2004، " استخدام المياه المعالجة في الأعمال الإنشائية بمملكة البحرين – منظور بيئي للتنمية المستدامة " ، البحرين، ص.ب. 54549 العدلية.
- أ.د. إبراهيم الدرويش و أ.د. عبد الوهاب عوض، 1986، " الخلطات الخراسانية" ، دار الكتب الجامعية، كلية الهندسة بجامعة الإسكندرية وجامعة بيروت العربية.

Proceedings of the 2nd Annual Gas Processing Symposium
Farid Benyahia and Fadwa T. Eljack (Editors)

Bio-ethanol from Municipal Solid Waste (MSW): the Environmental Impact Assessment

Majeda Khraisheh[1i], Aiduan Li[2]
[1]*University of Qatar, Department of Chemical Engineering, Doha P.O.Box 2713, Qatar*
[2]*University College London, Department of Civil, Environmental and Geomatic Engineering, Gower Street, London, UK*

Abstract
This paper discusses the potential impacts of using MSW as bio-ethanol resources from the prospects of both energy development and waste management. The investigation considered the waste produced in London as a case study. The use of MSW as biomass source for the production of bio-ethanol production was then investigated in the laboratories of University College London. The Experimental results showed that a conversion rate of around 85% was achieved, which is relatively high compared to other conventional biomass. This paper identifies the possible application of this technology on both energy production and waste management by providing valuable product to meet energy demand and protecting the environment from pollution. The potential impacts on related environmental issues, such as sustainable waste management, climate change, water issues, land use and biodiversity, are discussed.

Keywords: Waste management, bio-ethanol, municipal solid waste, bio-energy, waste management, environmental impacts

1. Introduction
In a context where energy security, food security and waste management are becoming some of the greatest challenges for many nations, bio-ethanol production from biodegradable municipal solid waste (BMSW) offers tremendous perspectives. This technique has a clear potential for commercialisation but the cost of this emerging process is currently too high to allow its implementation on an industrial scale. Indeed, a number of studies have attempted to identify specific technical opportunities to lower bioethanol production costs. In its simplest form, bioethanol production relies on three consecutive stages: pre-treatment, hydrolysis and fermentation (Fig.1). To reduce ethanol price and ensure food supply security, it is necessary to investigate alternative biomass sources; sometimes termed second generation biomass. These include forestry waste, agricultural waste, or municipal solid waste (MSW). Every year, each household in London produces a tonne of waste. Additionally, London Boroughs also collect some commercial and industrial waste, resulting in over 4.4 million tons of waste to be managed and costing over £30 million last year alone. Although London's recycling rate has increased over the past 5 years only 9% of waste is currently recycled. With this in mind, this project involves an experimental programme on the underlying processes of converting MSW into a valuable product (ethanol).

Ethanol is an organic solvent rated second to water in universal application. In 2001, 24.8 million tonnes of ethanol were produced worldwide. Of that, 92% was used in fuels. Ethanol can eliminate the use of methyl tertiary butyl ether, MTBE, an oxygenate currently used in fuel supplies. The reasoning behind this is that MTBE pollutes ground water resources, and can contaminate drinking water giving it a foul taste and odour. The European Union Directive (2003/30/EC) has set a target of 5.75% ethanol mixture with gasoline by 2010. This could result in an increase in the European ethanol demand. The great growth in ethanol demand is expected with an estimation of the ethanol market being worth up to 18.9 billion litres by 2012, approximately 150% greater than the market in 2004 (as represented on Figure 1).

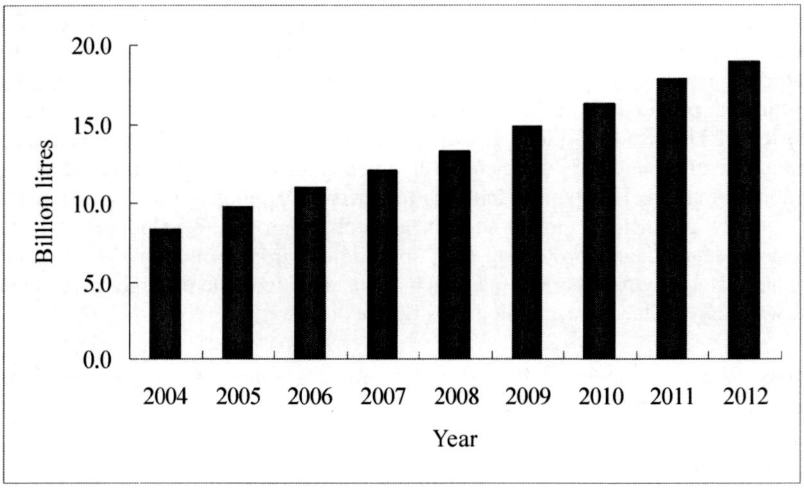

Figure 1 : EU Ethanol Demande

The process of converting BMSW to ethanol is shown in Figure 2. Pre-hydrolysis treatment and enzymatic hydrolysis of BMSW are of crucial importance during the bioconversion of BMSW-to-bio-ethanol, and thus their optimization will result in beneficial environmental and economic practices (Wyman, 1999). A previous study by the autors has investigated the impact of pre-hydrolysis treatments (i.e., dilute acid type and concentration, steam treatment, microwave) and enzymatic hydrolysis (i.e., enzyme type and concentration) on the bioconversion of BMSW-to-glucose at laboratory-scale (Li et al., 2007).

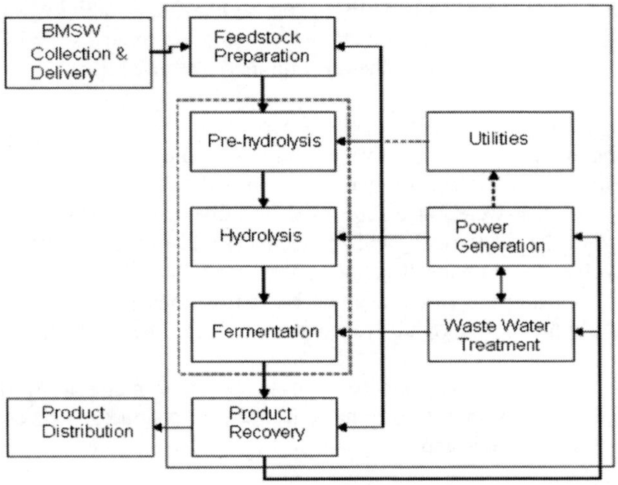

Figure 2: A typical ethanol production process

2. Materials and Methods

2.1. Sample and pre-treatment

The BMSW used in this study was selected from typical biodegradable waste, such as kitchen waste, garden waste and paper waste. The sample consists of 20% carrot peelings, 20% potato peelings, 20% grass, 20% newspaper, and 20% scrap paper. The selected waste was milled to small particles with size of 0.2mm-1.2mm, and then followed by pre-hydrolysis with sulphuric acid and steam treatment, and then enzymatic hydrolysis.

2.2 Pre-hydrolysis treatment

Biomass and dilute sulphuric acid (1%) were mixed at a solid/ liquid ratio of 0.5 g in 50 ml. The acid-soaked samples were then placed in an autoclave at a constant temperature 121°C for 60 minutes. Following steam treatment, the samples were washed with deionised water until the pH of the filtrate remained constant at an approximate value of 4.5. The pre-treated samples were drained and air-dried overnight to a final 45% moisture content

2.3. Enzymatic hydrolysis

Enzymatic hydrolysis of pre-hydrolysed substrate was carried out within the study temperature range in vials (50 ml) placed in an orbital agitator at 68 rpm for 24 h. The liquor pH was adjusted using 0.1 M citrate acid-sodium citrate buffer. The enzyme, *T. viride* (Sigma, UK) was used as received in combination with the enzyme *beta-glucosidase* (Sigma, UK). The cellulase activity of the enzymes was measured by the

Filter Paper assay developed by Mandels et al., (1979), and the activity was expressed in terms of Filter Paper Units (FPU). The b-glucosidase activity was measured by the PNPG assay (Paquot and Thonard, 1982), and reported as PNPG units (PNPGU). Samples were withdrawn from the reaction media after 24 h.

2.4 Analytical methods

Sugar concentrations were determined routinely from centrifuged samples (13,000 rpm, 5 min) using HPLC (Perkin-Elmer Ltd, UK). All compositional analyses were done in triplicate, and data were corrected to a 100% dry matter (DM) basis. Glucose yield as percentage of the theoretical yield [percentage digestibility, obtained from the equation which involves the transfer of cellulose to sugar $(C_6H_{10}O_5)_n + nH_2O = (C_6H_{12}O_6)n$] was computed by using the formula given by the National Renewable Energy Laboratory (Standard Biomass Analytical Procedures, NREL). The study range is summarised in Table 1.

The mass balances were calculated from the raw material to the pre-hydrolysed biomass, as well as to the enzymatic hydrolysis residues, with consideration of the weight loss during each step.

Table 1: Study experimental factors for the enzymatic hydrolysis process

Factor	Study range
A: Particle size	2-20 mm
B: Substrate concentration	5-15 g/l
C: Enzyme loading	10-100 FPU
D: Beta-glucosidase	10-100 PNPGU g–1
E: pH	3.7-5.0
F: Temperature	37-50 °C

2.5 Environmental Impact assessment

The environmental impact assessment was based on the results obtained from the laboratory work described above. The assessment covers a number of environmental issues such as waste management, carbon dioxide emission, water issue, land-use, and biodiversity etc, by comparing with traditional biomass such as corn and energy crops.

3. Results and Discussion

3.1 Conversion studies

Our preliminary study (Li et al., 2007) showed that BMSW as a feedstock produced about 52% glucose for the production of ethanol. Recent studies on hydrolysis optimization showed about 85% glucose yield can be reached under the optimal conditions (lower substrate concentration, lower beta-glucosidase loading, and temperature 50°C) (Figure 3).

The contribution of substrate concentration to the glucose yield was significantly higher (93.6%), than the contribution of temperature (3.2%) and beta-glucosidase loading (2.6%). Other factors such as particle size, enzyme loading, and pH have very little effect on the glucose yield within the study range. The results obtained by the model presented a satisfactory correlation to the actual values (R^2 = 0.90). Additionally, this study indicated that lower substrate concentration (5 g/l), higher temperature (50°C) and less beta-glucosidase produced the highest glucose yields. The significant contribution of substrate concentration and temperature at 50°C was in agreement with conventional biomass (Ballesteros et al., 2002). The reason of the lower value of beta-glucosidase may be due to enzymatic inhibition (Gregg and Saddler, 1996). This study indicated that BMSW fractions did not require higher cellulose loadings to obtain high glucose yields, which can reduce the cost of the process.

(a)

Factor	% Contribution
B - Substrate Concentration	93.6
D - Beta-glucosidase	2.6
F - Temperature	3.2

(b)

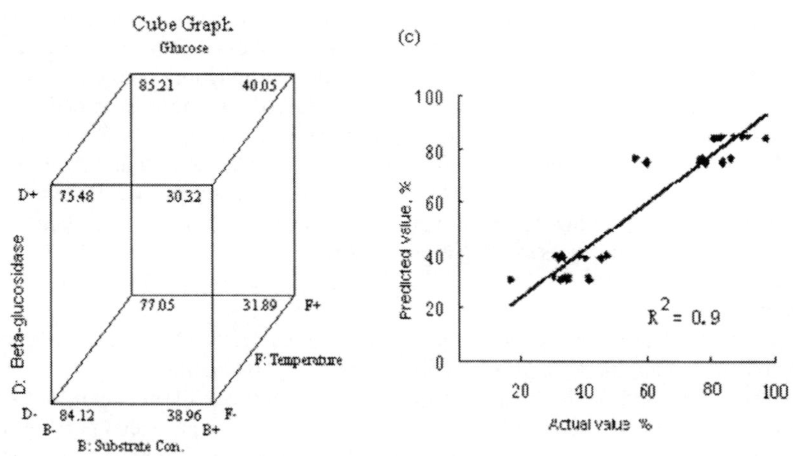

Figure 3: Factorial experimental design applied to BMSW fraction. (a) Factor effects on glucose yield, (b) cube graph of glucose yield, (c) comparison of actual value and predicted value

Results from the mass balance analysis revealed that about 0.25g biomass remained after the pre-hydrolysis step and 0.1g remained after the enzymatic hydrolysis process when 0.5g BMSW was used as feedstock. Consequently, around 80% of the biodegradable waste fractions were utilized in the BMSW-to-ethanol process.

3.2 Environmental impacts

The environmental impacts of the BMSW-to-ethanol process were briefly assessed for the following areas: waste management, climate change, water issues, land use, biodiversity and others.

3.2.1 Waste management

Using BMSW as resources for bio-ethanol production will reduce the biodegradable fraction from MSW going to landfill which is one of the requires of EU Landfill Directives. According to the mass balance from the laboratory work, the amount of waste will be massively reduced by 80% and the main composition of solid residues from the hydrolysis process are acid insoluble lignin, ash content, protein absorbed and dead cells, which can be used as compost. Thus, technology can be used as an alternative solution to the waste-to-landfill option. However, Sustainable waste management needs to involve different type of waste management methods in order to minimize the waste produced, and maximize energy/material recovery and to meet the needs of environmentally, economically and socially sustainable.

3.2.1 Climate change

Ethanol represents the closed carbon dioxide cycle as after burning, the released carbon dioxide is recycled back into plant material as they synthesize it to cellulose during photosynthesis cycle (Chandel et al., 2007). Ethanol production process only uses energy from renewable energy sources; no net carbon dioxide is added to the atmosphere, making ethanol an environmentally beneficial energy source. In addition, the toxicity of the exhaust emissions from ethanol is lower than that of petroleum sources. Apart from these, using BMSW as ethanol resources will avoid the biodegradable fraction going to traditional waste disposal plant such as landfill or incineration which often causes green gas emissions.

3.2.3 Water issues

Lange quantities of water are required for growing energy crops. One of the advantages is that irrigation can increase productivity surrounding plantations. For the ethanol process itself from both BMSW and energy crops, to produce 1 litre ethanol, 4-8 litres of water are required as well as 13 litres of sewage effluent are produced per 1 litre of ethanol produced as water is required to be the ground corn grain for the fermentation process. Water remains a major issue for bio-ethanol process from energy crops. The amount is reduced considerably when MSW is used to less than 30%.

3.2.4 Land use

Growing energy crop for bio-ethanol purpose required lots of land use. Harvesting corn stover may result in lowering soil organic carbon levels and soil nitrogen content, and may also increase soil erosion. If agricultural prices are increasingly linked to energy prices then impacts on food security will also be more volatility in the markets. World prices usually account for less than 25% for any variation at country level. It is currently hard to model effect of bio-fuels. Using BMSW as biomass can free the land used for landfill and growing energy crops.

3.2.5 Biodiversity

Input intensive energy crop production involves the movement of agrichemicals, especially nitrogen, phosphorus, and pesticides from farms to other habitats and aquifers. The use of BMSW for bio-ethanol purpose can avoid this problem.

Apart from the environmental impacts discussed above, this BMSW-to-ethanol system has its economic and social benefits. The end product: bio-ethanol is a valuable market product, which brings economic benefits from the non-valuable waste and meets the requirement of current energy market demands. However, this process requires the biodegradable waste fraction to be separated from non-biodegradable wastes from the households. The economic benefits with bio-fuel product will encourage waste separation.

3.2.6 Opportunities of MSW-to-ethanol as an alternative waste management

Waste management varies from developing nations to developed countries. Under different economic and social environment, different types of waste management methods from re-use, recycle, and energy/material recovery to disposal are used. Take the waste management system in London for example: if all London's BMSW available for new recovery options was used for bio-ethanol production an estimated 346 million Liters of ethanol could be produced by 2020, providing 14.6% of London's 2004 petrol car consumption. This evaluation should however be considered rough, and further consideration should to be made to a wider range of performance criteria and system requirements.

4. Conclusions

The assessments of the major environmental impacts (or sustainability indicators) of MSW have highlighted a range of environmental benefits to be gained from MSW energy recovery. These include:

- Reduced greenhouse gas emissions,
- Reduced acid gas emissions,
- Reduced depletion of natural resources (fossil fuels and materials),
- Reduced impact on water (leaching), and
- Reduced land contamination.

However, to realise the potential for bio-ethanol production requires performance across the lifecycle of the fuel chain against other fuels to be competitive within the end-product (bio-fuel and wider transportation fuel) market and the feedstock (waste resource) market. Net energy balance, greenhouse gas balance, and production costs are fundamental indicators for sustainable performance. There is a lack of published research examining this in regards to MSW-bio-ethanol by biochemical processes. Three important areas of research are required:

1. Lifecycle analysis of each of the competing MSW-ethanol technologies converting different BMSW feed-stocks;
2. A bio-fuel lifecycle comparative study which includes the optimum current and future MSW-bio-ethanol technologies; and
3. A waste management lifecycle comparative study which includes the optimum current and future MSW-bio-ethanol technologies

5. Acknowledgments

The authors acknowledge the financial support from RCUK Dorothy Hodgkin Postgraduate Award (UK) and University College London Graduate School Fellowship.

6. References

A. Li, B. Antizar-Ladislao, M. Khraisheh, 2007, Bioconversion of municipal solid waste to glucose for bio-ethanol production. Bioprocess and Biosystems Engineering. 30(3): 189-196.

A.K Chandel, E.S. Chan, R. Rudravaram, 2007, Economics and environmental impact of bioethanol production technologies: an appraisal Biotechnology and Molecular Biology Review, 2 (1), pp. 014-032.

C.E. Wyman, 1999, Biomass ethanol: technical progress, opportunities, and commercial challenges. Annual Review Energy Environment 24, 189–226.

D.J. Gregg, J.N. Saddler,1996, Factors affecting cellulose hydrolysis and the potential of enzyme recycle to enhance the efficiency of an integrated wood to ethanol process. Biotechnology and Bioengineering. 51 (4).

Geotimes, Weighing in on Renewable Energy Efficiency. August 2005.

M. Ballesteros, J.M. Oliva, P. Manzanares, 2002, Ethanol production from paper material using a simultaneous saccharification and fermentation system in a fed-batch basis. World Journal of Microbiology and Biotechnology 18 (6).

M. Mandels, R. Andreotti, C. Roche, 1976, Measurement of saccharifying cellulose. Bioeng Symp 6:21–33.

M. Paquot, P. Thonart, 1982, Enzymatical hydrolysis of regenerated cellulose. Holzforschung 36:177–181.

NREL (National Renewable Energy Laboratory). SSF Experimental protocols: Lignocellulosic Biomass Hydrolysis and Fermentation. 1995. http://www.devafdc.nrel.gov/pdfs/ 4691.pdf

Overseas Development Institute (ODI), 2007, Biofuels and poverty reduction:Is there a way through the maize? London, UK.

---

[i] Author to whom all correspondence should be addressed. Email: m.khraisheh@qu.edu.qa; Tel: 00974 4935403.

Proceedings of the 2nd Annual Gas Processing Symposium
Farid Benyahia and Fadwa T. Eljack (Editors)

Monoethanolamine biodegradation processes

Deshai Botheju[a], Jon Hovland[b], Hans Aksel Haugen[b], and Rune Bakke[a]

[a] *Telemark University College, Faculty of Technology, Porsgrunn, Norway*
[b] *Telemark Technological Research and Development Centre (Tel-Tek), Porsgrunn, Norway*

Abstract

Biodegradation of monoethanolamine is studied due to the potential relevance of biological MEA waste treatment. The suggested degradation pathways proceed through the formation of acetic acid and ammonium, under aerobic, micro-aerobic and anaerobic biodegradation conditions. Anaerobic and micro-aerobic digestion is identified as energetically favourable compared to aerobic treatment. A dynamic mass balance model for MEA biodegradation under complete anaerobic and micro-aerobic conditions is proposed based on published stoichiometric and kinetic parameters. The model is integrated in a standard anaerobic digestion model and implemented in a process simulation tool to predict and eventually test MEA wastewater treatment plant designs. The developed model is validated through a series of biodegradation experiments described elsewhere (Botheju *et al*, 2010a).

Keywords: Anaerobic digestion, biodegradation, micro-aeration, modeling, monoethanolamine

1. Introduction

Monoethanolamine (MEA) is often used as the solvent of choice in CO_2 scrubbing processes. Such scrubbing systems are increasingly being built (or designed to be built) during recent times due to the increased attention on carbon dioxide capture and sequestration (CCS) concept associated with CO_2 free power generation initiatives. Technical maturity of this solvent absorption system together with its unique features like high loading capacity for CO_2 at low partial pressures, fast reaction kinetics and high removal efficiencies (Goff, 2005) may ensure high use of this chemical for years to come.

A significant amount of MEA containing liquid wastes is generated due to degradation of the solvent in gas scrubbing. Land filling and incineration has been used for the disposal of such wastes in small quantities but such disposal methods may neither be economical nor environmentally sustainable for handling large quantities of such liquid wastes. Release of untreated or poorly treated amine wastes into the environment can harm the local ecosystems and public health. Eco-toxicity of different amines and health impacts of various chemical components present in waste amines are published (Eide-Haugmo *et al.*, 2008; Thitakamol *et al., 2007*).

1.1. Scope

The aim of this study is to incorporate published information on MEA degradation in a general model for anaerobic digestion of organic matter into methane rich biogas. The model is later implemented in a standard tool for process plant simulations (Botheju *et al*, 2010a). This article describes how monoethanolamine degradation pathways are implemented into a standard anaerobic digestion model, ADM1 (Batsone *et al.*, 2002). The model is developed based on amine biodegradation studies, mainly from a literature survey. An experimental study to validate the model is presented elsewhere (Botheju *et al.*, 2010a).

2. Biological Treatment of MEA wastes

The application of biological processes and biotechnology in the treatment of amine wastes generated by large CO_2 capture facilities has been investigated in few studies (Ohtaguchi *et al.*, 1995; Lai and Shieh, 1996; Ohtaguchi and Yokoyama, 1997; Liuzinas *et al.*, 2007). According to the study of Lai and Shieh (1996), MEA is highly biodegradable under the conditions of nitrate respiration using NO_3^- as an electron acceptor when sufficiently low ratio of total organic carbon to mixed liquor volatile suspended solids (TOC/MLVSS) is maintained. Ammonia, biomass and CO_2 are recognized as the major degradation end products under the condition of NO_3^- respiration.

Ohtaguchi *et al.* (1995) have demonstrated the possibility of aerobic degradation of amine wastes from CO_2 capture using pure cultures of *Escherichia coli* K 12. This microorganism is capable of cleaving monoethanolamine into ammonium ion and acetaldehyde. The generated acetaldehyde was further transformed into acetic acid while most of the ammonium was assimilated as a nitrogen source. Ohtaguchi and Yokoyama (1997) presented several different metabolic alternatives for MEA biodegradation used by a multitude of microorganism. They found that acetic acid is a main intermediate product in most of the aerobic metabolic pathways of MEA biodegradation. Therefore a possibility exists to redirect the process anaerobically towards CH_4 generation (biogasification) instead of CO_2 as the final aerobic end product. Other intermediate degradation products like ethanol, acetaldehyde, amino acids and H_2 are also candidates for further anaerobic degradation under methanogenic conditions.

2.1. Anaerobic biodegradation of MEA

Anaerobic digestion is energetically favourable in wet organic waste treatment, as energy can be recovered as methane. This is opposed to power consumption for aeration in aerobic processes. MEA conversion into mainly methane and ammonia has, however, not previously been demonstrated experimentally. Hongwei *et al.* (2005) theoretically recognized monoethanolamine as a readily biodegradable compound under anaerobic conditions. This was based on an analysis relating the anaerobic biodegradability of nitrogenous compounds to a set of three different molecular structure descriptors (second order molecular connectivity index (${}^2\chi^v$), third order path molecular connectivity index (${}^3\chi^v_p$) and energy of the highest occupied molecular orbital (EHOMO)). This method of biodegradability prediction can be a valuable tool as a screening method.

Speranza *et al.* (2006) summarized four different mechanisms to explain anaerobic biodegradation of alkanolamines (primarily MEA, DEA and TEA). All four mechanisms suggest that acetaldehyde and ammonia are products of anaerobic biodegradation of these alkanolamines. Acetaldehyde is easily converted to acetic acid, the main reactant in methane formation. Speranza *et al.* (2006) further illustrated that a strictly anaerobic gram-positive homoacetogenic bacterium (*Acetobacterium* sp., strain LuTria 3 – isolated from sewage sludge) was capable of converting triethanolamine (TEA) into acetate and ammonia. They suggested that the recognized bacterium can use a radical mechanism to process all three types of ethanolamines (MEA, DEA and TEA). The importance of adenosylcobalamin (coenzyme B12) is recognized to be a major co-factor initiating this kind of eliminase reactions (Speranza *et al.*, 2006).

Ethanolamine ammonia-lyase is recognized as a major type of enzyme catalyzing the biodegradation of ethanolamines. Numbers of microbes can produce this enzyme and use it to break down ethanolamines produced during breakdown of ethanolamine containing phospholipids (Abend *et al.*, 1999). Abend *et al.* (1999) further reported that several bacterial groups can use ethanolamines as the sole source of carbon and nitrogen. Monoethanolamine (MEA) seems to be the best substrate in this respect though other types of alkanolamines are also possible candidates (Fig. 1). It is also pointed out that the Ethanolamine ammonia-lyase belongs to a group of enzymes whose activity strongly depends on coenzyme B12.

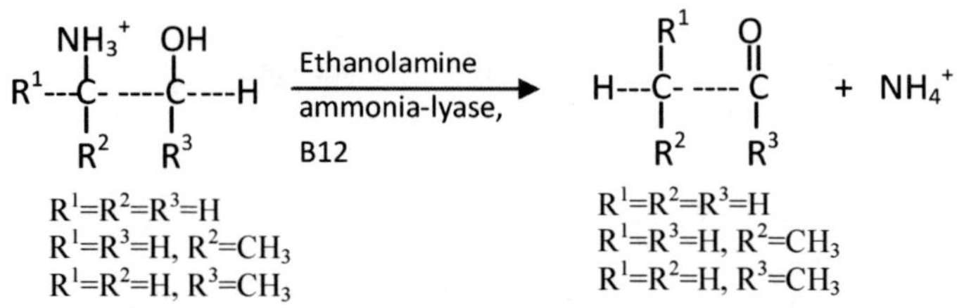

Figure 1: Ethanolamine ammonia-lyase and coenzyme B12 catalyses the biodegradation of alkanolamines (from Abend *et al.*, 1999)

2.2. Combined aerobic/ anaerobic treatment

The combined approach of anaerobic /aerobic treatment is sometimes used in treating recalcitrant nitrogen compounds like azo dyes and aromatic amines. Isik and Sponza (2008) reported that, in treating textile azo dyes, using an anaerobic step facilitate a rapid breakdown of azo compounds into aromatic amines but a subsequent aerobic stage is necessary to completely degrade these aromatic amine compounds. In general, azo compounds are degradable under highly reductive conditions like in an anaerobic environment, but the resulting aromatic amines need aerobic conditions in order to undergo further degradation (Georgiou and Aivasidis, 2006).

In the case of alakanolamines like MEA, an aerobic/anaerobic sequence can be more advantageous than the anaerobic/aerobic approach. This is proposed based on the observation that aerobic degradation of MEA can lead to rapid acetic acid formation (as described above). The subsequent anaerobic degradation stage can efficiently convert this into methane. Both the aerobic and the anaerobic steps can be carried out in a single reactor stage by the use of a micro-aerated anaerobic technique, to take advantage of both aerobic and anaerobic treatment (increased degradation rates and energy recovery). Micro-aeration is a technique where a limited amount of oxygen (usually as air) is injected to an anaerobic digester. Micro-aeration has been found to enhance the anaerobic digestability of wastes by enhanced hydrolysis (Zhu *et al.*, 2009; Johanson and Bakke, 2006; Juanga *et al.*, 2005).

2.3. Microbial toxicity

MEA can inhibit the growth of microorganisms, slow down or stop the degradation process due to its toxic effects. Microbial inhibitory concentrations suggested for MEA collected from the IUCLID (International Uniform Chemical Information Database) data sheets (available online on CPS&Q web pages) are presented in Table 1. These data sheets are based on the data given by industries producing large stocks of MEA. Note that there are large variations among the microbial toxicity results of different tests reported in the data sheets. This is probably due to the lack of good standard methods especially with respect to the inoculums, having different microbial diversity and adaptation. The lowest toxicity is found on activated sludge (AS) inoculums that have been exposed to the large variety of chemicals found in wastewater and where the microbial diversity can be high.

Table 1: Bacterial toxicity data for MEA (based on IUCLID data sheets)

Inoculum	Exposure time	Lethal – Effective conc. (mg/L)
Pseudomonas putida	17 hrs.	87 (EC 10) 110 (EC 50)
AS	30 min.	> 1000 (EC 10)
Nitrosomonas sp. (bact.)	2 hrs.	100 (EC 16) 12200 (EC 50)
Photobacterium phosphoreum	30 min.	13.7 (EC 50)
Other (bacteria)	24 hrs.	5 (EC 10) 15 (EC 50)
AS (not adapted)	30 min.	>1000 (EC 10)
AS (not adapted)	3 hrs.	>1000 (IC 50)
Consortium of bact.	6 hrs.	129 (IC 50)

2.4. Degredation rates

Aerobic degradation rates at different MEA concentrations can be estimated using the BOD consumption data obtained in a standard BOD test. Assuming a first order degradation rate constant, Thomas method (Abdelrasoul, 2001) can be adopted to determine the rate constants. This method is based on the following equation 1,

$$\left(\frac{t}{y}\right)^{1/3} = (2.3K_1L)^{-1/3} + \frac{K_1^{2/3}}{3.43L^{1/3}}t \tag{1}$$

Where, y is the BOD exerted at time t and K_I is the first order reaction rate constant (base 10) and L is the ultimate BOD. A linear graph of $(t/y)^{1/3}$ vs. t is used to determine K_I and L.

3. MEA degradation modeling

MEA degradation reactions are incorporated in to the (previously modified) standard Anaerobic Digestion Model no. 1, ADM 1 (Batstone *et al.,* 2002). The previous modification incorporated known effects of free oxygen to simulate micro-aerobic conditions as well as strictly anaerobic conditions. This ADM 1- Ox model is described elsewhere (Botheju *et al.,* 2009). The complete mass balance model incorporates the main physical, chemical and biological processes within the reactor chamber liquid phase and head space. It is developed to generalize the literature information presented above and to make it applicable for process simulations.

3.1. MEA degradation under anaerobic conditions

It is assumed that the organisms degrading amino acids (common structure $NH_2CHRCOOH$) also degrade monoethanolamine ($NH_2CH_2CH_2OH$), due to the close structural similarity. No new microbial species are therefore required in the model, apart from the 7 types of biomass originally presented in ADM 1. The reaction is shown in Eq. 2. This reaction can be considered as another acidogenic/acetogenic pathway according to the ADM 1 terminology.

$$NH_2CH_2CH_2OH \longrightarrow CH_3COOH + NH_4^+ + X_{aa} \qquad (2)$$

It is assumed that anaerobic MEA degradation rate is best described by a saturation type expression (Eq. 3), such as for all other organic reactants in anaerobic digestion (Batstone *et al.,* 2002). Maximum uptake rate (k_m), half saturation constant (k_s) and the yield coefficient (Y_{aa}) for anaerobic MEA degradation are assumed to have the same values as used for amino acid degradation in the ADM 1 model (50 kg COD_s/kg COD_x. d^{-1}, 0.3 kg COD /m^3 and 0.08, respectively).

$$r = k_{m,aa}\left(\frac{S_{mea}}{k_{s,aa} + S_{mea}}\right) X_{aa} I_1 I_{O_2} \qquad (3)$$

I_I is a combined inhibition term used to represent the inhibition due to unfavorable pH conditions and potential limitation of inorganic nitrogen. I_{O2} is the oxygen inhibition term.

3.2. MEA degradation under micro-aerobic conditions

The biological degradation of monoethanolamine to acetate in the presence of oxygen, shown in equation (4) (Ohtaguchi *et al.,* 1995) is energetically favourable as it generates a net positive amount of ATP.

$$NH_2CH_2CH_2OH + O_2 \longrightarrow NH_4^+ + CH_3COOH + ATP \quad (4)$$

With the inclusion of biomass growth the above reaction (4) can be uniquely balanced (Eq. 5).

$$14\,NH_2CH_2CH_2OH + 4\,O_2 \longrightarrow 9\,CH_3COOH + 2\,C_5H_7NO_2 + 12\,NH_4^+ \quad (5)$$

The biomass represented by $C_5H_7NO_2$ stands for the same X_{aa} syntrophic group which is responsible for amino acids degradation. The aerobic action of this and two other biomass groups (X_{aa}, X_{fa} and X_{su}) are already included in the ADM1-Ox model (Botheju *et al,* 2009). An aerobic biomass (X_{aa}) yield coefficient of 0.286 g COD_x/ g COD_{MEA} is determined from Eq. 5. This aerobic biomass yield is 3.5 times higher than the anaerobic value of 0.08, implying higher production of MEA degrading organisms with than without free oxygen available. Typically aerobic yield coefficients are 2- 10 times higher than the yields under anaerobic conditions (Tchobanoglous *et al.,* 2003). Further using equation 5, stoichimetric coefficients for inorganic nitrogen (IN), acetic acid and oxygen can be determined to be 0.0107 mol N/g COD_{MEA}, 0.514 g COD/g COD_{MEA} and 0.1143 g O_2/ g COD_{MEA}.

Rate of micro-aerobic MEA degradation is modeled by a double saturation function (eq. 6),

$$r = k_{m,mea,aer}\left(\frac{S_{mea}}{k_{s,mea,aer} + S_{mea}}\right)\left(\frac{S_{O2}}{K_{O2} + S_{O2}}\right)X_{aa}I_1 \quad (6)$$

Hales *et al,* (1996) fitted a saturation model to MEA degradation based on respirometric oxygen consumption data, using μ_m=0.152 hr^{-1} (maximum specific growth rate) and k_s=6.25 mg/L (half velocity constant). These experiments were carried out at 20 °C temperature. The values can be adjusted for other temperature conditions by using Vant-Hoff Arrhenius relation (Eq. 7).

$$K_T = K_{20}\theta^{(T-20)} \quad (7)$$

Temperature activity coefficient θ can be selected according to Tchobanoglous *et al.* (2003); θ for μ_m=1.07 and θ for k_s = 1.00. Accordingly, at the mesophilic condition of 35 °C, μ_m = 10 d^{-1} and $k_{s,mea,aer}$=0.008 kg COD/m^3.

The maximum specific MEA utilization rate ($k_{m,\,mea,\,aer}$=35.2 d^{-1}) is found based on the maximum specific growth rate μ_m using Eq. 8.

$$k_m = \mu_m / Y \quad (8)$$

The stoichiometric matrix for soluble components in the complete model (ADM 1-Ox including MEA degradation) is presented in Table 2. The modifications to the original ADM 1 model are shown highlighted.

4. Conclusions

Anaerobic degradation of MEA into mainly methane and ammonia as end products is possible and favourable for the energy balance of a biological MEA waste treatment plant compared to a completely aerobic process. Small quantities of free oxygen can, however, be supplied to speed up MEA degradation to acetate, while converting acetate to methane anaerobically. These processes are added to a standard anaerobic digestion model where free oxygen effects are also included. It is suggested that MEA degrades to acetate and NH_4^+ under both micro-aerobic and anaerobic biodegradation conditions. The proposed model presented in a stoichiometric matrix format includes relevant physical, chemical and biological reactions and can be used to simulate MEA waste treatment processes. Process parameters are estimated based on literature information. The model has also been validated using a series of biodegradation experiments described elsewhere (Botheju *et al.,* 2010a).

Table 2: The model in stoichiometric matrix format (the soluble components)

Components , i	i1	i2	i3	i4	i5	i6	i7	i8	i9	i10	i11	i12	i13	i14	Rate (ρ_j) [kg cod/m^3.d]
Processes, j	S_{su}	S_{aa}	S_{fa}	S_{va}	S_{bu}	S_{pro}	S_{ac}	S_{h2}	S_{ch4}	S_{IC}	S_{IN}	S_{O2}	S_I	S_{MEA}	
j1.Disintegration													$F_{sI,xc}$		$k_{dis}X_c$
j2. Hydrolysis of Carbohydrates	1														$k_{hyd,ch}X_{ch}X_{su}$
j3. Hydro. Proteins		1													$k_{hyd,pr}X_{pr}X_{aa}$
j4. Hydro. lipids	$1-F_{fa,li}$		$F_{fa,li}$												$k_{hyd,li}X_{li}X_{fa}$
j5. Uptake of sugars	-1				$(1-Y_{su})F_{bu,su}$	$(1-Y_{su})F_{pro,su}$	$(1-Y_{su})F_{fa,su}$	$(1-Y_{su})F_{h2,su}$		$-\Sigma c_i v_{i,5}$ i=1-9, 11-25	$-Y_{su}N_{bac}$				$k_{m,su}\frac{S_{su}}{K_{s,su}+S_{su}}X_{su}I_1I_{O2}$
j6. Uptake of amino acids		-1		$(1-Y_{aa})F_{va,aa}$	$(1-Y_{aa})F_{bu,aa}$	$(1-Y_{aa})F_{pro,aa}$	$(1-Y_{aa})F_{ac,aa}$	$(1-Y_{aa})F_{h2,aa}$		$-\Sigma c_i v_{i,6}$ i=1-9, 11-25	$N_{aa}-Y_{aa}N_{bac}$				$k_{m,aa}\frac{S_{aa}}{K_{s,aa}+S_{aa}}X_{aa}I_1I_{O2}$
j7. Uptake of LCFA			-1				$(1-Y_{fa})0.7$	$(1-Y_{fa})0.3$		?	$-Y_{fa}N_{bac}$				$k_{m,fa}\frac{S_{fa}}{K_{s,fa}+S_{fa}}X_{fa}I_2I_{O2}$
j8. Aerobic uptake of sugar	**-1**									**$C_1(1-Y_{su\_aer})$**	**$-Y_{su,aer}N_{bac}$**	-1.1			$k_{m,su}\frac{S_{su}}{K_{s,su,aer}+S_{su}}\frac{S_{o2}}{K_{o2}+S_{o2}}X_{su}I_1$
j9. Aerobic uptake of aminoacids		**-1**								**$C_2(1-Y_{aa\_aer})$**	**$N_{aa}-Y_{aa,aer}N_{bac}$**	-1.2			$k_{m,aa}\frac{S_{aa}}{K_{s,aa,aer}+S_{aa}}\frac{S_{o2}}{K_{o2}+S_{o2}}X_{aa}I_1$
j10. Aerobic uptake LCFA.			**-1**							**$C_3(1-Y_{fa\_aer})$**	**$-Y_{fa,aer}N_{bac}$**	-2.03			$k_{m,fa}\frac{S_{fa}}{K_{s,fa,aer}+S_{fa}}\frac{S_{o2}}{K_{o2}+S_{o2}}X_{fa}I_2$
j11.Uptake of valerate				-1		$(1-Y_{c4})0.54$	$(1-Y_{c4})0.31$	$(1-Y_{c4})0.15$			$-Y_{c4}N_{bac}$				$k_{m,c4}\frac{S_{va}}{K_{s,c4}+S_{va}}X_{c4}\frac{1}{1+S_{bu}/S_{va}}I_2I_{O2}$
j12.Uptake of butyrate					-1		$(1-Y_{c4})0.8$	$(1-Y_{c4})0.2$			$-Y_{c4}N_{bac}$				$k_{m,c4}\frac{S_{bu}}{K_{s,bu}+S_{bu}}X_{c4}\frac{1}{1+S_{va}/S_{bu}}I_2I_{O2}$
j13.Uptake of propionate						-1	$(1-Y_{pro})0.57$	$(1-Y_{pro})0.43$		$-\Sigma c_i v_{i,13}$ i=1-9, 11-25	$-Y_{pro}N_{bac}$				$k_{m,pro}\frac{S_{pro}}{K_{s,pro}+S_{pro}}X_{pro}I_2I_{O2}$
j14.Uptake of acetate							-1		$1-Y_{ac}$	$-\Sigma c_i v_{i,14}$ i=1-9, 11-25	$-Y_{ac}N_{bac}$				$k_{m,ac}\frac{S_{ac}}{K_{s,ac}+S_{ac}}X_{ac}I_3I_{O2}$
j15.Uptake of hydrogen								-1	$1-Y_{h2}$	$-\Sigma c_i v_{i,15}$ i=1-9, 11-25	$-Y_{h2}N_{bac}$				$k_{m,h2}\frac{S_{h2}}{K_{s,h2}+S_{h2}}X_{h2}I_1I_{O2}$
j16 – j22 Biomass decay															$k_{dis}X_i$ i=18-24
J23 Uptake of MEA							$(1-Y_{aa})$				$N_{mea}-Y_{aa}N_{bac}$			-1	$k_{m,aa}\frac{S_{mea}}{K_{s,aa}+S_{mea}}X_{aa}I_1I_{O2}$
J24 Aerobic uptake of MEA							0.514				$N_{mea}-Y_{mea,aer}N_{bac}$	0.1143		-1	$k_{m,mea,aer}\frac{S_{mea}}{K_{m,mea,aer}+S_{mea}}\frac{S_{o2}}{K_{o2}+S_{o2}}X_{aa}I_1$

Acknowledgement

The authors would like to thank Aker Clean Carbon AS, BioTek AS, Gassnova SF, NOAH AS, Norcem AS, Skagerak Kraft AS, StatoilHydro ASA and Østfold Energi AS for financing the study. Assistance given by Yuan Li, a post graduate student at Telemark University College at the time of this study, is appreciated.

References

Abdelrasoul, F.M., 2001, A comparative study of the BOD rate constant of industrial wastewater and sewage, Proceedings of the Sixth International Water Technology Conference, IWTC 2001, Alexandria, Egypt.

Abend, A., V. Bandarian, R. Nitsche, E. Stupperich, J. Retey and G.H. Reed, 1999, Ethanolamine Ammonia-Lyase has a "Base -on" Binding mode for Coenzyme B12, Archives of Biochemistry and Biophysics, vol. 370 (1), pp. 138-141.

Batstone, D.J., J. Keller, I. Angelidaki, S. Kalyuzhnyi, S.G. Pavlostathis, A. Rozzi, W. Sanders, H. Siegrist and V. Vavilin, 2002, Anaerobic digestion Model No.1, IWA publishing, London.

Botheju D, B. Lie and R. Bakke, 2009, Modeling free oxygen effects in anaerobic digestion, Proceedings of the MATHMOD 2009 - 6th Vienna International Conference on Mathematical Modeling, Vienna.

Botheju, D., Y. Li, J. Hovland, T. Risberg, H.A. Haugen, C. Dinamarca and R. Bakke, 2010a, Biogasification of waste monoethanolamine generated in post combustion CO_2 capture, Proceedings of the 2nd Annual Gas Processing Symposium, Doha, Qatar. – Accepted manuscript

Eide-Haugmo, I., O.G. Brakstad, K.A. Hoff, K.R. Sørheim, E.F. da Silva and H.F. Svendsen, 2008, Environmental impact of amines, Energy Procedia; Article in press.

Georgiou, D. and A. Aivasidis, 2006, Decolouration of textile wastewater by means of a fluidized bed loop reactor and immobilized anaerobic bacteria, J. Hazardous Materials, B135, pp. 372-377.

Goff, G.S., 2005, Oxidative degradation of aqueous monoethanolamine in CO_2 capture processes: Iron and Copper catalysis, Inhibition, and O_2 mass transfer, PhD. dissertation, The University of Texas, Austin, USA.

Hales, S.G., C.J. Philpotts and C. Gillard, 1996, A respirometer with improved sensitivity for ready biodegradation testing, Chemosphere, vol. 33, 7, pp. 1247-1259.

Hongwei, Y., J. Zhanpeng and S. Shaoqi, 2005, Biodegradability of nitrogenous compounds under anaerobic conditions and its estimation, Ecotoxicity and Environmental Safety, vol. 63, 299 -305.

Isik, M. and D.T. Sponza, 2008, Anaerobic/aerobic treatment of a simulated textile wastewater, Separation and Purification Technology, vol. 60, pp. 64-72.

IUCLID data sheets: Available on CPS&Q website, http://ecb.jrc.ec.europa.eu/

Johansen, J.E. and R. Bakke, 2006, Enhancing hydrolysis with micro-aeration, Water science and Technology, vol. 53(8), pp. 43-50.

Lai, B. and W.K. Shieh, 1996, Technical note; Batch monoethanolamine degradation via nitrate respiration, Water Research, vol. 30, No. 10, pp. 2530-2534.

Liuzinas, R., K. Jankevicius, M. Salkauskas, R. Jakubenas, A. Paskevicius and M. Mikalajunas, 2007, Biological method for the detoxification of spent monoethanolamine solutions, Proceedings of the Kalmar Eco-Tech' 07, pp. 853-860, Kalmar, Sweden.

Ohtaguchi, K., K. Koide and T. Yokoyama, 1995, An ecotechnology-integerated MEA process for CO_2 removal, Energy Conversion Management, vol. 36, no. 6-9, pp. 401-404.

Ohtaguchi, K. And T. Yokoyama, 1997, The synthesis of alternatives for the bioconversion of waste-monoethanolamine from large scale CO_2 removal processes, Energy Conversion Management, vol 38, Suppl., pp. S539-S544.

Speranza, G., C.F. Morelli, P. Cairoli, M. Britta and B. Schink, 2006, Mechanisms of anaerobic degradation of triethenolamine by a homoacetogenic bacterium, Biochemical and Biophysical Research Communications, 349, pp. 480 - 484.

Tchobanoglous, G., F.L. Burton and H.D. Stensel, 2003, Meltcalf & Eddy, Inc.'s Wastewater Engineering: Treatment, Disposal, and Reuse, 4th Edition, Tata McGraw-Hill, Publishing company Ltd., p. 580.

Thitakamol, B., A. Veawab and A. Aroonwilas, 2006, Environmental impacts of the integration of CO_2 capture unit into coal-fired power plants, 8th International conference on greenhouse gas control technologies, Trondheim, Norway.

Juanga, J. P., P. Kuruparan and C. Visvanathan, 2005, Optimizing combined anaerobic digestion process of organic fraction of municipal solid waste, Presented at International conference on integrated solid waste management in southeast asian cities, Siem Reap, Cambodia.

Zhu, M., F. Lu, L. P. Hao, P.J. He and L.M. Shao, 2009, Regulating the hydrolysis of organic wastes by micro-aeration and effluent recirculation, Waste Management, Article in Press.

Proceedings of the 2nd Annual Gas Processing Symposium
Farid Benyahia and Fadwa T. Eljack (Editors)

Environmental Sustainability and the Role of LNG in a Carbon Constrained World

Susan L. Sakmar, JD., LL.M.

Adjunct Professor of Law, University of San School of Law, 2130 Fulton Street, San Francisco, California, USA, 94117. Author contact: slsakmar@gmail.com

Abstract

The world is facing serious energy-related challenges that are further complicated by increasing concerns about global climate change. While energy is a key driver of economic progress throughout the world, the development, transport, and use of fossil fuels has contributed to an increase in greenhouse gas (GHG) emissions. These duel concerns have been brought to the top of the agenda in business, politics, and public debate and people around the world are searching for ways to meet rising energy demand in a cleaner and more sustainable manner. This non-technical paper provides an overview of the key issues related to the environmental sustainability of the gas processing industry with a particular emphasis on the role of liquefied natural gas (LNG) in a carbon constrained world. Section 1 of the paper analyzes whether LNG is a contributor to a sustainable energy future in the context of greenhouse gas emissions and LNG supply chain emissions. Section 2 of the paper provides an overview of some of the various ways in which the LNG industry can enhance environmental sustainability including methane mitigation and technological efficiencies.

Keywords: Liquefied natural gas, LNG, climate change, energy, environment, sustainability.

1. The Role of LNG in a Carbon Constrained World

Is LNG a contributor to a sustainable energy future? Initial analysis indicates that it may be. As the world grapples with issues related to climate change and carbon emissions, it has been widely recognized that natural gas is one of the cleanest burning fuels and produces relatively low carbon dioxide emissions. (CLNG Media Advisory) Nonetheless, there are two primary environmental considerations related to LNG. The first is the methane emissions that exist from all natural gas. The second consideration is the criticism that the energy needed to liquefy, transport, and regasify LNG (life-cycle emissions) diminishes any clean-burning benefits LNG might otherwise provide.

1.1 Natural Gas Methane Emissions

Although natural gas is a relatively low-carbon, clean-burning fuel, the principal component of natural gas is methane (CH_4). Methane is a potent greenhouse gas (GHG) and has 23 times the radiative forcing impact of CO_2 on a weight basis over a 100-year period. Methane, or CH_4, is the largest contributor to anthropogenic GHG emissions after carbon dioxide and accounts for about 16 per cent of the total on a CO_2 equivalent basis. This factor makes the control of CH_4 emissions an important component of any GHG emissions mitigation strategy. (IPIECA Workshop Report)

Methane emissions occur in all sectors of the natural gas industry, from drilling and production, through processing and transmission, to distribution. Emissions primarily result from normal operations, routine maintenance, fugitive leaks and system upsets. As gas moves through the system, emissions occur through intentional venting and unintentional leaks. Venting can occur through equipment design or operational practices, such as the continuous bleed of gas from pneumatic devices (that control gas flows, levels, temperatures, and pressures in the equipment), or venting from well completions during production.

In addition to vented emissions, methane losses can occur from leaks (also referred to as fugitive emissions) in all parts of the infrastructure, from connections between pipes and vessels, to valves and equipment. (Gas STAR).

1.2 Life-Cycle Emissions of LNG

Although LNG burns cleanly, concerns have been raised that the environmental impact and emissions associated with LNG production may nullify the clean-burning benefits of LNG. To date, there is limited independent research that analyzes the environmental impact of the entire life-cycle emissions of LNG and most environmental impact statements (EIS) tend to focus on just one aspect of the LNG supply chain, e.g. the emissions associated with the liquefaction process or import regasification terminal.

One recent study has suggested that the entire supply chain emissions from production through end-use of the delivered natural gas might be quite significant and should be considered in any environmental impact report (Heede, 2006).

In the Heede study, an analysis was conducted of the life-cycle emissions resulting from BHP Billiton's proposed Cabrillo LNG terminal off the coast of southern California. In it's permit application to the U.S. Coast Guard and the State of California, BHP estimated greenhouse gas emissions only from the operation of its proposed Cabrillo Deepwater Port. The Heede study was commissioned to estimate the entire life-cycle emissions of the project from the production platform offshore Western Australia and across the Pacific Ocean to Southern California, including combustion by end-users in Southern California (Heede, 2006).

The purpose of the Heede study was not to attribute the entire supply chain emissions to BHP but rather, to fully account for all the emissions attributable to the proposed project from start to finish – production to combustion. The study ultimately found that the "supply chain emissions from production through end-use of the delivered natural gas was equal 4.3 to 4.9 percent of California's total GHG emissions, and 5.3 to 5.9 percent of CO_2 emissions using EIA emissions data." (Heede, 2006).

The largest component of the supply chain emissions was the combustion of the natural gas delivered to the Southern Californian utility and it's end users. The emissions estimates for this segment ranged from 15.82 to 15.89 $MtCO_{2\text{-eq}}$ plus 0.58 to 0.72 $MtCO_{2\text{-eq}}$ of methane for an average total estimate of 16.50 $MtCO_{2\text{-eq}}$ per year, or 72% of the total emissions.

The most relevant findings for this paper are the emissions estimate for the processing segment and the transportation segment. The emissions estimates for the processing segment range from 1.97 to 3.17 $MtCO_{2\text{-eq}}$ for an average total of 2.69 $MtCO_{2\text{-eq}}$ per year, or 11.8% of the total. The emissions estimates for the transportation segment range from a low of 1.80 $MtCO_{2\text{-eq}}$ to a high of 2.37 $MtCO_{2\text{-eq}}$ for an average of 2.09 $MtCO_{2\text{-eq}}$, or 9.2% of the total (Heede, 2006).

A major limitation of the Heede Study is that it is based on estimates assuming industry best practices or in some cases, improvements over standard practice or industry benchmarks. The estimates were used since the facilities had not been designed or built and Heede did not have access to BHP engineering data other than limited information in the permit application. Nonetheless, the Heede study is instructive since the life-cycle analysis was used to support strong environmental opposition to BHP's proposed LNG facility – which was ultimately denied by the State of California (Sierra Club, 2007).

1.3 LNG versus Coal-Fired Power Plants

In much of the world, coal is a plentiful resource and therefore is the dominant fuel source for electrical power productions. Natural gas, and LNG as a supplement to domestic natural gas supplies, is increasingly playing a larger role in electrical power generation due to the perceived emissions benefits. At least two studies have accessed the GHG emissions from LNG versus coal-fired power plants and have reached different conclusions.

A study by researchers at Carnegie Mellon found that LNG imported from foreign countries to be used for electricity generation could have 35 percent higher lifecycle greenhouse gas emissions than coal used in advanced power plant technologies. (Jaramillo, 2007)

The Carnegie Mellon Study "analyzed the effects of the additional air emissions from the LNG/SNG life-cycle on the overall emissions from electricity generation in the United States." The study found that with current electricity generation technologies, natural gas life-cycle GHG emissions are generally lower than coal life-cycle emissions, even when increased LNG imports are included. However, "the range of life-cycle GHG emissions of electricity generated with LNG is significantly closer to the range of emissions from coal than the life-cycle emissions of natural gas produced in North America." The study also found that upstream GHG emissions of NG/LNG/SNG have a higher impact in the total life-cycle emissions than upstream coal emissions. (Jaramillo, 2007)

The Carnegie Mellon Study also analyzed advanced technologies and suggested that as newer generation technologies and carbon capture and sequestration (CCS) are installed, the overall life-cycle GHG emissions from electricity generated with coal, domestic natural gas, LNG or SNG could be similar. For SOx, the study found that coal and SNG would have the largest life-cycle emissions. For NOx, LNG would have the highest life-cycle emissions and would be the only fuel that could have higher emissions than the current average emission factor from electricity generation, even with advanced power design. (Jaramillo, 2007)

In contrast to the Carnegie Mellon Study, a study commissioned by the Center for Liquefied Natural Gas (CLNG), found that existing US domestic coal power plants produce two and a half times more emissions on a life cycle basis than that of LNG. (CLNG) LNG emissions were even lower when compared to advanced ultra supercritical coal (SCPC) power plants and integrated gasification combined cycle (IGCC) coal fired power plants (neither of which are commercially viable in the U.S.) The production and combustion emissions were greater in all of the coal cases but the processing and transportation segment emissions were greater in the LNG cases. (See Table 1).

Table 1: Summary of CLNG Results

	Production lbs CO2e/MWh	**Processing** lbs CO2e/MWh	**Transportation** lbs CO2e/MWh	**Combustion** lbs CO2e/MWh	**TOTAL** lbsCO2e/MWh
LNG	15	134	99	797	1,045
Coal IGCC	61	24	9	1,714	1,808
Advanced Ultra SCPC	61	24	9	1,773	1,868
Existing Coal Technology	76	30	12	2,614	2,731

Source: Pace, CLNG Report, all values in lbs CO2e/MWh

2. Enhancing Environmental Sustainability in the LNG Industry

Although there are numerous ways in which the LNG industry can enhance environmental sustainability, the most significant advancements can be achieved with methane mitigation and improved efficiencies in processing and transportation.

2.1 Benefits of Methane/CH_4 Mitigation

Reducing methane emissions has many important energy, economic, environmental, and safety benefits. Methane is second only to carbon dioxide (CO_2) as a GHG resulting from human activities and emissions of methane are expected to increase 23% to nearly 8 million metric tonnes (Mt) of carbon dioxide equivalent (CO_2-eq) by 2020, driven by growth in emerging economies, particularly in the natural gas and coal sectors (IEA, Methane, 2009).

Methane is 23 times as effective at trapping heat in the atmosphere as CO_2 and also has a relatively short atmospheric lifetime of approximately 12 years. These two characteristics make methane emissions reductions particularly effective at mitigating global warming in the next 25 years. (Methane to Markets) Since methane is the primary constituent of natural gas, the collection and utilization of methane provides a valuable, clean-burning and renewable energy source. Producing energy from recovered methane can also help to avoid the use of higher CO_2 and pollutant-emitting energy sources such as coal and oil.

Methane mitigation is also recognized as an important component under the Kyoto Protocol. The Kyoto Protocol's flexible mechanisms, the Clean Development Mechanism (CDM) and Joint Implementation (JI), can provide carbon credits for fuel switching or CH_4 emission reduction projects, However, as of the end of 2005, only 2 percent of the credits from approved or proposed CDM projects were for reduction of CH_4 emissions from oil and gas operations. Credits from the flexible mechanisms can be used in the EU's Emission Trading Scheme (EU ETS) or sold in the carbon market. Under the EU ETS fuel switching from coal-fired electricity production to natural gas-fired electricity production may be promoted where the incremental rise in costs is met or exceeded by the value of carbon credits. (IPIECA Workshop Report)

2.2 Barriers to Methane Mitigation

There are several barriers to reducing CH_4 emissions. In some instances there is a lack of information about emission levels and the value of the fuel being lost. Traditional industry practice, i.e. using high-pressure natural gas in pneumatic devises, can also be a barrier to considering lower emission alternatives. There may also be resistance from local communities to the building of the infrastructure (e.g. LNG terminals and pipelines) needed to transport recovered gas to market. (IPIECA Workshop Report)

Technical barriers to CH_4 emission reduction from oil and gas operation can arise from a lack of measurement techniques to characterize emissions and of familiarity with technical options and their benefits. Addressing technical barriers requires implementation of a phased approach to emission reduction. Initial steps require that all existing and potential sources of emissions are identified and inventoried. Following characterization of emission sources the alternative technological options for reducing emissions can be evaluated, and project feasibility studies developed. Reviewing and evaluating experiences, as well as information sharing through partnerships, enables these barriers to be addressed. (IPIECA Workshop Report)

Additionally, regulatory barriers can block production of otherwise economically recoverable natural gas resources. For example, much of the potential natural gas production capacity in the US is currently inaccessible because of regulatory restrictions. Removing these restrictions requires public, political and industry support.

To overcome the above barriers, the following partnerships and cooperative efforts have been formed:

Natural Gas STAR is a voluntary partnership between the US EPA and oil and gas industry partners with the goal of implementing cost-effective technologies and practices to reduce CH_4 emissions from oil and gas operations. The Natural Gas STAR partners have identified approximately 100 emission reduction techniques, which have led to cumulative emissions reductions of 460 Bcf.

Natural Gas STAR International works to promote technology transfer and provide assistance to international oil and gas companies to develop and implement cost-effective CH_4 emission reduction plans, and report on successes in controlling CH_4 emissions from their international natural gas operations.

The Methane to Markets Partnership is a cooperative agreement between countries designed to reduce CH_4 emissions from all anthropogenic sources, including the oil and gas industry. Private companies participate through a project network, which now includes more than 350 organizations. In 2005, the partner countries accounted for 56per cent of global methane emissions from the oil and gas industry.

The *Global Gas Flaring Reduction Partnership* is an agreement between countries, international oil companies, The World Bank, OPEC and the EU to support developing nation governments and the petroleum industry in their efforts to reduce gas flaring and venting. It has developed global standards for flaring and venting, regulatory best practices and a gas flaring data tool, and promoted carbon credit financing for flare reduction projects.

The *Asia-Pacific Partnership on Clean Development and Climate* involves six countries (Australia, China, India, Japan, Korea and the USA) and has the goal of promoting the development and deployment of clean energy technologies. One of its eight task forces is focused on cleaner fossil fuel technology, including greater use of LNG. Experience drawn from projects and initiatives highlights the importance of information in overcoming the technical barriers to CH_4 emission reduction, and the need for an improved regulatory environment and stakeholder support for projects to increase natural gas supply.

2.3 Methane Mitigation Opportunities in the Gas Industry

A wide variety of technologies are available to reduce CH_4 emissions from natural gas production and use. The Natural Gas STAR programme has identified almost 100 emission reduction techniques. Many of these technologies are relatively simple and cost effective such as vapour recovery for storage tanks and the use of electric motors rather than gas-fired engines to power wellhead equipment. (IPIECA Workshop Report).

Methane emission reduction opportunities generally fall in to one of three categories: 1) technologies or equipment upgrade that reduce or eliminate equipment venting or fugitive emissions, 2) improvements in management practices and operational procedures, or 3) enhanced management practices that take advantage of improved technology. In all cases, reducing methane emissions makes additional gas available for sale and use. (Methane to Markets).

2.4 LNG Methane Emissions Reduction Opportunities

Methane losses from LNG facilities are estimated at 4 Bcf and can be broken down as follows: (Conoco Phillips Nat Gas STAR Program)

Emissions from LNG liquefaction equipment

- Compressor seals
- CO_2 removal systems
- Dehydration systems
- Tank overpressure
- Ship loading displacement vapors

Emissions from LNG storage tanks

- Tank overpressure venting
- Leaks from pressure relief valves
- Vapor recovery compressors

Emissions from LNG Marine Terminals

- Fugitives
- Venting, if boil-off vapor can not be consumed as fuel

Emissions from LNG Tankers

- Flanges and fitting leaks during cool down
- Leaking vapor recovery systems, not operating
- Leaking cargo tank relief valves
- Cargo tank venting during delays

Emissions from LNG Sendout

- Vaporizer fuel system leaks
- Pressure relief valves

Methane savings in the LNG sector can come from the use of centrifugal compressor seals and by implementing a DI&M program. Other LNG emission prevention opportunities include improved connect/disconnect practices, improved tank pressure management, improved vapor recovery system maintenance and availability, and strict enforcement of ship venting rules. (Conoco Phillips Nat Gas STAR Program)

2.5 Efficiencies in LNG Liquefaction Processes

The liquefaction process transforms natural gas into LNG by cooling it to approximately -161° C, after which it is stored until it can be shipped on board LNG tankers to its final destination where it is regasified at the import terminal. There are a variety of proprietary processes marketed for large-scale LNG liquefaction plants which can be broken down into the following broad categories: (Tusiani, 2007).

- Pure-refrigerant cascade process;
- Propane-precooled mixed-refrigerant processes;
- Propane-precooled mixed-refrigerant, with back-end nitrogen expander cycle;
- Other mixed-refrigerant processes; and
- Nitrogen expander-based processes.

While a detailed technical discussion of the various methods is beyond the scope of this paper, the propane-precooled mixed-refrigerant (C3-MR) process is worth noting since it is considered the "work horse" of the LNG industry with over 80% of the world's completed trains utilizing a variation of this process due to its proven technology in a variety of process and environmental settings and high efficiency. (Tusiani, 2007)

In general, the C3-MR system uses a multi-component refrigerant to condense and evaporate natural gas in one cycle over a wider range of temperatures. The mixed refrigerant used is Air Products & Chemicals, Inc.'s (APCI) proprietary Multi-Component Refrigerant (MCR). High efficiencies are achieved by adding the propane pre-cooling stage for both feedgas and the mixed-refrigerant loop, allowing the MCR vaporization temperature curve to closely match the natural gas liquefaction curve. (Tusiani, 2007).

Since the LNG business is extremely capital intensive, economies of scale are critical, especially when it comes to the size of LNG "trains" which are the liquefaction facilities used to covert the natural gas into LNG. They are called trains because of the sequential arrangement of the equipment required to process the gas.

Over the past 10 years, LNG trains have doubled in size from 2 million tons per year to 5 million tons per year. The drive toward increased liquefaction economies of scale led APCI to develop a variant of their C3-MR process that increases the liquefaction train capacity from 5 to almost 8 MMt/y. The new AP-X process adds a third cycle of nitrogen expander (N_2) refrigeration to the back end of the C3-MR process's propane (C3) mixed-refrigerant cycles (MCR) cycles. This (N_2) cycle takes the LNG subcooling duty off the MCR cycle, increasing the natural gas capacity and reducing the refrigeration loads on the first two cycles. (Tusiani, 2007).

To gain a competitive advantage in the LNG market, ExxonMobil and RasGas began the development years ago of so called "mega" trains using the APCI AP-X process that will be able to process 7.8 MMt/y. These larger, more economical trains use state-of-the-art turbines, compressors and heat exchangers uniquely combined to result in greater efficiencies. (ExxonMobil).

In August 2009, it was announced that the first of the mega trains (RasGas Train 6) had produced it first LNG. It remains to be seen whether the production from Train 6 will result in significant efficiencies and further studies and research is need to determine whether the mega trains offer any net environmental benefit.

2.6 Efficiencies in LNG Transportation

For more than 30 years, the size of LNG ships remained virtually unchanged with capacity of about 140,000 cubic meters. In 2001, joint venture partners Qatar Petroleum and ExxonMobil wanted to expand beyond the primarily Asian market for Qatar's LNG and thus need to develop a better way to deliver Qatar's LNG to more distant ports. Since shipping accounts for about one-third the cost of LNG, a new class of carrier that was more efficient needed to be designed and built.

Over the course of several years, the design team settled on two similar ship platforms called the Q-Flex (Q for Qatar and Flex for the flexibility to access most LNG ports) and the slightly larger Q-Max (Max for the largest ship that can use the Qatar LNG terminal).

The new large LNG ship technologies include a number of industry breakthroughs and significant enhancements, including increased ship size, onboard reliquefaction units, slow-speed diesel engines, twin propellers and rudders, and the latest in hull and antifouling protection and improved fire-protection systems.

The Q-Flex carries 50 percent more LNG than the average carrier operating today while the Q-Max transports 80 percent more. The Q-Max carriers are longer than three football fields, tower twenty stories tall from keel to masthead and are equipped with the largest membrane containment tanks ever built. With a total capacity of up to 266,000 cubic meters, each ship carries enough natural gas to meet the energy needs of 70,000 U.S. homes for one year.

The innovative Q-Max ships carry up to 80 percent more cargo, yet require approximately 40 percent less energy per unit of cargo than conventional LNG carriers due to economies of scale and efficiency of the engines.

The end result of these new generation ships is a 20-30 percent reduction in transportation cost with improved efficiency and emissions reductions. Improved efficiency and emissions is key as the shipping industry is certain to encounter more stringent guidelines from the International Maritime Organization (IMO) going forward. In July 2009, the Marine Environment Protection Committee (MEPC) of the IMO agreed to disseminate a package of interim and voluntary technical and operational measures to reduce GHG emission from international shipping. (IMO MEPC).

In addition to increasing the size of the ship, a major initiative was undertaken to design, test and implement the on-board reliquefaction plant that re-liquefies natural gas that is vaporized during transit, re-injecting it as liquid into the cargo tanks rather than using is as vaporized gas to power the tanker itself – allowing for delivery of nearly 100 percent of the cargo. This is particularly beneficial for the long-haul voyages from Qatar to Europe and the Americas.

The on-board reliquefaction facilities created an opportunity to shift from steam boilers and turbines used for propulsion by conventional LNG ships to highly efficient slow-speed diesel engines. The Q-Max ships are equipped with two diesel engines driving twin propellers and rudders. This leads to more energy efficient, reliable and manoeuvrable ships, reducing fuel consumption by up to one-third. (ExxonMobil Corp News).

While the large tankers offer economies of scale and greater efficiencies, a limitation is that these tankers will require appropriate accommodation at the loading and unloading facilities. Unless existing LNG importers reconfigure and reinforce their terminals berthing facilities, the new generation LNG tankers will be tied to newly constructed import terminals designed specifically to accommodate the larger ships. (Tusiani, 2007).

2.7 Efficiencies in LNG Import Terminals

The LNG receiving or import terminal is the final link in the LNG chain and the point of connection to the consumers. Whereas the liquefaction plants serve as enormous refrigerators to cool natural gas into a liquid, the import terminals "regasify" or warm the gas back up so that it can be sent through the gas pipeline system. (Tusiani, 2007).

All baseload onshore LNG import terminals basically feature the following components: tanker berthing and unloading facilities, storage tanks, regasification system, facilities to handle vapor and boil-off gas, high-pressure LNG pumps, metering and pressure regulation station, gas delivery infrastructure, gas odorization, calorific value control and LNG truck loading facilities.

While there are many aspects of efficiencies to be gained at import terminals, the design of more efficient ambient air vaporizers, which heat the LNG into its gaseous state, has received the most focus. There are essentially two designs of ambient air vaporizers:

- Direct Ambient Air Vaporizers: This design transfers heat from the ambient air directly into the LNG through the heat transfer surface of a heat exchanger; and
- Indirect Ambient Air Vaporizers: In this design, heat from ambient air is transferred to an intermediate fluid which in turn transfers heat to LNG through a separate heat exchanger.

Although the final determination of the vaporization design is site specific, recent analysis suggests that direct ambient air vaporizers are a more efficient solution as they add some, if not all of the heat required to vaporize LNG under the range of operating conditions. Also, since the amount of supplemental heat that will need to be added (by natural gas fired heaters) will be lower in systems that use direct ambient air vaporizers, there is a corresponding reduction in emissions to the environment. (Oregon LNG).

3. Conclusion: Is LNG an Environmentally Sustainable Fuel for the 21st Century?

In the coming decades, the world must meet the challenge of producing more energy to meet growing worldwide demand while at the same time limiting and even reducing greenhouse gas emissions. This dynamic will create unprecedented challenges but also unprecedented opportunities for cleaner burning fuels and renewables. As a clean burning fossil fuel, LNG has the potential to play a key role in a carbon-constrained energy future.

Although long-term demand for LNG is projected to grow in all regions of the world, the application of technologies to create a more sustainable LNG industry must overcome an array of commercial, political, environmental and social barriers before its full potential can be realized. A crucial factor that may impact LNG's role in the global energy supply mix is the outcome of international climate change negotiations and resulting policy and regulatory actions. Whereas moderate climate change policies may stimulate natural gas and LNG consumption, more stringent climate change policies may favor carbon-free renewables over LNG and other fossil fuels.

References

Center for Energy Economics (CEE), (2007), *Introduction to LNG*, available at, www.beg.utexas.edu/energyecon/lng.

Conoco Phillips and EPA's Natural Gas STAR Program, Kenai, AK. May 25, 2006, Liquefied Natural Gas Emissions Reduction Opportunities, available at http://www.epa.gov/gasstar/documents/workshops/kenai-2006/lng-opportunities.pdf.

Energy Information Administration, Annual Energy Outlook 2008 (Early Release December 2007), (EIA AEO 2008) available at http://www.eia.doe.gov/oiaf/archive/earlyrelease08/production.html.

ExxonMobil Corp. News, *A Sea Change for LNG Carriers,* available at, http://www.exxonmobil.com/ corporate news_features_20070901_lngcarriers.aspx.

Federal Energy Regulatory Commission (FERC), "LNG – Laws and Regulations," available at http://www.ferc.gov/industries/lng/gen-info/laws-regs.asp.

Global Mitigation of Non-CO_2 Greenhouse Gases, available at www.epa.gov/climatechange/ economics/international.html.

Heede, Richard (2006) *LNG Supply Chain Greenhouse Gas Emissions for the Cabrillo Deepwater Port: Natural Gas from Australia to California.*

IPIECA Workshop Report (2006) *Natural Gas as a Climate Change Solution: Breaking Down the Barriers to Methane's Expanding Role.*

International Energy Agency (IEA), Energy Sector Methane Recovery and Use: The Importance of Policy (IEA, Methane 2009), available at http://www.iea.org/papers/2009/methane_brochure_pdf.

International Maritime Organization (IMO), 1978, MARPOL 73/78. International Convention for the Prevention of Pollution from Ships, 1973, as modified by the Protocol of 1978 relating thereto. London: IMO.

IMO MEPC, 59th Session, 13-17 July, 2009, *IMO Environment Meeting Issues Technical and Operational Measures to Address GHG Emissions from Ships*, available at, http://www.imo.org/ About/ mainframe.asp?topic_id=1773&doc_id=11579.

Jaramillo, P., Griffin, W.M., Matthews, H.S. (2007) *Comparative Life-Cycle Air Emissions of Coal, Domestic Natural Gas, LNG, and SNG for Electricity Generation,* Environ. Sci. Technol., 41 (17), 6290-6296, available at http://pubs.acs.org/cgi-bin/abstract.cgi/esthag/2007/41/i17/abs/es063031o.html.

Media Advisory, *LNG: An Essential Part of American's Clean Energy Mix* (April 24, 2008), Center for Liquefied Natural Gas, available at http://www.lngfacts.org/Media-Center/042408_media_advisory.asp.

Methane to Markets, *Global Methane Emissions and Mitigation Opportunities*, available at www.methanetomarkets.org/oil-gas/index.htm.

Natural Gas STAR Program, available at http://www.epa.gov/gasstar/.

Oregon LNG Import Terminal, *Vaporizer Alternatives Study*, prepared by CH-IV International, available at http://www.oregonpipelinecompany.com.

RasGas Train 6 produces its First LNG, Qatar News Agency, available at, http://www.qnaol.net/QNAEn/News_bulletin/Economics/Pages/09-08-12-2136_430_0083.aspx.

Sierra Club, *Huge Victory Against Offshore LNG Terminal,* available at http://www.sierraclub.org/ca/coasts/victories/victory2007-04-19.asp.

Severance, Craig, *Cap and Trade: Unexpected Friend to Gas Industry*, Energy Bulletin, June 21, 2009, available at http://www.energybulletin.net/node/49298.

Tusiani, Michael D. and Shearer, Gordan, (2007), *LNG A Nontechnical Guide*, PennWell Books.

Proceedings of the 2nd Annual Gas Processing Symposium
Farid Benyahia and Fadwa T. Eljack (Editors)

Industrial Water Discharge and Biocide Fate Simulations with Nonlinear Conversion

Johannes Lawen,[ac] Yu Huaming,[b] Patrick Linke,[a] Ahmed Abdel-Wahab[a]

[a]*Texas A&M University at Qatar, Texas A&M Engineering Building, Education City, 23874 Doha, Qatar*
[b]*Hamburg Institut of Oceanography , Bundestrasse 53 , 20146 Hamburg, Germany*
[c]*Hamburg University of Technology , Schwarzenbergstrasse 95 , 21073 Hamburg, Germany*

Abstract

Fouling in heat exchangers is a significant operational problem when using seawater for process cooling. It is associated with the formation of a biofilm due to the growth of micro organisms, the deposition of solid particles and organic matter as well as the growth of animals like mussels. Adding biocides to the inflowing cooling water can inhibit biofilm growth. If biocides like chlorine are added and the cooling circuit is fed with seawater the chlorine will finally be discharged into the environment at the circuit outlet. The primary biocide can undergo conversion to toxic components like halogenated organic matter within the cooling circuit itself and at the shore site environment. The environmental impact results out of both, primary biocide and forming toxic components. In different countries different boundary value regulations have to be met by industry. Therefore control of fouling using biocides has to be engineered and operated so as to satisfy process and environmental constraints. To predict whether the primary biocide and by-products exceed regulated boundary values or assess environmental fall out in general requires the simulation of transport and reaction. This work reports on the development of a plume simulation which incorporates the kinetics and fate of toxic by-products. The pollutant discharge simulation has been split into a hydrodynamic simulation and a superimposed reaction and flow field driven advection as well as dispersion system. The flow field computation has been carried out with the Finite Volume Coastal Ocean Model, FVCOM, whereas the transport and reaction simulation is based on ADI/ADE Finite Difference algorithm. The ADI/ADE transport simulation is bound to structured grids which could be curvilinear. The FVCOM utilizes a horizontally irregular mesh consisting of triangular cells with varying vertical layer thickness, σ-coordinates, along the coastal bathymetry.

Keywords: Finite Difference and Finite Volume Method, parabolic transport PDE, Navier-Stokes equation, biocide, unstructured and curvilinear grids, ADI/ADE algorithm

1. Introduction

Environmental outfall assessment at cooling water discharge sites does not only require the tracking of pollutant transport but also the prediction of forming byproducts. Initially discharge simulation tools such as Cormix and GEMSS have been considered (Huaming et al., 2009, Adenekan et al., 2009, Abdel-Wahab et al. 2009) and perceived as limited in their adoption of coastal geometries (Cormix) and process reaction kinetics beyond decay terms (Cormix and GEMSS) which does not allow to track the production of byproducts. This has led to the development of an endogenous simulation. The pollutant transport and fate simulation can be split into a hydrodynamic flow field computation and a system containing reactions and flow field driven advection as well as dispersion dependent on the influence of reactions on the hydrodynamic properties of the fluid (Gour-Tsyh Yeh et al., 2002). In the low orders of magnitude concerning biocide dosing and the corresponding kinetic constants the sensitivity of the system towards back coupling of reactions on the flow field is assumed to be negligible. Therefore the system is described by two models: A hydrodynamic model which computes a flow and dispersion field and a model which depicts the transport based on velocity and dispersion arrays supplied by the hydrodynamic model and the assumed kinetics (Versteeg et al., 2007). The transport model is set for each constituent of concern, each toxic component or reactant which can be transformed into a toxic component. The computational domain has three space and one time dimension. The achieved simulation can serve to compute the dynamic distribution of by-products due to first order models, multi-concentration-interval linearized kinetics and also due to mildly nonlinear reactions.

2. Modeling and Simulation

2.1 Modeling

The transport model is a balance of molar in and outflow of a species j along an infinite small control volume. Terms for dispersion mass flow, kinetic and convection are taken into account. In and Outflows at $r_i + dr_i$ are written at $\boldsymbol{r}$, the coordinate vector at the cell centre, by substituting them with a linear shift, $\dot{n}(r + dr) = \dot{n}(r) + \frac{d\dot{n}(r)}{dr} dr$. Linearized reactions l for each species j are notated with stoichiometric, m_{jl}, and kinetic matrix, k_{jl}.

$$\frac{\partial c_j(\boldsymbol{s})}{\partial t} = \sum_i \left(\frac{\partial}{\partial r_i} D(\boldsymbol{s}) \frac{\partial c_j(\boldsymbol{s})}{\partial r_i} \right) - \sum_i \frac{\partial \big(u_i(\boldsymbol{s}) c_j(\boldsymbol{s})\big)}{\partial r_i} + \sum_l m_{jl}\, k_{jl} c_j(\boldsymbol{s}) \ \ with\, \boldsymbol{s} = [\boldsymbol{r}, t]$$

The hydrodynamic model uses the shallow water assumption based primitive equation approximations in the set of momentum, mass and continuity balance. These equations may be denoted as it is common in the corresponding hydrodynamic models (Gour-Tsyh Yeh et al., 2002, D. Haidvogel et al., 1999) with the three velocity components u, v and w.

$$\frac{\partial u}{\partial t} + u\frac{\partial u}{\partial x} + v\frac{\partial u}{\partial y} + w\frac{\partial u}{\partial z} - fv = -\frac{1}{\rho_0}\frac{\partial P}{\partial x} + \frac{\partial}{\partial z}\left(K_m \frac{\partial u}{\partial z}\right) + F_u$$

$$\frac{\partial v}{\partial t} + u\frac{\partial v}{\partial x} + v\frac{\partial v}{\partial y} + w\frac{\partial v}{\partial z} - fu = -\frac{1}{\rho_0}\frac{\partial P}{\partial y} + \frac{\partial}{\partial z}\left(K_m \frac{\partial v}{\partial z}\right) + F_v$$

$\frac{\partial P}{\partial z} = -\rho_0 g$ and $\frac{\partial u}{\partial x} + \frac{\partial v}{\partial y} + \frac{\partial w}{\partial z} = 0.$

2.2 Kinetics

Reaction rates are either introduced as first order models or in the form of linearized multi-interval approximations of measurement data. In case of concentration interval dependent kinetics after each time step for each cell the kinetic constant which corresponds to the local concentration interval is assigned to a cell. This approach has been favored in order to generalize the treatment of nonlinear or mechanistically not well understood kinetics. The applied kinetic constants are assigned based on concentrations of the preceding time level and therefore explicit. In case of multi-interval kinetics algorithms which potentially allow high Courant number are restricted to small times steps corresponding to the explicit control of the kinetic constants.

2.3 Outfall Assumption

Cooling water outfall geometries have been observed on industrial sites to have a variety of, partially above the sea surface located, outlets. Above surface discharge is preferable not modeled by continuum models which rely on well defined fluid phase geometries if no extensive re-meshing is utilized. Since the objective of the study is to evaluate whether boundary values are exceeded at the mixing zone boundary any assumption which introduces less mixing will not pose a threat to underestimate pollutant levels beyond the imminent discharge. Therefore the embedded subsurface discharge is considered to depict an equivalent scenario except possibly lower pollutant level compared to a highly mixing fountain like discharge.

2.4 Domain Geometry, Flow Field and Dispersion Arrays, Boundary Conditions, Source Terms and Tidal Influences

The Transport model relies on a fully cell-discretized topography whereas the hydrodynamic model depicts the topography curvilinear with σ-coordinates. Dynamic velocity component and Eddy diffusivity arrays are handed over from the hydrodynamic to the transport model. At the boundaries of the considered water body with the Arabian Gulf open boundary conditions are employed which assume no dispersive flux. At solid-fluid interfaces no-flux boundary conditions are embedded. Discharges drive the model by a spatial source term array. Besides first order kinetics also zero order source term like conversion terms are possible. Tidal variations apply to the surface layer as far as no σ-coordinates are employed. In case of σ-coordinates each layer varies in thickness with tidal oscillations.

The mesh of the transport model has a resolution of just 1 m throughout the whole domain. The triangle Finite Volume mesh has cell sizes ranging from 1 meter close to the discharge up to about two logarithmic orders inflated cells in the bulk region of the water body.

2.3 Simulation

The choice of the algorithm reflects to recognize the system's convective mass flow as dominating and thereby stability limiting. Furthermore Horizontal convective fluxes are logarithmic orders of magnitude higher than vertical convection. It can be therefore of advantage to apply possible implicit notation on the horizontal mass flow terms reducing the implicit-wise dimensionality from 3 to 2.

These dominating horizontal terms have been treated with an Alternating Direction Implicit-Alternate Direction Explicit algorithm and pure ADI algorithms. These methods are bound to structured grids. Beneficial stability properties of the ADI algorithm can be granted only for rectangular geometries. If σ-coordinates are used the rectangular geometry is violated only by the land mass but not the sea topography. A combined ADI-ADE algorithm has been applied to reduce this remaining stability dependency with better information propagation within the computational domain compared to a pure ADI algorithm. For comparison also a 2D (horizontal) ADI + 1D (vertical) explicit algorithm has been applied as well as a fully 3D-ADI Douglas and Brian based second order accurate algorithm. The algorithm, boundary condition, source term, as well as domain geometry specifications have been programmed purely in array fashion. Per fractional time step including all modifications only one array operation is required allowing a practically arbitrary degree of parallelization.

The FVCOM Finite Volume hydrodynamic model gains its efficiency by the volume size flexibility of the unstructured triangular grid reducing the cell number of the entire domain sufficiently to compute the fluxes iteratively. This allows approximating the system arbitrarily implicit allowing high Courant number. The unstructured grid for the hydrodynamic model has also the benefit of inflating volume sizes (like also horizontal curvilinear grids) with increasing distance from the shore. This is necessary to allow comparatively small cell size in near-shore shallow water to fulfill the shallow water condition of the utilized primitive equations.

For the ADI/ADE algorithms von-Neumann and heuristic stability analysis has been carried out. Not considering the violations of rectangular domain geometries the ADI approximations allows unconditional stability. The practically reduced stability has been partially regained by the constructed mixed ADI-ADE algorithm.

Beyond this error considerations have been undertaken. The error analysis considers quasi steady state. It is assumed that for a fully developed plume the error contribution from the slight dynamics due to the comparable slow tidal movements are small versus the error of the total process simulation. Besides that an error analysis of time series, especially time series for spatial systems, could not have been incorporated in an available general statistical error theory as it is given for time invariant processes with

$$\Delta c_j(x,y,z) = \sum_i \left|\frac{\partial c_j(x,y,z)}{\partial P_i(x,y,z)} \Delta P_i(x,y,z)\right|.$$

Therefore the uncertainty of c has to be evaluated numerically by running the simulation with slightly different Parameter P_i. The parameter is disturbed around the magnitude where the simulation is operated since the partial derivative in the operational region determines the error weight. This gives a 3 dimensional error array for the steady state and requires disturbing the velocity and dispersion array, the kinetic constant matrix and the discharge specifications. Thus also errors of velocity and dispersion arrays have to be computed. Therefore partial error for each nod at x,y and z will be obtained in later comparative and case sensitivity studies. For the hydrodynamic simulation it can be proceeded analogously where the environmental parameter, depth distribution, shoreline, radiation, precipitation, in and outflow of the Arabian Gulf and the discharge specifications can be disturbed.

2.6 Shoreline Approximation

First a coarse, thereupon a with ~0.5 m resolution accurate shoreline have been used for the simulations. The resulting horizontal domain characteristics have been casted in a structured mesh for the transport-reaction model and a triangular unstructured mesh for the hydrodynamic model as depicted in figure 1. Close to the discharge increased attention has been given recognizing the higher velocities with finer cells.

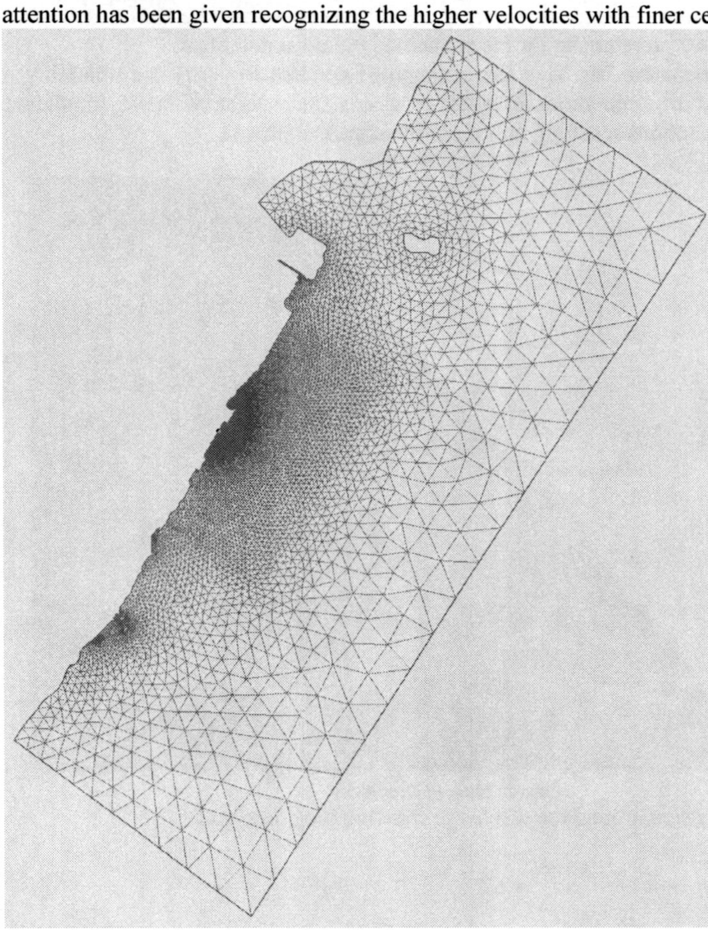

Figure 1: Finite Volume mesh at Mesaieed coast

3. Results

Two different simulations with a coarse and finer shore line and varied outflow rates triggered the same south-western heading plume shape to develop along the shore.

Superimposed on the hydrodynamic simulation the transport-reaction simulation has been carried out as depicted in figure 2 close to the discharge.
According to the decoupled step-wise approach in a first step the flow field has been computed as depicted in figure 3.
Due to the correlating plume orientation in case of different shore line data quality, discharge flows and different bathymetry data the case appears to be well posed which is considered to be a prerequisite for the concerned plume simulations.
This well posed behavior has been the question of concern to verify the reliability of shallow water outfall simulations in order to assess the weight of made simulations compared to distributions resulting out of pollutant measurements.

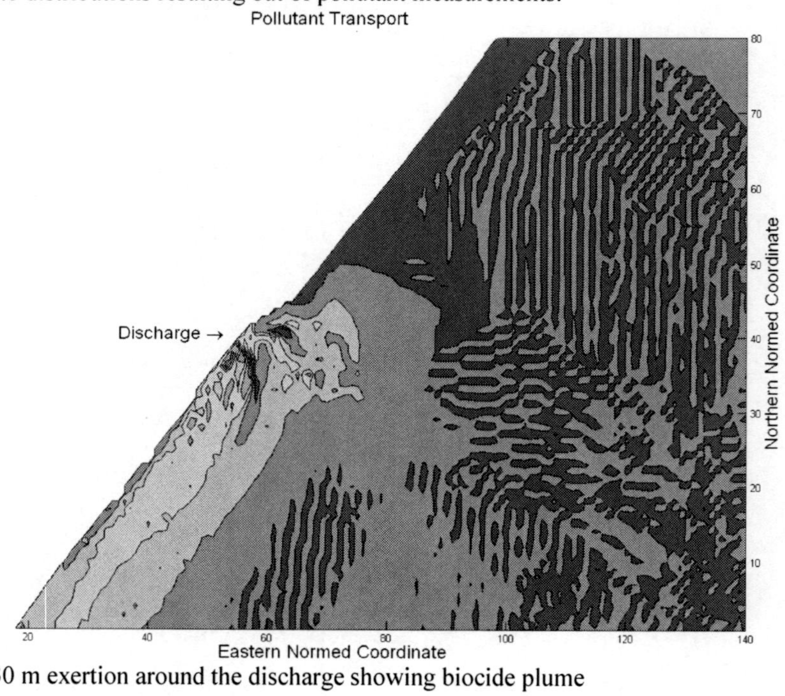

Figure 2: 30 m exertion around the discharge showing biocide plume

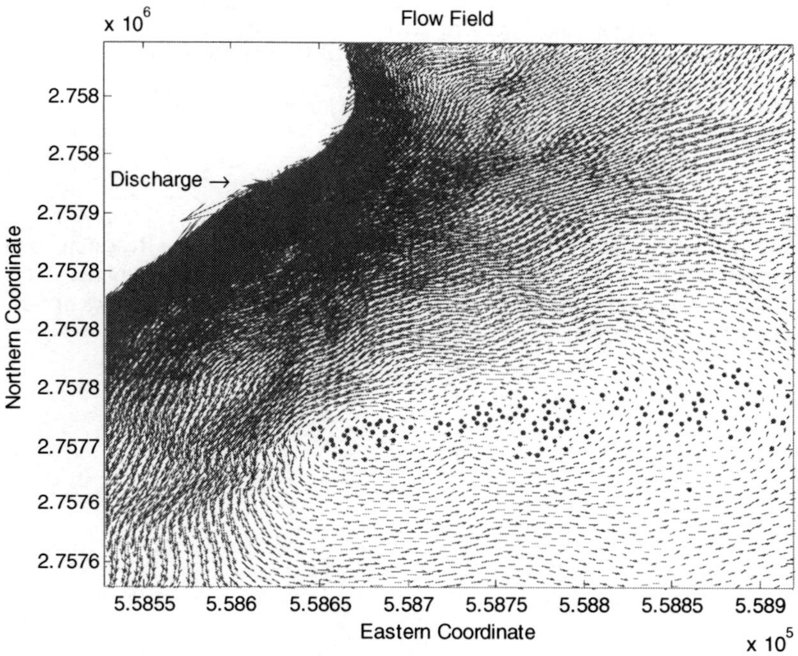

Figure 3: Flow field input for transport and reaction simulation

4. Conclusion

The separate transport model allows more freedom to effectively incorporate multi-interval linearized and mildly nonlinear conversion terms. Hydrodynamic and transport-reaction simulations have been successfully carried out in a split manner with two different meshes though we recognize that using the same mesh is preferable. The problem is well posed which further justifies discharge simulations.

References

Y. Huaming et al., 2009, A Study on Optimum Schemes for Reclamation in a Narrow Bay using the Finite Volume Method, China Academic Journal Electronic Publishing House

A.E. Adenekan, V.S. Kolluru, J.P. Smith, 2009, Transport and Fate of Chlorinated By-Products Associated with Cooling Water, Proceedings of the 1st Annual Gas Processing Symposium, pp 340-353

A. Abdel-Wahab, P. Linke, H. Alfadala, M. El-Halwagi, B. Batchelor, 2009, Towards a Holistic Approach to the Sustainable Use of Seawater for Process Cooling, Proceedings of the 1st Annual Gas Processing Symposium, pg. 330-339

Gour-Tsyh Yeh, Ming-His Li, M. D. Siegel, 2002, Fluid Flows and Reactive Chemical Transport in Variable Saturated Subsurface Media, Environmental Fluid Mechanics Theories and Applications, pg. 207-255

H. Versteeg, W. Malalasekra, 2007, The finite volume method for convection-diffusion problems, Introduction to Computational Fluid Dynamics: The Finite Volume Method
L. Lapidus, G.F. Pinder, 1999, Numerical Solution of Partial Differential Equations
A. Date, 2005, Introduction to Computational Fluid Dynamics
D. Haidvogel, A. Beckmann, 1999, The hydrostatic Primitive Equations, Numerical OceanCirculation Modeling, pg.19-21

Acknowledgments

This publication was made possible by a grant from the Qatar National Research Fund under its national Priority Research Program (NPRP). Its contents are solely the responsibility of the authors and do not necessarily represent the official views of the Qatar National Research Fund.

Proceedings of the 2nd Annual Gas Processing Symposium
Farid Benyahia and Fadwa T. Eljack (Editors)

Sustainable Water Management: A Systems Integration Approach

S. Atilhan[a,b], P. Linke[a], A. Abdel-Wahab[a], A. Bin Mahfouz[b], and M. El-Halwagi[b],

[a]*Texas A&M University at Qatar, Chemical Engineering Department, Doha, QATAR*
[b]*Texas A&M University, Artie McFerrin Chemical Engineering Department, TX, USA*

Abstract

Water production by desalination is a key solution to satisfying future water demands in many countries with scarce fresh water resources. Desalinated water is needed in numerous industrial operations for purposes such as cooling, washing, quenching, and absorption. In order to meet the industrial needs of desalinated water, there are usually tradeoffs between using centralized facilities with a distribution system or using decentralized facilities on or near the using sites. In the Gulf region, water management has a high priority because of the limited water resources, the population growth, and the industrial and agricultural expansion. This work develops an optimization-based methodology for water management. This task requires the identification of network configuration composed of desalination facilities and the assignment of water sources to users. A new source-desalination plant-sink (user) representation is developed to manage water needs in the region. Then, a mathematical formulation is developed and solved to minimize the total annual cost of the desalinated water including desalination plant cost, desalinated water operation, piping and pumping. A case study is solved to illustrate the convenience of the developed procedure.

Keywords: Desalinated water; Industrial usage of water, Water management; Optimization.

1. Introduction

Production of fresh water via desalination is an important option for satisfying water needs in industrial processes in countries with limited supply of fresh water such as the Gulf area. Developing a water management strategy is a complex task because of the availability of various sources of water, the possible use of different desalination technologies and different plant sizes, the competition between larger centralized facilities and smaller decentralized facilities, and the need to transport desalinated water to the industrial users. As such, a systems integration approach provides very attractive features that enable cost-effective and sustainable decision making for water management.

The cost of seawater desalination, which depends on location of the desalination plants, energy usage, transportation and other costs; has recently decreased due to improvements and advancements in desalination technologies (Kim et al., 2009, Mehdizadeh, 2006, Hawlader et al., 2000). According to recent figures, desalinated water production and energy costs are aimed around \$0.5 per m3 of water and \$0.02 per kWh(Awerbuch, 2002). Techno-economic analysis of different desalination techniques has also been also studied and compared to select the most appropriate technology (Fiorenza et al., 2003, Manolakos et al., 2008, Borsani and Rebagliati, 2005).

In this work, we introduce a systems integration approach that can be used as the basis for developing design and planning decisions for desalination and distribution networks on a macroscopic scale (region, country, etc.). A simplified version of source-interception-sink representation is developed and presented to account for the potential configurations of interest. The design problem is formulated as an optimization problem subject to the supply, demand, mass conservation, technical performance, and economic constraints. We also use LINGO (Schrage, 2006) for the solution of the formulated optimization problem. Output results are used to determine the selection, design, capacity, and location of the desalination technologies and how the various desalination plants are best integrated via a distribution network to serve a region.

2. Problem Statement

The following is a statement of the addressed problem. Given is a process with:
A set of saline water sources is equal to $\{i|i=1, 2, \ldots, N_{sources}\}$. Sources are saline water sources that can compose of process streams, which are available to be allocated. Given components are $\{p|p=1, 2, \ldots, N_{component}\}$. Each source has a maximum allowable saline water flowrate, G_i and composition $y_{i,p}$.

$$G_i^{\min} \le G_i \le G_i^{\max}$$

$$y_{i,p}^{\min} \le y_{i,p} \le y_{i,p}^{\max}$$

A set of process sinks is equal to $\{j|j=1, 2, \ldots, N_{sinks}\}$. Sinks are process units that can accept the sources. Each sink requires a given flowrate, F_j and composition $y_{j,p}$.

$$F_j^{\min} \le F_j \le F_j^{\max}$$

$$y_{j,p}^{\min} \le y_{j,p} \le y_{j,p}^{\max} \qquad j \in \text{ SINKS, } p \in \text{ component}$$

A set of interception units is equal to $\{k|k=1, 2, \ldots, N_{Int}\}$. Interception units are called desalination plants. Each interception has a given flowrate, Q_k and composition $y_{k,p}$.

$$Q_k^{\min} \le Q_k \le Q_k^{\max}$$

$$y_{k,p}^{\min} \le y_{k,p} \le y_{k,p}^{\max}$$

3. Approach

The following hierarchical procedure is proposed:

i. Minimizing the total annual cost for sustainable water management problem
ii. Targeting maximum production of desalinated water and minimum brine discharge;
iii. Optimizing to distribution of saline water sources to sinks to achieve the goal;
iv. Selecting the most effective desalination technology for this problem.

First a schematic representation of the problem is developed as shown in Figure 1 to embed potential configurations of interest. It is composed of sources (saline water), interception (desalination technologies), and sinks (industrial units using fresh water). This is based on the source-interception-sink representation used in mass integration(El-Halwagi, 2006).

The design problem is formulated as a nonlinear programming problem whose objective is to minimize the total annualized cost of the system. The constraints include performance of the desalination units, splitting of the saline sources to the desalination plants, assignment of desalinated water to industrial users, and disposal of brine streams.

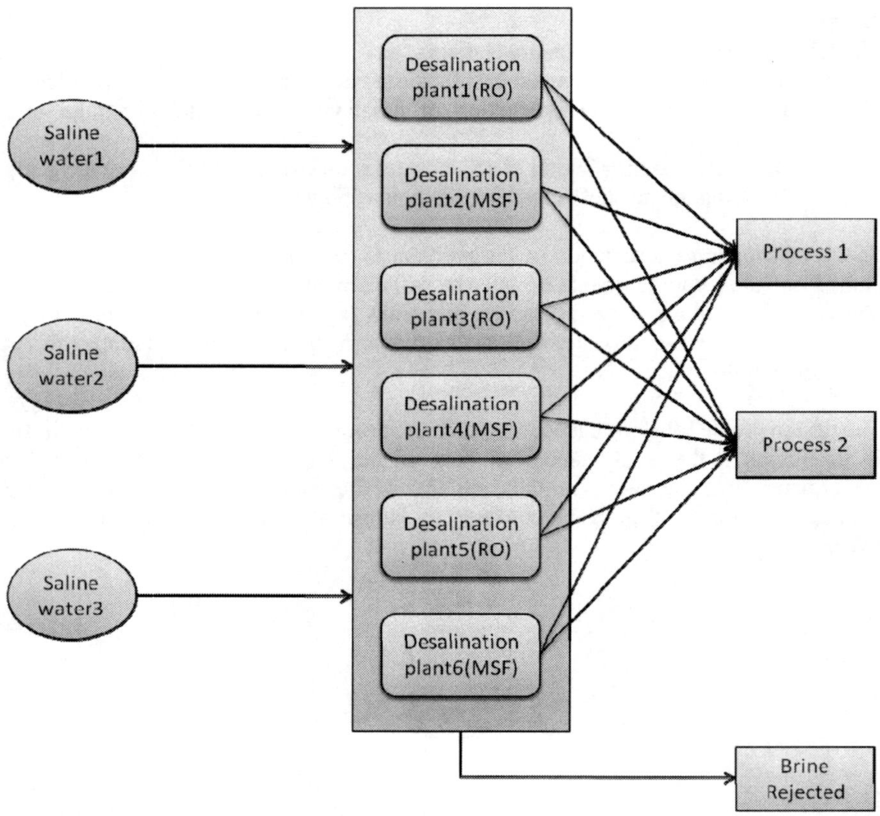

Figure 1: Scheme of water management problem used in this work.

4. Case Study

Consider a water management problem with three saline water sources and two processes as shown in Figure 2. The data for the saline water sources and the processes are given in Tables 1 and 2. The distances between saline water sources and the processes are shown in Table 3. Six desalination plants (interceptors) are considered for the desalination of saline water. The cost and flow data for the interception technologies are detailed in Table 4. The cost of piping is assigned to be 0.018 $/(m3. mile)

Table 1: Allowable flow rate and saline concentrations for process 1 and 2

Process	Minimum water required, m^3/day	Maximum allowable saline concentration, ppm
1	167,000	267
2	120,000	134

Table 2: Allowable flow rate and saline concentrations for saline water sources

Saline water source	Maximum allowable saline water rate, m^3/day	Saline concentration, ppm
1	267,000	30,000
2	240,000	13,400
3	40,000	3,340

Table 3: The distances between the saline water sources and process 1 and 2

Saline water source	Process	Distance (miles)
1	1	3
1	2	73
2	1	76
2	2	30
3	1	68
3	2	25

The aim of this case study is to minimize the total annualized cost of the system under suitable process constraints. Non-linear programming (NLP) formulation is developed and solved. The global optimum solution is found by using the Global Solver of software LINGO 10.1(Schrage, 2006). Results are shown in Figure 2. The minimum annual cost of water management system is determined to be $52.9MM/yr. Cost and flow rate results of water management problem are feasible. These results are shown that this case study is applicable for sustainable water management problems.

Table 4: Costs for each technology and sources

Technology	**Saline water source**	**Outlet saline concentration (ppm)**	**Water recovery (m^3desalinated water per m^3 feed water)**	**Annual fixed cost function (\$/yr)**	**Operating cost (\$/m^3 desalinated water)**
RO	1	267	0.45	$2215*Q^{0.75}$	0.167
MSF	1	234	0.58	$1655*Q^{0.65}$	0.133
RO	2	134	0.53	$777*Q^{0.75}$	0.120
MSF	2	54	0.69	$1485*Q^{0.65}$	0.160
RO	3	67	0.68	$235*Q^{0.55}$	0.053
MSF	3	54	0.75	$575*Q^{0.6}$	0.073

*Q is flow rate of desalinated water in m^3/day

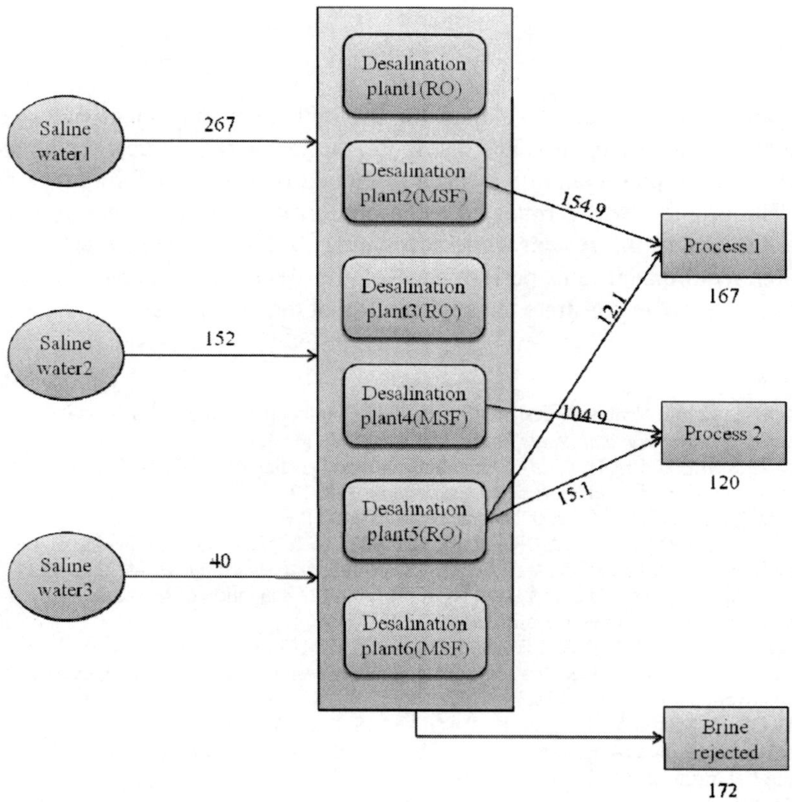

Figure 2: Global optimum solution of the problem solved by software LINGO 10.1(All flow rates are in terms of 10^3 m^3/day)

5. Conclusions

A systematic approach was developed for the optimal management of water resources involving desalination and distribution to industrial sources. A source-interception-sink representation was used to embed potential configurations of interest. The problem was formulated as an optimization program that aims to minimize the cost of the system while accounting for the qualities of seawater, fresh water requirement, and performance of the desalination plants. A case study was solved to demonstrate the applicability of the procedures.

References

AWERBUCH, L. (2002) Vision for desalination-challenges and opportunities. *IDA World Congress on Desalination and Water Reuse.* Manama, Bahrain.

BORSANI, R. & REBAGLIATI, S. (2005) Fundamentals and costing of MSF desalination plants and comparison with other technologies. *Desalination,* 182, 29-37.

EL-HALWAGI, M. M. (2006) *Process Integration,* Academic Press.

FIORENZA, G., SHARMA, V. K. & BRACCIO, G. (2003) Techno-economic evaluation of a solar powered water desalination plant. *Energy Conversion and Management,* 44, 2217-40.

HAWLADER, M. N. A., HO, J. C. & CHUA KOK, T. (2000) Desalination of seawater: an experiment with RO membranes. *Desalination,* 132, 275-280.

KIM, Y. M., KIM, S. J., KIM, Y. S., LEE, S., KIM, I. S. & KIM, J. H. (2009) Overview of systems engineering approaches for a large-scale seawater desalination plant with a reverse osmosis network. *Desalination,* 238, 312-332.

MANOLAKOS, D., MOHAMED, E. S., KARAGIANNIS, I. & PAPADAKIS, G. (2008) Technical and economic comparison between PV-RO system and RO-Solar Rankine system. Case study: Thirasia island. *Desalination,* 221, 37-46.

MEHDIZADEH, H. (2006) Membrane desalination plants from an energy-exergy viewpoint. *Desalination,* 191, 200-209.

SCHRAGE, L. (2006) LINGO User's Guide. Chicago, IL, LINDO System Inc.

Proceedings of the 2nd Annual Gas Processing Symposium
Farid Benyahia and Fadwa T. Eljack (Editors)

Overview of Conversion of Greenhouse Gas Carbon dioxide to Hydrocarbons

S. M. Javaid Zaidi
Center of Excellence in Renewable Energy and Chemical Engineering
King Fahd University of Petroleum & Minerals, Dhahran-31261, Saudi Arabia

Abstract

Conversion of green house gas carbon dioxide to abate global warming is of interest to save the planet from the catastrophe. Carbon dioxide can be used for the synthesis of important petrochemicals like methanol, ethylene etc. Various aspects of CO_2 conversion have been extensively studied in both aqueous and non-aqueous medium. The process requires an efficient catalyst and energy and strongly depends on electro-catalyst used for the reduction products. Both in aqueous and non aqueous environments and at higher over potentials the selectivity towards desired products is still relatively low. The major limitation is the low solubility of carbon dioxide in electrolyte and high over potential at which this reaction proceeds. In this presentation, various approaches for the reduction of CO_2 to useful petrochemical products will be discussed with particular emphasis on the solid-state electrochemical method. In addition different parameter affecting the conversion of CO_2 and the optimum conditions will be discussed.

Keywords: Electrochemical reduction, Carbon dioxide, Electro-catalyst, SPE, Hydrocarbons.

1. Introduction

Carbon dioxide is a greenhouse gas and the concentration of CO_2 in the atmosphere has increased significantly since the beginning of the industrial revolution. There are three main sources of CO_2 emissions: stationary, mobile and natural sources. Within the industrial sectors, there are two principle routes of CO_2 formation. CO_2 accumulation in the atmosphere is widely blamed as a main contributor to global warming and cause of extreme weather (Tryk et al. 2001). This situation has stimulated worldwide research directed at reducing CO_2 emission. More stringent laws are coming into existence on carbon dioxide tolerance and pressure on industries as well as automobile manufacturers to curb their CO_2 emission and use eco-friendly fuels and processes is mounting. To meet stringent climate change targets, such as stabilizing CO_2 concentrations below 550 ppm, or limiting the global temperature rise to 2°C above pre-industrial levels requires drastic CO_2 reductions of 60–80% by 2050 compared to 1990 emissions level, which is a big challenge. This will require a portfolio of technologies and mitigation activities across all sectors such as improving energy efficiency, carbon capture and storage and the use of renewable energies or nuclear power.

2. CO_2 Conversion and Utilization

CO_2 conversion and utilization should be an integral part of carbon management. As an example of utilization of CO_2, approximately 110 million metric tons per year of carbon dioxide are used as a raw material for the production of urea, methanol, polycarbonates, cyclic carbonates and specialty chemicals (Arakawa et al. 2001). The chemical industry has pledged an industry wide goal of reducing its greenhouse gas intensity (ratio of net greenhouse gas emissions to production) by 18% to 1990 levels by 2012. A detailed breakdown and a total of 68 million metric tons for carbon equivalent was reduced in 2001 of which 50 million metric tons were from direct reduction, 16 million metric tons from indirect reduction and 2 million tons were sequestered. A potential upper limit of 650 million metric tons of CO2 use as a raw material has been estimated [Song et al. 2002] , which included traditional processes for urea and methanol in addition to plastics, fibers, rubber and other uses. This tonnage is comparable to carbon dioxide emissions from all US fossil fuel power plants.

3. Challenges for CO_2 Utilization

The costs involved for CO_2 capture from a manufacturing process and energy requirements for CO_2 conversion are some of the main challenges being faced for the CO_2 utilization (Song et al. 2002) . The total emissions in 2010 will be 44% above the Kyoto target, and these emissions will be 62% above the target in 2020. The emissions from electric utilities and transportation were 523 and 473 million metric tons of carbon equivalents. The total carbon dioxide emission in U.S. in the year 1997 was 1500 million metric tons of carbon equivalents. These emissions were high enough and will still put the net emissions well above the Kyoto target. Thus, the need for the introduction of the new technology and the change of infrastructure are desired (Flannery 2000). The rate at which these new technologies would be developed is also an equally important issue compared to the development of these technologies. The important issue would be how to introduce new technology in a small scale and then get them to grow into widespread commercial use. No single new technology will solve the entire problem. There should be an emergence of a number of promising new technologies that could contribute to the carbon dioxide emission reductions. All of them have to overcome challenges of economics, performance, and associated environmental impacts, which are some of the barriers for new technology into widespread commercial use. (Flannery 2000)

4. Research Strategies for CO_2 Utilization

Carbon dioxide can be used as a reactant or co-feed in various chemical processes. . An array of chemicals can be manufactured from Carbon dioxide. It can be used as the whole molecule in reactions, and it can be used as a carbon source or as an oxygen source (Li, 2006) . The synthesis of urea from ammonia and carbon dioxide, and the production of salicylic acid from phenol and carbon dioxide are good examples of the large-scale production processes where carbon dioxide is utilized as a raw material. These are some of the possible ways to expand the utilization of carbon dioxide in chemical industry. Possible processes that may be used for CO_2 conversion include the radio-chemical, photo-chemical, bio-chemical, thermo chemical, electrochemical and photo-electrochemical methods. Most of these processes are still research subjects in the laboratories and few have reached large scale production (Song et al. 2002, Scibioh and Vishwanathan, 2004).

The electro-reduction of CO_2 (ERC) is of great interest in the fields of theoretical and applied electrochemistry, as reflected in the number of papers on ERC published since 1982, i.e. 441 (Scibioh and Vishwanathan, 2004). However, the key argument against ERC is that, if it came from the combustion of fossil fuels, the electric energy used in such a process would increase CO_2 emissions. Therefore, ERC will have a future only when it can be carried out with high efficiency on an industrial scale and renewable energy is adopted as the source of electricity. There are several issues in the research of ERC that should be dealt with in the development of a process that will be interesting at the industrial scale (with reproducibility, long-term stability, high efficiency, low cost etc.

The major issues related to the ECR are the mass transfer constraints, reactor capacity and lack of engineering research. For industrial electro-chemical processes, the superficial current density and current efficiency should be respectively at least 1 kA m^{-2} and 50 %. However, the relatively low solubility of CO_2 in aqueous solutions (ca. 70 mM at STP), coupled with the CO_2 (aq)/HCO_3/ CO equilibria, creates a mass transfer constraint on the reduction of CO_2 that limits the primary current density to a maximum value of the order 0.1 kA m^{-2} Several devices have been suggested to relieve the CO_2 mass transfer constraint, including operation at super-atmospheric pressure and/or sub-ambient temperature, using a gas-diffusion cathode (GDE) or using a fixed-bed cathode while providing a "3-phase interface" for the reaction by sparging the cathode chamber with CO_2 gas (Mahmood et al. 1987). Apart from the constraint of the CO_2 mass transfer limiting current density, a practical reactor for ERC must be able to handle volumetric gas feed rates in the order of 1 m^3 STP per hour per kA.
Research on the engineering aspects of ERC should be initiated to bridge the gap between the previous laboratory work and industrial reality. Such engineering research

includes the design and scale-up of continuous electro-chemical reactors, together with the conception, design and economic projections for complete ERC processes.

4.1. System development for ERC

Procedures for electro-reduction of CO_2 can be classified according to both the nature of the electrodes (cathodes) and the media used for the catholyte (aqueous and non-aqueous). If the reduction of carbon dioxide to liquid fuels is to be accomplished through electrochemical means, the deployment of efficient electrocatalyst will be essential for the development of practical industrial processes. Identification of electro catalytic activity can be seen easily in cyclic voltammetry (CV)). For this reason, ECR has been studied on different electrode materials such as Hg, Pb, Sn, In, Au, Ag, Ni, and Cu and their reduction products are given in Table 1. ERC has been carried out on a variety of electrode surfaces, including most of the metals in the periodic table, a number of alloys, metal oxides and chemically modified electrodes [Qu et al. 2005].

Table 1: Various products from metal electrodes

Groups	Metals	Products
A	Cu	Hydrocarbons and alcohols
B	Au, Ag, Zn, Pd, and Ga	Carbon monoxide
C	Pb, Hg, In, Sn, Bi, Cd, and Tl	Formate/formic acid
D	Ni, Pt, Fe, Co, Rh, Ir, and W	Multi-products; only exhibit useful catalytic activities at high pressure and low temperature

4.2. Electrode Design

Various types of 3-D electrode with three-phase (G/L/S) operation have been used for ERC. Gas diffusion electrodes (GDE), based upon PTFE (polytetrafluoroethylene) as a binder, have been developed to overcome mass transport limitations with gaseous reactants in fuel cells. When employed in ERC, GDE's operated in batch mode, provided much higher superficial current densities than those obtained with 2-D metal plate electrodes [Mahmood et al. 1987]. ERC was carried out by Koleli et al. in 2004 [Koleli and Balun 2004] in an un-divided fixed bed reactor where Pb and Sn granules in a batch mode operation. Only one product, "formic acid", was supposedly produced, with the maximum current efficiency of 90% at a superficial current density of 0.008 kA m-2. Also, ERC was carried out in a batch reactor at the 3-phase (G/S/L) interface on Cu and Pt mesh electrodes at ambient pressure and temperature. In this system CO2 was supplied to the electrode from the gas phase by bubbling CO_2 through the mesh, aiming to maintain a high concentration of CO_2 at the cathode surface.

4.3. Electrolyte for ERC.

Because of the influence of the solvent on the nature of the products, there have been two main routes for CO_2 electro-reduction based on whether the catholyte is aqueous or non aqueous. Aqueous media have long been a popular selection for ERC research because the water provides a proton source and have higher electric conductivity than that of non-aqueous electrolytes. A lot of studies have been done in aqueous media on almost all the electrodes. Non-aqueous media present certain advantages for CO_2 reduction: higher CO_2 solubility than that in water, the suppression of H_2 evolution, and the possibility to work at low temperature, e.g., below the freezing point of water.
Although the main products from CO_2 reduction in non-aqueous media have been found to be quite similar to those in aqueous media, i.e., carbon monoxide, oxalic acid and formic acid, the reaction mechanisms can follow pathways different from those in aqueous media [Ikeda et al. 1987]. The electrolytes used in non-aqueous system are different from those in aqueous media. KOH and $TBABF_4$ (tetrabutylammonium tetrafluoroborate) were both employed as the electrolyte for non-aqueous media, but the latter was more frequently used. A variety of combinations of alkali metal cations and various anions, such as PO_4^{-3}, HCO_3^-, SO_4^{--}, ClO_4^-, HPO_4^- have been used as the electrolyte in aqueous media for ERC.

5. ERC to Hydrocarbons at Solid Polymeric Electrolyte

The limiting problem with the electrochemical reduction of CO_2 is the low solubility of CO_2 in aqueous solution which limits the current density to 0.2 kAm-2 Hori et al. 2003].This problem by metal- coated ion exchange membrane electrode (occasionally called "solid polymer electrolyte" or SPE) [Cook et al. 1990, Dewulf and Bard 1988]. Ion-exchange membrane coated with porous catalyst metal can provide gas phase electro-reduction of carbon dioxide. The use of ion-exchange membrane as electrolyte medium is convenient so as to limit cross over, resulting in a decrease in current efficiency of the electrochemical cell. SPE electrodes were applied to CO_2 reduction as previously reported by some workers.

There are two components of such SPEs, the ion-conducting membrane and electrocatalyst coating. Polymer electrolyte membrane (PEM) is recognized as the key element for an efficient ERC system. There are basically two types of membranes used for ERC applications. One is cation exchange membranes (CEM) and other is anion exchange membrane (AEM). Till now only, proton conducting perfluorinated ionomers based DuPont Nafion® has been used as a membrane material. Although strictly speaking not lying in the category of SPE, yet some relevant works were done using membrane as an integral part of the ERC system.

Acknowledgements:

The support of the King Fahd University of Petroleum and Minerals for this work is highly appreciated. The author also would like to thank Mr. Safdar Hossain for help in the literature review.

References

Arakawa H, M., Aresta, J.N. Armor, Chemical Reviews, Vol.101, (2001), pp953-958

Cook , RL, RC. MacDuff, AF. Sammells (1988) Journal of Electrochemical Society, Vol.137, No.1, pp187-189

Dewulf, DW, AJ. Bard (1988) Catalyst Letters, Vol.1, No.1-3, (1988), pp73-79

Flannery B.P, Workshop Report to the Chemical Sciences Roundtable, (2000), pp44-59

Gonçalves MCMR, JAD. Condeço TCD. Pardal , DM. Roncero, DMF. Santos, CAC. Sequeirac (2007) ECS Transactions, Vol.3, No.18, p 49.

Hori Y, I. Takahashi , O. Koga , N. Hoshi (2003) Journal of Molecular Catalysis A: Chemical, Vol.199, No.1-2, (2003), pp39-47

Ikeda. S, T. Takagi, K. Ito, The Chemical Society of Japan, Vol.60, (1987), pp2517-2522

Koleli F, D. Balun (2004) Applied Catalysis A: General, Vol.274, No.1-2, pp237-242

Li H (2006) Ph.D. thesis, Chemical & Biological engineering, University of British Colombia, p. 156

Mahmood MN,D. Masheder , C.J. Harty (1987) Journal of Applied Electrochemistry, Vol.17, No.6, pp1159-1170

Qu J, X. Zhang , Y. Wang , C. Xie (2005) Electrochimica Acta, Vol.50, No.16-17, (2005), pp3576-3580

Scibioh AM, B. Vishwanathan, Proceedings of Indian National Science Academy, Vol.70A, No.3, (2004), pp407-462

Song C., A.M. Gaffney , K. Fujimoto , An American Chemical Society Publication.(2002)

Tryk DA, V. Yamamoto , K. M, Hirota, K.M. Hashimoto, Okawa , Journal of Organometallic Chemistry Vol.15, (2001), pp113-120

Proceedings of the 2nd Annual Gas Processing Symposium
Farid Benyahia and Fadwa T. Eljack (Editors)

Sulphur Sustainable Applications: Initial Field Monitoring and Performance of Shell Thiopave Trial Road in Qatar

Marwa Al-Ansary[a], Eyad Masad[b], and David Strickland[c]

[a]*Qatar Shell Science and Technology Centre, Doha, Qatar*
[b]*Texas A&M University at Qatar, Doha, Qatar*
[c]*Shell Sulphur Solutions, United Kingdom*

Abstract

The estimated sulphur output from Qatar is around 4 Mtpa by 2012, which is mainly from the gas processing operations in Qatar. The technology of using sulphur in roads was developed in the 1970s and was originally known as sulphur extended asphalt (SEA). However, the change in sulphur and hydrocarbon economics as well as environmental concerns related to the handling and safety of molten sulphur, rendered SEA mixes an unattractive option. Nevertheless, in the early 2000's, interest in the use of SEA mixes was renewed by innovations in the technology that eliminated hazards associated with molten sulphur use.

Shell Thiopave[1] is a technology for sulphur-enhanced asphalt mix developed by Shell Sulphur Solutions since 2003. Shell's technology enables a portion of the bitumen in the asphalt mix to be replaced with sulphur and special additives, resulting in a pavement that has enhanced mechanical properties such as increased stiffness and significantly reduced permanent deformation in the asphalt layer of a pavement particularly at high ambient temperatures. A trial two-lane section of roadway of asphalt mix containing Shell Thiopave was constructed in October 2007 at Pearl Village, the worker accommodation for the Pearl GTL project, which is under construction by Qatar Petroleum and Shell in Ras Laffan Industrial City, Qatar. The roadway section has been periodically evaluated by Texas A&M University at Qatar. This paper reports the initial results of the field evaluation, and it provides a summary of the findings from laboratory characterization of mixtures and industrial hygiene monitoring during construction.

Keywords: Gas Processing, Sulphur, Shell Thiopave, SEA, Road Construction, Sustainability.

---

[1] ***Shell Thiopave is a trade mark of the Shell Group of Companies***

1. Introduction

Most of the current reserves of gas and oil are richer in H_2S and other sulphur components than previously developed fields. An increasing proportion of currently produced elemental sulphur is generated from oil and gas processing operations. Moreover, with the recent introduction of more stringent environmental legislation primarily in the USA and Europe, more sulphur has to be removed from fuels than previously. The development of Qatar's large gas resources, particularly the North Field, will increase its sulphur production significantly (Sulphur magazine, March/April 2009). The North Field is only moderately sour (0.5-1.0% H_2S) but combined with the sheer scale of operations planned, the forecast is a volume of 4 million tons per annum by 2012 (equivalent to almost 12,000 tons/day of sulphur) (Sulphur magazine, March/April 2009).

Shell will contribute to sulphur production in Qatar via two projects with Qatar Petroleum; viz. the Pearl GTL (Gas-to-Liquids) project and the Qatargas 4 LNG (Liquefied Natural Gas) facility. Figure 1 presents a simplified process integration scheme for the Pearl GTL and Qatargas 4 LNG facilities, located in Ras Laffan Industrial City (RLIC), Qatar. From Figure 1, it can be shown that in both the GTL and LNG processes, sulphur has to be removed in the feed gas processing section. In the GTL process, which is a technology that converts natural gas via a chemical conversion over a catalyst into a range of products such as naphtha, kerosene, diesel and lubricant oils, the catalyst is very sensitive to sulphur poisoning and virtually all sulphur needs to be removed in the gas treating step. In the LNG process, the sulphur has to be taken out to avoid freezing out of H_2S and consequently blocking the main cryogenic heat exchangers. Also, the aluminium linings of LNG carriers can be quickly corroded by H_2S.

One of Shell's main goals is to create a virtuous circle of sulphur management; removing sulphur where it adds little value (e.g. in fuels), and then using this recovered sulphur to add value, for example as a component of new end-products which have the potential for large scale commercialisation with significant benefits. Hence, Shell is using leading-edge technological expertise to utilise sulphur or Thio (Greek word for sulphur) in various sustainable applications such as in roads (Shell Thiopave), in concrete (Shell Thiocrete[2]) and in fertilisers (Shell Thiogro[3]).

There is a combination of factors that suggest that Shell Thiopave could offer an important paving solution in Qatar. First, the construction industry is growing, resulting in a high demand for construction materials. This growth creates an opportunity for construction materials based on Qatari natural materials as long as these comply fully with Qatar Construction Standards. Second, due to the high temperatures in Qatar, especially during the summer season, roads in Qatar are susceptible to permanent deformation in the asphalt layer of the pavement. Permanent deformation, also known

---

[2] ***Shell Thiocrete is a trade mark of the Shell Group of Companies***

[3] ***Shell Thiogro is a trade mark of the Shell Group of Companies***

as rutting, occurs primarily due to reduction in the shear strength of an asphalt mixture at high temperatures; and consequently, the loss of the mixture's ability to resist shear stresses exerted by repeated traffic loading. Figure 2 highlights an example of permanent deformation in roads in Qatar. Shell Thiopave could be applied in Qatar, as it will use sulphur, a locally produced material, to improve shear strength of the asphalt mix and construct superior roads with enhanced resistance to permanent deformation. Both Qatar Shell New Business Development (NBD) and Qatar Shell Research and Technology Centre (QSRTC) are providing commercial and technical support, respectively, for the development of Shell Thiopave in Qatar.

This paper will discuss the results of the first field trial of Shell using Shell Thiopave modified asphalt mixes in Qatar. The trial road stretch was located at Pearl Village, the worker accommodation for the Pearl GTL project in Ras Laffan Industrial City, Qatar. The test sections were constructed in October 2007 and periodically evaluated by experts from Texas A&M University at Qatar.

2. Use of Sulphur in Road Construction

The technology of using sulphur in roads was developed in the 1970s and was originally known as sulphur extended asphalt (SEA). There is a wealth of information published about the properties of sulphur modified asphalt road sections constructed in the 1970s and early 1980s (Beatty et al., 1984). Laboratory and theoretical studies from the US Federal Highway Administration indicated that the addition of sulphur to asphaltic pavements could produce better, more economic pavements (Pickett, 1978). In spite of the improvements in road mechanical properties, the change in sulphur and hydrocarbon economics as well as environmental concerns related to the handling and safety of molten sulphur meant that SEA became an unattractive option. In the early 2000's, however, interest in the use of SEA mixes was renewed by innovations in sulphur technology that eliminated hazards associated with molten sulphur use.

Shell has been developing a technology called 'Shell Thiopave' for sulphur-enhanced road construction since 2003. Shell Thiopave is a patented asphalt mix additive based on sulphur technology. It is added to the asphalt mix as solid pellets with a diameter less than 5 mm. It is added at ambient temperature and quickly melts in the mix to form part of the binder as it comes into contact with the hot asphalt mix. Figure 3 shows the Shell Thiopave pellets. In general, bitumen contains a very small amount of natural sulphur, however, some of the added Shell Thiopave acts as a bitumen extender, whilst the other portion of the added Shell Thiopave forms a lattice structure that gives the improved mechanical properties.

Therefore, there are many advantages for the addition of Shell Thiopave to the asphalt mix in order to obtain superior roads characteristics. For example, the addition of Shell Thiopave to the asphalt mix increases the overall mix strength as demonstrated by stiffness modulus and a higher resistance to permanent deformation. The benefits of using Shell Thiopave are realized in terms of extended road life and/or thinner pavements. It also works effectively across a wide range of climatic conditions. In cold climates, where cracking due to brittleness is an issue, Shell Thiopave allows the use of softer asphalt binders with improved resistance to cracking at low temperatures, without sacrificing the high temperature performance. In hot climates, the increased high-temperature stiffness and shear strength of the Shell Thiopave asphalt mix improves resistance to permanent deformation. These enhanced properties of roads constructed using Shell Thiopave have been demonstrated under real field conditions in case studies in the USA, Canada, China and Saudi Arabia.

Shell Sulphur Solutions conducted laboratory characterization and analysis on Shell Thiopave modified mixtures for several field projects in USA, Canada, China and Saudi Arabia prior to the one in Qatar (Gilbert, 2008). The main conclusions from these studies were:

- Shell Thiopave acts as an asphalt mix modifier, it lowers the binder viscosity and increases its ductility and improves the asphalt mix performance properties.
- The asphalt mixture stability increases with the use of Shell Thiopave.
- The resistance to permanent deformation with the use of Shell Thiopave was enhanced.
- The temperature for production of asphalt mixtures with Shell Thiopave is 135 - 145°C. As the viscosity of the binder is reduced with Shell Thiopave, this allows the aggregate to be coated and workable at lower temperatures compared to conventional asphalt mixtures. The temperature manufacturing window is also low enough to avoid almost all gaseous emissions during production.

3. Shell Thiopave Trial Road in Qatar

3.1 Construction of Shell Thiopave-Modified Asphalt Road in RLIC

The overall positive experiences discussed above with the use of Shell Thiopave have encouraged construction of a trial section of Shell Thiopave-modified asphalt mixture in Qatar. The mass ratio of bitumen to Thiopave in this mixture was 40 to 60. The Thiopave-modified asphalt mixtures had about 25% less bitumen mass than the conventional mixture. This percent reduction in bitumen is the result of designing the Thiopave-modified mixture such that the volume of Thiopave and bitumen is the same as the volume of bitumen in the conventional mixture and the fact that the Thiopave has about twice the density of bitumen (Bailey et al., 2009).

The trial road stretch was located at Pearl Village, the worker accommodation for the Pearl GTL project in Ras Laffan Industrial City (RLIC), Qatar. The construction commenced in October 2007. Pictures of the construction process are shown in Figure 4. A 600-m section was constructed using Shell Thiopave asphalt modifier followed with a 300-m section of conventional asphalt mixture. Sections of the roadway were constructed using both material: Shell Thiopave and conventional, consisting of two lanes with a base course thickness of 8 cm and a wearing course thickness of 5 cm. A schematic of the location of the test section in Pearl GTL facility is shown in Figure 5. The red line shows the approximate location of the Shell Thiopave section (about 600 m long) while the blue section shows the conventional mixture section (about 300 m) constructed in series with the Shell Thiopave section. It is estimated that these test sections will experience about 1,500 bus passes daily in each direction. For a design period of 10 years, the estimated traffic loading is between 0.7-1.5 million Equivalent Single Axle Loads (ESAL).

3.2 Field Monitoring of trial road in RLIC

The field surveys of the road surface were conducted on December 2008 and May 2009. The visual surveys of the road section started at the south end of the test section and covered about 900 m (Shell Thiopave and conventional asphalt mixture sections). Overall, the 900 m road section was in a very good condition. No signs of road distresses such as longitudinal cracking, transverse cracking, permanent deformation or potholes were observed. The first bus stop area at about 250 m from the south end was paved using the Shell Thiopave modifier mixture. Conventionally, these areas tend to be prone to permanent deformation under braking forces and slow traffic. However, this bus stop was in a noticeably good condition and free of any visible distresses, which is also the case for bus stops in the sections paved with the conventional mixture. The plan is to continue the field monitoring during different climatic seasons over the next few years.

3.3 Laboratory Characterization Results

Shell Sulphur Solutions has conducted laboratory characterization (Gilbert, 2008) of the Shell Thiopave-modified and conventional mixtures used in the Shell field trial in the GTL village. The results from the laboratory characterization have shown that the Shell Thiopave-modified mixture had more resistance to permanent deformation than the conventional mixture. This can be observed in Figure 6, where the permanent deformation caused by applying 8000 cycles using the Asphalt Pavement Analyzer device. This laboratory device applies load cycles in the form of a tire on the top of a tube that is placed at the surface of an asphalt mixture specimen. The pressure in the tube is controlled to be similar to the tire pressure in the field. In most cases, the maximum depth of permanent deformation in specimens of Shell Thiopave-modified mixtures was less than half of permanent deformation in the conventional mixtures. The laboratory characterization has also involved conducting indirect tensile stiffness measurements at 10°C, 20°C, 30°C, and 40°C. The test was conducted according to "Indirect tensile Stiffness Modulus of Bituminous Mixtures Draft: DD213:1993". The Shell Thiopave mixture exhibited a higher indirect tensile stiffness at all temperatures (Figure 7a), but the difference between stiffness of conventional and Shell Thiopave mixtures increased with an increase in temperature reaching about 65% at 40°C (Figure 7b). These results confirm the improvements in mechanical properties of asphalt mixtures due to the use of Shell Thiopave especially at high temperatures.

3.4 Industrial Hygiene Monitoring

The industrial hygiene study focused on measuring gas emissions during processing and paving operations (Froes, 2008). At the mixing plant site, the initial concentrations of SO_2 were elevated. However, the SO_2 concentration dropped significantly when the method of mixing the binder and Shell Thiopave was altered such that the asphalt binder was added first followed by Shell Thiopave (Froes, 2008). The SO_2 emissions at the plant and paving sites were below the acceptable exposure limits (2 ppm for 8-hour average and 5 ppm for 15-minute rolling average). In fact, most of the measurements were below the measuring method detection limit.

The H_2S concentrations at the mixing plant and paving site are also below acceptable exposure limits (10 ppm for 8-hour average and 15 ppm for 15-minute rolling average). Again, the H_2S emissions were most of the time below the detection limits of the measurement methods. An important conclusion from the industrial hygiene monitoring was to maintain the temperature during the mixing process to less than 145°C in order to ensure that the SO_2 and H_2S concentrations are less than the exposure limits.

4. Summary of Findings

The field study results indicated that:

1. The total road section (Shell Thiopave and conventional) in the Pearl GTL village constructed using Shell Thiopave asphalt modifier and conventional mixtures are in a very good condition.
2. The total road section (Shell Thiopave and conventional) was free of any moderate or major distresses such as longitudinal cracking, transverse cracking, permanent deformation or potholes.
3. The laboratory characterization of Shell Thiopave-modified and conventional mixtures showed that the Shell Thiopave-modified mixture exhibited better resistance to permanent deformation and higher stiffness than the conventional mixture. The increase in stiffness suggests that the Shell Thiopave-mixture would have better resistance to fatigue cracking when it is used in relatively thick asphalt pavement sections (thicker than 125 mm).
4. The industrial hygiene monitoring study has shown that SO_2 and H_2S emissions remain below the maximum limits when the temperature is controlled to be less than 145°C. Mixing of Shell Thiopave-modified mixtures can be achieved at temperatures below 145°C because the addition of Shell Thiopave decreases the binder viscosity and improves its ability to be mixed with and coat the aggregates.

The construction trial emphasized the following advantages associated with Shell Thiopave use, relative to conventional asphalt:

1. Saving in bitumen content of about 25%, by mass.
2. Higher mix stability and a significant reduction in permanent deformation in wheel-tracking tests at relatively high temperatures.
3. An energy saving at the asphalt hot-mix plant from a 17°C lower mixing temperature; thus reduced CO_2 emissions.
4. Conservation of hydrocarbon resources.

5. Recommendations

After this field trial, new and improved Shell Thiopave technology was developed. This new Shell Thiopave technology allows the asphalt mix to be manufactured at even lower temperatures of 125 - 145°C, which falls into the 'Warm Asphalt Mix" category. The use of low temperatures in road construction means reduction in energy consumption during construction and reduction in emissions due to burning fuels. Thus, future trials in Qatar will incorporate this improved Shell Thiopave formula.

This paper reports the results from the initial monitoring of the road section in RLIC. The monitoring of this section will continue during different environmental seasons in the next few years. The plan is to continue the field monitoring of the test sections under different climatic conditions over the next few years to investigate whether the expected performance benefits suggested by Shell Thiopave's superior performance in laboratory-based comparisons with conventional mix, materialise in the field.

It is recommended that another field section be constructed in an area with high traffic loading, document the energy consumption in constructing the new section in comparison with conventional sections, characterize the materials using laboratory mechanical tests for fatigue and permanent deformation, and monitor this new section using field characterization devices such as the Falling Weight Deflectometer. The data from the new section will be needed in order to conduct detailed analysis of the benefits of adding Shell Thiopave in terms of reduction in energy consumption, improvement of asphalt pavement performance, and reduction of the total cost for the whole life of the pavement (initial construction cost and maintenance cost).

References

H. K. Bailey, R. Allen, D. Strickland, J. Colange and K. Gilbert (2009). *"Innovative Sulphur Technology Applied to European Asphalt Mixtures."* Mairepav6: The sixth international conference on maintenance and rehabilitation of pavements and technological control, 8-10 July 2009. Politecnico Di Torino, Torino, Italy.

T. Beatty, K. Dunn, E. Harrigan, K. Stuart, and H. Weber (1984). *"Field Evaluation of Sulfur-Extended Asphalt Pavements."* Transportation Research Record 1115, Transportation Research Board, Wshington DC.Z.

D. Froes (2008). *Industrial Hygiene Monitoring of Sulfur Extended Asphalt Mix (SEAM) trials-Doha Qatar*, Shell International B.V., the Hague, Netherlands.

K. Gilbert (2008). *An evaluation of the performance properties of the Asphalt mix containing SEAM TM Asphalt mix modifier paved at the Pearl GTL Village site.* Shell Sulphur Solutions.

D. Pickett (1978). *Extension and replacement of asphalt cement with sulfur: executive summary.* Federal Highway Administration Report no. FHWA-RD-78-160: Dept. of Transportation, Federal Highway Administration, Offices of Research & Development; Springfield, Washington.

Sulphur magazine (March/April 2009). *"Sour gas in the Middle East."* BC Insight Ltd, London, UK

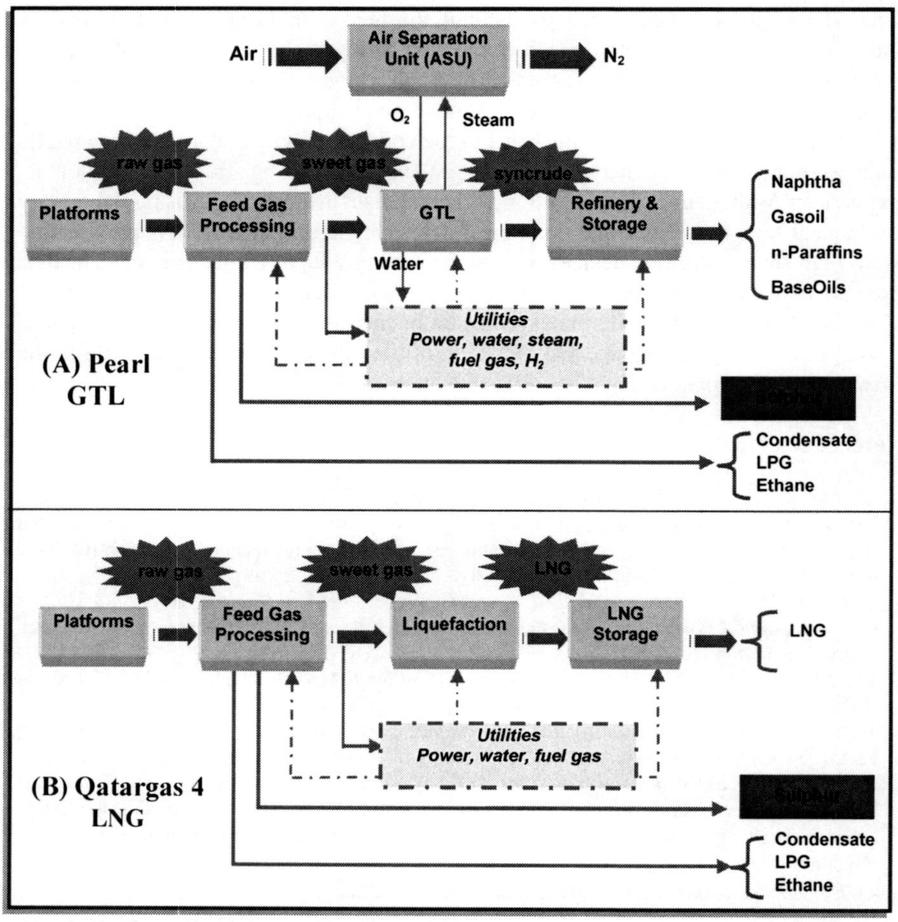

Figure 1: Simplified Process Integration Scheme for
(A) Pearl Gas-to-Liquid (GTL) plant, Ras Laffan, Qatar and (B) Qatargas 4 LNG.

Figure 2: Permanent deformation in a road in Qatar.

Figure 3: Shell Thiopave pellets.

Figure 4: Shell Thiopave asphalt mix laydown in RLIC on October 30th 2007.

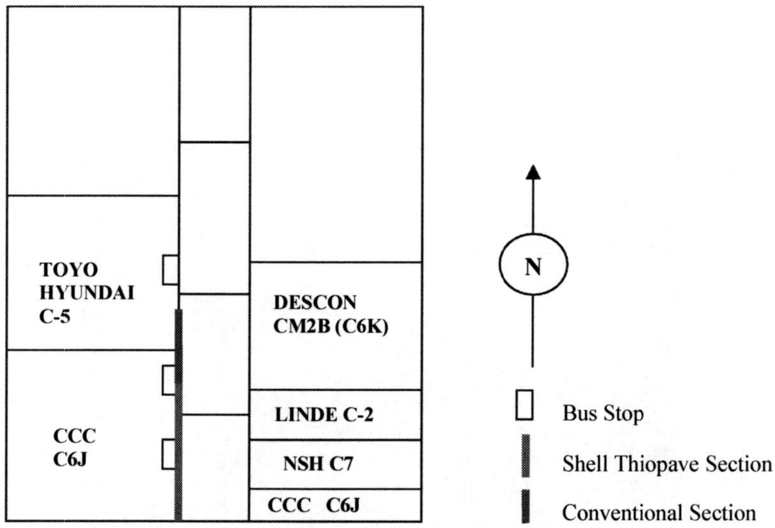

Figure 5: A schematic of the location of the test sections in Pearl GTL facility in RLIC.

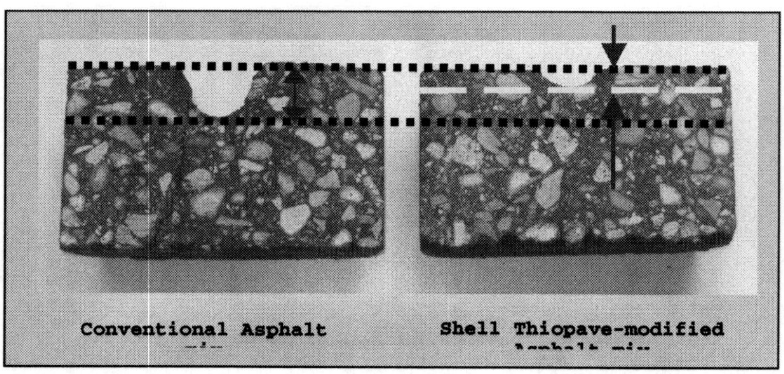

Figure 6: Typical appearance of permanent deformation (rutting) resistance test specimens After 8000 wheel load applications.

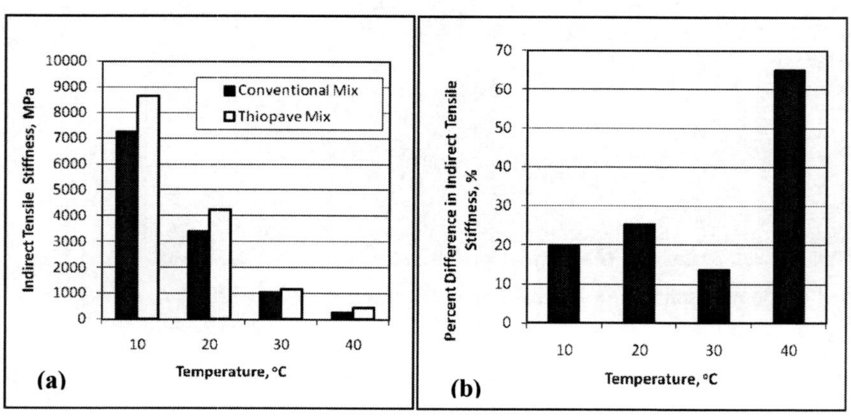

Figure 7: A comparison of stiffness between a Shell Thiopave modified mixture and a conventional mixture.

Proceedings of the 2nd Annual Gas Processing Symposium
Farid Benyahia and Fadwa T. Eljack (Editors)

Utilization of Byproduct Sulfur for Chromium Waste Treatment

Ahmed Abdel-Wahab[a], Ali Dashti[b], Ali Anany[b], Khaled Elsaid[a]

[a]*Chemical Engineering Program, Texas A&M University at Qatar, P.O. Box 23874,Doha, Qatar*
[b]*RasGas Company Ltd, P.O. Box 24200, Doha, Qatar*

Abstract

The ongoing increase in the production of natural gas results in the production of huge amounts of sulfur as byproduct from natural gas processing. Hydrogen sulfide (H_2S) existing in natural gas extracted from Qatar's North Field is converted to elemental sulfur using conventional Claus process. Managing byproduct sulfur from natural gas processing is a key aspect of economic development and environmental protection in Qatar. Therefore, new markets must be found for sulfur to avoid disposal crises. One attractive use of byproduct sulfur is to treat hazardous wastes. Sulfur cement can be applied as a primary binder in solidification/stabilization (S/S) treatment of hazardous wastes. Furthermore, elemental sulfur and lime can be combined to form calcium polysulfide which can be used to immobilize toxic metals by chemically reducing them to inert forms and to degrade toxic chlorinated organics by converting them to non-chlorinated compounds. The goal of this research was to investigate use modified sulfur products to immobilize hexavalent chromium (Cr(VI)) wastes. Cr(VI) was reduced to Cr(III) using calcium polysulfide (CaS_x) as the reducing agent. CaS_x was prepared by mixing sulfur with lime and water using mass ratios of 1:1.05:7.48, respectively for two hours at 95 °C. The mixture was then cooled to room temperature, filtered and the supernatant was used in Cr(VI) reduction experiments. Results of kinetic experiments showed that Cr(VI) concentration decreased with time and the reduction of Cr(VI) was almost complete after reaction time of 12 hours. The effect of CaS_x dose on Cr(VI) reduction was evaluated.

Keywords: chromium, sulfur, stabilization, polysulfide.

1. Introduction

Qatar's proven reserves of natural gas are now estimated at almost 22 trillion cubic meters (750 trillion cubic feet), the third largest in the world behind Russia and Iran. At this point, Qatar's strategy of emphasizing development of its natural gas resources is being aggressively and effectively pursued. This results in the production of huge amounts of sulfur as byproduct from natural gas processing. Managing byproduct sulfur is a key aspect of economic development and environmental protection in Qatar.

Hydrogen sulfide (H_2S) existing in natural gas extracted from Qatar's North Field is converted to elemental sulfur by means of a conventional Claus sulfur process that comprises one thermal and two catalytic Claus reactor stages. Large quantities of hydrogen sulfide are produced as a byproduct of natural gas processing and must be disposed. Finding beneficial uses for this material is important for economic development and environmental protection. One approach is to convert hydrogen sulfide to elemental sulfur. Although this reduces toxicity and mobility, there is still a need to dispose of the solid sulfur product. Elemental sulfur has been shown to be effective in encapsulating waste materials and substantially reducing their tendency to leach contaminants. Elemental sulfur can be combined with alkali or alkaline metal hydroxides or sulfides to form polysulfide which can be used to immobilize toxic metals and to destroy toxic chlorinated organics by converting them to non-chlorinated compounds. Calcium polysulfides can be formed by reaction of sulfur and lime. These compounds have been used as agricultural amendments (Smilanick and Sorenson, 2001) and as a method of immobilizing toxic metals in waste materials (USEPA, 1981, Alkemade and Koene, 1996). For immobilization, they are typically used in combination with conventional solidification/stabilization processes that combine Portland cement with the waste material to produce a solid that is easier to handle and safer to dispose. Polysulfides have also been found to be effective reagents for treating metals contamination (Yan, 1991, Lan et al., 2005) and for degradation of organic pollutants (Guo and Jans, 2006). Furthermore, polysulfides can also act as agents to reductively dehalogenate halogenated organic compounds (Hwang, 2000). Additionally, modified sulfur products can be used for encapsulating hazardous wastes in substitution of cement (Kalb et al., 1996).

Recently, sulfur-based treatment technologies have been the focus of an increasing number of research studies and commercial applications for treating soils and groundwater contaminated with toxic metals. Calcium polysulfide reduces and precipitates some highly soluble metals such as arsenic, lead, copper, and zinc to form metal sulfides, which often are much less soluble than metal hydroxides and remain insoluble within a wider range of pH values (USEPA, 1981).

In addition to their ability to immobilize and degrade hazardous wastes, modified sulfur products can be used as binders in solidification/stabilization treatment of hazardous wastes. One of the modified sulfur products that can be used for that purpose is called plasticized sulfur. Plasticized sulfur is made by heating elemental sulfur to the point where it becomes a viscous fluid. At that point hydrocarbon plasticizers are added to maintain a semi-solid consistency of the sulfur as the temperature returns to normal levels. Plasticized sulfur has recently emerged as a possible alternative binding and stabilizing agent for the solidification and stabilization of hazardous, low-level radioactive and mixed wastes (Kalb and Colombo, 1997, Darnell, 1996). Plasticized sulfur has been used as a binder for solidification/stabilization of wastes contaminated with heavy metals, elemental mercury waste (Mattus, 1998) and a variety of mixed wastes (Darnell, 1996).

Solidification/Stabilization (S/S) is the most commonly used remedial alternative for containment of hazardous wastes. Conventional S/S has largely been used to immobilize inorganic contaminants at high pH using binders such as Portland cement, fly ash, and lime. S/S has been used to treat a variety of wastes such as nuclear wastes, municipal ash, contaminated soils/sediments, and wastewater (Means et al., 1995). Portland cement has been the most widely used binder in conventional S/S and has been applied to a variety of hazardous materials. However, conventional S/S processes are not effective for encapsulation of soluble organic contaminants. Degradative solidification/stabilization (DS/S) is a modification of solidification/stabilization (S/S) that combines degradative processes for organic contaminant destruction with the low cost of conventional S/S. DS/S is attractive because many contaminated sites contain both inorganics and organics. Reduced forms of sulfur are among the most abundant electron donors in natural environments that are reducing (Stumm and Morgan, 1996). Therefore, applications of sulfur-containing compounds for DS/S are well supported by theory. Sulfide, polysulfide (Sx^{2-}), and dithionite ($S_2O_4^{2-}$) are among the sulfur-containing compounds that are known to reduce chlorinated organics in the environment (Rodriguez and Rivera, 1996).

The purpose of this research is to study the feasibility of utilizing byproduct sulfur from natural gas processing for the reduction of hexavalent chromium [Cr(VI)].

2. Materials and Methods

All solutions were prepared by using deionized (DI) water from Barnstead EasyPure RoDi. Cr(VI) stock solution was prepared by dissolving 2.94 g of primary standard $K_2Cr_2O_7$ (ACS grade) in 1000 mL DI water; this resulted in a Cr(VI) concentration of 1.04 g/L. Solutions at the desired concentration were then obtained by successive dilutions. CaS_x was prepared following Hajjatie, et al. (2006) by mixing of sulfur, analytical reagent lime, and deoxygenated water with mass ratios of 1:1.05:7.48 respectively. The mixture was heated to 95°C and allowed to mix for 2 hours in an anaerobic environment. Then, the unreacted sulfur was allowed to settle and removed. The solution was then filtered using 0.45 μm membrane filters and the filtrate was used for Cr(VI) reduction experiments. 0.2 N H_2SO_4 and 0.1 N NaOH solutions were used to adjust pH. The pH values were measured before and after the reaction using pH meter calibrated with buffer solutions at pH 2, pH 7, and pH 12.45. Experiments were performed in 20-mL sealed Pyrex vials and solutions were mixed continuously with magnetic stir bars coated with Teflon. Cr(VI) concentration was determined using the colorimetric method with UV-VIS spectrometer at 540 nm wavelength.

3. Results and Discussions

3.1. Kinetics of Cr(VI) reduction using CaS_x

Kinetic experiments were conducted to evaluate the kinetics of Cr(VI) reduction using CaS_x. Initial Cr(VI) concentration of 10 mg/L and a CaS_x dose of 1.50 g/L were used in these experiments. Samples were taken at times 0.5, 1, 2, 4, 8, 12, and 24 hours from the initiation of the reaction. At the end of each time period, the mixture was filtered using 0.45 µm filters and the filtrate was analyzed for Cr(VI) using the colorimetric method.

Fig. 1 shows the results of kinetic experiments and indicates that Cr(VI) concentration decreased with time until it reaches its minimum concentration at reaction time of 12 hours. Between 12 and 24 hours, less than 5% change in Cr(VI) concentration occurs. This indicates that the reaction is almost complete after 12 hours of reaction time. Results of these kinetic experiments were used to set the time for equilibrium experiments. The reaction time for equilibrium experiments was chosen to be 24 hours.

3.2. Effect of CaS_x dose on Cr(VI) reduction

Effect of CaS_x dose on Cr(VI) reduction was studied by adding various doses of CaS_x to a solution of 10 mg/L Cr(VI). This was achieved by successive dilutions of stock CaS_x solution. Doses of CaS_x used in these experiments ranged from 0.0748 to 2.24 g/L. Figure 2 shows the effect of CaS_x dose on the reduction of Cr(VI) and indicates that the final concentration of Cr(VI) decreased with increasing CaS_x dose. The minimum dose of CaS_x required to achieve complete removal of Cr(VI) was 0.36 g/L when the initial Cr(VI) concentration was 10 mg/L.

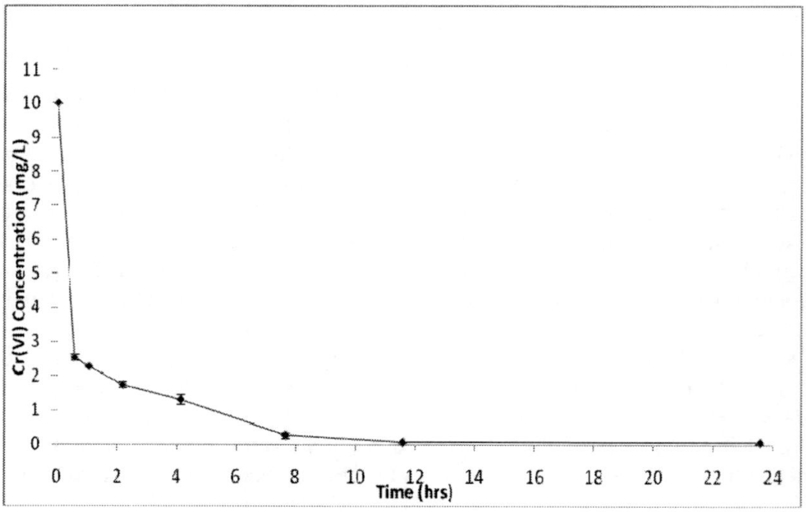

Figure 1: Kinetics of Cr(VI) reduction using CaS_x.

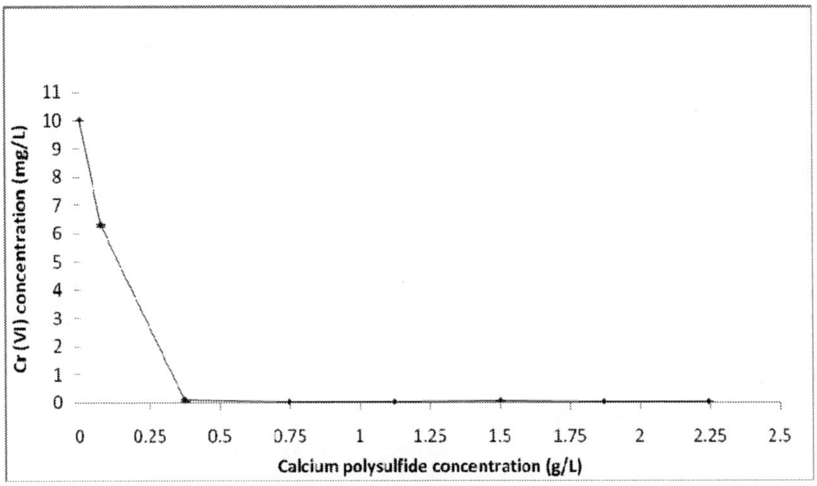

Figure 2: Effect of CaS_x dose on Cr(VI) reduction.

3.3. Effect of pH on Cr(VI) reduction

Effect of pH on Cr(VI) reduction was studied by varying the initial pH of the reaction from 2 to 12 using initial Cr(VI) concentration of 10 mg/L and CaS_x dose of 0.75 g/L . Fig. 3 shows the effect of pH on Cr(VI) reduction and it shows that maximum Cr(VI) reduction was achieved at pH around pH2 and at pH values above pH9 while minimum removal was achieved at pH around pH4. The high Cr(VI) reduction at low pH could be attributed to the formation of hydrogen sulfide which was the active reducing agent at this low pH value. At high pH values above pH9, the increase of Cr(VI) reduction could be due to the formation of chromium hydroxide solid according to the following reaction.

$$2CrO_4^{2-} + 3CaS_x + 10H^+ \rightarrow 2Cr(OH)_3\,(s) + 3xS\,(s) + 3Ca^{2+} + 2H_2O \quad (1)$$

3.4. Effect of initial Cr(VI) concentration on Cr(VI) reduction

Effect of initial Cr(VI) concentration on Cr(VI) reduction was studied by varying initial Cr(VI) concentration from 1 to 100 mg/L. The dose of CaS_x was fixed at 0.75 for this set of experiments. Figure 4 shows the final Cr(VI) concentrations afected by initial Cr(VI) concentrations using 0.75 g/L CaS_x. Final Cr(VI) concentration increased with increasing initial Cr(VI) concentration. However, 0.75 g/L CaS_x had capacity to remove more than 10 mg/L Cr(VI). For instance, at an initial Cr(VI) concentration of 20 mg/L, 0.75 g/L CaS_x was able to remove 16.5 mg/L Cr(VI). Similarly, at initial Cr(VI) concentrations of 50 and 100 mg/L, CaS_x dose of 0.75 g/L was able to remove 20.4 and 56.6 mg/L Cr(VI), respectively.

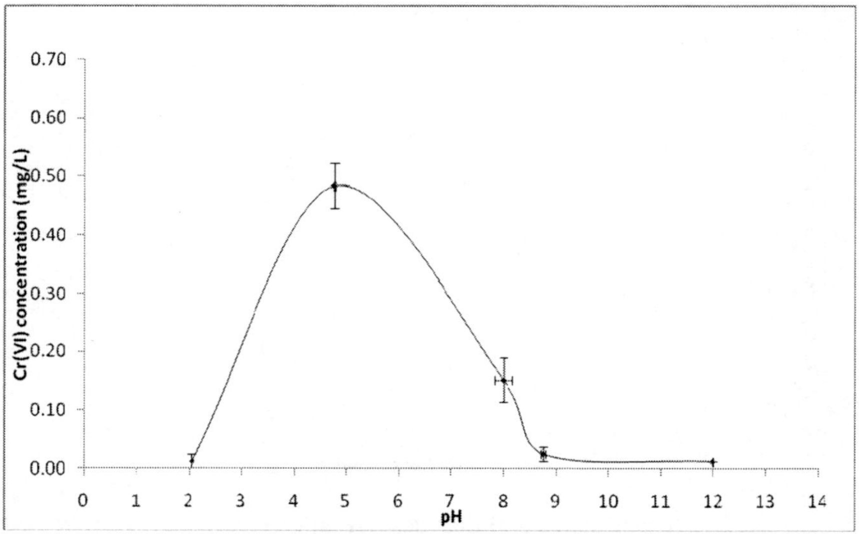

Figure 3: Effect of pH on Cr(VI) reduction.

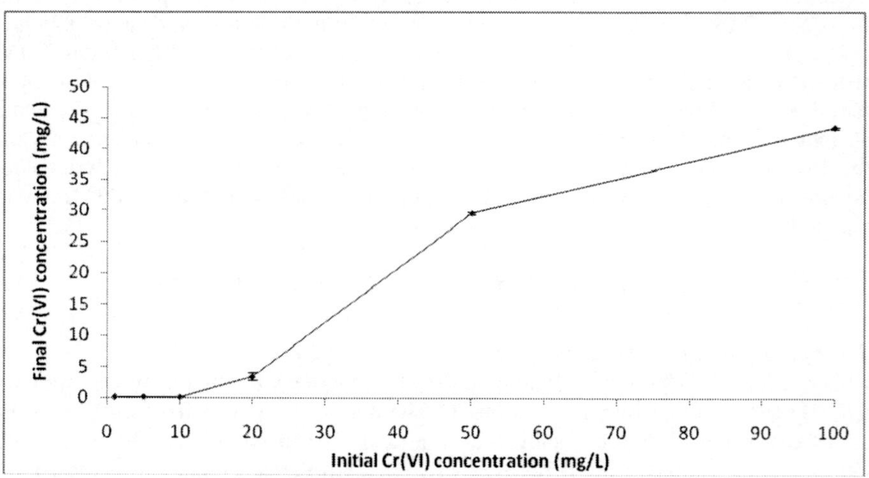

Figure 4: Effect of initial Cr(VI) concentration on Cr(VI) reduction (CaS_x dose = 0.748 g/L).

3.5. Use of industrial byproduct sulfurfor Cr(VI) reduction

Additional experiment was conducted using byproduct sulfur from the QatarGas plant (Ras Laffan City, Qatar). In this experiment, 0.10 g/L sulfur and 0.10 g/L lime were added directly to a solution of 10 mg/L Cr(VI) solution. The mixture was allowed to react for 24 hours at room temperature. Samples were taken and filtered and the filtrate was analyzed for Cr(VI). Final Cr(VI) concentration was under the detection limit of 0.05 mg/L and the final pH was pH 12.4. The mechanism of Cr(VI) removal in these conditions is currently under investigation.

4. Conclusions

The feasibility of Cr(VI) reduction and immobilization using byproduct sulfur was investigated. The results of laboratory experiments demonstrated that Cr(VI) can be reduced with high efficiency using calcium polysulfides. Kinetics of Cr(VI) was rapid and complete Cr(VI) was achieved after 12 hours of reaction time at room temperature. Equilibrium experiments showed that final Cr(VI) concentration decreased with increasing polysulfides dose and it reached its minimum concentration when polysulfides dose was 0.35 g/L at initial Cr(VI) concentration of 10 mg/L. pH was found to have significant effect on Cr(VI) reduction. High pH values are desired for optimum Cr(VI) reduction. Actual byproduct sulfur was used with lime for and it resulted in achieving high reduction efficiency of Cr(VI).

References

Alkemade M.M.C., J.I.A. Koene, 1996, The Useful Application of Sulphur-Bound Waste Materials, Waste Management, 16, No 1-3, pp. 185-188.

Darnell, G.R., 1996, Sulfur polymer cement, a final waste form for radioactive and hazardous wastes, in: T. Michael Gilliam, C. Carlton, Wiles (Eds.), Stabilization and Solidification of Hazardous, Radioactive, and Mixed Wastes: ASTM STP 1240, vol. 3, American Society for Testing and Materials, West Conshohocken, PA.

Guo X., U. Jans, 2006, Kinetics and Mechanism of the Degradation of Methyl Parathion in Aqueous Hydrogen Sulfide Solution Investigation of Natural Organic Matter Effects, Environ. Sci. Technol., Vol. 40. No. 3, pp900-906.

Hajjatie, M.M., M.D. Aspenqren, M.P. Clarkson, H.C. Kominski, C.L. Lockhart, 2006, Process for Preparing Calcium Thiosulfate Solution, US Patent No. 6984368,

Hwang I., 2000, Fe(II)-Based Reductive Dechlorination of Tetrachloroethylene in Soils Treated by Degradative Solidification/Stabilization, Ph.D. Dissertation, Texas A&M University, College Station, Texas.

Kalb, P.D., J.W. Adams, M.L. Meyer, H.H. Burns, 1996, Thermoplastic encapsulation treatability study for a mixed waste incinerator offgas scrubbing solution, in: T. Michael Gilliam, C. Wiles Carlton (Eds.), Stabilization and Solidification of Hazardous, Radioactive, and Mixed Wastes: ASTM STP 1240, vol. 3, American Society for Testing and Materials, West Conshohocken, PA.

Kalb, P.D., P. Colombo, 1997, Composition and process for the encapsulation of radioactive hazardous and mixed wastes, United States Patent 5,649,323.

Lan Y., B. Deng, C. Kim, E.C. Thornton, H. Xu, 2005, Catalysis of Elemental Sulfur Nanoparticles on Chromium(VI) Reduction by Sulfide under Anaerobic Conditions, Environ. Sci. Technol., Vol. 39, No. 7, pp2087-2094.

Mattus, C.H., 1998, Sulfur polymer cement for macroencapsulation of mixed waste debris, in: Proceedings of the International Conference on Decommissioning and Decontamination and on Nuclear and Hazardous Waste Management, Denver, Colorado, September 13–18.

Means, J. L.; L.A. Smith, K.W. Nehring, S.E. Brauning, A.R. Gavaskar, B.M. Sass, C.C. Wiles, C.I. Mashni, 1995, The Application of Solidification/Stabilization to Waste Materials. Lewis Publishers: Boca Raton, Florida.

Rodriguez, J. C.; M. Rivera, 1996, Reductive Dehalogenation of Aqueous Carbon Tetrachloride by Sodium Dithionite. *Abstracts of Papers of the American Chemical Society*; American Chemical Society: Washington, DC, 212, 97-ENVR

Smilanick J.L., D. Sorenson, 2001, Control of Postharvest Decay of Citrus Fruit with Calcium Polysulfide, Postharvest Biology and Technology, Vol. 21, No. 2, pp. 157-168.

Stumm, W., J.J. Morgan, 1996, Aquatic Chemistry, 3rd ed.; John Wiley & Sons, Inc.: New York, New York.

United States Environmental Protection Agency (USEPA), 1981, US EPA Chart Comparing Sulfide and Hydroxide Solubilities, EPA Publication, EPA-600/2-82-OIIC.

Yan T. Y. (1991) Reaction of Trace Mercury in Natural Gas with Dilute Polysulfide Solutions in a Packed Column, Industrial and Engineering Chemistry Research, Vol. 30, pp 2592 – 2595.

Acknowledgments

This publication was made possible by a grant from the Qatar National Research Fund under its Undergraduate Research Experience Program. Its contents are solely the responsibility of the authors and do not necessarily represent the official views of the Qatar National Research Fund.

Proceedings of the 2nd Annual Gas Processing Symposium
Farid Benyahia and Fadwa T. Eljack (Editors)

Removal of Acid gases from Natural Gas Streams by Membrane Technology

S. M. Javaid Zaidi
Chemical Engineering Department and Centre of Excellence in Renewable Energy
King Fahd University of Petroleum & Minerals, Dhahran-31261, Saudi Arabia
e-mail: zaidismj@kfupm.edu.sa

Abstract

Membrane-base gas separations are getting popular due to the environmental and economic advantages. It can successfully be used for the natural gas purification to remove acid gas such as CO_2 and H_2S. It competes directly against amine absorption process for removal of carbon dioxide. Even though there are a large number of other potential applications for gas separation using polymeric membranes, only few of them have been adopted in practice. Progress in gas separation studies using membranes was greatly accelerated and this technique has emerged as a commercially viable method in the 1980s due to the development and refinement of novel synthetic polymers. Since then, progress has been made continuously in improving membrane formation processes, chemical and physical structures and different modular configurations for specific end applications. In order to ensure the performance of a membrane, the scientific factors, such as processing of new materials and behaviour of membranes with respect to various feed mixtures under practically realistic operating conditions must be considered. To sum up, permeability and selectivity of a membrane are the deciding factors in separating gaseous mixtures. Molecular sieve and selective surface flow membranes also represent the "revolutionary" sorption-diffusion materials, that could ideally open up many new opportunities for higher performance and convenient separations that are not achievable with today's membrane forming materials. One or more of the above approaches are expected to be useful for separating acid gases from natural gas with desired high selectivity and permeability. The aim of this presentation is to review the status of the membrane-based natural gas purification and propose the applicability of membrane technology for the natural gas industry in Saudi Arabia

Keywords: natural gas, membranes, acid gases, permeability.

1. Introduction

Gas separation through membranes has emerged as an important unit operation offering specific advantages over conventional separation methods such as cryogenic distillation and adsorption. Over the years, membrane-based technology has contributed tremendously to the growth of new research frontiers such as carbon dioxide capture, which serves as an efficient tool for natural gas purification. Saudi Arabia is one of the leading producers of Natural Gas. Before selling natural gas or feed it to petrochemical plants, it is necessary to purify it from acid gas such as CO_2 and H_2S. The current existing technology (Absorption by Amine solvents) has a thermodynamic limitation, plus it is not attractive from environmental point of view. Therefore, there is a need for an alternative efficient and environmentally sound "solvent-less" process. Membrane Separation can fulfil these two criteria.

Membrane separations can be efficiently and successfully used for the purification of natural gas replacing the amine absorption process, which dominates the natural gas industry. It has benefits such as lower energy consumption and flexibility in operation. Membrane technology can be effectively applied for the removal acid gases such as CO_2 and H_2S and water vapor present in the natural gas streams. The membrane technology has proven to be technically feasible and has impacted the market by its many advantages over the amine processes, which are listed below: (Bernardo et al, 2009; Bake and Lokhandwala, 2008)

1. Smaller and lighter streams
2. Simultaneous removal of CO_2, H_2S and water vapor
3. Reduced energy consumption
4. Lower operating and capital cost
5. High process flexibility
6. Ease of installation
7. Ability to treat gas at wellhead.

For example, with the membranes modules of hollow fiber or spiral wound design carbon dioxide in natural gas can be reduced to less than 2%. Since carbon dioxide removal membranes are also selective to hydrogen sulfide, H_2S can be removed in the same membrane stage along with carbon dioxide. Membrane processes compete with the glycol absorption process for natural gas de-hydration. The disadvantages of glycol process are loss of glycol due to contamination with aromatic hydrocarbons and maintenance requirements.

The acid gases (CO_2, H_2S) and water vapor should be removed from the natural gas before its use in industry in order to increase the heating value of natural gas, reduce corrosion, prevent atmospheric pollution, and prevent water from freezing in pipelines and preventing the distribution of natural gas. The effective method of purifying natural gas to remove acid gases is by selective permeation of its components through the nanoporous membranes [Stern et al 1986]. Membrane processes are well suited for this purpose as natural gas is typically available at high pressures (400-1200 psig) and the rate of gas permeation through the membranes increases with increasing pressure. Carbon dioxide, H_2S and water vapor permeate faster than CH_4 through most polymer membranes, and thus CH_4 is concentrated in the retentate (high pressure) stream without a significant pressure drop of the sweetening gas. The gas selectivities of several glassy polymers are sufficiently large to produce pipeline quality natural gas (Bhide et al, 1993). Membrane processes are potentially energy efficient. They offer advantages such as ease of scale-up, environmental safety, flow rate variation flexibility, pressure and composition of feed streams (Koros et al 1987).

2. Membrane Modules and Process Configurations

Various modules design can be used depending on the purity requirement such as single-stage membrane system and multi-stage membrane systems. Most membrane-based separations are accomplished with asymmetric hollow fiber or spiral wound modules, which provide a large amount of separation area per unit volume (Juaied, 2006). Asymmetric fibers are composed of a thin 500-2000 °A separating skin layer, which is supported on a second porous layer that is usually 50-200 μm thick. Ideally, the porous support has no effect on the permeation properties of the membrane (Spillman, et al 1989, Kesting et al 1993, Porter 1990, Paul et al 1994). It has been shown in various studies (Bhide at al 1993) that the performance of the membrane-based separations is superior to the amine processes at higher pressures and concentrations, and the membrane system cost 25% less than the conventional system. A number of studies (Babcock et al 1988) have been carried out to evaluate the membrane processes for acid gases removal from natural gas

Bhide and Stern (1993) studied membrane processes for the removal of acid gases from natural gas streams. They carried out detailed investigation on process configurations and optimization of operating conditions. In the second part of their report, they studied the effects of operating conditions, economic parameters, and membrane properties. The costs of membrane-based separation and those of amine process are also compared in their study. In the first part of their study, they identified the optimum membrane process configurations for reducing the concentration of CO_2 in natural gas to 2 mole % or less, as required for the pipeline-quality gas. The flow rate of the natural gas feed stream was 35 MMSCFD at 800 psia containing 5-40 mole % CO_2 using asymmetric cellulose acetate membrane. Membranes made from cellulose acetate are more permeable to CO_2 and H_2S than CH_4. So, CO_2 and H_2S and water vapor present in the feed are concentrated in the permeate side, whereas CH_4 is permeated in the retentate. Seven membrane process configurations consisting of one, two, or three permeation stages in series or cascade arrangement with recycle of a fraction of either the retentate or permeate product streams. It was found in their study that the cost of removing CO2 from natural gas is the lowest for a three-stage process configuration. This configuration consisted of a single permeation stage in series with a two-stage permeation cascade with retentate recycle.

A number of studies have also been carried out to compare the economics of membrane processes with those of conventional separation process for the removal of acid gases from natural gas (Spillman 1989, Babcock et al 1988). These studies found that membrane processes are more economical than DEA gas absorption process over the CO_2 concentration range of 5-40 mole % using three-stage membrane process configurations. The economic advantage of membrane processes increases with decreasing flow rate, increasing feed pressure and increasing CO_2 concentration in the feed. When the feed also contains H_2S, the cost of removing CO_2 and H_2S was reported to increases with H_2S concentration (Bhide et al 1993). In cases where membrane processes alone are not economically competitive, the separation cost could be significantly reduced by using hybrid membrane processes (Puragason and Houston, 1984, Ryan and Anderson, 1988). In such cases the bulk of CO_2 and H_2S are removed from the natural gas with membrane, and final purification is performed by means of suitable gas-absorption process. Such hybrid processes have already been considered for the upgrading of natural gas as well as for the separation of CO_2 in enhanced oil recovery applications (Ryan and Anderson, 1988).

3. Polymer Membranes for Natural Gas Purification

For the membrane process to be successful it requires the membrane with high selectivity for the acid gases. The separation cost as well as the power requirements of the membrane processes can be lowered by the synthesis of membrane forming polymers with higher permeability and selectivity for CO_2 and H_2S relative to CH_4, which will reduce membrane area as well as hydrocarbon losses. Research is under way in various academic and industrial laboratories to develop new type of polymers and the membrane to improve the efficiency of membrane process for natural gas purification and reduce the overall cost.

Generally glassy polymers are used for the removal of acid gases from the natural gas. The most common material used for the natural purification is the cellulose acetate. These materials show relatively low CO_2/CH_4 selectivities under typical operating conditions in the field, but have fair-to-good tolerance for contaminants such as benzene, toluene, and xylene in the natural gas streams (Wind et al 2004, Schell 1988, Lee 1989). The decline in selectivity is primarily is due to swelling-induced plasticization of the membrane by CO2 and heavy hydrocarbons. The development of stable membrane material with high CO_2/CH_4 selectivities will enhance the competitive position of membrane relative to gas absorption. In particular polyamides have excellent CO_2/CH_4 separation properties and have robust mechanical properties to withstand high-pressure natural gas feeds. Most of the reported permeability properties for polymers are from the dense film samples. These films provide that can be considered as upper limit for the selectivity that can be obtained with a membrane made from the polymer. In the specific application to carbon dioxide recovery many cases exists where the apparent performance of a material depends on the exact testing conditions (White et al 1995). Despite their excellent properties polyimides also causes decline in the performance due to plasticization effect, which negatively impacts process economics and reliability. If the polymer shows very high values for permeability and selectivity with dense films then it may be a good candidate for development of practical membrane. For commercial applications the active separation layer of the membrane should have an equivalent dense film thickness of the order of Angstroms. If thin coherent films of the polymer cannot be prepared then useful membrane devices will not result. In order to improve the properties of the polyimide membranes, these polymers have been modified by cross-linking with propanediol monoester cross-link (PDMC). Cross-linking has been shown to be an effective approach to stabilize properties of polymeric membranes used in carbon dioxide removal from natural gas mixtures (Hillock et al 2007).

Polyimides modified with several different cross-linking agents were examined by Wind et al 2004 to assess the dependence of permeation properties on the size of cross-linking modifiers. The cross-linked PDMC membranes were reported to have significantly higher CO_2 permeability and slightly higher selectivity for CO_2/CH_4 separation than uncross-linked membranes. The cross-linked membranes not only outperformed the uncross-linked with their enhance transport properties but also provide vastly superior plasticization resistance against aggressive CO_2 streams (Hillock et al 2007). Recently, new polymer membrane based on polyethylene oxide (PEO) crosslinked hydrogel (Freeman et al 2006). It has been demonstrated by Freeman et al 2006 that highly branched cross-linked PEO rubbery membrane materials exhibit high CO_2 permeability, high CO_2/CH_4 mixed gas selectivity, and excellent stability to contaminants in natural gas due to high gas diffusivity and high CO_2/CH_4 solubility selectivity. Finally, the membrane device must be able to function under the environmental conditions and constraints where natural gas is processed. It is often a challenge to truly duplicate the working environment of a membrane in the laboratory.

4. Conclusions

Membrane technology has a strong potential and promise for the removal of acid gases from natural gas. A number of studies have demonstrated the use of membrane processes for natural gas purification as feasible and economical and environmental friendly. It is time now for the natural gas industry in Saudi Arabia and Gulf Region to start looking at and apply this technology.

Acknowledgement:

The support of the King Fahd University of Petroleum and Minerals for this work is highly appreciated.

References

Babcock, R.E., R.W. Spillman, G.S. Guddin, T.E. Cooley, Energy Progress, 8 (1988)135-142.

Babcock, R.E., R.W. Spillman,C.S. Goddin, T.E. Cooley, Energy Progress, 8(3) (1988) 135-42.

Baker, R.W., K. Lokhandwala, , Ind. Eng. Chem. Res. 47 (2008), 2109-2121

Bernad P., E. Drioli and G. Golemme, Ind. Eng. Chem. Res. 48 (2009), 4638-4663.

Bhide, B.D., S.A. Stern, Journal of Membrane Science 81 (1993) 209-237.

Bhide, B.D., S.A. Stern, Journal of Membrane Science 81 (1993) 239-252.

Freeman, F.D.S. Matteucci, H. Lin, S. Kelman, V. Kusuma, Proc: Tobago Gas Technology Conference 2006, "Directions in Gas Conversion", may 22-24, 2006, Trinidaa, Tobago.

Freeman, F.D.S. Matteucci, H. Lin, S. Kelman, V. Kusuma, Proc: Tobago Gas Technology Conference 2006, "Directions in Gas Conversion", may 22-24, 2006, Trinidaa, Tobago.

Hillock, A. M.W and W.J. Koros, Macromolecules, 40 (2007), 583-587.

Juaied, M, W.J. Koros, Journal of Membrane Science 274 (2006) 227-243.

Kesting, R.E., A.K. fritzsche, Polymer Gas Separation Membranes, John Wiley & Sons Inc., New York, 1993.

Koros, W.J, R.T. Chen, in R.W. Rousseau (Eds.), Handbook of Separation Process Technology, Wiley Interscience, New York 1987, pp 862-963.

Pauk, D.R. (Eds.) Polymeric Gas Separation membranes, CRC Press, Boca Raton, FL, 1994.

Porter, M.C. (Eds.) Handbook of Industrial Membrane Technology, Noyes Publications, Park Ridge, NJ, 1990.

Puragason, R.S., C.D. Houston, Energy Progress, 4(1) (1984) 54-56.

Schell, W.J., C.J. Wensely, M.S. K. Chen, J.A. Stuart, Recent Advances in Cellulose membranes for gas separation and Pervaporation, Gas Sep. Purif., 3 (1989) 162-169.

Spillman, R.W, T.E. Cooley, Membrane gas treating, in proceedings of the sixty-eight Gas Processors Association Annual Convention, 1989.

Spillman, R.W., Chem. Eng Prog., 85 (1989) 41-62

Stern, S.A., New Developments in membrane processes for gas separations, in M.B. Chenoweth (Eds.), vol 5, Harwood Academic Publishers, New York 1986, pp 1-37.

White, L.S., T.A. Blinka, H.A. Kloczewski, I. Wang, Journal of Membrane Science 103 (1995) 73-82.

Wind, J.D, D. R. Paul, W.J. Koros, Journal of Membrane Science 228 (2004) 227-236.

Proceedings of the 2nd Annual Gas Processing Symposium
Farid Benyahia and Fadwa T. Eljack

Catalytic Dry Reforming of Methane Using Ni/MgO-ZrO$_2$ Catalyst

Teddy G. Monroy[a], Leonila C. Abella[a], Susan M. Gallardo[a], Hirofumi Hinode[b]
[a]*De La Salle University, 2401 Taft Ave., Manila, Philippines*
[b]*Tokyo Institute of Technology, 2-12-1 O-okayama, Meguro-ku, Tokyo,152-8552, Japan*

Abstract

Methane dry reforming is one of the important routes in natural gas processing. Here, CH_4 and CO_2 are converted into syngas CO and H_2 which later can be used as feedstock for the processing of other chemicals such as CH_3OH and NH_3. Methane dry reforming requires the use of catalysts, often nickel-based, to increase the reaction rate. To date, the dry reforming of methane has limited commercial application due to the rapid deactivation of the catalysts. The reactants contain carbon species that has a high potential of blocking active sites on the catalyst surface, deactivating them.
This study focuses on the use of 15% Ni/MgO-ZrO$_2$ as a catalyst for CH_4 dry reforming where the high basicity of MgO and the mobile oxygen species provided by ZrO_2 are expected to provide high catalyst activity and stability. The catalysts were prepared using the impregnation method where a 1:1 mole ratio of MgO to ZrO_2 was achieved. Surface characterization of the catalysts was conducted using advanced instrumentation such as the AAS for nickel loading, BET for surface area, pore volume, and average pore size, SEM_EDX for surface morphology and composition, XRD for bulk crystal structure, and TPD for the number of acidic and basic sites.
The catalysts were placed inside a U-tube quartz reactor (7mmϕ) and inserted into a tube furnace. Together with the carrier gas He, CH_4 and CO_2 were injected into the reactor at 100 ml/min total flow rate. The CH_4 dry reforming reaction was conducted at 850°C reaction temperature. The exit gases went into an on-line gas chromatograph for the determination of gas compositions.

Keywords: methane, reforming, catalyst, magnesia, zirconia

1. Introduction:

Methane dry reforming is one of the most important processes used in the production of syngas (H_2 and CO). CH_4 and CO_2, two of the most potent greenhouse gases, are reacted at high temperatures in the presence of a metal catalyst. This process is not widely-used in the gas processing industries because of rapid catalyst deactivation due to carbon deposition.
Methane is the major component in natural gas and is also commonly present as a by-product from landfill wastes. It was estimated that the Philippines has about 3.8 trillion cubic feet of proven natural gas reserves, with significant production starting in 2001. Meanwhile, around 819.50 tons of methane emissions are released daily from MSW dumpsites around the country (MMSWMP, 2003). These gases may be used as feedstock in the production of syngas through methane dry reforming. Subsequently,

syngas may be further processed to produce important chemicals like NH_3, CH_3OH, and other oxygenated chemicals.

The dry reforming of methane is a process that requires the use of catalysts such as the noble metals Rh and Pt, supported on metal and nonmetal oxide support materials such as Al_2O_3, MgO, TiO_2, ZrO_2, and CeO_2 (Topalidis, et. al., 2007). Although these noble metals yield high activity and selectivity, they are relatively expensive. As such, transition metals are used instead which include nickel, cobalt, and copper (Rostrup-Nielsen and Hansen, 1993, Gadalla and Sommer, 1989).

Rapid catalyst deactivation caused by carbon deposition on the catalyst surface during CH_4 dry reforming is particularly rampant at lower reaction temperatures. The carbon formation on the catalyst surface is generally attributed to the decomposition of methane and the Boudouard reaction. Past researches have shown that the effect of carbon deposition on the activity and stability of the catalyst may be minimized by choosing the right support material. Support materials that are basic in nature are generally preferred. These include MgO, K_2O, and Ca_2SiO_4. The high basicity of these support materials (Horiuchi, et. al., 1998, Hu and Ruckenstein, 2004) and the presence of mobile oxygen species in supports like in ZrO_2, (Razaei, et. al., 2008), discourages the formation of inactive carbon deposits on their surfaces. The mobile oxygen is readily available to react with inactive carbon to form CO, minimizing carbon deposits (Therdthianwong, et. al., 2008). A combination of a highly basic support material and one that can provide mobile oxygen species such as MgO-ZrO_2 may greatly improve catalyst activity and stability during methane dry reforming especially at low reaction temperatures where crucial financial savings may be achieved through lower energy consumption. This study is focused on the investigation of the catalytic activity of 15% Ni/MgO-ZrO_2 in the dry reforming of methane.

2. Materials and Methods

2.1 Catalyst Preparation

The 15% Ni/MgO and 15% Ni/MgO-ZrO_2 catalysts were prepared using the dry impregnation method. MgO (Sigma Aldrich) and ZrO_2 (JRC-ZRO-3) from the Tokyo Institute of Technology in Japan were mixed in a 1:1 mole ratio to make up the dual support. $Ni(NO_3)_2{\cdot}6H_2O$ (Bayer) was used as precursor and was impregnated onto the surface of the support materials. $Ni(NO_3)_2{\cdot}6H_2O$ solution was added to the support material drop-wise until a well-mixed paste is formed. This mixture was dried at 40°C for 24 hours to allow the moisture to slowly evaporate and the precursor to impregnate the support material. After which, the partially dried paste was dried even further at 120°C for 12 hours to completely remove the water. The dried powder was calcined inside a temperature-controlled furnace set at 850°C for 16 hours. The calcined catalysts were pelletized, crushed, and sieved until the desired particle size of +20 mesh (0.841 mm) and – 40 mesh is achieved (0.420 mm). The catalysts were then reduced under a 100 ml/min stream of 20% H_2/He flow for 2 hours at 850°C. All the prepared catalysts were stored inside a dessicator to prevent moisture from being adsorbed. Catalysts with varying nickel loading were also included in this study as a basis for comparison.

2.2 Catalyst Characterization

The catalysts prepared were characterized using advanced instrumentation and techniques to determine their surface characteristics and morphology: These include:

- surface area, pore size, and pore volume – Autosorb - I Analyzer
- crystal structure – XRD (Phillips PW1050/80)
- morphology and elemental composition – SEM-EDX (JEOL JSM 5310)
- acid/basic sites – Temperature Programmed Desorption with GC

2.3 Activity Test

The reactant gases methane and carbon dioxide were allowed to flow into the flow reactor system (Figure 2) until they reached a quartz microreactor where the dry reforming reaction is expected to take place at 850°C. Both feed gases were delivered, using helium as the carrier fluid, directly into the reactor using separate lines. The flow was controlled by calibrated rotameters. The CH_4, CO_2, and He streams met at the entrance of the reactor. Once inside, the gases passed through a fixed bed of catalysts where they underwent the dry reforming reaction. Since the reaction is highly endothermic, a tube furnace was used to heat up the reactor to the desired reaction temperature. The furnace was controlled by a programmable temperature controller.
The exit gases then left the reactor and entered the molecular sieve column to affect component separation. After which, the separated gases entered a thermal conductivity detector (TCD) where the gas composition was determined.

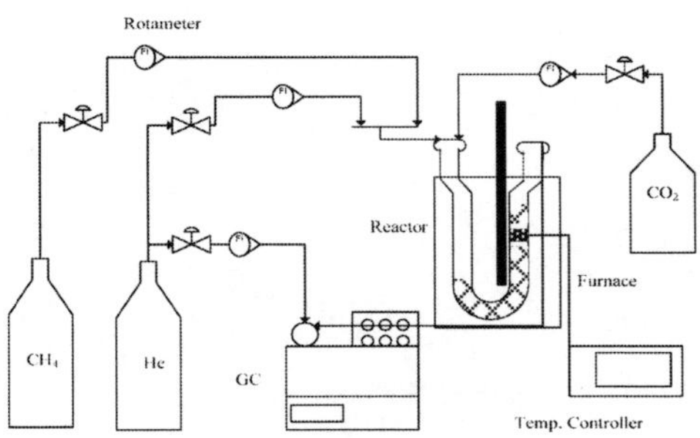

Figure 2: Methane Dry Reforming Experimental Set-up

3. Results and Discussion

3.1 Catalyst Surface Area, Pore Size, and Pore Volume

Table 1 summarizes the measurements done by the Autosorb – I Analyzer for the catalysts that were calcined at 850°C and then reduced at 850°C. In general, as seen in Figure 3, the 15% Ni/MgO-ZrO_2 catalysts exhibited a decrease in surface area after calcination and after reduction. The decrease in the surface area for these catalysts may be due to sintering that occurred during the calcination and reduction processes.

The decreasing trend was not seen in the 15% Ni/MgO catalysts. The different behavior exhibited by the catalyst using MgO as support can be attributed to the fact that MgO has high thermal stability and is not prone to sintering. This was also reported by Ruckenstein and Hu (Ruckenstein and Hu, 1996). The increase in surface area for these catalysts can be due to the addition of nickel and the formation of NiO-MgO solid solutions whose ensemble sizes lead to higher surface area. This was not seen in the catalysts containing dual support MgO-ZrO_2 since the sintering effect on the ZrO_2 support material may have been more prominent.

Table 1: Surface Area, Pore Size and Pore Volume of Test Catalysts

Sample	Average SA (m^2/g)	Pore Volume (cc/g)	Pore Width (Å)
MgO	32.465	0.0132	17.2
NiO/MgO	32.92	0.0088	15.28
Ni/MgO *	39.845	0.014	14.155
MgO/ZrO_2	73.69	0.0296	19.33
NiO/MgO-ZrO_2	13.22	0.0046	17.975
Ni/MgO-ZrO_2 *	15.385	0.0045	16.6

Reduction at 850°C under a 100 ml/min stream of 20% H_2/He flow for 2 hours

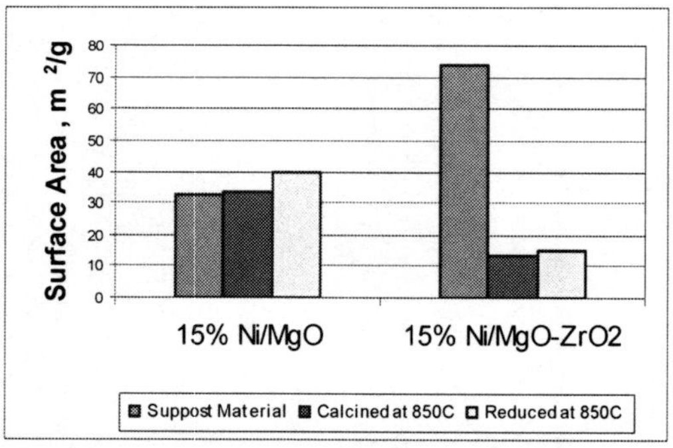

Figure 3: Effect of Calcination and Reduction on Catalyst Surface Area

3.2 Surface Morphology

As seen in the SEM pictures presented in Figure 4 where the catalyst surface was magnified 5,000 times, there is evidence that, consistent with the results from the Autosorb, catalyst surface area for 15% Ni/MgO-ZrO_2 decreased upon being subjected to elevated temperatures during calcination. As the calcination temperature is increased, surface area decreased, which also led to lower pore volume and pore width.

Reduction in surface area was not very prominent in the 15% Ni/MgO catalyst as seen in Figure 5. Again, this is consistent with the data produced by the Autosorb.

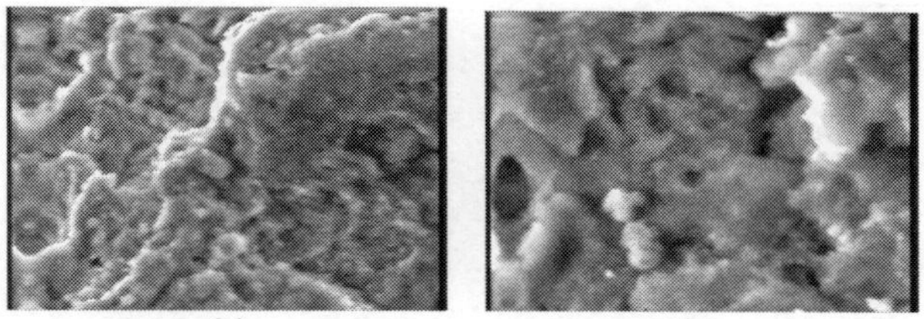

(a) (b)

Figure 4: SEM (5000x) - (a) 15% NiO/MgO-ZrO_2 calcined at 500°C (40.27 m^2/g) (b) 15% NiO/MgO-ZrO_2 calcined at 850°C (13.22 m^2/g)

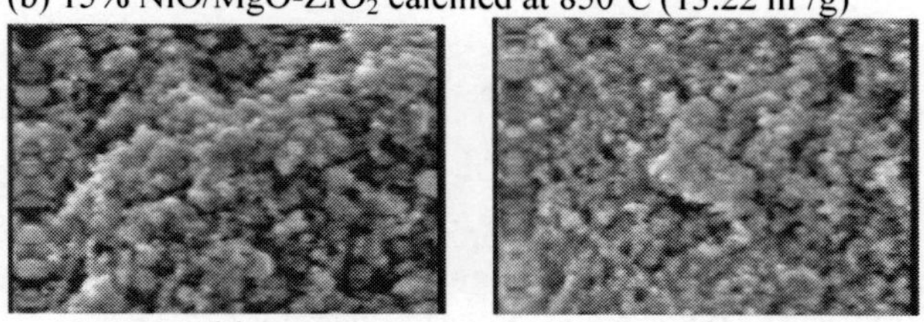

(a) (b)

Figure 5: SEM (5000x) - (a) 15% NiO/MgO calcined at 500°C (36.37 m^2/g) (b) 15% NiO/MgO calcined at 850°C (32.92 m^2/g)

3.3 Catalyst Crystal Structure

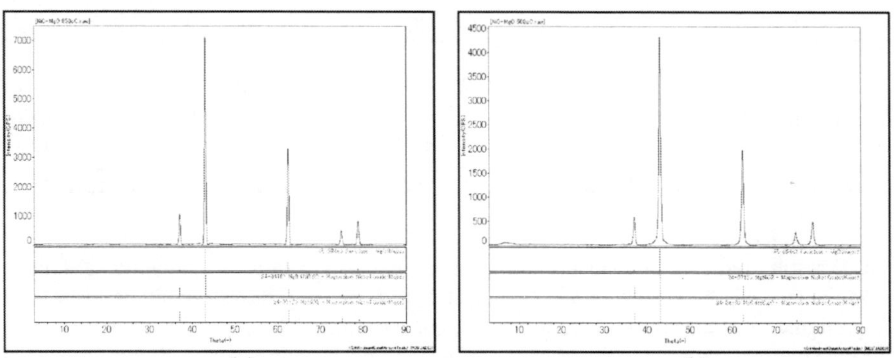

(a) (b)

Figure 6: XRD (a) Ni/MgO calcined at 500°C (b) Ni/MgO calcined at 850°C

Based on Figure 6, solid solutions (Djaidja, et. al. , 2006 , Chen, et. al. , 1994 , Yamazaki, et. al., 1992) of NiO-MgO were present in the catalysts after calcination at 850°C. The solid solution was also present for 15% NiO/MgO calcined at 500°C, which is why these were very resilient against sintering and loss of surface area. Also, the intensity of the peaks was higher for the catalyst calcined at 850°C versus the one calcined at 500°C.

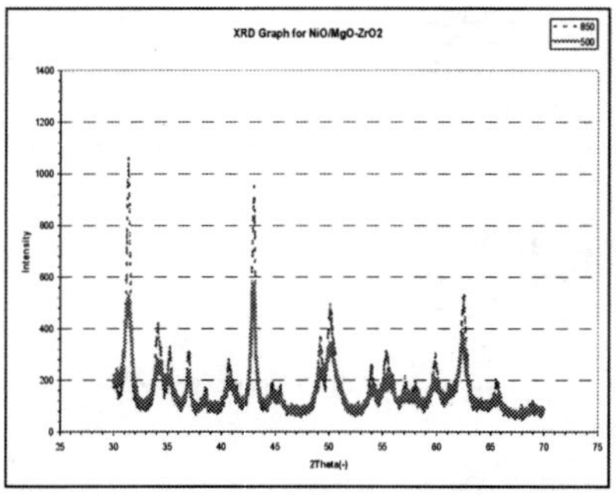
(a)

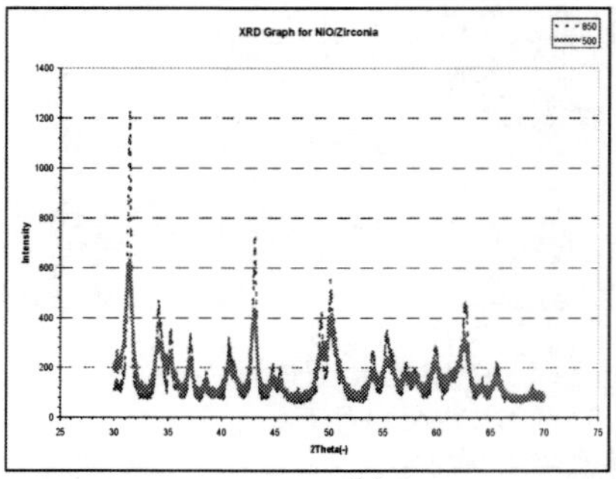
(b)

Figure 7: XRD (a) NiO/MgO-ZrO_2 (b) NiO/ZrO_2

Even with a 1:1 molar ratio of MgO to ZrO_2, the XRD profile of Ni/MgO-ZrO_2 appears to behave more closely like Ni/ZrO_2 than Ni/MgO (Figure 7). This is consistent with the fact that unlike Ni/MgO, Ni/MgO-ZrO_2 is highly prone to sintering.

3.4 Surface Composition from EDX

Table 2: EDX Surface Composition (% Element)

Sample	% Element				% Atomic				
	Mg	Ni	O	Zr	Mg	Ni	O	Zr	Mg:Zr
15% Ni/MgO	41.86	12.25	45.89	-	35.88	4.35	59.77	-	-
15% Ni/MgO-ZrO_2	8.89	11.70	36.49	42.91	11.03	6.01	68.78	14.19	0.777
5% Ni/MgO-ZrO_2	11.92	3.82	42.46	41.80	13.37	1.77	72.36	12.49	1.071
10% Ni/MgO-ZrO_2	9.31	7.24	37.22	46.23	11.46	3.70	69.67	15.17	0.755
20% Ni/MgO-ZrO_2	8.24	21.68	37.24	32.84	9.98	10.87	68.54	10.60	0.942

EDX results indicate that surface composition is close to target at roughly 12% nickel and about 0.77:1 Mg:Zr ratio. Catalysts with varying nickel loadings were also investigated. The deviations from target composition may be due to inadequate mixing of the support materials. Overall, the surface composition indicates that there is still an opportunity to improve the catalyst preparation method.

3.5 Acid and Basic Sites

Based on the TPD results shown in Table 3, 15% Ni/MgO has the most number of basic sites. 15% Ni/MgO-ZrO_2 has a lower number of basic sites which could be because it has a smaller surface area to begin with. Furthermore, for Ni/MgO-ZrO_2, the number of basic sites seems to be decreasing with increasing nickel loading. In contrast, the number of acid sites appears to be increasing. More nickel on the surface of the support may be preventing the basic sites from being fully exposed.

Table 3: Acid and Basic Sites from Temperature Programmed Desorption Experiments

Catalyst	Basic Sites	Acid Sites
15% Ni/MgO	1.02×10^{21}	-
15% Ni/MgO-ZrO$_2$	2.08×10^{20}	2.63×10^{20}
5% Ni/MgO-ZrO$_2$	2.36×10^{20}	1.20×10^{20}
10% Ni/MgO-ZrO$_2$	1.85×10^{20}	1.69×10^{20}
20% Ni/MgO-ZrO$_2$	1.78×10^{20}	2.68×10^{20}

3.6 Activity Tests

At 850°C, using 15% Ni/MgO-ZrO$_2$ led to an average % methane conversion of 91.34% and a H_2/CO ratio of 1.05. Theoretically, the equilibrium conversion of the methane dry reforming reaction alone as shown in Equation 1 is 86.90% with an H_2/CO ratio of 1. The results indicate that since the H_2/CO ratio is close to 1, very little methane decomposition occurred and thus, less carbon formation on the surface of the catalyst. This could be due to the presence of mobile oxygen species from zirconia that is available to bond with the surface carbon forming carbon monoxide, thus minimizing carbon deposition on the catalyst surface.

$$CH_4 + CO_2 \leftrightarrow 2H_2 + 2CO \quad ---Eq1 \qquad \Delta H^o{}_{298} = 261kJ/mol$$

$$CH_4 \rightarrow C + 2H_2 \quad ---Eq2 \qquad \Delta H^o{}_{298} = 261kJ/mol$$

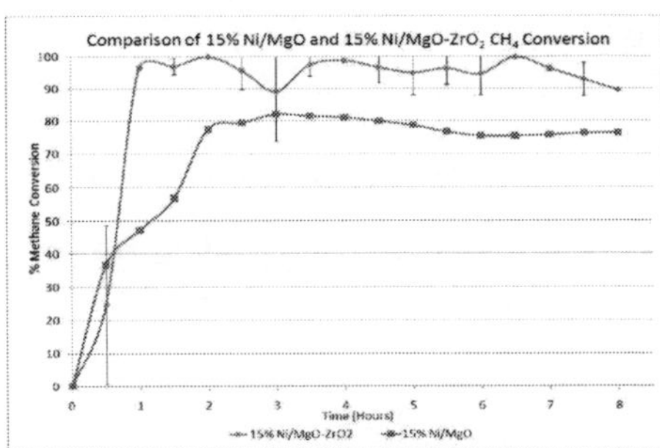

Figure 8: Time Course Activity of 15% Ni/MgO-ZrO$_2$ at 850°C

In comparison, at 850°C, using 15% Ni/MgO led to an average % methane conversion of 77.64% and an H_2/CO ratio of 0.18. This low value of H_2/CO ratio may indicate the occurrence of methane decomposition (Equation 2) that is more prevalent than the one experienced when using 15% Ni/MgO-ZO$_2$. This also means that more carbon may have deposited on the surface of the catalyst.

4. Conclusions and Recommendations

The addition of ZrO$_2$ changed the surface properties of the traditional catalyst Ni/MgO. 15% Ni/MgO-ZrO$_2$ has a lower surface area than 15% Ni/MgO. It also has a lower number of basic sites. However, despite of the relatively lower number of basic sites, it appears that using 15% ni/MgO-ZrO_2 leads to higher % methane conversion and potentially lower carbon deposition. The presence of mobile oxygen species from zirconia may explain this behavior. It is recommended to conduct activity tests for longer periods of time to verify catalyst stability in the presence of carbon deposits.

In addition, the Ni/MgO-ZrO_2 catalyst had a crystalline structure closely similar to that of Ni/ZrO_2 indicating that the NiO-MgO solid solution may have not been formed. This may lead to lower catalyst activity in spite of the high resistance to carbon deposition. Increasing the MgO:ZrO_2 ratio is also recommended.

Finally, the catalyst preparation method can be improved in order to make the surface composition achieve values that are closer to target. Other preparation methods such as the sol-gel and co-impregnation methods should also be investigated.

References

Chen, Yang-guang, and J. Ren. 1994., Catalysis Letters, 29, pp. 39 - 48

Djaidja, A., S. Libs, A. Kiennemann, and A. Barama. 2006., Catalysis Today, 113, pp. 194 – 200

Gadalla, A., and M. Sommer. 1989., Journal of American Ceramic Society, Volume 72, No. 4, pp. 683 – 687

Horiuchi, T., H. Hidaka, T. Fukui, Y. Kubo, M. Horio, K. Suzuki, and T. Mori. 1998., Applied Catalysis A, General 167 pp. 195 –202

Hu, Y and E. Ruckenstein. 2004., Advanced Catalysis, 48, pp. 297 – 345

Metro Manila Solid Waste Management Project, 2003

Rezaei, M., S. M. Alavi, S. Sahebselfar, P. Bai, X. Liu, and Z. Yan. 2008., Applied Catalysis B, Environmental, 77, pp. 346 – 354

Rostrup – Nielsen, J. and J. Hansen. 1993., Journal of Catalysis, Volume 144, Issue 1, pp. 38 – 49

Ruckenstein E. and Y. Hu 1996., Journal of Catalysis, Volume 162, pp. 230 – 238

Therdthianwong, S., A. Therdthianwong, C. Siangchin, and S. Yongprapat. 2008., International Journal of Hydrogen Energy, No. 33, pp. 991 – 999

Topalidis, A., D. Petrakis, A. Ladavos, L. Loukatzikou, and P. Pomonis. 2007., Catalysis Today, 127, pp. 238 – 245

Yamazaki, Osamu, K. Tomishige, and K. Fujimoto. 1992., Chemistry Letters, 21, pp. 1953 – 1954

Proceedings of the 2nd Annual Gas Processing Symposium
Farid Benyahia and Fadwa T. Eljack (Editors)

Smart Leak Detection and Repair at Q-Chem

Homssi, Ahmad M., Environmental Manager, Qatar Chemical Co. ltd., P. O. Box 24646, Doha, Qatar

Abstract

Industrial Cities in Qatar included the requirements for fugitive emissions monitoring in their land lease agreements with industries in their individual areas. Although there are some differences in each city's guidelines, the overall requirement is for a Leak Detection and Repair (LDAR) program to be in use to ensure early response to leaks and minimization of fugitive emissions.

Recently, inclusion of a LDAR program became a State of Qatar environmental requirement, although not in the executive Environmental Regulations as yet. The State authority had enforced LDAR program in some industries' consent to operate, when Q-Chem was only required to monitor hydrocarbon tanks. Q-Chem had chosen to be in full compliance with our contractual agreement with Mesaieed Industrial City (MIC) and Ras Laffan Industrial City (RLIC) for our operated plants at these cities. For that, Q-Chem decided to build and implement a "Smart LDAR Program", incorporating the use of Infra Red Optical Gas Imaging Cameras (IR Camera) for qualitative monitoring of leaks and traditional VOC “Sniffers” for the quantitative measures of identified leakers. Moreover, this “Smart LDAR Program” is predicated on the use of P&IDs and PFDs and populating a database that lists and tracks all fugitive sources, their annual leaking values, and repair histories without physical tagging of individual sources in the field.

Unlike the initial LDAR US-EPA Method 21, this Smart LDAR, which has recently been approved by US-EPA, requires much less manpower; is much faster; is less costly; has less potential for missing leakers; reduces the number of difficult to monitor (DTM) or Unsafe to monitor (UTM) sources; quickly locates the highest leaking components without having to monitor every individual piping component in the plant. ; and has > 90% of fugitive emissions come from only about 0.13% of the piping components, it is most cost effective to concentrate and direct efforts and resource towards those emitting > 500 ppm using scanning IR camera.

In this paper, Q-Chem will present the information collected from facilities operated by Chevron Philips Chemicals (CPChem) and its joint ventures to develop LDAR programs and compare traditional and “Smart” programs. Additionally, this paper will outline efforts to secure RLIC and MIC consent for use of this new technology.

This paper will also discuss how this Smart LDAR program had maximized emission reduction while minimizing the program cost, and getting work done faster in locating and repairing the leakers, especially that 99% of the fugitive sources do not leak appreciably and therefore do not contribute significantly to the mass emissions.

Past experience in building the LDAR at Q-Chem will be shared with its positives and lessons learned as we continue building this program for all Q-Chem facilities as they are commissioned and started up.

Keywords: VOC, Leak detection, Hazardous pollutants, Monitoring

1. Leak Detection and Repair

1.1. Scope and Purpose

Although the VOC emission from a single fugitive emission source, like a pressure relief device, a flange, a bellows valve, a control valves, a block valve, a check valve, a line connection, or an open-ended pipe might be in ppb unit only, it is a fact that when all of such fugitive sources that are in thousands in a plant are added up, the total emitted volumes of the VOCs included those considered being HAPs are significantly high unless controlled and reduced to the minimum possible.

For that, LDAR program is implemented for checking the fugitive sources on gas or liquid phase streams of VOC and HAP (e.g., Ethane, Benzene, Butane, Cyclohexane, Ethylene, Hexene, and Isobutene). This LDAR program is being regulated by the US-EPA for past 25 years using Method 21 – Determination of volatile organic compound leaks, 40 C.F.R. Pt. 60, App.A-7 (2006). The program aims at finding the larger mass rate leakers sooner and repairing them more quickly to offset the smaller mass rate leakers.

For full implementation of US-EPA Method 21, an organization has to first identify each and every regulated fugitive source, tag physically, draw sampling routes, flag the sources that are unsafe to monitor (USM) and difficult to monitor (DTM) for alternative monitoring (i.e., correlation), utilize skilled manpower to monitor the leakers and address the repair requirements, follow up on corrective/preventive actions, and report annual emissions including the advancement in controls every passing year.

LDAR program therefore is known to reduce green house gases, hazardous air pollutants, loss of raw materials or products through leaks, enhances air quality, and reduces losses, while being in compliance with regulatory requirements and posturing continuous improvement in resources conservation and pollution prevention.

In search for a faster monitoring program that is less costly, more practical, more efficient, can eliminate the UTM and DTM sources, and still meet regulatory requirements, the "Smart" LDAR program was recently introduced, using Infra Red Optical Gas Imaging Cameras and acknowledged by US-EPA.

1.2. LDAR Legal Requirements

LDAR program was introduced many decades back in USA and Europe. For the past 25 years, the US-EPA was regulating the implementation of LDAR programs through Method 21 and enhancing it over the years. The latest version was introduced in 2006 and known as "Determination of volatile organic compound leaks, 40 C.F.R. Pt. 60, App.A-7".

Locally, Law 30 of year 2002 included some requirements for leaks limits, detection and repairs. This is more specifically in the Executive Regulations Bylaw, Annex 3/1st/ Bullets # 5 & 6. This bylaw failed to specify the key requirements of what VOCs and

HAPs to be monitored, at what concentrations the actions are to be taken, and the reporting requirements. The authority was more specific on their requirements when issuing site-specific consents to operate based on case-by-case evaluation. Hence, not all operational facilities are found having same LDAR requirements.

At industrial cities level, each industrial city has its list of LDAR requirements in their land lease agreement's environmental guidelines' section. Although the requirements of the different industrial cities are close, there are still some requirements that are found unique for each one. These requirements are mostly considered for being contractual requirements rather than law enforced ones, yet subject for compliance.

Table 1: Comparison of LDAR Legal Requirements

Agency	Program	Citation	Regulatory Leak Definition (ppmv)					
			Valves	Pumps	Compressors	Connectors	PRVs[1]	OELs[2]
MIC	Proposed to Q-Chem	n/a	≥ 500 (OHAP) ≥ 10,000 (VOC)	≥ 500 (OHAP) ≥ 10,000 (VOC)	≥ 500 (OHAP or VOC)	≥ 500 (OHAP only)	≥ 500[3] (OHAP or VOC)	≥ 500 (OHAP) ≥ 10,000 (VOC) <u>and</u> Design Standard[4]
RLC	Applicable to Q-ChemII	n/a	≥ 10,000 (VOC or H_2S[11])	AVO[10] and ≥ 10,000 (VOC or H_2S[11])	Design Standard[7] or NDE[8]	No Standard	NDE[8] (VOC or H2S[11])	Design Standard[4]
U.S. EPA	NSPS for SOCMI facilities	40 CFR Part 60, Subpart VV	≥ 10,000[5]	AVO[10] and ≥ 10,000[6,7]	Design Standard[7] or NDE[8]	AVO[10] observation of fluids dripping	≥ 500[3]	Design Standard[4]
U.S. EPA	HON MACT	40 CFR Part 63, Subpart H	≥ 500[5]	AVO[10] and ≥ 1,000[6,7,9]	Design Standard[7] or NDE[8]	≥ 500[5]	≥ 500[3]	Design Standard[4]
Texas - TCEQ	Regulation 115 - Fugitives for Refinery and SOCMI facilities	30 TAC Chapter 115, Subchapter D, Division 3	≥ 500[5]	≥ 10,000[6,7]	Design Standard[7] or ≥ 10,000	AVO[10] observation of fluids dripping	≥ 500[3]	Design Standard[4]

Notes:
1. PRV = Pressure Relief Valves
2. OEL = Open-Ended Lines
3. PRVs in gas/vapor service only. PRVs in liquid service are exempt from instrument monitoring except after a release.
4. Install cap, blind flange, second valve, or other barrier on the end of each OEL.
5. Components in light liquid or gas/vapor service only. Heavy liquid components do not require instrument monitoring. (MIC and RLC proposals do not exclude heavy liquids.)
6. Pump seals in light liquid service only. Heavy liquid pump seals are exempt from instrument monitoring. (MIC and RLC proposals do not exclude heavy liquids.)
7. Pump (light liquid) and compressor (gas/vapor) seals meeting a design standard of a barrier fluid system, dual mechanical seals, or other "leakless" design are exempt from instrument monitoring. (MIC proposal does not have design standard alternatives to monitoring. RLC proposal has design standard for compressors, but not for pumps.)
8. NDE = No Detectable Emissions. Defined as an instrument monitored concentration < 500 ppmv.
9. If material is a polymerizing monomer, threshold is 5,000 ppmv.
10. AVO = Audible, Visual, or Olfactory
11. H_2S cannot be measured by the same organic vapor analyzer (OVA) that is required for VOC monitoring and thus component monitoring will require 2 instruments, calibrations, etc. Also, H_2S concentrations of > 200 ppmv are considered lethal. Thus a 10,000 ppmv threshold for H_2S is impractical.

Table 1 above compares some of the local industrial cities' requirements with those in the international ones and more specifically in the USA.

To be in full compliance with regulatory and contractual requirements, Q-Chem who operates Q-Chem and Q-Chem II plants at Mesaieed Industrial City (MIC) and RLOC at Ras Laffan Industrial City (RLIC) decided to build and implement a "Smart LDAR Program", incorporating the use of Infra Red Optical Gas Imaging Cameras (IR Camera) for qualitative monitoring of leaks and traditional VOC "Sniffers" for the quantitative measures of identified leakers with the concurrence of MIC and RLIC, such that tagging and flagging are only done on the P&IDs and P&FDs and reporting is done from a populated database that lists and tracks all fugitive sources at each location.

1.3. Standard LDAR Versus Smart LDAR

Method 21 of US-EPA 40 C.F.R. Pt. 60, App.A-7 is known for being the standard LDAR regulating guidelines. If to implement this method fully, a medium size petrochemical industry for example would need 4 or more skilled skillful technicians

with 2 VOC monitoring equipment working 8 hours per day around the year to monitor the expected 50,000 sources on site. This is known for being a very expensive and complicated program and in most cases contracted to specialized companies.
The USA American Petroleum Institution (US-API) published Figure-1 below describing that more than 90% of fugitive emissions come from only about 0.13% of the piping components and hence it is most cost effective to concentrate and direct efforts and resource towards those emitting > 500 ppm.

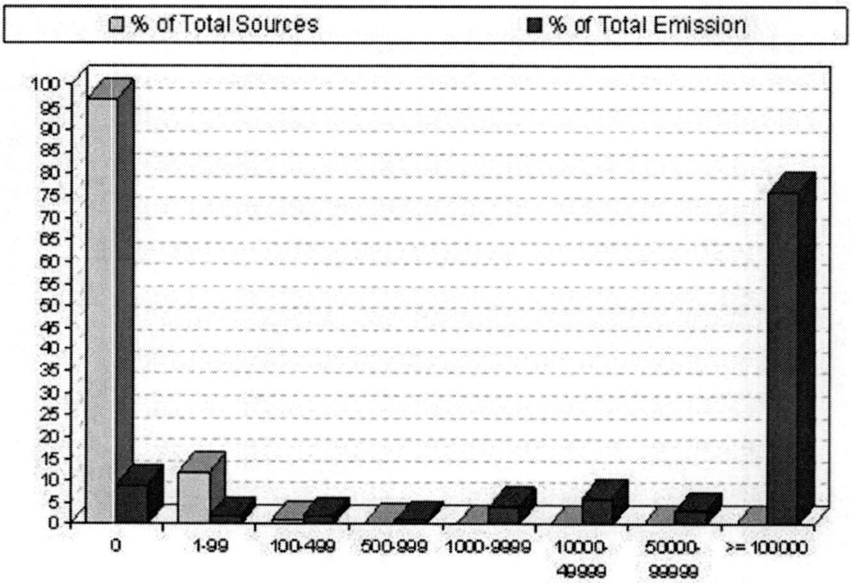

Figure 1: Fugitive emissions (American Petroleum Institution)

Significant reductions in costs and emissions can be realized when Smart LDAR is implemented using IR Optical Gas Imaging Camera, as it can locate the very high leaking components without having to monitor every individual piping component in the plant using EPA Method 21.
In 2006, the US-EPA acknowledged the need for a more efficient method. In describing the "Smart" LDAR using IR Camera, US-EPA announced "The current work practice requires the use of a monitor which meets required performance specifications. This work practice is based on 25-year-old technology. New technology has been developed which we believe provides equal, or better, environmental protection than that provided by the current work practice. This action proposes a voluntary alternative work practice (AWP) for finding leaking equipment using optical gas imaging."

US-EPA had then sponsored this API survey, realizing these facts, and had approved this alternative method on December 15, 2008. The US-EPA had issued a final amendment to the leak detection and repair requirements allowing the use of optical gas imaging technology to locate emission leaks. This amendment had modified about 40 USA national rules requiring facilities to find and repair leaks from equipment, including pumps, valves, and connectors, from refineries, chemical production plants, and bulk liquid storage facilities.

Table 2 below compares between Method 21 and the IR Camera practices and try to justify why IR Camera surveys is becoming the promising tool for successful LDAR.

Table 2: Comparision Between Method 21 and IR Camera Practices

LDAR Using Method 21	IR Optical Gas Imaging
Team of 3-6 people	Team of 2 people
Excessive Capital Cost	Relative Lower Capital Cost
High Operational Cost	Low Operational Cost
100 Sources / Day	> 2000 Source / Day
Many DTM & UTM Sources	No DTM & UTM Sources
Technicians Quit Job	Technicians Enjoy Job
Tracking Based on Tags	Video Taping for Tracking
Major Cost on Monitoring	Minimum Cost on Monitoring
Less Monitoring Frequently	High Monitoring Frequently
All Intrinsically Safe	Not Intrinsically Safe Yet
US-EPA Approved	US-EPA Accepted

LDAR program based on IR Optical Gas Imaging should maximize emission reduction while minimizing the program cost. Also, this more cost effective emission-reduction approach would emphasize the location and repair of the very high leakers in place of monitoring all of the piping components, 99% of which do not leak appreciably and, therefore, do not contribute significantly to the mass emissions. Locating very high leakers can be done more frequently at reasonable costs. More frequent repair of the very high leakers can result in lower overall emissions than would a program with less frequent repair including the low level leakers.

1.4. Smart LDAR Tests and Applicability

Derek Reese and Charles Melvin from ExxonMobil Chemical Company, Baton Rouge, LA, and Wayne Sadik from ExxonMobil Chemical Company, Baytown, TX, had issued a two-year survey and conclusively shown that using optical imaging in a Smart LDAR program for fugitive emissions control results in lower emissions compared with the current Method 21-based regulatory required procedures. Also, small concentration leaks were shown to not generally increase over time. This study demonstrated that finding the larger mass rate leakers sooner and repairing them more quickly offset the smaller mass rate leakers that would be not have been detected using the AWP. With

regard to monitoring efficiency, fewer personnel will be required. Using the AWP, they will be able to monitor a facility in a fraction of the time that would have been required using the CWP. The combination of all these benefits demonstrates that optical imaging should be allowed to replace Method 21 for fugitive emission control.
According to Mr. Eric Anderson from ERM, a consulting partner with Leak Surveys Inc. (LSI), there have been many contracts in place recently by a number of companies to provide additional training once a camera is purchased and on-site in the Middle East.
In Q-Chem, difficulties and limitations were of most interest as we were building the Smart LDAR. To mention just few:

1. Eliminating the possible removal of tagged sources, either for maintenance or replacement, without notifying LDAR Technician or not putting the tags back on the removed source.
2. Avoiding the requirement of assistance from field operators, while their management is reluctant to spare them to assist in building and verifying sources.
3. The limited manpower within the organization for full-time LDAR implementation and to avoid sources being identified and tagged in assigned employees spare times.
4. Not going to fully-automated system which that need management approval for capital expenditure, which might not be justified, as it's not legally required to conduct full-scale LDAR.

Q-Chem Environmental Engineering Team had to justify the use of the Smart LDAR Program to stakeholders within the organization, the land lords at MIC and RLIC, and the Ministry of Environment.

2. Smart LDAR Program at Q-Chem

2.1. Q-Chem LDAR Program Establishment

2.1.1. Regulatory Compliance

Q-Chem is consented to only monitor fugitive emissions from the hydrocarbons storage tanks. In the land lease agreement, environmental guidelines, Q-Chem, like all land lease holders at Mesaieed Industrial City (MIC), has a list of fugitive sources to be monitored as well. Nonetheless, MIC did not enforce this requirement; possibly knowing how much cost and efforts it need.
Chevron Philips Chemicals (CPChem), the 49% shareholder of Q-Chem, used their Operations Excellence Protocol to audit Q-Chem in November 2007, when a gap was found in meeting the contractual requirement in the land lease agreement for fugitive air emission sources monitoring and leak repairs. This had driven Q-Chem management to send two engineers to visit with CPChem headquarters in Houston, TX, USA, and some industrial facilities for LDAR familiarization. Following the visit, Q-Chem Environmental Engineering evaluated the requirements for LDAR full implementation, including its new facilities, Q-Chem II at Mesaieed Industrial City (MIC) and RLOC at Ras Laffan Industrial City (RLIC).

There were number of questions raised after this evaluation, as the requirements of MIC and RLIC are a pit different in definitions and implementation requirements. Q-Chem had hence invited MIC and RLIC for a round-table brainstorming discussions and agreed to start Smart LDAR using Infra Red Optical Gas Imaging Cameras (IR Camera) for qualitative measures and TLV 1000B VOC sniffers for the quantitative measures of the sources detected by the camera to be leaking. The agreement was also to identify the fugitive sources on the P&IDs and populate a database that list of these source and for generating annual reports, provided that this program is enhanced in the following years, working together with MIC and RLIC.

The plan was to complete Q-Chem sources identification and conduct the first survey by end of 2009 (i.e., baseline survey), then complete Q-Chem II in 2010 and RLOC in 2011. The Ministry of Environment was copied on this program and is being updated every quarter on the progress of this Smart LDAR Program. Q-Chem Consent to Operate issued in January 2009 acknowledged this and it is Q-Chem expectation that its Consent to Operate due for renewal in January 2010 would include Smart LDAR as a new requirement that was started voluntarily to meet contractual requirement.

2.1.2. Recruitment, Equipment Purchase and Training/Certification

Q-Chem Smart LDAR program is based on the usage of two full-time hired LDAR technicians, who have reliability or maintenance past experience and can read P&IDs and are trained and certified to identify fugitive sources (on lines containing non-methane and non-ethane hydrocarbons by marking on P&IDs) and certified for the use of the LDAR equipment, populate these sources in an in-house built LDAR Database, verify these sources by site visits and conducting QA/QC sessions using the process area specialists and before they finally use the two IR Cameras and two TLV 1000B VOC Sniffers for the measurements without any third party consultancy help.

These Smart LDAR surveys with IR Cameras are conducted by one LDAR technician (primary/Camera User) while the other LDAR technician works with him as (backup/Safety Watch and Gas Tester).

To ensure a state-of-the-art program, Q-Chem had consulted with many facilities like Saudi Chevron in KSA; Pasadena and Cedar Bayou in USA; and Ras Gas and Dolphin in Qatar. In conclusion, Q-Chem decided to purchase the equipment shown on Figure 2 below.

	ThermaCAM® GasFindIR® HSX ThermaCAM GasFindIR Camera: 320x240 InSb High Performance FPA Fixed 25mm Lens with Lens Cover Black and White Viewfinder Auto Gain Control (AGC) (2) Intelligent Battery Chargers Battery, 3 each, NiMH Video Cables BNC S-Video

	Neck Strap Shipping/Carrying Case Operating Manual Personal Video Recorder (PVR), Battery, and Accessories
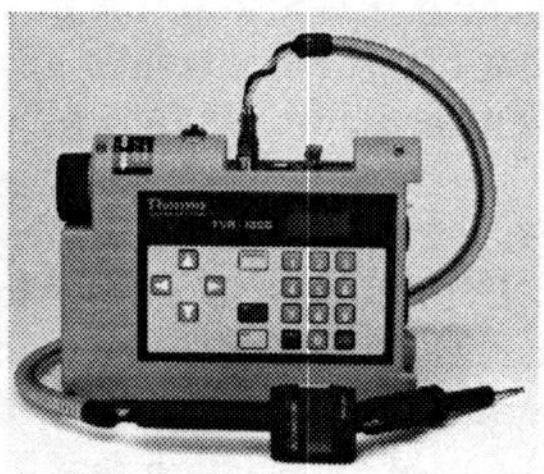	<u>TVA-1000B Toxic Vapor Analyzer</u> Thermo Electron Model TVA-1000B (Both FID/PID) Filter Kit (10 Pcs) Filter Cup (Package of 5) Calibration Kit for FID; Includes zero gas and span gas (methane) cylinders, regulator, tubing Calibration Kit for PID; Includes zero gas and span gas (Isobutylene) cylinders, regulator, tubing Frame Backpack Hydrogen tank Battery Pack Battery Charger, 230V Annual Maintenance Service Contract

Figure 2: Smart LDAR Equipment

The detailed Smart LDAR Program is built and based on IR Camera's vendor's operations and maintenance procedure and training/certification of LDAR technicians. This Program document and its supporting Standard Operating Procedures (SOPs) for the monitoring and inspection are incorporated into Q-Chem Environmental Management System. Q-Chem believes that its Smart LDAR program is robust and yet practical, given the size of the three plants (Q-Chem, Q-Chem II and RLOC) with an anticipation of more than 250,000 fugitive emission sources, the complexity of operations, the physical locations of the three sites, and the potential for this program to be mandated in future within Qatar.

2.2. Q-Chem LDAR Database

2.2.1. Source Identification

The Q-Chem two LDAR technicians had successfully developed an effective LDAR monitoring and inspection procedure, with inputs from Safety, Industrial Hygiene, Operations, and Maintenance. LDAR technicians target accessible fugitive VOC emissions from valves (control valves, block valves, bellow valves, and check valves), pump seals, compressor seals, pressure relief valves, connectors and tube fittings, open-ended lines in compliance with MIC and RLIC regulations and guidelines. This was done with the following in mind:

1. Q-Chem identifies fugitive emission sources on P&ID drawings, including types of components, the counts, hydrocarbon streams containing more than 5% VOC (under maximum production conditions "MPC"), but exclude fuel gas, sales gas, methane, and ethane.

2. P&ID drawings are obtained to complete component counts for components in light liquid, heavy liquid and vapor services. All components are identified on P&IDs for future references.
3. Finally Q-Chem prepares a report documenting the methodology used, monitoring surveys done and results of the fugitive monitoring results and repair plans.
4. Q-Chem will partially implement the conditions listed in US-EPA method 21 for VOC monitoring (e.g., tagging & flagging are not physically done in the field), while all monitoring and reporting requirements set by LLAs are met and per this program.

2.2.2. Source QA/QC Program and On-Site LDAR Source Verifications

The marked P&ID's and P&FD's are studied carefully by the LDAT Technicians and a process area specialist and filtered using design-base listed sources, site verifications, and outdoor walk-downs to:

1. Ensure if all sources are captured or not
2. Ensure if the captured sources are existing on site or not
3. Mark if sources are on open-ended pipes or not
4. Mark if all sources are accessible or not and if safe to access or not
5. Mark if any of these sources require special accessing arrangements
6. Verify line numbers in the field and log the sources' "Major" and "Minor" equipment they are connected to, and
7. Verify the route from previous point and route to next point

2.3. Q-Chem LDAR Surveys

2.3.1. Flagging Leakers

Leaking fugitive sources with VOC more than 10,000 ppm or HAP more than 500 ppm are physically tagged with LDAR tags for reporting and repairing and are logged in the database for tracking. Leaking HAP from fugitive sources with more than 10,000 ppm shall be fixed within one month. All leaking sources logged in the database for tracking are monitored every month up till the leak is repaired and then for a minimum of three monthly readings to confirm that; all of these readings are logged in the database.

2.3.2. Repairs of Flagged Leakers

Fugitive sources found to be leaking more than 10,000 ppm of VOC or more than 500 ppm of HAP during these Smart LDAR program surveys will be repeatedly surveyed every month up till that source is fixed and for a minimum of three monthly readings after the fix to confirm that. The flagging is done using the two flags shown in Figure 3 below.

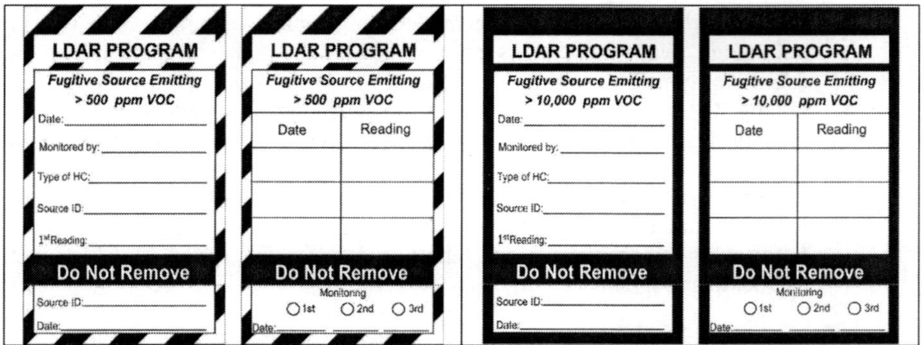

Smart LDAR Tags (Front)	Smart LDAR Tags (Back)

Figure 3: Q-Chem Smart LDAR Tags

2.4. Q-Chem LDAR Reports

This Smart LDAR program results in annual reports generated to the individual industrial cities for the accessible VOC sources that had HAPs leaks exceeding 500 ppm and VOCs leaks exceeding 10,000 ppm only. Hence, this Smart LDAR program is based on annual surveys at each plant site.

The database is built to log and track the leakers of Volatile Organic Carbons with leaks greater than 500 ppm. The database does not include the monitoring each and every valve, flange, pump, etc, but rather a track of the main sources found leaking, assigning them unique numbers, and following up on their repairs. This is because a big portion of the cost of setting up and maintaining an LDAR program is for tracking all the tags and database management.

References

American Petroleum Institution (US-API), Taback et al., 1997, API Publ 310, "Analysis of Refinery Screening Data", 1997.

Derek Reese and Charles Melvin from ExxonMobil Chemical Company, Baton Rouge, LA, and Wayne Sadik from ExxonMobil Chemical Company, Baytown, TX, Smart LDAR: Pipe Dream or Potential Reality?, January, 2008, Page 18.

Eric Anderson, ERM, a consulting partner with Leak Surveys Inc. (LSI), Meeting with Chevron Philips Chemicals Company, Meeting Agenda, April 2008.

Ministry of Environment, State Environmental Law 30 of Year 2002, Qatar, 2002.

Ministry of Environment, Q-Chem Consent to Operate, January, 2009.

Mesaieed Industrial City, Environmental Guidelines, Qatar, 1998.

Ras Laffan Industrial City, Environmental Guidelines, Qatar, 1998.

United States of America, Environment Protection Agency, Code of Federal Regulations, 40 CFR, Appedix A-7, Method 21 – Determination of volatile organic compound leaks, 2006.

United States of America, Texas Health and Safety Code, Chapter 382, (also known as the Texas Clean Air Act), May 2005.

Proceedings of the 2nd Annual Gas Processing Symposium
Farid Benyahia and Fadwa T. Eljack

Gas processing and Integrated Environmental Management

Per Gerhard Grini[a], Jostein Pettersen[a], Henrik Solgaard Andersen[b]

[a]*Statoil R&D Centre Trondheim, 7005 Trondheim. Norway*
[b]*Statoil R&D Centre Porsgrunn, 3908 Porsgrunn, Norway*

Abstract

From the start of petroleum production off the coast of Norway in the late 1960s the country has practised strict environmental regulations for the gas processing industry. In 1991, a CO_2 tax of approx. 60 USD/tonne was introduced. These regulations have led to the development and implementation of innovative technologies that minimize energy consumption and greenhouse gas emissions, as well as minimizing discharges to sea. At the same time, high levels of plant and process regularity have been achieved.
Examples of implemented technologies includes energy optimization on the Åsgard B Platform, gas compression with power from shore to the Troll A platform, application of "best available technologies" in the Hammerfest LNG plant and Snøhvit field development, and reduced environmental footprint and increased hydrocarbon recovery by the development of subsea gas compression. In addition, CO_2 capture from natural gas and geological storage of the CO_2 gas are applied on the Sleipner offshore facility, on the onshore facilities at Hammerfest (northern Norway), and in the In Salah gas processing plant (Algeria).
A strong base of expertise has been built up to support these projects, including fundamental developments in natural gas liquefaction process technology and cryogenic heat exchangers, investigations of phase behaviour of natural gas and relevant fluid mixtures, development of high pressure gas/liquid separation knowledge, and improved understanding of acid gas removal from high pressure natural gas
The combination of fundamental technology and competence development has strongly supported projects in the early phases as well as during operation, and as a result there is increased efficiency and regularity of the plants.

Keywords: Gas processing, Energy optimization, CO_2 capture and storage, LNG.

1. Introduction

Since the pioneering period of the late 1960s, the development of the Norwegian Continental Shelf has taken place under increasingly stricter environmental regulations. An example of this is the ban on the flaring of associated gas and the establishment of the principle that all offshore oil and gas exploration should take place respecting the fisheries and protecting nature and the environment. These two regulations were part of the so-called "10 commandments" for the oil industry that were passed by the Norwegian parliament in 1971.

Greenhouse gas emissions have had a strong focus in Norway since the 1980s and in 1991 a CO_2 tax of 350 NOK/tonne (approx 60 USD/tonne) was introduced for the petroleum industry. From 2008, this tax was changed to a combination of CO_2 tax and quotas, while maintaining the cost level of CO_2 emissions. The high taxation has further increased the focus on reducing CO_2 emissions.

Statoil operates the Sleipner gas processing platform which was the world's first CO_2 capture and underground storage facility in 1996. By 2009, 11 million tonnes of CO_2 had been permanently stored there. In a more recent example; during the period 2005 to 2010 Statoil will have completed energy efficiency modifications on existing platforms reducing the CO_2 emissions by 370 000 tonnes annually. In 2006, the average CO_2 emission per produced barrel of oil was 7.8 kg/bbl from the Statoil assets off Norway. Here, the world average is 19 kg/bbl.

There is a strong political push in Norway for the oil and gas processing industry to capture and store all CO_2. This is not only the CO_2 removed from the natural gas, but also CO_2 emitted from energy production and industrial sources. In order to get concessions for new onshore power plants based on fossil fuel, carbon capture and storage (CCS) must be included. As an example the CO_2 permits for the combined heat and power plant (CHP) under construction today at Mongstad requires that a CCS solution is in operation 3 years after the CHP start-up.

The methanol plant at Tjeldbergodden on the west coast of Norway came into production in 1997 and had a design that was focused on reduced energy consumption (Hansen and Olsvik 1999). Extensive heat integration, selection of energy efficient processes (e.g. synthesis gas production with two-step reforming), and continuous improvements, have developed a very energy efficient plant. The energy consumption is now approx. 2000 kWh/tonne produced (7.2 GJ/tonne produced); placing Tjeldbergodden among the most energy efficient methanol plants in the world.

Regarding discharges to sea there has been a shift during the recent years from the existing discharge limits on a single component to a holistic quantification of the total environmental impact of all components in the discharge. This includes dissolved and dispersed hydrocarbons, and chemicals added during the separation and processing. A new way of quantifying the environmental impact of such discharges was introduced on the Norwegian sector in 2000 and is called the Environmental Impact Factor. This method is a computerized tool that calculates the environmental impact from each of a number of chemical components present in the discharge (Johnsen et al. 2000).

For the last 5 years, Statoil has been among the industry leaders on the Dow Jones Sustainability Index.

This paper will give examples of how new innovative technologies and fundamental knowledge have been applied to minimize energy consumption and greenhouse gas emissions, as well as minimizing discharges to sea, while maintaining high levels of plant and process regularity.

2. Cases of energy optimization in gas processing

This section presents successful energy optimization cases in our gas processing plants. This includes the efficiency optimized Åsgard B platform, power from shore to the Troll A platform, and the development of the Hammerfest LNG plant. In addition, R&D activities on subsea compression will be briefly outlined.

2.1. Åsgard B Platform: 30% reduction in CO_2 emissions vs. conventional design

The Åsgard B semi-submersible gas processing platform located in the Norwegian Sea was put in operation in 2000. By choosing energy efficient processes and new technology, the energy requirement was reduced by 30% compared to conventional design. The energy optimization gave a reduced emission of 184 000 tonnes CO_2/year (not including the contribution from the flare system and amine plant described below). Key elements in the design included (IEA 1999):

- Condensate stabilization and heat integration: Heat is used instead of pressure for condensate stabilization. This saves re-compression power of 10 MW. In addition, extensive heat integration has been installed to reduce stabilization heating and product cooling requirements. The integrated preheating of condensate prior to stabilization provides 37 MW of heat which would normally be provided by fired heaters.
- Gas export inter-stage cooling in the compressor train: Such cooling is usually only installed on export compressors with higher compression ratios than existed on Åsgard B. By the installation of an additional compact inter-stage cooler and scrubber, the export compression power requirement was reduced by approx. 10 MW.
- Warmed coolant used to heat colder stream: The cooling water return is used to provide pre-heating for the gas production, saving 32 MW of heat which would normally be provided by fired heaters.
- Optimized seawater lift pumps: Optimized location of the seawater lift pumps provided savings of 70 m pump head requirement. This reduced the pump power requirement by 2.2 MW.
- Flare gas reduction: The plant uses a closed flare system. A High Integrity Pressure Protection System (HIPPS) has been included to further reduce flaring and other gaseous emissions. The HIPPS shuts down production to avoid flaring during process upsets, as would be the case when using a traditional pressure relief system.
- Use of selective amine: A selective amine is used to remove H_2S from the natural gas. This significantly reduced the CO_2 emissions from the regeneration process compared to a non-selective amine.

It should be mentioned that the reference case in these calculations is an efficient plant with high efficiency aero-derivative gas turbines with waste heat recovery from the exhaust stream.

2.1.1. Troll A platform; Gas Compression with power from shore

The Troll A platform is located 65 km offshore the western coast of Norway and receives all electric power from shore, thereby eliminating the need for offshore gas turbines. Power is supplied from the onshore electric grid. The gas is transported together with the condensate and condensed water in a multiphase flowline to the onshore gas processing plant at Kollsnes. From the start up in 1996 the reservoir pressure has been gradually reduced through depletion, and in 2005 two compressor units were installed at Troll A to increase the capacity and maintain the receiving pressure at Kollsnes. The new compressors get their power from shore through a new high voltage (60 kV), DC transfer system supplying 80 MW to the compressor motors. This reduced the CO_2 emissions from the Troll A platform by 230 000 tonnes/year compared to an alternative gas turbine installation. In total, electrification of the Troll A platform and the receiving facility Kollsnes has reduced CO_2 emissions by approx 1.5 million tonnes/year (Norwegian Government report 2006).

Power from shore will also be applied for the semi-submersible Gjøa platform that will start production in 2010. The platform is located at 380 m water depth and will be the first floating platform receiving power from shore. It should be emphasised that for power from shore to be environmentally beneficial, its generation has to be either from renewable sources (most of Norway's electricity is from hydro power), or from fossil fuels with carbon capture and storage.

2.2. Hammerfest LNG plant and Snøhvit field development example

The Statoil-operated Hammerfest LNG plant in the Barents Sea in northern Norway receives unprocessed feed from wells 143 km offshore, mainly comprising the Snøhvit gas condensate field. The onshore plant, which started production in 2007, has a capacity of 4.3 million tonnes LNG (equal to 5.7 billion Sm^3) annually in addition to condensate and LPG.

Onshore facilities were designed based on a philosophy of "Best Available Techniques" (BAT). Systems involving emissions were evaluated and selected based on environmental performance as well as economic robustness in life cycle cost comparisons involving potential quota purchases.

The most significant BAT contributions in the design of the Snøhvit LNG chain are (Lillesund et al. 2003), (Heiersted et al. 2004):

- Snøhvit is the first land-based industry development that uses technology for CO_2 sequestration involving offshore reservoirs. The design CO_2 re-injection rate is 700 000 tonnes a year. Details are explained elsewhere in the paper.
- The Snøhvit reservoir area comprises a subsea field development and the gas wells are connected to shore by a multiphase pipeline. No harmful discharges or emissions occur from the offshore system. All wellstreams are routed to the land-based facility for processing.
- The power and heat system is designed to provide a reliable, fully self-sufficient supply to the offshore transportation system and the onshore processing and liquefaction plant. The national grid is providing a certain backup of electricity to the LNG plant, but is not robust enough to supply all the power needed. Aero-derivate gas turbines were selected for their high efficiency (41%), and the balance with heat demand in the plant, resulting in an excellent overall thermal efficiency for the LNG plant.
- The efficient refrigeration process (discussed elsewhere in the paper) and the cold Arctic climate help to boost the energy efficiency of the liquefaction plant. This in turn cuts emissions substantially in comparison with similar facilities.

- Electric motors, provided with variable speed features, driving the refrigeration compressors, are part of BAT and contribute to optimal capacity balancing. The availability of the all-electric concept is approximately ten onstream days more a year than for a mechanical drive, industrial heavy-duty based concept.
- With the selection of the gas treatment processes, using MDEA® for the CO_2 removal, the overall heat demand is optimized for minimum emissions to air and sea. The gas turbine selection primarily addresses high mechanical (electrical) efficiency, as this is the constraint relative to fuel gas consumption. At the same time the waste heat from the flue gas meets all process heat demands through a hot oil system.
- Dry low NOx burners for the gas turbines were applied to guarantee an overall emission level of < 25 ppm while maintaining high combustion efficiency.
- Fuel gas to the gas turbines is taken from the gas feed to the LNG facility instead of taking processed gas or LNG as fuel. This gives an increase in plant capacity due to reduced gas processing and cooling duty demand.
- As an industry first, the Hammerfest LNG plant processes vapour return from loading of the LNG and LPG carriers at the facility.

The Hammerfest LNG plant has been through its run-in and early operational phase since autumn 2007, including shut-downs to repair leaks in seawater-cooled heat exchangers and to conduct minor modifications. Efforts are now underway to prepare for necessary modifications to ensure full capacity utilization of the plant in 2009. In this run-in period, the know-how built up in Statoil on liquefaction process and cryogenic heat exchangers has proven to be of great importance for troubleshooting and optimized operation. One example is the reduction in warm start-up flaring which has been significantly reduced due to improved operational procedures, see Figure 1.

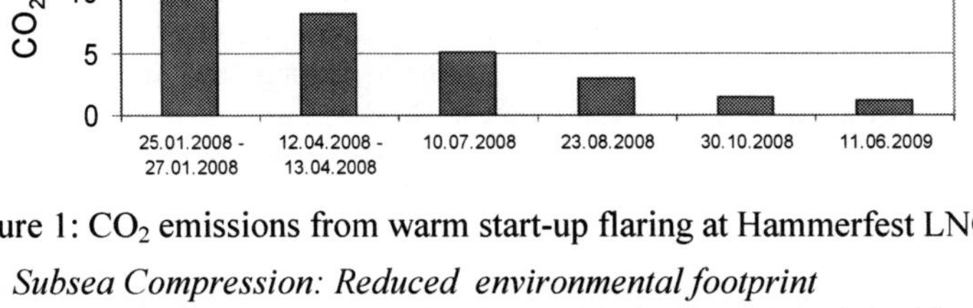

Figure 1: CO_2 emissions from warm start-up flaring at Hammerfest LNG (Vist 2009)

2.3. Subsea Compression: Reduced environmental footprint

Subsea compression is a new technology with the potential of significantly reducing the environmental footprint of offshore gas production, as it eliminates the need to build a platform to accommodate the compressors on, and since power from shore can be applied.

This option is highly relevant on the Norwegian Continental Shelf (NCS) where subsea fields with wet wellheads are routing the wellstream directly to shore. The Ormen Lange subsea field with 120 km multiphase transport to shore is the most well-known field considering this option. But other fields on the NCS such as Snøhvit, Åsgard and Gullfaks are also evaluating this technology, and have taken active steps to qualify subsea compressors from several suppliers (Brenne et al. 2008).

Two different concepts are developed: subsea dry gas compression and subsea wet gas compression. The dry gas compression is basically a marinization of the compressors currently used in gas processing plants, where the gas and liquid is separated and the liquid is routed to a pump, and the gas to the compressor. The wet gas compressors are designed to accept a relatively large fraction of liquids, typically up to 5 vol% of the feed, at the cost of somewhat lower polytropic efficiency. The range of subsea compressors under consideration covers the gas volume rates from 5 to 15 million Sm^3/d, pressure ratios up to 3, and power consumption from 4 to 12 MW.

3. CO_2 capture, transport and storage

When discussing CO_2 capture, transport and storage, it is important differentiate between two areas of CO_2 management:

- Handling of CO_2 from CO_2-rich gas fields, i.e. so-called acid or sour gas
- Handling of CO_2 from energy production

Many natural gas reservoirs contain a high amount of CO_2 and in order to meet a market specification or downstream process requirements the CO_2 has to be removed to a specific level. In normal field development the cost of removing CO_2 is an integrated part of the project development. Statoil is currently operating several gas fields where CO_2 is removed from the natural gas. In these processes the CO_2 is produced in a clean form and only needs to be compressed and possibly dried before being sent to transport and storage. On a worldwide basis, CO_2 processing, transport and geological storage solely from an environmental perspective (i.e. not using the CO_2 for enhanced oil recovery) have been adapted in four cases: Sleipner, In Salah, Snøhvit, and Weyburn-Midale. Statoil is involved in the first 3 and this paper summarise the experience from these fields.

The transport and storage part of a carbon capture and storage chain for managing CO_2 emissions related to energy production is basically identical to handling CO_2 from a gas field. The CO_2 capture part contains the greatest difference and the technology readiness, cost and complexity are significantly more challenging.

Figure 2 outlines current and future Statoil carbon capture and storage projects.

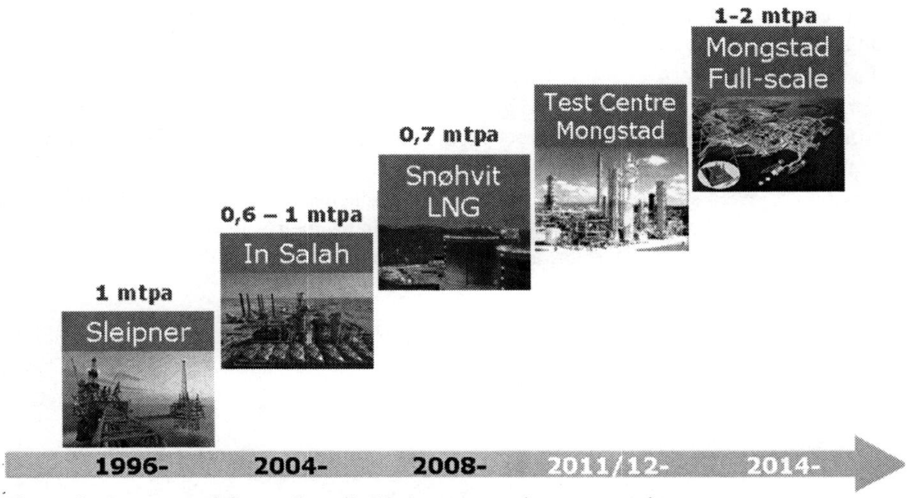

Figure 2: Current and future Statoil CO_2 capture and storage projects

3.1. Sleipner CCS project

The Statoil operated Sleipner field is a large gas and gas-condensate production unit in the Norwegian sector of the North Sea. The daily gas export capacity is 20 million cubic metres and 20 000 barrels of gas condensate. The 9% CO_2 content in the natural gas had to be reduced to meet the sales gas specification of maximum 2.5%. The technical experts came up with the unprecedented idea of capturing the CO_2 offshore and injecting it into a saline formation beneath the Sleipner installations. In this way, the Sleipner asset would minimize CO_2 emissions – the prime motive – while avoiding environmental taxes.

An amine-based absorption process was selected for CO_2 capture, because it was deemed more compact than competing systems. One of the greatest challenges, however, was to scale down the process plant sufficiently so that it could be accommodated on an offshore platform. Even so, the 'miniaturized' version of the CO_2 capture module weighed 8 200 tonnes.

The Sleipner field came onstream in October 1996, and included the world's first large-scale offshore CO_2 extraction plant; and world's first industrial facility for CO_2 injection and storage. By 2009, 11 million tonnes of CO_2 had been stored. A comprehensive programme has been established to evaluate if the CO_2 is safely stored. This has included 7 seismic surveys. No indication of CO_2 leakage into the overburden has been detected, and the surveys have shown how the CO_2 has spread inside the Utsira aquifer (Hansen et al. 2005).

3.2. Snøhvit CCS project

Statoil and partners started the world's second offshore CO_2 storage at the Snøhvit field off northern Norway in 2007. The production area extends across the Snøhvit field itself and the Albatross and Askeladd satellites. The natural gas in these fields contains 5 to 8% CO_2, and this has to be reduced to less than 50 parts per million to avoid CO_2

freezeout in the LNG process. Thus 700 000 tonnes of CO_2 will be captured each year in an amine absorption plant.

To avoid emissions, the captured CO_2 is transported back offshore and stored. The 153 km, 8 inch diameter offshore pipeline transporting CO_2 is the first of its kind in the world. New modelling tools and procedures have had to be developed for safe operation of the CO_2 pipeline and subsea injection well.

Having reviewed several disposal options, it was decided that the CO_2 will be injected into the Tubåen Formation – a sandstone formation 60 metres beneath the Snøhvit field 2 600 metres below the seafloor, and below the main natural gas reservoir, as shown in Figure 3.

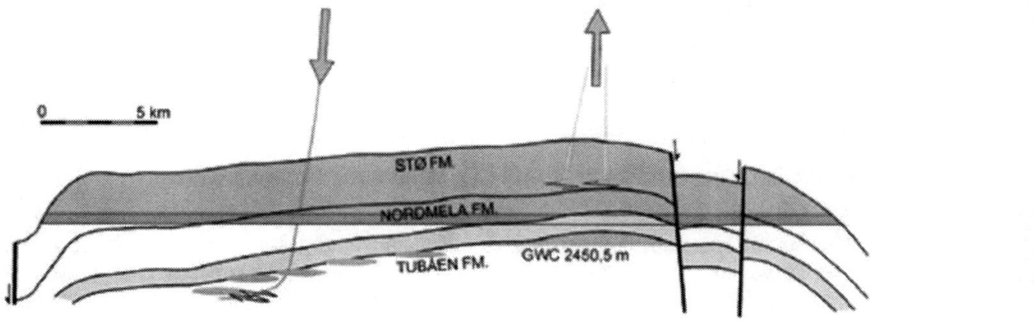

Figure 3: Geological cross-section showing the path of the injection well.

The Tubåen Formation has a thickness varying from 47 to 75 m, and good reservoir properties. The formation is sealed by shale cap rocks, and has capacity to cover the storage need of 23 million tonnes of CO_2 during the 30-year lifetime of the Snøhvit project.

Injection started in April 2008. It is being monitored by wellhead and downhole pressure and temperature gauges. Seismic baseline data have been gathered, and seismic and gravimetric surveys will be carried out later to monitor the behaviour of the CO_2 in the storage reservoir.

3.3. In Salah CCS Project

The CO_2 injection is part of the jointly operated Sonatrach-BP-Statoil In Salah gas field in Algeria. The development covers eight gas discoveries in the central Saharan region of the country, and delivers 9 billion cubic metres of natural gas per annum.

The gas fields contain CO_2 with concentration ranging from 1 to 9%, whereas export gas specifications require a CO_2 concentration $< 0.3\%$. To achieve this target and avoid emissions, the capture and storage of 1.2 million tonnes of CO_2 annually is necessary. The CO_2 is stripped off from the gas stream using an amine process and injected into 3 wells in a saline formation surrounding one of the gas fields – Krechba. According to the field development plan, the injected CO_2 should slowly migrate upwards and reach the production reservoir after it has been depleted.

Several R&D activities are initiated and novel well data has been acquired to further investigate and verify the nature of the cap rock seal, surface and downhole gases have been measured and a range of geophysical monitoring methods are being tested. The In Salah project complements the pioneering projects at Sleipner and Snøhvit and thereby stimulates the worldwide adoption CO_2 storage.

3.4. European CO_2 Technology Centre Mongstad

Statoil is a partner in the "European CO_2 Technology Centre Mongstad" (TCM). TCM is based on an agreement between the Norwegian government and Statoil and it is planned to build facilities to test two post-combustion CO_2 capture technologies. TCM is a first step of implementing CO_2 capture at the Mongstad refinery. The other TCM partners are Gassnova (representing the Norwegian State) and Shell. The test facilities has a design capacity to capture 100 000 tonnes of CO_2 annually. Other ambitions of the TCM project are:

- Development of technologies for CO_2 capture that are capable of wide national and international deployment
- Reducing the cost and technical and financial risks related to large-scale CO_2 capture
- Testing, verification and demonstration of CO_2 capture technology owned and marketed by vendors
- Encourage the development of such technology
- Reduce the environmental risk related to large-scale CO_2 capture

Two different CO_2 absorption technologies (amine and chilled ammonia) are to be tested in two parallel plants. The plants will be able to capture CO_2 from two different flue gas sources with 3.5 and 13 mole% CO_2, respectively, as shown in Figure 4. These CO_2 sources provide a unique flexibility to carry out tests on flue gases with relevance to different fuels.

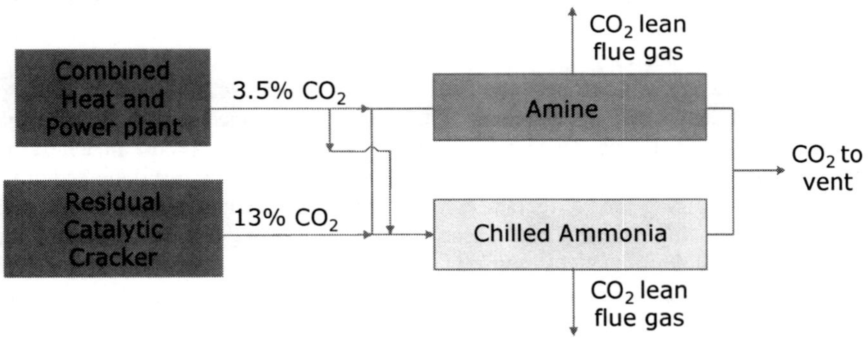

Figure 4: Opportunities for TCM with two different flue gas sources and two different technologies.

The amine technology is considered to have a moderate risk, since this technology has been used in similar gas processing applications for decades. However, there is still

potential for improvements and the need to qualify certain components of the process before implementation in a full-scale plant.
The chilled ammonia technology (Kozak 2009) is still under development and therefore represents a higher risk. The aim at Mongstad is to verify it for large-scale use. An important part of the test is to verify the HSE performance of the different CO_2 capture technologies in particular the local emissions and discharges.

4. Examples of the need for fundamental expertise and know-how

Statoil has developed a strong base of expertise in order to facilitate and support developments which focus on environmental management through reduced energy needs and increased regularity. The base of expertise has provided fundamental developments in several areas, and the following are given as examples:

- Natural gas liquefaction process technology and cryogenic heat exchangers
- Phase behaviour of natural gas and relevant fluid mixtures, where new equations of state have been evaluated vs. own experimental data
- High pressure gas/liquid separation where improved understanding and new theoretical models have been developed based on results from small- to full-scale experiments
- Acid gas removal from high pressure natural gas

4.1. Natural gas liquefaction process and cryogenic heat exchangers

In 1984 Statoil established long-term R&D cooperation with the research and university environment at SINTEF and the Norwegian University of Science and Technology (NTNU) in Trondheim, Norway, focusing on advanced liquefaction process modelling and optimization as well as laboratory experiments and modelling of cryogenic heat exchangers. Efforts were focused on the main spiral-wound heat exchanger (SWHE). Detailed results from experimental measurements with improved fluid property data and models led to the development of an advanced simulation tool for spiral-wound heat exchangers and liquefaction processes.
An LNG technology alliance between Statoil and Linde was established in 1996 to pursue joint research and development. The alliance developed a new liquefaction process using three separate mixed refrigerant cycles. In this way a close match to the idealized cooling curve was obtained, giving flexibility for optimization, and high efficiency. By adapting the refrigerant compositions, pressure/temperature levels, and load distribution, this Mixed Fluid Cascade (MFC®) process can be optimized for varying site conditions such as ambient temperature, gas composition and driver configuration. Especially in a cold climate, the use of a refrigerant mixture gives more flexibility for loading up the precooling stage than with pure propane as the refrigerant.
Another important task for the technology alliance was to conduct exhaustive tests on a prototype spiral-wound heat exchanger to qualify Linde as a supplier of SWHE for base-load natural gas liquefaction, and provide the confidence needed to manufacture commercial versions. In May 1997, the alliance installed an industrial-scale prototype of a SWHE in the LNG plant at PetroSA's GTL refinery in Mossel Bay, South Africa, and a successful qualification programme was completed over the next years.

The Statoil-Linde liquefaction technology and heat exchanger technology was selected for the Hammerfest LNG plant in 2002. The history of Statoil LNG research and technology development shows the value of long-term systematic work together with academic, research and industry partners, starting with fundamental studies, experimental investigations and model developments, and finalizing with the industrialization, qualification and implementation/follow up and verification of technology (Pettersen et al. 2008).

As a result of this long-term technology focus Statoil has developed in-depth competence on LNG technology which has proven valuable in the planning, start-up and troubleshooting at the Hammerfest LNG plant.

4.2. Phase behaviour of natural gas

Much of the gas produced on the Norwegian continental shelf is transported as partially processed rich gas at dense phase conditions in pipelines up to 42" and 700 km length. This transport option is profitable as the capacity is high, but it is also a very positive option in terms of environmental management as moving a larger fraction of the gas processing onshore facilitates the use of larger and more energy efficient processes, as well as utilizing power that is generated from renewable sources. The offshore environmental footprint is also reduced.

It has been observed that malfunctioning scrubbers and incorrect thermodynamic models in process design can lead to off-spec conditions and/or capacity limitations in the gas/dense phase transport systems. In order to avoid liquid hydrocarbon in the pipeline and at the receiving facilities, it is critical to operate above the cricondenbar.

In order to get a better estimation of the phase envelope of hydrocarbon mixtures, and thereby improve the prediction of the conditions in the transport systems, Statoil initiated an R&D project to evaluate the equations of state (EoS) used to simulate the phase envelope of such mixtures. The Soave-Redlich-Kwong, Peng-Robinson, GERG-2004 (Kunz et al. 2003) and PC-SAFT (Von Solms et al. 2003) EoS were compared (Rusten et al. 2008).

Hydrocarbon dew point measurements were performed in our laboratories for both synthetic hydrocarbon gas mixtures and real natural gases. The experimental data were compared to simulated dew points based on compositions given by chromatographic gas analysis. Significant deviations between measured and predicted dew points were observed for the synthetic gas mixtures and the real natural gases. For real natural gases, characterization of the C7+ fraction was shown to be important.

In general, none of the EoS evaluated determined the whole phase envelope correctly. Even when a detailed composition of the C7+ fraction was available, a maximum deviation at 8°C for the cricondentherm and 6 bar for the cricondenbar was observed between the simulated and experimental dew points. The EoS to be recommended will depend on the pressure and temperature of interest as the models are currently not satisfactorily accurate to predict the whole phase envelope for real natural gases.

4.3. High pressure gas liquid separation

The performance of gas/liquid separators is crucial in order to avoid problems with liquid sensitive equipment like compressors and adsorbers/absorbers and fulfil the hydrocarbon and water dew point specifications of treated gas. In the past decade scrubbers have been designed to be more compact and the focus today is very much on compact equipment for subsea separation. With more compact equipment the complexity increases as there are malfunctioning scrubbers due to the lack of fundamental knowledge about fluid behaviour at real conditions.

Statoil has built several laboratory test rigs with different scales and capabilities to establish more reliable and robust scrubber design as well as keeping the scrubber size acceptable (Gjertsen et al. 2007). The goal was to acquire knowledge about the performance of separator internals under real operating conditions. These rigs facilitate testing with live hydrocarbon fluids instead of the model fluids (often air/water/kerosene) regularly used for testing at the vendors.

The test rigs have capabilities to operate with both model systems and real hydrocarbon systems in the pressure range from 1.5 bar to 150 bar. The gas and liquid rate can be individually varied, and fluid composition has varied from less than 0.1 vol% to above 3 vol% liquid. The test units have the possibility to give both the overall scrubber performance as well as the performance of the individual units within a scrubber; i.e. inlet section and demisting section.

The testing has shown that the hydrocarbon fluid properties at elevated pressure significantly reduce the separation efficiency compared to low pressure air/water/-kerosene experiments. The test facilities have been actively used in the qualification of new technologies and equipment, as well as in optimizing our existing facilities.

4.4. Acid gas removal from high pressure gas

During operation of the alkanolamine (MDEA) CO_2 removal plant at the Statoil operated Sleipner gas platform it was observed that the plant did not operate as designed. It removed less CO_2 than expected at a given circulation rate (Hansen et al. 2005). Research was therefore initiated to identify why and find ways to better understand and predict the CO_2 capture capacity of alkanolamines at high pressure (De Koeijer, Solbraa 2004). The CO_2 content and hydrocarbon solubility in the amine solution were measured under different operational conditions (T, p) in a high-pressure equilibrium cell. The measurements were performed with natural gas and amine sampled at the field, and recombined at pressures and temperatures relevant for the conditions at Sleipner.

It was experimentally observed that varying pressure had a significant and non-linear effect on the absorption capacity of the amine. The absorption capacity of the Sleipner amine decreased with increasing pressure at constant CO_2 partial pressure. This is a non-desirable property that was often not anticipated in design, and can lead to a lack of driving force for CO_2 capture in the absorber column. The experiments also showed that the addition of an activator to the Sleipner amine would have a limited, or even negative, effect on the cyclic CO_2 removal capacity of the plant. The positive effects of enhanced capacity and kinetics obtained by adding an activator can be negligible compared to a decreased desorption efficiency in processes utilizing flash regeneration. However, for deep CO_2 removal with full regeneration of the amine, the addition of an activator is crucial.

The experimental data were used in combination with advanced thermodynamic models to establish a process model of the plant. Finally, the model was used to optimize the

CO_2 capture plant on the Sleipner platform, leading to more stable operation and higher plant availability.

5. Conclusions

The paper presents real life examples of:

- Energy optimization on the Åsgard B Platform resulting in 30% reduction in CO_2 emission vs. conventional design
- Gas compression with power from shore on the Troll A platform
- Application of "best available technologies" in the Hammerfest LNG plant and Snøhvit field developments
- Reduced environmental footprint and increased hydrocarbon recovery by development of subsea gas compression

CO_2 capture from natural gas and geological storage of the CO_2 gas is applied on the Sleipner offshore facility, on the onshore facilities at Hammerfest (northern Norway), and in the In Salah gas processing plant (Algeria). The operation of these plants has been a success. One example is the quantity of CO_2 from the Sleipner facility stored in the Utsira aquifer which is presently more than 11 million tonnes.

A strong base of expertise has been developed in order to enable and support the above developments with focus on environmental management through reduced energy needs and increased regularity. The base of expertise has provided fundamental developments in:

- Natural gas liquefaction processes and cryogenic heat exchangers
- Phase behaviour of natural gas and relevant fluid mixtures, where new equations of state have been evaluated vs. own experimental data
- High pressure gas/liquid separation where improved understanding and new theoretical models have been developed based on the results from small to full scale experiments
- Acid gas removal from high pressure natural gas

The fundamental technology and competence development has significantly supported projects in their early phases as well as during the operative phases. This has resulted in increased efficiency and availability of the plants.

References

Brenne L., Bjørge T., Bakken L.E., Hundseid Ø., 2008, Prospects for subsea wet gas compression. Proceedigs of ASME Turbo Expo 2008, Berlin Germany.

De Koeijer G., Solbraa E., 2004, High pressure Gas Sweetening with Amines for Reducing CO_2 Emissions. Preceedings (Elsevier) from IEA GHGT-7 Vancouver 2004.

Gjertsen L.H., Rusten B.H., Austrheim T. Knowledge of Real Fluid Behaviour- Key to Solve High Pressure Separation Challenges. 2007, Procedings from 86th Annual GPA Convention, San Antonio, USA.

Heiersted, R. S., Lillesund, S., Nordhasli, S., Owren, G. A. and Tangvik, K. 2004. The Snøhvit design reflects a sustainable environmental strategy. 14th LNG Conference, Qatar

Hansen H., Eiken O., Aasum T.O., 2005, Tracing the path of carbon dioxide from a gas – condensate reservoir, through an amine plant and back into a subsea acquifer. Case study: The Sleipner area, Norwegian North Sea, SPE 96742, paper presented at Offshore Europe 2005.

Hansen R., Olsvik O.,1999, Experiences and Opportunities for the Statoil Mid Norway Methanol Plant. Proceedings of AICH 1999 Spring National Meeting.

IEA 1999, Retrofit of offshore oil and gas installations for grenhous gas abatement, IEA Greenhouse gas R&D programme, International Energy Agency, 1999.

Johnsen S., Frost T.K., Hjelsvold M., Utvik T.R., 2000, The Environmental Impact Factor, a proposed tool for produced water impact reduction, management and regulation. SPE 61178, Paper presented at the SPE HSE int. conference 26-28 June 2000, Stavanger, Norway.

Kozak F., Petig A, Morris E., Rhudy R., Thimsen D., Chilled ammonia process for CO2 capture, Energy Procedia, Volume 1, Issue 1, , Pages 1419-1426, 2009.

Kunz, O., Klimeck, R., Wagner, W., Jaeschke, M., "The GERG-2004 Wide Range Equation of State for Natural Gases and Other Mixtures", GERG TM15, 2007.

Lillesund, S., and Heiersted, R. S. Concept selection for Norway's first LNG Plant. Research RIPI 12th Oil, Gas & Petrochemical Congress, February 24-26, 2003, Tehran, Iran.

Norwegian Government report (white paper) to the Parliament 2006, Norwegian Climate Policy, Report no. 34, 2006-2007.

Pettersen, J, Bredesen, A. M. and Nekså, P. Research on natural gas liquefaction technology for the Hammerfest LNG plant. Cryogenics and Refrigeration - Proceedings of Fourth International Conference on Cryogenics and Refrigeration (ICCR'2008), Shanghai April 2008.

Rusten B.H., Gjertsen L.H., Solbraa E., Kirkerød T., Haugum T., Puntervold S., 2008, Determination of the Phase envelope- Crucial For Process Design and Problem Solving, Preceedings from GPA annual meeting, Texas.

Vist, S., Lessons Learned from Start Up of the Snøhvit LNG Plant. Energy Frontiers International

Gas-to-Market & Energy Conversion Forum, Washington, D.C., 21 - 23 September 2009

Von Solms, N., Michelsen, M.L., Kontogeorgis, G.M., Computational and Physical Performance of a Modified PC-SAFT Equation of State for Highly Asymmetric and Associating Mixtures, Ind.Eng.Chem.Res., Vol. 42, p. 1098, 2003.

Proceedings of the 2nd Annual Gas Processing Symposium
Farid Benyahia and Fadwa T. Eljack (Editors)

Dehydration of Acid Gas Prior to Injection

Eugene Grynia [a], John J. Carroll [a], Peter J. Griffin [a]

[a] Gas Liquids Engineering, #300, 2749 – 39th Avenue N.E., Calgary, Alberta T2W 3Y2

Abstract

Acid gas is a mixture of hydrogen sulfide and carbon dioxide, with small amounts of other components, and usually saturated with water, which is the by-product of the sweetening of sour gas. Acid gas injection is an environmentally friendly and economically viable option for dealing with unwanted acid gas and in particular small quantities of acid gas. Basically acid gas injection involves the compression of the gas to a sufficient pressure, transport of the compressed gas and injection into a disposal well.

In many injection schemes the gas does not require dehydration beyond that achievable by compression and cooling alone. In some circumstances additional dehydration is required. There are a number of methods commonly used to dehydrate the acid gas. The most popular is absorption of water with triethylene glycol (TEG). This paper discusses the design considerations for application of both methods of dehydrating acid gas: by compression and cooling alone, and by TEG.

Keywords: acid gas, compression, dehydration, injection

1. Introduction

Acid gas is a mixture of hydrogen sulfide and carbon dioxide with a small amount of lower molecular weight hydrocarbons, mainly methane. Acid gas is most often produced in sour natural gas sweetening units, during regeneration of the sweetening chemical, usually an amine. Acid gas liberated from rich amine leaves the amine regeneration tower at a low pressure, typically below 200 kPa, and at about 50°C, the typical temperature downstream of the regeneration tower overhead condenser. In addition, the acid gas is saturated with water. This is not produced water, which contains dissolved salts, which can cause additional problems but rather this is condensed water which should contain virtually no dissolved solids.

If acid gas is to be injected into a subsurface formation such as an aquifer or a depleted reservoir, it must be compressed and sent to an injection well through a pipeline. Water present in acid gas can contribute to corrosion of the compression equipment and the pipeline, and it may facilitate hydrates formation. All main constituents of acid gas: H_2S, CO_2 and methane, are hydrate formers. Hydrates can form in acid gas without the presence of free water; however free water is required in the acid gas to cause corrosion of carbon steel. It is recommended to keep the relative water saturation of the acid gas mixture considerably below 100% in order to avoid condensation of water in the piping, facilities, pipeline or well.

In some cases acid gas dehydration by compression and cooling alone is sufficient to avoid corrosion and hydrate issues. If this is insufficient, an additional dehydration process is required. The most popular method of dehydrating acid gas prior to injection, other than compression and cooling is the use of triethylene glycol (TEG).

2. Acid Gas Dehydration

As mentioned in the Introduction, the presence of water in the acid gas poses several problems, including corrosion and hydrate formation. There are a number of ways water can be removed to various degrees from acid gas, including:

- Compression and cooling alone
- TEG dehydration
- Refrigeration
- Mole sieves
- Thermodynamic phase separation, Dexpro process, patent pending (McKay, 2009)

Dehydration of acid gas by compression and cooling is always used since the acid gas needs to be compressed and cooled anyway. If additional dehydration is required then water absorption with TEG is the most popular method.

2.1 Compression and Cooling Alone

Acid gas is usually compressed from a suction pressure of less than 200 kPa to a discharge pressure dictated by the reservoir pressure, injection well depth and the length of the acid gas injection pipeline. The number of compression stages depends on the required injection pressure, but usually does not exceed five. If more stages are required, acid gas is typically condensed after the last stage of compression and a pump is used to increase the acid gas pressure to the required value.
Figure 1 shows a four-stage compression curve superimposed on saturated water curves at 48.9°C (120°F). The compression ratio for each stage is approximately 3, including inter-stage cooler pressure drops.

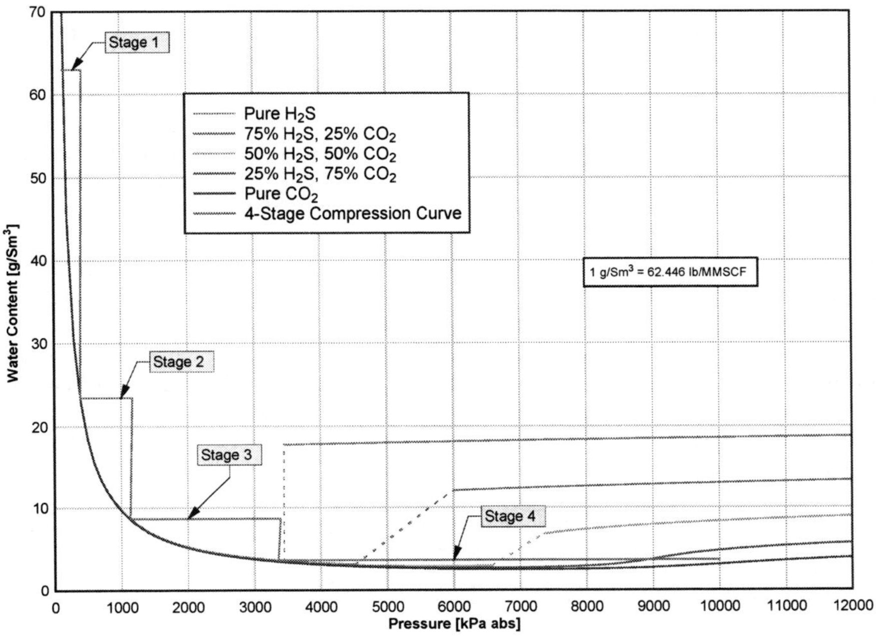

Figure 1: Dehydration by compression and cooling to 48.9°C (120°F)

In this particular example, acid gas was dehydrated from 63 g/Sm^3 to approximately 3.73 g/Sm^3. At 10000 kPa saturated water content and relative saturation after compression of acid gas mixtures shown in Figure 1 are as follows:

Acid Gas	Saturated Water Content at 10000 kPa and 48.9°C g/Sm^3	Relative Water Saturation %
Pure H_2S	18.62	20.0
75% H_2S, 25% CO_2	13.05	28.6
50% H_2S, 50% CO_2	8.34	44.7
25% H_2S, 75% CO_2	4.88	76.4
Pure CO_2	3.20	116.6 *

* Oversaturated, free water present.

Pure carbon dioxide cannot be dehydrated to relative water saturation below 100% if the 3rd stage discharge pressure is around 3400 kPa, as in the above example. However, if the next to last compression stage discharge pressure is increased to 5000 kPa, at which pressure the saturated water content is 2.80 g/Sm3, the relative water saturation of this particular acid gas decreases to 87.5%.

If this acid gas contains 1% methane, the relative water saturation at 10000 kPa increases to 90.6%. To increase the pressure from the suction pressure of 142.7 kPa (6 psia) to 5000 kPa in three stages, the compression ratio would have to be around 3.4, including pressure drops across the interstage coolers.

When setting the interstage pressures to optimize acid gas dehydration by compression and cooling alone, one needs to check whether the interstage discharge temperatures do not exceed the limits recommended by the compressor vendor. If the limit is exceeded, the compression ratio needs to be lowered.

Based on the above analysis, a rule of thumb can be proposed which says that if the H2S content of acid gas is 25 mole% or more, dehydration by compression and cooling alone is sufficient to prevent corrosion. Otherwise additional dehydration may be required.

The above rule of thumb can be used for short injection pipelines and excludes situations when acid gas injection is suspended and acid gas sits in the pipeline for a long period of time, reaching the ground temperature. If the temperature of the acid gas falls to 2°C, assumed to be the ground temperature, the relative water saturation increases as follows:

Acid Gas	Saturated Water Content at 10000 kPa and 2°C g/Sm^3	Relative Water Saturation %
Pure H_2S	6.63	56.3
75% H_2S, 25% CO_2	4.80	77.7
50% H_2S, 50% CO_2	3.27	114.1*
25% H_2S, 75% CO_2	2.16	172.7*
Pure CO_2	1.42	262.7*

* Oversaturated, free water present.

2.2 Acid Gas Dehydration with TEG

Triethylene glycol (TEG) is commonly used in dehydration of natural gas due to its relatively low vapor pressure. However, above 48.9°C (120°F) in the contactor, the TEG losses to the dry gas are high enough to justify cooling the inlet gas to or below 48.9°C. Another possibility would be to use tetraethylene glycol (TREG). It is more expensive than TEG but its losses to the dry gas are lower at high contactor temperatures due to its low vapor pressure. On the other hand pumping costs are higher because TREG is more viscous than TEG (Carroll, 2009).

The lower the temperature of the inlet gas at constant pressure the smaller the amount of water in the gas and smaller the required diameter of the contactor, since the density of the gas and consequently its actual velocity in the contactor is lower. However, the limiting factors in the selection of the gas temperature are the availability of the cooling media and hydrate formation, with the most important: the avoidance of acid gas condensation in the compressor interstage coolers.

The higher the pressure of the inlet gas at constant temperature the smaller the amount of water in the gas and smaller the required contactor diameter since the actual velocity of the gas in the contactor is lower. However, the acid gas feed pressure is determined earlier, in the design of the compressor inter-stage pressures.

Figure 2 shows a typical gas dehydration equipment configuration.

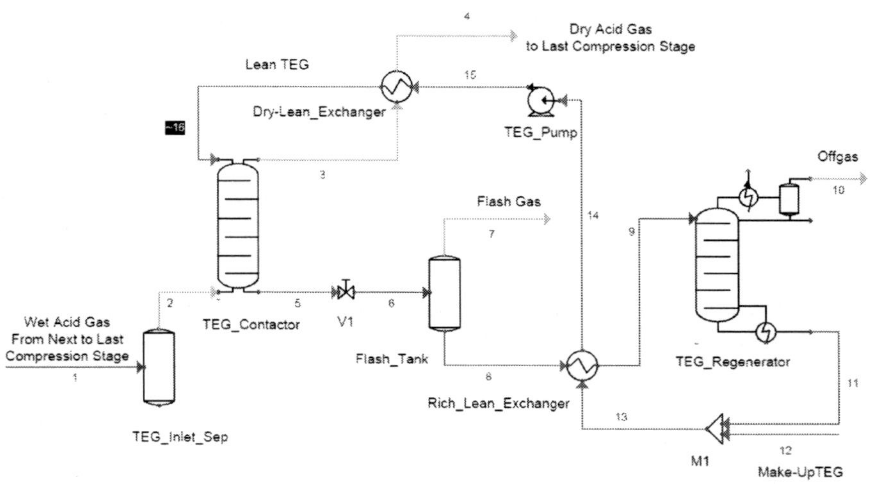

Figure 2: Acid gas dehydration with TEG

The acid gas from the stage of compression prior to last flows through the Inlet Separator and enters the TEG Contactor at the bottom. It then flows upward through a number of trays or a layer of random or structured packing. The gas comes in contact with TEG flowing from the top of the Contactor. The dried gas leaves the Contactor at the top, cools down the lean TEG in the Dry/Lean Exchanger, and flows to the suction scrubber of the last stage of compression. The TEG, rich in water, leaves the Contactor and flows into the Flash Tank, where the pressure is lowered, usually to approximately 345 kPag (50 psig) or lower, to flash off some of the gas that was absorbed at high pressure in the Contactor. The TEG is then directed to the Rich/Lean Exchanger (glycol-glycol preheater) before being fed to the Still Column on top of the Reboiler, operating at the pressure close to atmospheric. The Still Column and the Reboiler are shown on Figure 8 as the TEG Regenerator. Sometimes, the rich glycol is preheated in the Rich/Lean Exchanger first, and then it is sent to the Flash Tank.

The number of trays or the height of the packing depends on the required degree of dehydration. A typical number is 6 to 12 trays (real trays, not ideal stages). Since the acid gas does not have to be dehydrated to the same level as sales gas, a smaller number of trays might be sufficient. It is worth noting that the number of trays can have a

greater effect on the amount of water removed than the circulation rate. The concentration of the TEG depends on the required degree of dehydration. If deeper dehydration is required, stripping gas is used to increase the concentration of lean glycol. In the case of acid gas dehydration there is no need for deep dehydration and stripping gas is not normally used in TEG units for acid gas dehydration.
The higher the glycol circulation rate at the constant gas flow rate the more water is removed from the gas. Normally, the glycol circulation rate is given as gallons of TEG per pound of water removed. A typical design is 2 to 7 gallons of glycol per pound of water to be removed, or, to simplify the design, per pound of water in the wet feed gas. Less than two gallons of glycol per pound of water may affect the gas-glycol contact and more than 7 gallons of glycol per pound of water may decrease the reboiler temperature and lean glycol concentration, which in turn decreases the glycol water absorption capacity.

Since the acid gas feed to the TEG Contactor is still wet, the feed lines are usually made from a corrosion resistant 316 stainless steel. The vessels, equipment, and piping on the TEG rich side are usually made from carbon steel; however there are specific areas where gas breakout is possible. In these instances 316 stainless steel or 316 clad carbon steel is usually specified. Carbon steel is the material of choice for the lean side (Williamson, 2009). If the feed is pre-dehydrated with glycol upstream of the TEG Contactor then the contactor itself can be made of carbon steel (Carmody, 2000).

VMGSim, the trademark software of Virtual Materials Group, Calgary, Alberta, was used to simulate the dehydration of acid gas with TEG. Three theoretical stages were used in the TEG Contactor; this is equivalent to approximately 12 real trays. The flow rate of acid gas is 1 million SCMD on a dry basis, TEG circulation assumed to be the minimum recommended for good gas-glycol contact, which is 2 gallons (US) of glycol per pound of water in the feed gas. It was also assumed that rich glycol is heated up to 300°F in the Rich/Lean Exchanger before being fed to the Still Column. TEG is heated to 400°F in the Regenerator. The concentration of lean TEG is 98.7 wt. %.
Table 1 summarizes the results of the simulation.

The flash gas from the Flash Tank and the offgas from the Still Column have a high concentration of hydrogen sulfide. Instead of sending the two streams to the flare or incinerator, they can be recycled to the first stage of compression to minimize emissions.

Figure 3 presents the bottom part of the compression curve to better show how much TEG dehydration lowers the water content of acid gas compared to dehydration by compression and cooling alone.

Table 1: Results of the simulation of TEG dehydration for three compositions of acid gas.

	Acid Gas Composition

		50% H_2S 50% CO_2	25% H_2S 75% CO_2	0% H_2S 100% CO_2
Water In	g/Sm^3	3.577	3.520	3.447
Hydrate Temp.	°C	30.1	23.9	10.8
Water Out	g/Sm^3	0.265	0.238	0.223
Hydrate Temp.	°C	-39.5	-34.9	-32.3
Reboiler Duty	kW	191.2	197.7	201.2
TEG Pump Power	kW	3.22	3.25	3.23
TEG Make-Up	kg/h	0.56	0.38	0.30
Flash Gas	SCMD	9377	4612	1996
Offgas	SCMD	7732	5924	4556

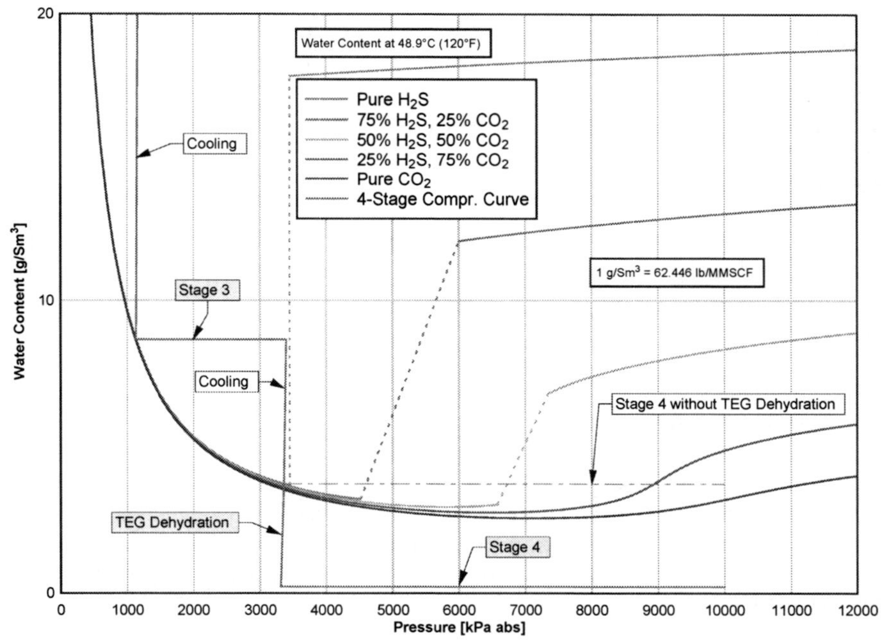

Figure 3: TEG dehydration after the 3rd stage of compression

The water content of acid gas shown in Figure 3 is for 48.9°C (120°F). When the acid gas is transported in a long injection pipeline to the injection well, or if the injection operation is suspended for a period of time, the temperature of the acid gas may drop to the ground temperature. Since the saturated water content of acid gas decreases with decreasing temperature, the relative water saturation of the acid gas in the pipeline increases.

Table 2 compares relative water saturation of acid gas dehydrated with TEG after the third stage of compression. Again, the acid gas pressure after the third stage is 3400 kPa, the compressor discharge pressure is 10000 kPa. The comparison is for 48.9°C and for 2°C, assumed to be the ground temperature.

Table 2: Relative water saturation of TEG dehydrated acid gas at 48.9°C and 2°C

		Acid Gas Composition		
		50% H_2S 50% CO_2	25% H_2S 75% CO_2	0% H_2S 100% CO_2
Water in Dehydrated Acid Gas at 3400 kPa	g/Sm^3	0.26	0.24	0.22
Saturated Water Content at 10000 kPa and 48.9°C	g/Sm^3	8.34	4.88	3.20
Relative Water Saturation	%	3.1	4.9	6.9
Saturated Water Content at 10000 kPa and 2°C	g/Sm^3	3.27	2.16	1.42
Relative Water Saturation	%	8.0	11.1	15.5

Acid gas dehydrated with TEG remains under-saturated with respect to water when the temperature of acid gas drops from 48.9°C to the ground temperature of 2°C.

3. Conclusions

Acid gas has to be compressed prior to injection. The process of compressing and inter-stage cooling is often sufficient to dehydrate acid gas below 100% relative water saturation thus avoiding corrosion problems. Otherwise other means of dehydrations are required, most often by using tri-ethylene glycol (TEG). TEG units are installed before the last stage of compression. No stripping gas is required. TEG circulation rate can be kept to a minimum but to ensure good contact between the gas and the glycol, the circulation rate has to be at least 2 gallons (US) of glycol per pound of water in the acid gas being fed to the TEG contactor.

There are cases where dehydration of acid gas by means other than compression and cooling alone should be considered even if the relative water saturation of the acid gas is below 100%. Additional dehydration is required if the acid gas injection pipeline is long enough to allow the acid gas to cool to ground temperature or the temperature below hydrate formation, or if the operating philosophy of the acid gas injection scheme allows for suspension of injection with the acid gas sitting in the injection pipeline until the acid gas temperature reaches ground temperature or hydrate formation temperature.

References

Carroll, J.J., 2009, Natural Gas Hydrates, Gulf Professional Publications
Carmody, P., 2000, Glycol Pre-Dehydration of Gas before a Contactor, Proceedings of the 50th Laurance Reid Gas Conditioning Conference, Norman, Oklahoma
McKay, W., 2009, CO2 – Properties and Dehydration, EPRI – CO2 Capture & Storage Technical Meeting, Edmonton
Williamson, S., 2009, Ammonite Corrosion Engineering, personal communication

Proceedings of the 2nd Annual Gas Processing Symposium
Farid Benyahia and Fadwa T. Eljack (Editors)

Adsorption Thermodynamics of Natural Gas Storage onto Pitch-Based Activated Carbons

Kazi Afzalur Rahman[a], Wai Soong Loh[a], Anutosh Chakraborty[a], Bidyut Baran Saha[a], and Kim Choon Ng[a]

[a]*Department of Mechanical Engineering, National University of Singapore, 9 Engineering Drive 1, Singapore 117576, Singapore*

Abstract

Adsorbed natural gas (ANG) has recently become competitive to compressed natural gas (CNG) for storage and transportation purposes with high energy density and at much lower pressure than CNG. The adsorption characteristics of adsorbents are important information for uptake capacity and thermal management of the adsorptive gas storage. This article presents the adsorption characteristics of methane, which is the major component of NG, on two types of pitch-based activated carbons namely Maxsorb III and ACF (A-20) in the temperature ranges from 5 to 65 °C and pressures up to 2.5 MPa. The uptakes have been experimentally measured by a volumetric technique. The results are regressed with various popular isotherm models used in the literature for physisorption and are compared with the data of other researchers for similar adsorbent - adsorbate pairs. The heat of adsorption has also been extracted from the measured data. The thermodynamic study has been accomplished on the performance of adsorbed natural gas storage system for charging and discharging along with the thermal effects.

Keywords: Adsorbed natural gas, Activated carbons, Adsorption isotherm, Heat of adsorption.

1. Introduction

In recent years, natural gas has potentially become an alternative energy source in many sectors due to the instability in oil markets and the increase in environmental concerns. The substantial reduction of pollutant (COx, NOx, and SOx) emissions and clean combustion are main benefits of natural gas, whose main ingredient is methane (85-95%). It is widely being popular as a transportation fuel because of relatively lower cost than the oil-derived fuels and prevalent availability in many parts of the world. A good proportion of vehicles employ compressed natural gas (CNG) at very high pressure (20 ~ 30 MPa). This implies high manufacturing and filling costs and also represents a safety concern. Liquefied natural gas (LNG) is another alternative storage and transportation technology, which requires huge energy for cryogenic cooling of the gas (as a boiling liquid to approximately −163 °C). Adsorbed natural gas (ANG) is an intermediary option to CNG and LNG by the way of reduction in pressures and near atmospheric temperature, reported by numerous researchers. Natural gas can be stored at relatively low pressure (2 ~ 4 MPa) in a lightweight cylinder filled with porous adsorbent using sorption technique which allows a good design flexibility in tank configuration and placement. The storage tank pressure is easily reachable with an inexpensive single-stage compressor which is another attractive feature of ANG.

To date, the most promising adsorbents for NG storage are highly microporous activated carbons with high packing density (Quinn and Macdonald, 1992). Among the practical solids used in industries, pitch based activated carbons are the most versatile because of their extremely high surface area and micropore volume. Moreover, their bi-disperse pore size distribution provide good access of adsorbate molecules to the interior (Ruthven, 1984) and good thermal conductivity allow for improved thermal management of the adsorptive gas storage [Saha et al., 2008]. For this purpose, an adsorbed natural gas storage up to 2.5 MPa is proposed to analyze the performance using two types of activated carbons. The uptake values have experimentally been measured by a volumetric technique. The results are evaluated through regressions to fit with the Langmuir and the Tóth adsorption isotherm models. The results are compared and found in good accordance with the previous studies (Himeno et al., 2005; Lozano-Castello′ et al., 2002) for similar adsorbent - adsorbate pairs. The heat of adsorption is extracted from the isotherm data which is functional for the reproduction of thermal properties during charging and discharging of the adsorptive gas storage system.

2. Materials and Experimental Methods

2.1. Activated Carbon and the Methane

The ultra pure methane sample supplied by SOXAL, Singapore is used and the purity grade is greater than 99.9%. All properties of methane used in this paper have evaluated using the generalized equation of state proposed by Setzmann and Wagner (1991). The Maxsorb III sample, supplied by Kansai Coke and Chemicals Co. Ltd., Japan, is in powder form, which is highly micro-porous and the ACF (A-20) sample, supplied by Osaka Gas Co. Ltd., Japan, is fibrous type. Figures 1(a) and 1(b) show the scanning electron microscope (SEM) photographs of Maxsorb III and ACF (A-20), respectively. The surface structure is observed to be flake-like layers with porous volumes entrenched in between in case of Maxsorb III and cylindrical shapes with uniform surface diameter for ACF (A-20). The porous properties such as the BET surface area, the pore size, the pore volume, and the skeletal density of the assorted adsorbents are listed in Table 1.

2.2. Experimental Apparatus

The schematic of the experimental apparatus is shown in Figure 2, and measurement procedures were described in detail by Loh et al. (in press); hence only a brief description is given here. The experimental apparatus consists mainly of a stainless steel (SS 304) adsorption cell and a charging cell with internal volume of 62.78 ml and 1026.15 ml respectively.

Table 1: The thermo-physical properties of Maxsorb III and ACF A-20

Adsorbent	Surface Area $[m^2/kg]$	Total Pore Volume $[m^3/kg]$	Average Pore Diameter [Å]	Skeleton Density $[kg/m^3]$	References
Maxsorb III	3.14×10^{6}	20.1×10^{-4}	20.08	2200	Saha et al., 2008
ACF(A-20)	1.9×10^{6}	10.28×10^{-4}	21.6		El-Sharkawy, 2006

(a) (b)

Figure 1: Scanning electron micrographs (SEM) photo of (a) Maxsorb III at magnifications 3700 and (b) ACF (A-20) at magnifications 1000.

The samples were loaded into the adsorption cell which is connected to the charging cell through 1/4″ nominal stainless steel plumbing and a set of Swagelok fittings (valves, T's, and reducers). Both the adsorption and charging cells were immersed in a constant-temperature water bath. The pressure readings of methane were measured using a 0-5 MPa range Kyowa pressure transducer (PGS-50KA) with an uncertainty of 0.1% of full scale in measurement. The temperatures of the adsorption and charging cells were recorded using class-A Pt 100 Ω RTDs with estimated uncertainty of ±0.15 K. The adsorption cell RTD was contact with the activated carbon to enable direct temperature measurement. Hence the adsorption cell temperature was used as the isotherm temperature. All the temperatures and pressure readings were logged into an Agilent data logger to enable real time monitoring of the system. The readers may approach to the article by Loh et al. (in press), which has described the adsorption characteristics for activated carbon/HFC-134a pair, for the detailed experimental procedures.

2.3. Data Reductions

The raw data were the pressure and temperature of both adsorption cell and charging cell. After evacuation of the entire assembly for 24 hours, the charging cell was pressurized with methane from its source (with Ball valve 2 being closed). The initial readings were recorded when the pressure (P_{Charge}) and temperature (T_{Charge}) was stabilized. Thus, the initial amount of methane (m_i) was determined from the (ρ_i) density at P_{Charge} and T_{Charge} and the charging cell volume, $V_{Charge\ Cell}$. Then, the methane gas was released into the adsorption cell and the final mass of methane (m_f) in the charging side was determined from the recorded pressure and temperature after stabilization. Therefore, the amount of the methane transferred to the adsorption cell is ($m_i - m_f$) which was partly adsorbed by the adsorbent and it took about an hour to reach the adsorption equilibrium state. The left over methane in the adsorption cell ($m_{adsorption\ cell}$) was calculated from the density ($\rho_{adsorption\ cell}$) at $P_{adsorption}$ and $T_{adsorption}$ and the void volume in the adsorption cell (V_{void}).

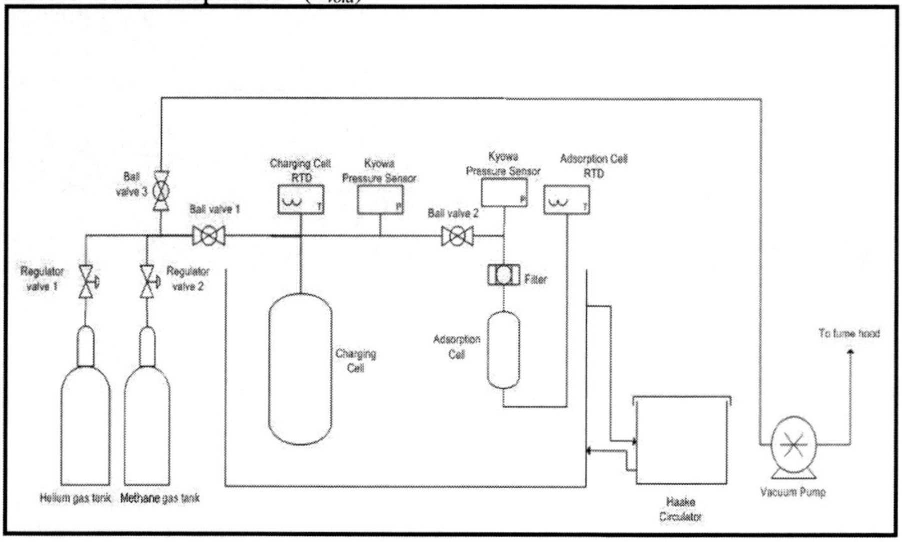

Figure 2: The schematic diagram of the adsorption isotherm experimental apparatus

As adsorption occurs in the micro pores, the void volume (V_{void}) in the adsorption cell is calculated as follow:

$$V_{void} = V_{ads\_cell} - m_{ac} / \rho_{solid} - v_{\mu} m_{ac} \quad (1)$$

where V_{ads_cell} is the adsorption cell volume, m_{ac} is the mass of activated carbon sample in the adsorption cell, ρ_{solid} is the solid density of activated carbon, and v_{μ} is the micro pores volume of activated carbon. Hence, the amount of methane adsorbed ($m_{adsorbed}$) can be estimated from ($m_i - m_f - m_{adsorption\ cell}$) which gives the final the specific uptake, $C = m_{adsorbed} / m_{ac}$. Subsequently, the charging cell is charged to next higher pressure with water bath temperature remain constant. Measurements were made up to 2.5 MPa and the same procedures were repeated with different isotherm.

3. Adsorption Thermodynamics

The Langmuir and the Tóth models are used to fit the adsorption equilibrium data. The Langmuir model presumes homogeneous surface of adsorbents where adsorption energy is constant over all sites. This model has also assumed that the adsorption on surface is localized and each site can accommodate only one molecule or atom (Do, 1998). The Langmuir model is written as

$$\frac{C}{C_o}=\frac{k_0\exp(h_{st}/RT)p}{1+k_0\exp(h_{st}/RT)p} \tag{2}$$

where C is the amount adsorbed, C_0 is the limiting adsorbed amount, P is the equilibrium pressure, k_0 is the equilibrium constant, h_{st} is the isosteric heat of adsorption and R stands for gas constant.

The Langmuir model has the limitation to fit at relatively higher pressure region and for material heterogeneity. The Tóth model is commonly used for heterogeneous adsorbents such as activated carbon because it predicts correctly the behaviour at both low and high pressure conditions. The Tóth model can be represented by

$$\frac{C}{C_o}=\frac{k_0\exp(h_{st}/RT)p}{\left\{1+\left(k_0\exp(h_{st}/RT)p\right)^t\right\}^{1/t}} \tag{3}$$

where t is the parameter that indicates the heterogeneity of the adsorbent. The Tóth model becomes identical to the Langmuir model when the heterogeneity parameter (t) becomes unity.

The isosteric heat of adsorption (h_{st}) is the evolution of heat during adsorption process since the adsorbate molecules are more stabilized on the adsorbent surface than in the bulk phase. This parameter is useful for the thermal management of any adsorption system and also functional for the reproduction of thermodynamic properties such as enthalpy (h), entropy (s), and internal energy (u). Recently, Chakraborty et al. (2009) has formulated the thermodynamic properties of an adsorbent–adsorbate system from the rigor of classical thermodynamics and the extensive entropy quantity is written as,

$$s(P,T,C)\approx s(P,T,C=0)+\int_{T_o}^{T}\left\{\frac{c_{p,s}}{T}+\frac{h_{st}}{T}\left(\frac{1}{T}-\frac{1}{v_g}\frac{dv_g}{dT}\right)\right\}dT+\int_{0}^{C}\left\{s_g-\frac{h_{st}}{T}\right\}dC \tag{4}$$

where $c_{p,s}$ is the specific heat capacity of the adsorbents, v_g indicates the specific volume of the gaseous phase.

4. Results and Discussion

Figure 3 shows the uptake data for the adsorption of methane on Maxsorb III and ACF (A-20) for temperatures 5, 25, 45, and 65 °C and pressures up to 2.5 MPa. The experimental data have been regressed with both Langmuir and Tóth models which are shown in Figure 3(a) for Maxsorb III, and Figure 3(b) for ACF (A-20). The regression values agree within ±5% of experimental data for the Langmuir model and in case of the Tóth model the regression results fit the experimental data within ±3% due to the account of heterogeneity parameter (t). The numerical value of the parameters C_0, k_0, h_{st} and t are listed in Table 2.

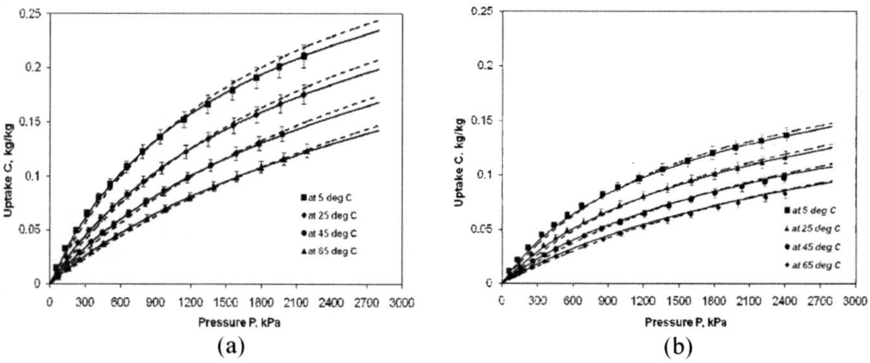

Figure 3: Adsorption isotherm of methane on (a) Maxsorb III, (b) ACF (A-20). Broken lines indicate the Langmuir model and the solid lines indicate Tóth model.

The equilibrium uptakes of methane on the both assorted adsorbents have been compared with the data cited from the literature of Himeno et al. (2005) and Lozano-Castello′ et al. (2002) and these are shown in Figure 4. The present adsorption uptake data are slightly higher than those of Himeno et al. (2005) where the uptake results of methane onto Maxsorb sample were presented. The porous properties of Maxsorb are comparable but slightly lower than Maxsorb III. Another reason for observing deviation between the present uptake data and the data obtained by Himeno et al. (2005) could be the taking of bath temperature as the isotherm temperature whereas the adsorbent temperature was considered as the isotherm temperature in the present study. Nevertheless, the present experimental data on ACF (A-20) are fairly matched with the isotherm at 25 °C for the same sample which has been carried out in a high-pressure microbalance by Lozano-Castello′ et al. (2002). Figure 4 also proclaims the principle benefit of ANG storage that the volumetric capacity on the adsorbents (for packing density of 500 kg/m^3) is significantly higher than the compressed natural gas (CNG) for the same vessel volume.

Table 2 The numerical values of C_0, k_0, h_{st} *and* t

Parameters	Maxsorb III		ACF (A-20)	
	Langmuir	**Tóth**	**Langmuir**	**Tóth**
C_0 (g/g)	0.402	0.439	0.232	0.277
h_{st}/R (K)	1550	1610	1440	1453.6
k_0 x 10^6 (kPa^{-1})	2.10	1.97	3.50	3.65
t	1.0	0.780	1.0	0.742
Error of Regression (%)	4.2	1.0	4.4	2.6

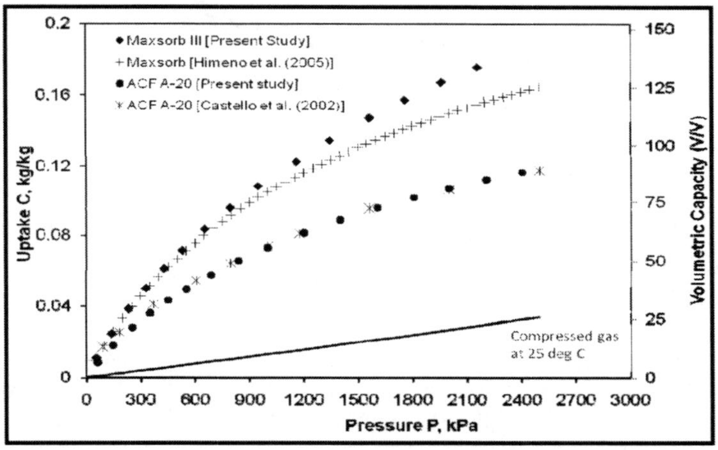

Figure 4: Comparison of isotherm data for 25 ^{0}C with cited data from the literature.

The entropy plot is shown in Figure 5 for methane adsorption onto the two assorted adsorbent samples as a function of pressure and temperature. The entropy values at adsorbed phase has been calculated from Equation (4) where the specific heat capacity of the adsorbents ($c_{p,s}$) has considered as constant.

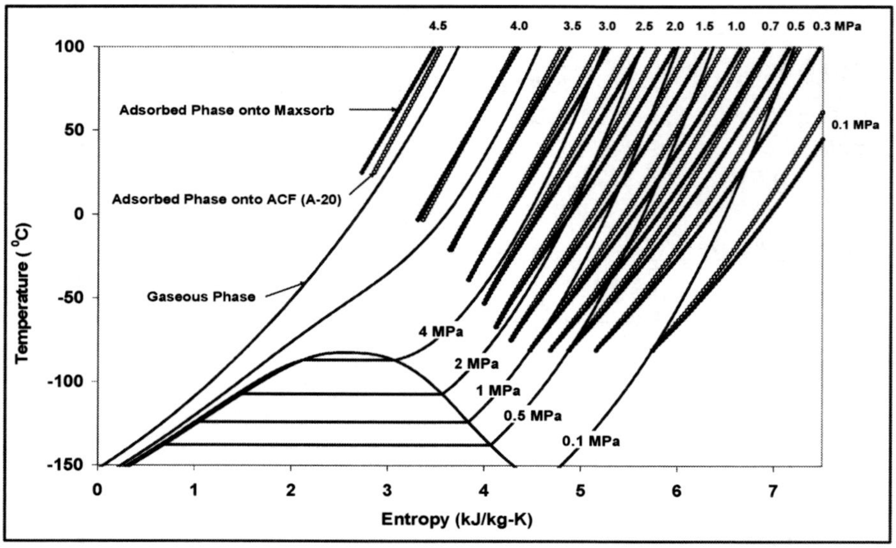

Figure 5: Temperature–entropy maps for methane adsorption onto Maxsorb III (closed symbol) and ACF A-20 (open symbol) as a function of P and T.

Here, the solid lines denote the entropy of pure gaseous phase for various pressures. A value of 1.375 kJ/kg.K for Maxsorb III and 1.05 kJ/kg.K for the ACF (A-20) have been used. This plot indicates that the amount of entropy at adsorbed phase (s_a) for both adsorbent-adsorbate pairs is higher than that of the gaseous phase when pressure is below 2 MPa along with relatively higher temperatures. However, the adsorbed phase entropy (s_a) is very close to the gaseous phase when pressure is 2 MPa and even lower for higher pressures. Thus the storage of methane found to be suitable at pressure near but above 2 MPa for ambient temperature condition.

5. Conclusion

The adsorption characteristics are the basic parameters of any adsorbent-adsorbate system. In this study, the experimental isotherms have been derived for two different activated carbons namely, Maxsorb III and ACF (A-20) with methane which are useful for the charging and discharging analysis of natural gas storage system. The Maxsorb III sample shows higher storage capacity due to its high specific pore volume than the ACF (A-20) sample. However, the heat of adsorption is lower in case of the ACF (A-20) sample which accelerates the adsorption rate and thus minimizes the charging time. Moreover, the cylindrical shape of the fibre enhances the adsorption rate. Temperature-entropy maps of the assorted adsorbent-adsorbate pairs have been plotted which are useful in designing an adsorbed natural gas storage system.

References

Chakraborty, A., Saha, B. B., Ng, K. C., Koyama, S., and Srinivasan, K., 2009, Theoretical Insight of Physical Adsorption for a Single-Component Adsorbent + Adsorbate System: I. Thermodynamic Property Surfaces, Langmuir, 25 (4), 2204-2211

Do, D. D., 1998, Adsorption Analysis: Equilibria and Kinetics, Imperial College Press, London.

Himeno, S., Komatsu, T., Fujita, S., 2005, High pressure adsorption equilibria of methane and carbon dioxide on several activated carbons. J. Chem. Eng. Data, 50, 369-376.

Loh, W.S., Kumja, M., Rahman, K.A., Ng, K.C., Saha, B.B., Koyama, S., El-Sharkawy, I.I., In Press, Adsorption parameter and heat of adsorption of activated carbon/HFC-134a pair, Heat Transfer Engineering.

Lozano-Castello´, D., Cazorla-Amoro´s, D., and Linares-Solano, A., 2002, Powdered Activated Carbons and Activated Carbon Fibers for Methane Storage: A Comparative Study, Energy & Fuels, 16, 1321-1328

Quinn, D. F., and Macdonald, J. A., 1992, Natural Gas Storage, Carbon. 30 (7), 1097-1103.

Ruthven, D. M., 1984, Principles of adsorption and adsorption processes, John Wiley & Sons, New York.

Saha, B. B., Chakraborty, A., Koyama, S., Yoon, S. H., Kumja, M., Yap, C. R., and Ng, K. C., 2008, Isotherms and thermodynamics for the adsorption of n-butane on Maxsorb III, International Journal of Heat and Mass Transfer, 51, 1582-1589.

Setzmann, U., And Wagner, W., 1991, A new equation of state and tables of thermodynamic properties for methane covering the range from the melting line to 625 K at pressures up to 1000 MPa, J. Phys. Chem. Ref. Data, 20 (6), 1061-1151.

Proceedings of the 2nd Annual Gas Processing Symposium
Farid Benyahia and Fadwa T. Eljack (Editors)

Opportunities for Selective Control of Fischer-Tropsch Synthesis Hydrocarbons Product Distribution

Nimir O. Elbashir

Chemical Engineering Program, Texas A&M University at Qatar
P.O. Box 23874, Doha, Qatar

Abstract

This paper reports a comparison between the hydrocarbon product distribution of Fischer-Tropsch synthesis (FTS) over three catalysts (cobalt-based (15% Co/Al_2O_3), ruthenium-based (10% Ru/ SiO_2, areogel), and iron-based (RLS 4.55 Si/isocyelan)) under near-critical and supercritical solvent operating conditions. The Anderson-Schultz-Flory (ASF) model is used to describe the chain growth mechanism and the hydrocarbon product distribution from the three catalysts as obtained from a fixed-bed reactor. The data presented in this paper showed interesting phenomenon within the middle hydrocarbon product range of the distributions indicating a unique deviation from ASF model, specifically for the cobalt catalyst and the ruthenium catalyst. According to the enhanced α-olefin incorporation model the extent of the deviation from ASF distribution could be correlated with the capability of the middle distillate olefins to incorporate in the chain growth process (Elbashir and Roberts, 2005). The experimental data represented by the ASF plots of both cobalt catalyst and the ruthenium catalyst were found to be in agreement with the aforementioned model; however, the iron catalyst data showed standard ASF plots. Preliminary explanation of this difference is that FTS reaction mechanism of the iron catalysts favor oxygenates as intermediate product not α-olefins as the case for cobalt and ruthenium catalysts. For the former catalysts surface reaction mechanism is favored and the primary products α-olefins have the capability to re-adsorb and affect the chain growth process.

Keywords: Fischer-Tropsch synthesis, Supercritical Solvents, Hydrocarbon Product Distribution

1. Introduction:

Fischer-Tropsch synthesis (FTS) continues to receive attention as an alternative for the production of ultra-clean transportation fuels, chemicals, and other hydrocarbon products through heterogeneous catalytic conversion of the synthesis gas or syngas (a mixture of H_2 and CO) derived from readily available natural resources, such as coal, natural gas, or biomass. Diesel fuels from FTS have unique characteristics that include very low sulfur and aromatic contents, high cetane index and exceptional clean burning in compression-ignition engines. Compared to crude oil derived diesel, FTS diesel fuel has been shown to reduce emissions of carbon monoxide, nitrogen oxides, hydrocarbons and other particulates. Three types of FTS reactors found their way to commercialization including tubular-fixed-bed reactor, fluidized-bed reactors and slurry reactor. Detailed comparison between the different FTS reactor technologies have been covered elsewhere (Elbashir, et al. 2009a). The shortcomings in industrial gas-phase FTS (fixed-bed-reactor) and liquid-phase FTS (slurry-reactor) have led to research efforts towards the application of supercritical fluid solvents in order to improve catalyst activity and selectivity (Elbashir et al. 2009b). Conducting FTS under supercritical fluid (SCF) solvent conditions (later referred to as SCF-FTS) affords unique opportunities to manipulate the FTS reaction environment, and to enhance production of liquid fuels and value-added chemicals (such as α-olefins) from the syngas. Supercritical phase combines the desirable properties of gas-like diffusion with liquid-like heat transfer and solubility, thus overcoming several of the current industrial limitations. Fine adjustments in the operating conditions near the critical point can result in superior control of the FTS process, including tunable product distributions (Elbashir and Roberts 2005).

Previous studies in this field attributed performance enhancement of cobalt-based and iron-based catalysts in SCF-FTS reaction performance relative to conventional gas-phase FTS and liquid-phase FTS to the followings (Elbashir, et al. 2009b and references therein): (1) in situ extraction of heavy hydrocarbons from catalyst pores due to high solubility in the supercritical phase (2) elimination of inter-phase transport limitations, thus promoting reaction pathways toward the desired products; (3) enhanced desorption of α-olefins that promotes the chain growth process prior to secondary reactions, and (4) superior heat transfer compared to gas-phase reaction, resulting in more long chain products.

2. Selective Control of Hydrocarbon Product Distribution in Fischer-Tropsch Synthesis

One of the most well known models to describe the chain growth mechanism in FTS reaction is the Anderson-Schultz-Flory (ASF) model. According to this model, the polymerization process in FTS is assumed to initiate on the surface of the catalyst by a monomer that contains one carbon atom, while chain growth takes place by the addition of one monomer at a time. As soon as chain growth is terminated, by either addition or subtraction of hydrogen molecule, the product molecule desorbs from the surface of the catalyst and creates a vacant site for further reaction. The mathematical form of the ASF model is as follows:

$$\ln\left(\frac{W_n}{n}\right) = n \ln \alpha + \ln\left(\frac{(1-\alpha)^2}{\alpha}\right) \qquad (1)$$

where W_n represents the weight fraction of all hydrocarbon products with carbon number n, and α is the chain-growth probability (represents the rate of growth divided by the sum of growth and termination rates).

According to the ASF model a plot of $\ln(W_n/n)$ should yield a straight line for all hydrocarbon products whereby the α-value can be determined from the slope of that line. Higher α-value indicates higher selectivity towards heavier hydrocarbon production and vice versa.

Numerous studies have reported deviations from the standard ASF product distribution in FTS yielding different curvatures for the relation between $\ln(W_n/n)$ vs n (Inoue et al. 1987). Many researchers have targeted such deviations as a mean to control the product distribution in FTS in order to maximize the production of hydrocarbons within specific carbon number ranges (e.g. C_5-C_{11} for gasoline fraction, C_{12}-C_{19} for diesel fraction, or heavier hydrocarbons (C_{20+}) for wax range). Many kinetic and reaction route models have been developed to explain the observed deviations from the standard ASF model. Mass transfer limitations inside the catalyst pores due to the condensation of heavy hydrocarbons was one of the reasons suggested for enhanced α-olefin (primary products) re-adsorption thereby increasing the production of heavy hydrocarbons (Madon et al. 1993). Huff and Satterfield also suggested a model that accounts for two-catalytic sites with two different growth factors (iron promoted catalysts) that yield a positive deviation from the standard ASF model (Huff and Satterfield 1984). Pore size effects (molecular sieve effects) (Iglesia et al. 1992) and concentration gradients as well as temperature gradients in fixed bed reactors (Matsumoto and Satterfield 1989) were also suggested as causes for non ASF distributions. In a series of studies, Puskas and his coworkers argue that multiplicity in the chain growth probability is the reason for positive deviations from the ASF model (Puskas et al., 1993). In addition, deviations from the ASF model have also been attributed to experimental artifacts, non steady state operation (Davis 2003), and transient holdup of higher molecular weight products in the oil phase surrounding the catalyst (Dictor and Bell 1983).

Several studies reported enhancements in chain growth probability and increase of high value added chemicals (e.g. α-olefins) selectivity in SCF-FTS compared to conventional gas-phase FTS and liquid-phase FTS (Elbashir, et al. 2009b and references therein). Higher productivity of α-olefins, the primary product in the FTS chain growth process, indicates suppression of secondary reactions such as hydrogenation to paraffins, isomerization to isomers, and/or oxygenation. Deviations from the ASF product distributions in SCF-FTS were reported in several studies. Tsubaki, et al. (2002) reported an interesting non-ASF product distribution in the supercritical pentane phase at a reaction temperature of 200 °C over a silica supported cobalt catalyst. The authors reported a hydrocarbon product distribution where the formation rate of hydrocarbons from C_7-C_{28} was totally independent of the carbon number, with α-value (chain growth probability) higher than 0.96. They suggested that the high ability of the olefins (C_nH_{2n}) formed to readsorb in the supercritical phase and initiate new chain-growth processes acts to increase the formation rate of heavy hydrocarbons.

In previous studes we reported the hydrocarbon product distribution in SCF-FTS on cobalt-based catalysts (15% Co/Al_2O_3) over wide range of operating conditions (Elbashir and Roberts, 2005, Elbashir et al. 2005, Elbashir et al. 2009b). The conclusion we drawn from the results reported in the aforementioned papers is that operating FTS in the supercritical phase is not sufficient by itself to produce deviations from the standard ASF product distributions. Rather, deviation from the standard ASF distribution in SCF media is a result of its influence on both the thermophysical characteristics of the reaction media and FTS reaction kinetics (i.e. chain growth mechanism and rates of different reactions take place on catalyst surface). Enhanced incorporation of primary products α-olefins in the chain growth mechanism was modeled as the cause of the deviations from standard ASF in SCF-FTS (Elbashir and Roberts, 2005). In this paper we extended our study to investigate the hydrocarbon product distribution of SCF-FTS over the other known catalysts for this reaction; ruthenium-based catalyst and iron-based catalyst. The objective behind these investigations is to see whether deviations from the standard ASF product distribution observed on cobalt-based catalyst under SCF-FTS are also exists over a ruthenium-based catalyst and an iron-based catalyst over wide range of operating conditions.

3. Experimental

The FTS high pressure reactor unit that used to measure the reaction performance of these catalysts has been described in details elsewhere (Elbashir, et al., 2005). The reactions experiments were continued until steady state was reached in terms of both catalyst activity (CO and syngas conversion) and hydrocarbon product distribution. The reported results in this study represent samples that were collected after at least 50-hr of time-on-stream (TOS). The analysis of reactants and products were conducted by two online gas chromatographs. The finger prints of the peaks were used for the selectivity and conversion calculations. A Varian 3300 GC with capillary column (DB-5) and a FID detector was used for the analysis of C_2-C_{35} hydrocarbons and oxygenates. A Varian CP-3800 GC with a packed column (Hayesep-DB100/120) and TCD were used for the analysis of the permanent gases (H_2, CO, N_2, CH_4, CO_2, C_2H_4, and C_2H_6). Quantitative analysis from Varian CP-3800 provides the necessary data for the calculation of catalyst activity and the selectivity. GC-MS analysis (using HP5890 Series II with 0.25um column) was also conducted on different liquid samples collected after gas-phase FTS and supercritical hexanes FTS (SCH-FTS) to quantify isomers and straight hydrocarbons ratio, and oxygenates content for specific hydrocarbons. The alumina supported catalyst used in this study is a 15% Co/Al_2O_3 that purchased from the United Catalysts Company. The ruthenium catalyst is a 10% Ru/SiO_2 (aerogel) obtained from Dr. Richard Ernst research team at the University of Utah. The iron-based catalyst (RLS 4.55 Si/isocyelan) is a FTS catalyst purchased from the United Catalysts Company.

4. Results

4.1 Deviations from the Standard ASF Distribution over the Cobalt-Based Catalyst in SCF-FTS:

Figure 1 represents the hydrocarbon products from gas phase FTS and supercritical hexane (SCH)-FTS over a cobalt-based catalyst at 240 °C and partial pressure of syngas ($CO+H_2$) ca. 20 bar (for SCH the partial pressure of the supercritical solvent is kept at ca. 45 bar). The light hydrocarbons fraction in Figure 1 is represented by C_1-C_4, while gasoline fraction is represented by C_5-C_{11}, diesel fraction by C_{12}-C_{19}, and wax fraction by C_{20+}. As shown in the figure, the overall product distribution shifts towards heavier products in SCH- FTS compared to gas phase FTS. Considering the gas-phase reaction as the base line case, Figure 1 shows that there is a significant reduction in light hydrocarbons selectivity coupled with significant improvement (100% increase) in diesel fraction selectivity in the SCH-FTS compared to the gas phase-FTS. The enhanced selectivity in the supercritical phase can be attributed to the liquid-like density that provides better temperature management of hot spots formed on the catalyst surface due to the highly exothermic nature of FTS reaction (see Huang and Roberts 2003). On the other hand, in the SCF media the solubility of heavy compounds is increased to

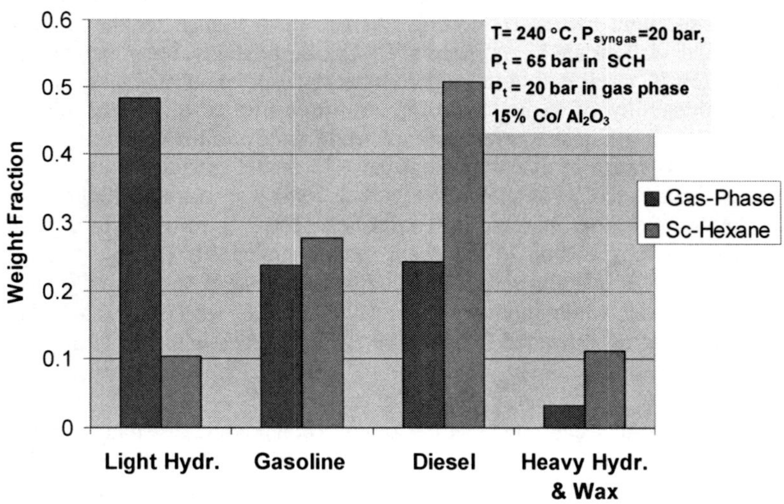

Figure 1: Hydrocarbons and fuel fractions distributions over 15% Co/Al_2O_3 in gas-phase FTS and SCH-FTS environment at 240 °C. H_2/CO feed ratio is 2/1.

enhance the in situ extraction of heavy hydrocarbons that occupy catalyst's active sites, thereby freeing up more active sites for primary products to incorporate in the chain growth process.

In order to better understand the causes of these interesting results, we conducted detailed analysis of the hydrocarbon distribution of alumina supported cobalt catalyst over a wide range of operating conditions in near critical and supercritical FTS conditions. Typical results are shown in Figure 2. The data represented in Figure 2a-d were collected after more than eighty hours TOS, after which both the activity and the selectivity of the cobalt catalyst showed steady performance with time. The degree of deviation from the ASF model was found to vary from near-critical to supercritical and from liquid-like density to gas-like density as illustrated in the figure. It is noteworthy to mention here that at 230 °C the reaction mixture exists as a compressed liquid phase as evidenced by the experimental measurement of the critical properties of the reaction mixture (Elbashir and Roberts, 2005). On the other extreme, operation at 260 °C results in a reaction mixture that is well within the supercritical region (i.e. the temperature is well above the mixture critical temperature). The overall hydrocarbon distribution closely followed the ASF model at 230 °C with a chain growth probability α-value of 0.8. Increasing the temperature to 240 °C, whereby the reaction mixture is maintained near the critical point, the overall chain growth probability is maintained at 0.82 as shown in Figure 2b. The product distribution in Figure 2b follows closely the ASF model in the middle distillate and heavy hydrocarbon range whereas marked deviations are observed in the light hydrocarbon range. As shown in Figure 2b, the deviations extend to C_8 (i.e. deviations are in the range of C_2-C_8). Surprisingly, upon increasing the temperature to 250 °C (well within the supercritical region), no substantial drop in the chain growth probability of heavy hydrocarbons (α_{hp}-value of 0.78) was observed despite the increase in the reaction temperature while the overall α-value increased to 0.87. In addition, the range of deviation from the ASF model extended up to C_{14} (i.e., deviation range from C_2 to C_{14}) as shown in Figure 2c (which agrees with Tsubaki, et al. (2002) ASF plots). A further increase in temperature (260 °C), resulting in a gas-like density, produced a drastic drop in the chain growth probability ($\alpha_{overall}$ = 0.66) as shown in Figure 2d. The overall product distribution of hydrocarbons up to C_{20} closely followed the ASF model, while only very slight deviation in C_2 selectivity is seen in the product distribution at this elevated temperature (which agrees with Shi et al. (2005) ASF plots).

The aforementioned results suggested that the hydrocarbon product distribution in SCH-FTS could be represented by more than one chain growth probability for different ranges of hydrocarbons (typical model was proposed by Elbashir and Roberts (2005)). The high chain growth probability within middle distillate hydrocarbons (i.e. C_5-C_{15}) (α-value = 0.95) showed in Figure 2c can be attributed to the enhanced α-olefin incorporation in the chain growth process. The enhanced olefin incorporation range can be considered as unique contribution of the supercritical solvent on the chain growth process in near-critical and supercritical FTS conditions. The main assumption of Elbashir and Robert's (2005) model is that only α-olefins have the capability to readsorb and incorporate in the chain growth process. The presence of the supercritical solvent enhances the solubility of these primary products (α-olefins) and thus affects their adsorption/desorption equilibrium inside catalyst pores. As a result, they have better opportunity to readsorb and incorporate in the chain growth process by either initiating a new chain or becoming a part of a growing one, which will result in higher selectivity than the one obtained by the standard ASF model (the details of this models is described in Elbashir and Roberts (2005)). It is important to mention that the concept of α-olefins incorporation is not new since Smith, et al. (1930) has shown that in mixture with carbon monoxide and hydrogen, ethylene can react on a on a cobalt based catalyst to form large quantities of higher hydrocarbons and oxygen-containing compounds. Several other studies have also shown that α-olefins can incorporate in the chain-growth process and behave similarly to the FTS building blocks, specifically on cobalt catalysts (Snel and Espinoza 1987, Patzlaff et al., 1999).

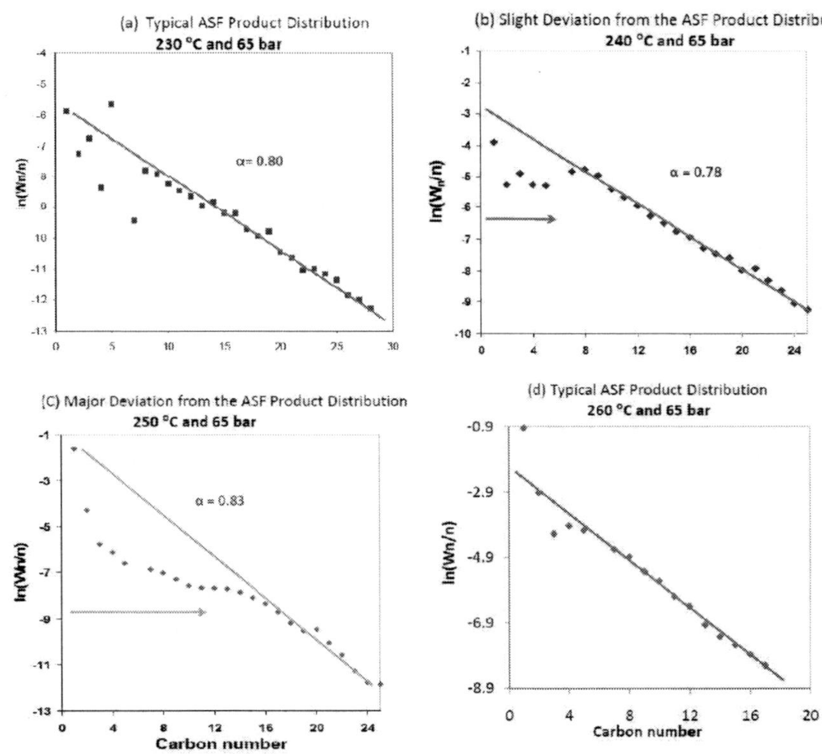

Figure 2: Hydrocarbon product distribution of 15% Co/Al_2O_3 from SCH-FTS at 65 bar and four temperatures: 230°C, 240 °C, 250 °C, and 260 °C (α-value at 260 °C = 0.63). Reaction conditions: 50 sccm/g_{cat} syngas flow rate, 1.0 mL/min hexane flow rate, and H_2/CO feed ratio of 2. α-value represents the overall chain growth probability.

4.2 Deviations from the Standard ASF Distribution over the Ruthenium-Based Catalysts in SCF-FTS:

Prior to the evaluation of the ruthenium catalyst performance under typical SCF-FTS conditions, we have conducted stability tests at relatively long TOS. The conditions for stability tests are similar to the one reported previously for the cobalt-based catalyst (Elbashir et al. 2005). The stability performance of the ruthenium catalyst is illustrated in Figure 3 and it shows that decreasing the temperature from 250 °C to 230 °C resulted in lower syngas conversion; however, CO conversion showed slightly decreased with temperature upon returning to the original condition (250 °C and 60 bar). In general, the outcome of these tests indicates that the activity of the ruthenium catalyst is reasonably

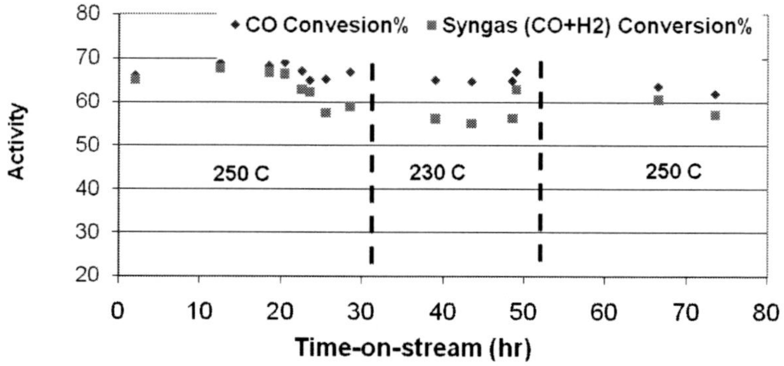

Figure 3: Stability of 10% Ru/SiO2 activity (syngas conversion (%) and CO conversion (%)) with TOS in SCH-FTS at different reaction conditions (total pressure is 60 bar and two temperatures 230 °C and 250 °C, P_{syngas} = 20 bar, syngas flowrate 50 sccm/g_{cat}, and hexane/syngas molar ratio is 3).

stable under the test conditions for relatively long TOS. The important part of the catalyst performance is presented in Figures 4a and 4b that show the hydrocarbon product distribution under near-critical and supercritical FTS conditions. It is noteworthy to mention here the combination of 35 bar and 230 °C (conditions of Figure 4a) represents near-critical conditions based on experimental measurements of the critical properties of FTS mixture. The aforementioned condition is equivalent to the low temperature FTS with overall α-value of 0.74 (Figure 4a). However, at 250 °C and pressure c.a. 45 bar, which is equivalent to high-temperature FTS, the hydrocarbon product distribution showed significant deviation from ASF distribution within the middle distillate hydrocarbons (Figure 4b). More importantly, the overall α-value has also significantly increased to 0.83. This trend is in agreement with the ASF distribution of the cobalt catalyst that reported earlier in Figure 2c.

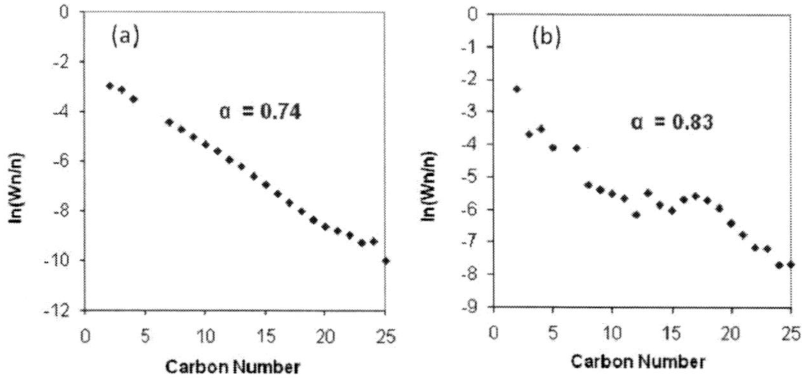

Figure 4: Hydrocarbon product distribution in SCF (hexanes)-FTS of 10% Ru/SiO_2 catalyst at the following conditions: (a) 230 °C and (b) 250 °C and 45 bar. Reaction condition: Reaction conditions: 50 sccm/gcat syngas flow rate, 1.0 mL/min hexane flow rate, and H_2/CO feed ratio of 2. α-value represents the overall chain growth probability.

4.3 Deviations from the Standard ASF Distribution over the Iron-Based Catalysts in SCF-FTS:

The hydrocarbon product distribution of the iron-based catalyst (RLS 4.55 Si/isocyelan) was found to be different from that of the cobalt-based catalyst (reported in Figure 2) and the ruthenium catalyst (reported in Figures 4). The hydrocarbon product distribution was found to be in a reasonable agreement with the standard ASF product distribution for both near-critical conditions and supercritical FTS conditions. Figure 5a shows typical hydrocarbon product distribution in near-critical hexane-FTS (230 °C and 45 bar). The α-value was found to be relatively high at 0.86 and no deviation from ASF was observed in the middle distillate hydrocarbons. At higher temperature 250 °C and 45 bar (Figure 5b) no significant changes in the overall α-value was observed (α=0.84) with a typical ASF product distribution. It is noteworthy to mention that syngas conversion of the iron catalyst is ≈33% lower than that of cobalt and ruthenium catalysts at the prescribed conditions. In addition, the main products over the iron-based catalysts are olefins, oxygenates, and isomers with low amount of linear paraffins.

There are several mechanistic models that have been developed to predict the kinetics and the chain growth process over iron-based catalysts. The hydroxycarbene and CO insertion mechanisms whereby [CH_2O-s] and [CO-s] are the corresponding monomers for the chain growth mechanism is one of these mechanisms. Another one is the CO insertion mechanism where alcohols are formed by the successive hydrogenation of the acyl intermediates (Teng, et al. 2005). In the aforementioned mechanisms neither olefin incorporation nor re-adsorption is playing a role in chain growth mechanism like the case in the cobalt and ruthenium catalysts.

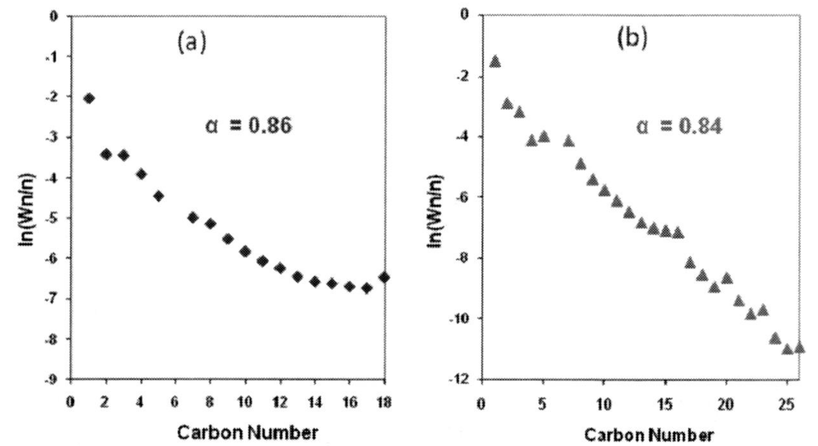

Figure 5: Hydrocarbon product distribution in SCF (hexanes)-FTS of iron catalyst (RLS 4.55 Si/isocyelan) at total pressure of 45 bar and at two temperatures: (a) 230 °C (b) 250 °C.

5. Conclusions:

This paper reported a comparison between FTS hydrocarbon product distribution and ASF plots of three catalysts (cobalt-based, ruthenium-based, and iron-based). The objective is to investigate opportunities for selective control of the hydrocarbon products in near-critical and supercritical FTS conditions. Non-ASF plots of the hydrocarbon product distributions are not the innovation of this study, since numerous studies have reported the existence of more than one α-value. The data presented in this paper showed significant deviations from the ASF within the middle hydrocarbon product range of the ASF plots indicating a unique deviation from ASF model, specifically for the cobalt catalyst and the ruthenium catalyst. According to the previously developed enhanced α-olefin incorporation model (Elbashir and Roberts, 2005), the extent of the deviation is supposed to be correlated with the capability of the middle distillate olefins to incorporate in the chain growth process. An important finding is that deviations from the standard ASF in near-critical and supercritical FTS can be correlated to the thermophysical properties of the reaction mixture that influenced by reaction temperature and pressure. Nevertheless, benefits of liquid-like density and heat capacity in the SCF media have to be balanced by the effect of temperature on kinetics (see results at 260 C- Figure 2d). However, in the case of the iron catalyst no deviation from the standard ASF model is observed even over wide range of operating conditions in near and supercritical FTS. This has been attributed to the reaction mechanism over the iron catalyst that known to favor oxygenates intermediate during the synthesis process. Even though oxygenates might readsorb over the catalyst surface (Teng, et al. 2006) they are mainly known to take part in the corresponding secondary reactions and not towards formation of primary olefins that would cause deviations from the standard ASF distribution. The possible contribution of oxygenate intermediates in chain growth mechanism over the iron-based catalyst will be investigated at equivalent syngas conversion of that obtained for the cobalt-based catalyst.

Acknowledgement

The author would like to acknowledge the Consortium of Fossil Fuel Science and the DOE under grant (DE-FC26-99FT40540) for the finical support of the first phase of this study by and to Qatar National Research Funding under grant (NPRP 08-261-2-082) for funding the second phase of the study. The appreciation also extended to Dr. Christopher Roberts invaluable discussions.

References

B. Davis, Anderson Schulz Flory product distributions-Can it be avoided for Fischer-Tropsch Synthesis. American Institute of Chemical Engineers (AIChE) 2003 Spring Meeting, New Orleans, LA, 2003.

R. A. Dictor, A. T Bell (1983). An Explanation for Deviations of Fischer-Tropsch Products from a Schulz-Flory Distribution. Ind. Eng. Chem. Process Des. Dev., 1983, 22, 678-681.

M. E. Dry (2001). High quality diesel via the Fischer-Tropsch process-a review. J. Chem. Technol. Biotechnol. 77 (1), 43-50.

N.O. Elbashir, D.B. Bukur, E. Durham, C.B. Roberts. 2009. Advancement of Fischer Tropsch Synthesis via Utilization of Supercritical Fluid Reaction Media. AIChE J. in press.

N. O. Elbashir Bao B., El-Halwagi MM. 2009. An Approach to the Design of Advanced Fischer-Tropsch Reactor for Operation in Near-Critical and Supercritical Phase Media. In Alfadalla H. E., G.V. Reklaitis, M. M. El-Halwagi Advances in Gas Processing: Proceedings of the 1st Annual Symposium on Gas Processing Symposium (volume 1). Elsevier, Amsterdam, 2009; 423-433.

N.O. Elbashir, C. B. Roberts (2005) Enhanced Incorporation of α-olefins in the Fischer-Tropsch Synthesis Chain-growth Process over an Alumina Supported Cobalt Catalyst in Near-critical and Supercritical Hexane Medium. Ind. Eng. Chem. Res.; 44: 505-521.

N. O. Elbashir, P. Dutta, A. Manivannan, M. S. Seehra, C. B. Roberts (2005). Impact of Cobalt-Based Catalyst Characteristics on the Performance of Conventional Gas-Phase and Supercritical-Phase Fischer Tropsch Synthesis. Appl.Catal. A: Gen. 285, p. 169.

X. Huang, C. B. Roberts (2003). Selective Fischer–Tropsch synthesis over an Al2O3 supported cobalt catalyst in supercritical hexane. Fuel Process. Technol.,. 83: p. 81-99.

E. Iglesia, S. L Soled,. R. A. Fiato (1992). Fischer-Tropsch synthesis on cobalt and ruthenium. Metal dispersion and support effects on reaction rate and selectivity. J. Catal., 137 (1), 212-224.

M. Inoue, T. Miyake, T. Inui (1987) Simple criteria to differentiate a two-site model from a distributed-site model for Fischer-Tropsch synthesis. J. Catal. 105, 266-269.

R. J.Madon, E.Iglesia, S. C. Reyes (1993). Non-Flory product distributions in Fischer-Tropsch synthesis catalyzed by ruthenium,cobalt, and iron. ACS Symp. Ser. 517, 383-396 (Selectivity in Catalysis).

D. K.Matsumoto, C. N. Satterfield, Effects of temperature and hydrogen/carbon monoxide ratio on carbon number product distribution from iron Fischer-Tropsch catalysts. Energy Fuels 1989, 3 (2). 249-254.

J. Patzlaff, Y.Liu, C.Graffmann, J. Gaube (1999). Studies on product distributions of iron and cobalt catalyzed Fischer-Tropsch synthesis. Appl. Catal. A: Gen.186: p 109.

I. Puskas, R. S. Hurlbut, R. E. Pauls, Telomerization Model for Cobalt-Catalyzed Fischer-Tropsch Products. J. Catal., 1993, 139, 591-601 (and references therein).

B. Shi, G. Jacobs, D. Sparks, B. Davis (2005). Fischer–Tropsch synthesis: 14C labeled 1-alkene conversion using supercritical conditions with Co/A12O3. Fuel. 84: 1093-1098.

D. F. Smith, C. O.Hawk, P. L. Golden (1930). The mechanism of the formation of higher hydrocarbons from water gas. J. Am. Chem. Soc. 1930, 52, p. 3221-3232.

R.Snel, R. L. Espinoza (1987). Secondary reactions of primary products of the Fischer-Tropsch synthesis: Part 1. The role of ethene. J. Mol. Catal. 43(2): p 237-247.

B.T. Teng, C.H. Zhang, J. Yang, D.B. Cao, J. Chang, H.W. Xiang and Y.W. Li (2005). Oxygnate kinteics of Fischer Tropsch syntheis over an industerial Fe-Mn catalyst. Fuel 7–8, p. 791.

B. T. Teng, J. Chang, C.H. Zhang, D. Cao, J. Yang, Y. Liu, X. Guo, H. W. Xiang, and Y. W. Li (2006). A comprehensive kinetics model of Fischer–Tropsch synthesis over an industrial Fe–Mn catalyst. Appl. Catal. A: Gen. 301, p. 39-50.

N. Tsubaki,; K.Yoshii, K. Fujimoto (2002). Anti-ASF Distribution of Fischer-Tropsch Hydrocarbons in Supercritical-Phase Reactions. J. Catal. 207, p. 371-375.

Proceedings of the 2nd Annual Gas Processing Symposium
Farid Benyahia and Fadwa T. Eljack (Editors)

Infrared (IR) Thermography for Condition Monitoring at Q-Chem

Salahuddin Ghori

Qatar Chemical Company Limited(Q-Chem), Qatar

Abstract

Infrared (IR) Thermography, among other applications, reveals and measures heat generation in machines and equipments. It visualizes overheated components and detects and prevents "creeping" breakdowns. It has become a familiar and accepted technique for predictive maintenance. Reliability Professionals worldwide are gradually switching towards Risk Based Scheduling of equipment maintenance to optimize their reliability programs. Online inspection tools like IR Thermography supplements visual inspection and help to plan repair/ maintenance jobs in advance and acquire material beforehand. The main advantage of Infrared application is to avoid unscheduled shutdowns, which normally incur 5-10 times greater cost and take 25-30 % longer time to rectify than scheduled ones. Thermography also allows us to realize the time/cost difference between unscheduled outage and planned work.

Today, IR thermography has become a part of reliability improvement program in Oil &Gas and petrochemical industries worldwide. Realizing the benefits of this technology, many industries in the world are in the process of incorporating IR Thermography as a part of their reliability and safety management programs. Q-Chem, one of the leading High Density Polyethylene(HDPE) manufacturers in the region has managed to develop a successful IR inspection program which it would like to share with industry professionals through this paper.

This paper will discuss the application of this technology at QChem's operating units and the wealth of benefits use of this technology has brought to Q-Chem's Reliability Improvement Program. This paper will discuss the initial challenges and limitations faced when Q-Chem embarked upon making IR Thermography part of their best in class condition monitoring program of electrical and process equipment. This paper will share the experience of Q-Chem's first in the region use of Infrared View Ports and Spy Glass lens to carry out inspections of energized Electrical equipment which were rather not possible or extremely difficult due to stringent NFPA 70E requirements. This paper will also share QChem success stories where IR Thermography has helped preventing potential equipment breakdowns by identifying anomalies that could trigger equipment breakdowns and process interruptions. This paper will also emphasize on the necessary certifications and trainings a Thermographer should be equipped with before such a program is implemented.

Keywords: Infrared, Thermography, Condition Monitoring, Q-Chem, View Ports, SpyGlass, Switchgears

1. Introduction:

The use of Infrared (IR) Thermography has revolutionized the world of condition monitoring by providing a means of non contact, non destructive testing to identify imminent failures which have always been categorized as 'Hidden' and have always had the potential to hurt the safety and reliability statistics of any process industry. Since first becoming commercially available in the 1970's, IR thermography technology has saved millions of dollars by avoiding unscheduled downtime and detecting hidden failures in their early stages. With the advent of new technology, high resolution thermal imagers and user friendly software are available today, as a result infrared thermography is being used widely as an effective condition monitoring tool for Electrical and Process equipment in many process industries.

This technology has numerous uses in petrochemical industry. Q-Chem, one of the leading high density polyethylene manufacturers in Qatar took yet another 'first step' by adopting this technology as part of their best in class condition monitoring program. In the first part of this paper, the discussion will be focused on what thermography is all about and how this amazing technology fits into the spectrum of reliability, and how industry can maximize the potential of this highly visual, intuitive condition-monitoring technique? The remaining part of this paper will explore the application of this technology at Q-Chem, the methodologies used to set up an effective IR Inspection program followed by an account of challenges and successes associated with the program.

2. Why Thermal Imaging?

Temperature is one of the first observable parameters that indicate a change in the operation of equipment. Every equipment has one thing in common, it dies a death of heat. When failing, this equipment may have an increase in work, it may vibrate, it may make noise, but it always heats up. Heat, or thermal energy, is a byproduct of all types of 'work' whether electrical, mechanical, or chemical. The thermal behavior of equipment can be a powerful clue to diagnose problems and predict the reliability of industrial equipments.

3. What is Infrared Thermography?

Infrared Thermography is the technique that uses an infrared imaging and measurement camera to "see" and "measure" invisible infrared energy being emitted from an object. In simple terms it's a technology for making 'invisible' Infrared radiations 'visible'.

All objects which have a temperature above absolute zero emit heat energy in the Infrared portion of spectrum called Infrared radiations. As an object increases in temperature, it radiates more energy, and the wavelength gets shorter. Our eyes are unable to see these radiations unless the object temperature exceeds 500degree Celsius. Thermal Imaging cameras convert infrared energy into a visual two dimensional image allowing us to see the thermal energy associated with the equipment under inspection without actually coming in contact with it.

Infrared radiation, visible light and ultraviolet light are all forms of energy in the electromagnetic spectrum. The only difference is their wavelength or frequency. The wavelength regions most industrial and commercial IR cameras typically use are 'Short Wave' (2-5 μm)and 'Long Wave'(8-14 μm).

3.1. Thermogram:

An IR camera produces a thermal image or thermogram, much like a photograph as shown in Fig1. It displays the thermal radiation from an object based on radiance variations. The standard image is a grey tone picture displaying hot objects as White, descending in Grey tones to Black, indicating the coolest temperature within the scene. When appropriate, this image can be enhanced by overlaying a false color palette. Typical palettes use white, red and yellow to depict higher energy levels and violet, blue and black to display lower energy levels.

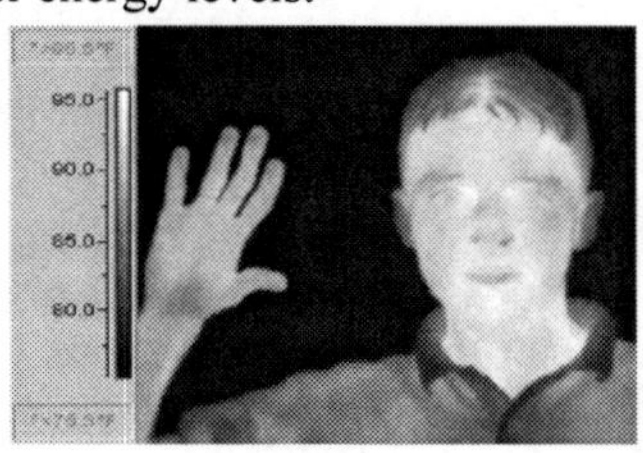

Figure 1: Example of a Thermogram

4. Thermography as Condition Monitoring Tool:

Thermography can be used as a complimentary Condition Monitoring technique alongside more commonly used techniques like Vibration Monitoring and Oil analysis, as it adds a new dimension to any good reliability program.

It may be appropriate to define these two terms:

4.1. ***Condition Monitoring*** - the continuous or periodic measurement and interpretation of data to indicate the condition of an equipment or component and determine the need for maintenance.

4.2. ***Condition Based Maintenance*** - maintenance initiated as a result of knowledge of the condition of an equipment or component from routine or continuous condition monitoring.

Big advantage of thermography is that large quantities of equipment could be scanned while in operation, hence production operation needs not to be interrupted.

5. Applications of Thermography in Process Industry:

Wherever a difference in radiated energy may indicate an abnormal condition, there is an application for Infrared Thermography. With this capability, plant maintenance personnel have recognized infrared thermography as one of the most versatile and effective condition monitoring tools available today. Effective application of condition monitoring technique, such as IR thermography, serves to reduce unscheduled emergency repairs, which tend to have a higher frequency of safety incidents and cost implications than works that are planned and scheduled.

5.1. Electrical Application:

Although there are numerous applications of IR Thermography in industry that but by far the most common use of thermography in process industry is within the Electrical field. The reason is obvious; virtually every industry, process, building, or facility in the industrialized world requires electrical energy to operate. It is a well documented fact that infrared surveys can be used as one of the most effective Condition Monitoring tools to survey electrical equipment ranging from high voltage switchgears, right down to low voltage battery systems. Infrared thermography allows a technician to test and help detect faulty connections in early stages, so that repairs may prevent possible future breakdowns that would be very costly. The investment in an infrared program can be paid back from the discovery of one major electrical fault.

5.1.1. The principle:

The principle driving the use of thermography in electrical applications is very simple. Current flowing through any electrical system will generate a small amount of heat because of electrical resistance. With time, the components and contact surfaces of the electrical system will begin to deteriorate. Any loose connection in the electrical system will also exhibit a point of high resistance. As an electrical connection deteriorates or if there is any loose connection, the resistance to the flow of electrical energy increases. This electrical energy is converted into thermal energy, which shows up as increased temperature. Wherever there is increased resistance caused by a condition such as a poor connection or overload, it will show up in the Infrared image as an anomaly.

5.2. Other Applications:

One of the benefits of thermography is its versatility, having the capability to assess the condition of a variety of applications, including process equipment and sometimes even the process itself. In addition to electrical inspections, infrared thermography can be used to discover anomalies in number of other applications.

Table 1 below summarizes some of the most common industrial applications of this technology.

Table 1: Applications of Infrared Thermography

Key Equipments	Identified Conditions
Motors, Pumps, Compressors, IC Engines, Fans etc	Overheated bearings, high discharge temperatures, unbalanced loads, overheating/overloading, blocked cooling passages.
Boilers, Furnaces, Ovens	Refractory/insulation breakdown, hot gas leaks.
Storage tanks, vessels, spheres	Product levels, debris build-up, inadequate insulation
Pipelines, steel, plastic, lined pipe spools	Blockages and restrictions, integrity of lining
Heat exchangers, plate and shell and tube	Operation, blockages
Steam Systems	Valve or joint leakage, insulation effectiveness, underground system leakage, restricted or blocked condensate return lines.
Refrigeration Systems	Air leaks, clogged condenser /heat exchanger tubes.

6. Thermography Program at Q-Chem :

Many businesses and organizations today are pursuing performance goals that are defined in terms of operational excellence. Q-Chem being one of them, recognizes safety performance as an integral element of its business success and understands that an effective condition monitoring program has the potential to provide even more benefits to an organization pursuing operational excellence.

Q-Chem's IR thermography program was mainly established with an objective of enhancing the condition monitoring of critical 'Electrical' equipments, realizing that failure of such can significantly impact the performance of an organization in pursuit of Operational Excellence. An incident resulting in disruption of electrical energy or control to hazardous chemical process could result in a process safety event, waste of raw materials, loss of production and damage to facilities and equipment, all of which having the potential to adversely affect the two most important assets; people and environment. It was also envisaged that once this program will go through its inception, acceptance and growth stages, other applications such as Mechanical and Process will also be made part of this condition monitoring program.

7. Path to Reliability:

Once it was established what is required, the next challenge was to define how to achieve it. To start off on the right foot, the program implementation team at Q-Chem developed a comprehensive implementation plan that defined the path of how and when each individual milestone will be achieved. Following paragraphs will briefly discuss the major steps that were undertaken by the implementation team and the challenges that were faced.

7.1. Selection of Thermal Imaging Equipment:

The first step towards establishing an IR Inspection program was the selection of thermal imaging equipment that would satisfy all the needs of intended application and the environment where it will be used. The proper selection of thermal imaging equipment is one of the crucial factors that success of any IR Thermography program depends on.

Since major application of thermography at Q-Chem was Electrical, the equipment selection criteria were made accordingly. The electrical equipment inventory comprised of an array of Electrical Switchgears, Motor Control Centers(MCCs), Power Transformers, AC Drives, Power and Lighting distribution boards, Battery Banks, UPS etc. The voltage levels involved were 132kV all the way down to 24V.

Based on the application requirements and after a lot of market research Mikron M7800 Thermal Imager with a temperature range of -40 to 500°C was selected. This temperature range was good enough to monitor all the above mentioned electrical equipments as well as mechanical and process equipments except furnaces. Another important selection criteria was capability of IR camera to work in conjunction with SpyGlass lens and View Ports, allowing closed door inspections of live electrical equipments, which will be discussed shortly.

7.2. Building up Equipment Inventory:

The next step was determination of what would be inspected i.e. development of a database. Equipment data base is the backbone of any condition monitoring program and must be established before onset of any inspection program. Without a database, a thermographer cannot tell what was inspected and what was not, cannot schedule an inspection and for the worst part cannot keep track of failures. A piece of equipment can go for years without being tested if no inspection record is kept. Realizing importance of a data base management tool, the implementation team opted for 'InspecTrend' which is a standalone Infrared inspection Management tool with numerous features to make life of a thermographer easier. InspecTrend manages everything from inspection inventory, scheduling, problem tracking and trending to report generation.

7.3. Frequency of Inspection:

Once the inventory was built up, the next step was development of inspection schedules in 'InspecTrend'. Frequency of inspection was driven by a number of factors, including safety, the criticality of the equipment, the expense of a failure, and the frequency with which problems impact production and/or maintenance. To begin with, an infrared survey and assessment was performed on each equipment with an initial frequency of 3 months. Based on the results of initial inspections and criticality, it was decided that this frequency will be stretched to 6 months for critical equipments and 12months for non critical ones. However there was always a room left for more frequent surveys and assessments to be carried out in special cases based on severity of operation and/or equipment frequency of failure.

7.4. Training:

Receiving the proper training is the best way to assure not only best return on investment and efforts but also keeping personnel safe. If maximum benefits are to be achieved from investment in this powerful, cost reducing technology, a commitment must be made to provide proper training for the Thermographer. Training is an essential component of IR Thermography program and should not be compromised with. One might perceive IR Thermography to be a relatively easy and straight forward task, i.e. survey the equipment and find the hot spot, but unfortunately this is not the case. Regardless of the application of an IR camera, its effectiveness is dependent not only on the person's skill in operating the camera, but also on knowledge and understanding of the component being inspected as well as the science involved in the assessment of the thermal evidence. Only a trained and experienced thermographer knows that not every hot or cold spot represents a problem, but may in fact reflect a component's normal operation, performance or location in the structure. Backed up by management support, training had always been considered a vital component of the program implementation at Q-Chem. All the thermographers received the necessary Level1 training and certification from a recognized training provider before any inspections were initiated in the field.

8. NFPA 70E Requirements:

For obvious reasons, Thermal imaging of any electrical equipment is only performed when it's energized and loaded. Removing or opening live electrical panels for conducting an IR survey would always pose a risk of Arc Flash especially when higher voltages are involved. NFPA (National Fire Protection Agency) 70E is the standard for safe electrical work practices which defines flash protection boundaries as limited, restricted, and prohibited.

NFPA 70E recommends that only "qualified" personnel donning high energy arc flash protective suits and specialized Personal Protective Equipment(PPE) be allowed to perform work inside the flash protection boundary. These stringent NFPA requirements and even more stringent Q-Chem Safety Procedures, forced the program implementation team to re evaluate their plans for initiating thermographic inspection program especially for energized electrical equipments.

9. IR View Ports and SpyGlass Lens-An innovation:

In search of alternative methods of conducting safer, standards compliant inspections we learnt about IR Windows or Viewports. An IR view port is basically an infrared transparent material with a holder/mounting body. The viewport uses only a 0.5 inch aperture that is covered by a sealed, screw-on cap to maintain the integrity and safety rating of the cabinet when not in use. These viewports can be easily installed as a retrofit in the field on any electrical cabinet provided safe distances from Live parts could be maintained as per IEEE C37.20.2 table A.3.

The SpyGlass lens is a specially designed wide angle lens, with its wide field of view (53° horizontal by 40°vertical, or 66° diagonal), which allows easy scanning of the interior of the electrical cabinet through the viewport, while providing a temperature measurement accuracy of ±3°C. View ports eliminate many of the risks associated with live inspections since they enable an Infrared camera fitted with spy glass lens, direct view of live electrical components without the need to open electrical enclosures which is a great advantage and adds a new safe and convenient dimension to the field of IR Thermography.

9.1. Important Considerations:

- It is of critical importance to choose an IR window/viewport that has been tested and approved by certification bodies such as UL, IEEE for the application.
- One must also ensure that material of IR window is suitable for the environment in which they will be used.
- An incorrect IR window or wrong installation of it can invalidate the safety rating of the switchgear as well as result into measurement inaccuracies.

Keeping in view the above factors, before venturing into the option for installing view ports on electrical switchgears, Q-Chem consulted the manufacturers of respective Switchgears and IR Viewports to ensure that installation of these view ports will not harm the safety rating and integrity of switchgear cabinets. Only after seeking approval from switchgear manufacturers, a total of 120 IR Viewports were installed in the cable compartments of 33KV, 11KV and 6KV switchgears, to facilitate closed door IR inspections of Medium Voltage (MV) cabinets. These view ports were strategically positioned on the cabinets to provide maximum view of cabinet interior to the thermographer. It is important to mention here that while carrying out inspections through IR viewports, only standard Personal Protective Equipment (PPE) is required thus eliminating the need for 'NASA Suit PPE'.
Fig 2 shows a 33KV Switchgear panel at Q-Chem with installed view ports. Fig 4 shows how IR scanning is carried out through View Port/SpyGlass combination without opening the panel covers. Fig 5 shows an inside thermogram of Medium Voltage cable compartment when scanned through SpyGlass and View Ports.

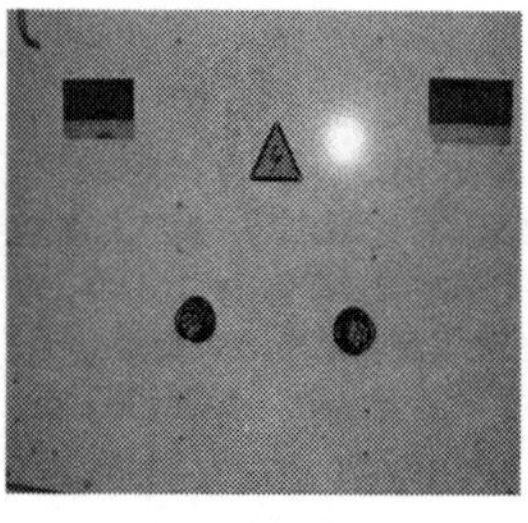

Figure 3: View Ports installed on 33KV Switchgear

Figure 4: Thermal Scanning through View Port/SpyGlass

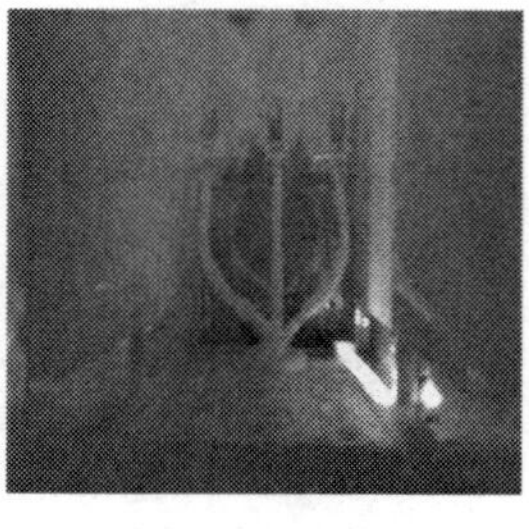

Figure 5: Inside Thermal View of 33KV Cable Compartment

10. Program Success Stories:

Q-Chem's Thermography program is a manifestation of the fact that with strong management commitment and support, combined with the assembly of a technically competent and dedicated team, significant value can be added to the Reliability statistics of the plant. Since its inception in November 2007, IR thermography program has proven its worth by identifying numerous hidden problems which had a very high likelihood of affecting plant availability if left undetected. This program provided a quick return on investment by identifying potential problem areas and avoiding many costly unplanned outages and process interruptions. Fig 6 statistically indicate the proportion of different types of problems identified by Thermography program at Q-Chem.

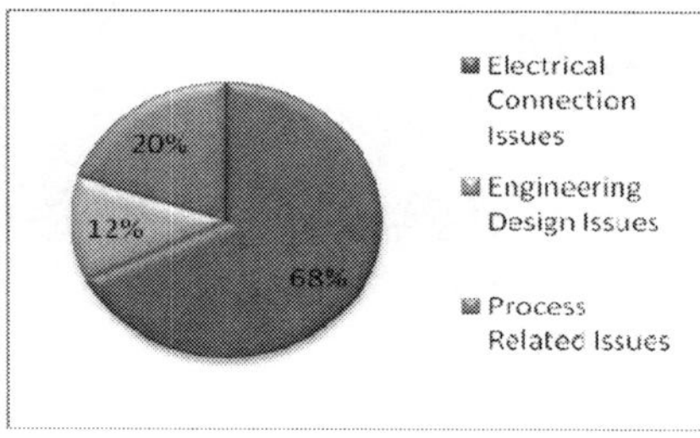

Figure 6: Different types of Problem Identified through IR Thermography

Some of the success stories are also being shared here.

Case 1: Ref Fig.7

Equipment: 2000KVA, 6KV/415V Distribution Transformer in Ethylene Substation

Detected Anomaly: High Temperature on middle phase cable termination (Transformer Primary).

Remedial Actions: Transformer taken off line. Found loose middle phase termination. All the Terminations re tightened. Carried out Thermal Imaging again to ensure normal operation.

Digital Image of Transformer

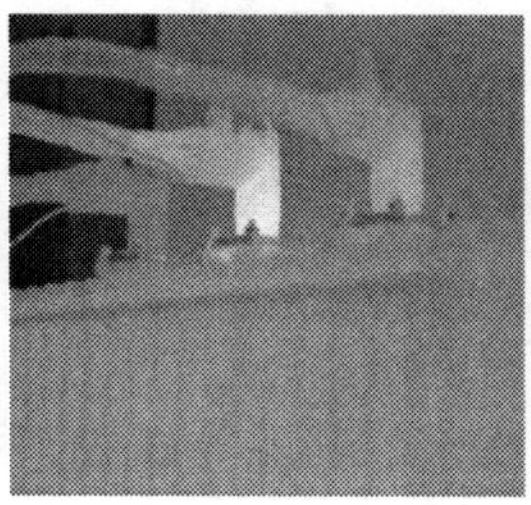

Thermal Image-Anomaly Identified

Thermal Image-Anomaly Repaired

Figure 7: Case 1 A 2000 KVA Transformer

Case 2: Ref Fig.8

Equipment: 75KW Critical Motor Feeder in Poly Ethylene Substation.
Detected Anomaly: High Temperature observed on one phase of motor feeder cable, most probably due to loose connection or inadequate lug crimping.
Remedial Actions: Physical inspection revealed looseness on the terminals. All the cable terminations were re tightened. Carried out Thermal Imaging again to ensure normal operation.

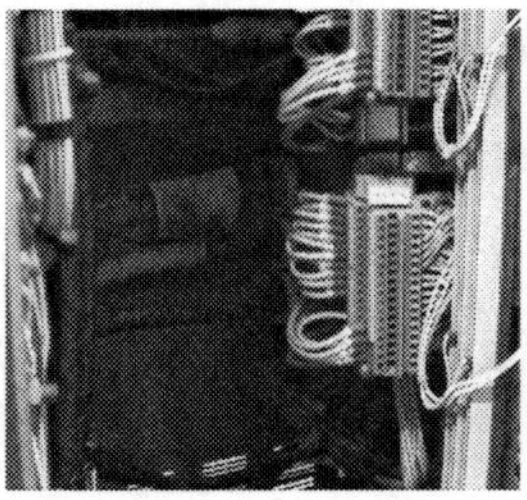

Digital Image of Motor feeder cable Compartment

Thermal Image-Anomaly Identified

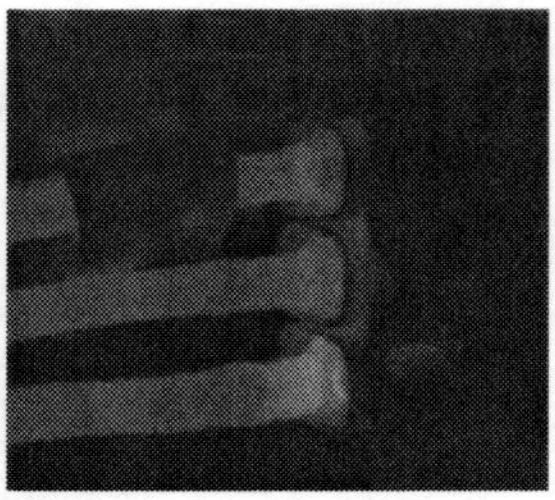

Thermal Image-Anomaly Repaired

Figure 8: Case 2 A 75KW Motor Feeder

Case 3: Ref Fig 9
Equipment: Sea water Electro chlorination Unit
Detected Anomaly: One of the 20 cells was observed with abnormally high temperature due to blockage caused by salt deposits, thus affecting the efficiency of the unit.
Remedial Actions: Blockage removed after dismantling the cells and unit was put back in operation.

Digital Image of Electro Chlorination unit

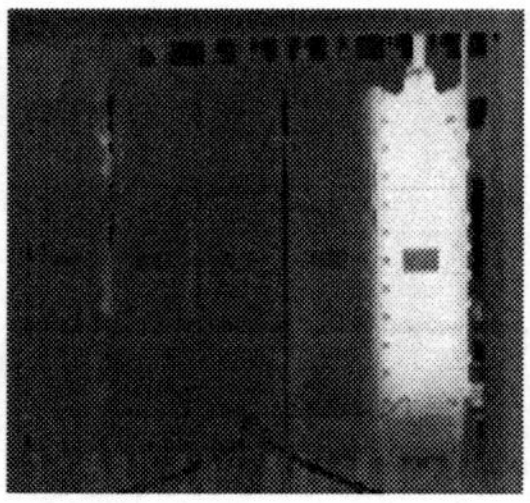

Thermal Image-Anomaly Identified

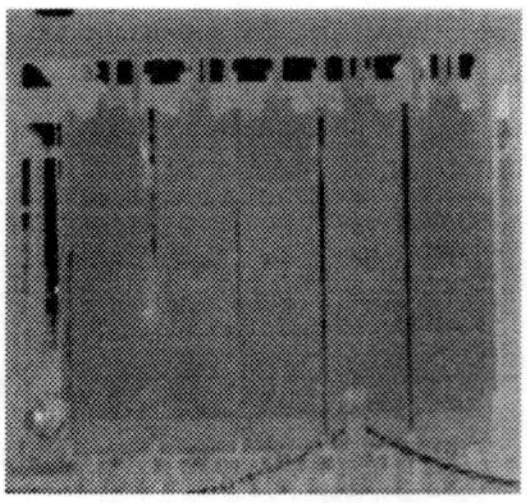

Thermal Image-Anomaly Repaired

Figure 9: Case 3 Sea Water Electro chlorination Unit

11. Conclusion

Use of Infrared Thermography as condition monitoring technique is poised to leap forward owing to its versatility of applications in process industry. A properly implemented and maintained, Infrared condition monitoring program has all the potential to enhance reliability and improve operating profits of any organization. Q-Chem's thermography program has witnessed significant success in a short span of time and has brought in wealth of benefits to its safety and reliability statistics. Despite the fact that there are efforts involved in establishing a successful program such as this, the benefits are well worth it.

Acknowledgments:

The author wants to greatly acknowledge the support of Q-Chem management in developing this paper.

References

Austin Dunne, 2004, "*Thermography Assists In Maintaining Equipment Reliability*"

Ron Newport, 2002, "*Electrical System Reliability utilizing Infrared Thermography*"

Scott Cawlfield, 2003, "*Infrared Inspection Methodologies and Data Collection Techniques*"

Martin Robinson, 2009, "*Windows Open the Door to Savings*", Uptime Magazine, Dec/Jan09.

Robert Kern, 2006, "*IR: The Next Generation*", Uptime Magazine, June 2006

"*Level 1 Infrared Certification*" Training Manual Doha, 2007, Institute of Infrared Thermography.

Mikron Infrared, "*SpyGlass_ViewPort_DataSheet Rev. 120406*"

Proceedings of the 2nd Annual Gas Processing Symposium
Farid Benyahia and Fadwa T. Eljack (Editors)

Industrial Operation of HySWEET®, a New Hybrid Solvent for Improved Mercaptan Removal

Renaud Cadours[a], Virenkumar Shah[a] and Claire Weiss[a], Damien Roquet[b], François Lallemand[c]

[a]*Total S.A., 2 Place Jean Millier, 92078 Paris La Défense, France*
[b]*Total EP France, Route de Bayonne, 64170 Lacq, France*
[c]*Total S.A., Avenue Larribau,64018 Pau, France*

Abstract

Removing mercaptans from sour natural gas has always been considered as a challenge. With the global trend towards more stringent specifications for commercial gases, it is becoming an even more important issue.

Solvent-based deacidification processes are very well-suited to the removal of H_2S and CO_2 from large quantities of sour gases, but generally require an additional treatment step to achieve the specification for total sulphur content in the exported gas. Amines have been extensively used because of their ability to meet the most severe H_2S and CO_2 specifications and their very high acid-gas selectivity over hydrocarbons, but their performances as regards mercaptan removal are extremely limited. Hybrid solvents are more efficient in removing mercaptans, but have the disadvantage of poor acid-gas selectivity over hydrocarbons, with the result that some of the hydrocarbons are lost with the separated acid gases.

On the strength of its extensive know-how and experience in acid gas removal with amine mixtures, Total has developed a new hybrid solvent formulation that offers simultaneous absorption of acid gases and mercaptans, with limited co-absorption of hydrocarbons. The solvent was selected at the laboratory stage, focusing specially on operational constraints such as cost, corrosion, foaming, degradation, etc. It was then tested on a pilot rig to validate its performances in effectively removing acid gas and mercaptans.

The opportunity of a first industrial demonstration then arose at the Lacq sour gas processing plant, in South-West France, where, with the decline in the sour gas production flow rate, one of the three amine units was due for definitive shutdown in July 2007. It was decided to run field trials on the new hybrid solvent before the unit was shut down.

With the success of this industrial demonstration, the new HySWEET® process was then implemented, in the first quarter of 2008, in one of the two remaining units of the Lacq plant. Half the gas production is now treated with the hybrid solvent, and the plant has achieved a higher global rate of mercaptans removal.

This paper fully documents the results of both the 2007 campaign and the current operation, and shows that the newly developed process is plainly a good contender for achieving the increasingly stringent commercial gas specifications to be met in the development of new sour gas fields.

Keywords: HySWEET® process, acid gas, mercaptan, thiodiglycol

1. Introduction

Total's expertise, which started with the production of the Lacq field in South-west France in the late 1950s, has evolved and improved continuously (Figure 1) to reduce the cost and energy consumption involved in sweetening sour natural gases. Today, the AdvAmine™ technology (Streicher et al., 2004) is grounded in considerable industrial experience, with more than 110 units built, about one fifth of which are, or were formerly, operated by companies in the Total Group. This knowledge spans not only design "know-how" but also all the practical operational issues downstream, such as corrosion control (Bonis et al., 2004), foaming and degradation. Other specific improvements have also been made, for instance for high COS removal (Cadours et al., 2006).

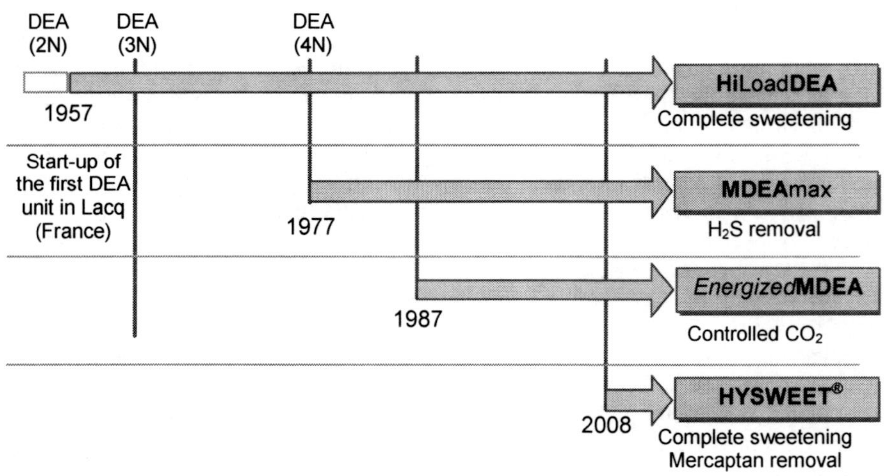

Figure 1: History of the AdvAmine™ technology developed by Total

Whereas deep removal of CO_2 and H_2S is now perfectly mastered, mercaptan (R-SH) removal from a sour gas is still considered a challenge. For the treatment of a sour gas containing mercaptans, the usual CO_2 and H_2S specifications can be achieved by amine units but additional treatments are then necessary to remove the mercaptans. Three main options are generally considered for this.

This first option consists of implementing a finishing stage, for example molecular sieves, downstream of the amine treating process. The main drawback in this case is posed by the regeneration gas used in the sieves, which contains mercaptans and must in turn be treated. The succession of treatment stages entails higher investment and operating costs. In the second option, mercaptans are concentrated in the liquid hydrocarbon cuts, which are usually sweetened by the use of caustic soda. This option can be entertained only if LPG recovery is envisaged. Moreover, as with the first option, the succession of treatment stages increases the capital and operating costs. The third option consists of removing the mercaptans and acid components simultaneously, in a single deacidification unit, by using a hybrid solvent process. However, the current hybrid technologies suffer the disadvantage of significant hydrocarbon co-absorption, which has important consequences for the downstream Claus unit.

Taking advantage of its extensive know-how and experience in gas sweetening, Total has improved its solvent technology by developing the HySWEET® process, which offers simultaneous removal of CO_2, H_2S, and mercaptans, while preserving low hydrocarbon co-absorption. A meticulous methodology was used for the development of the new hybrid solvent process, starting with the screening of molecules, followed by laboratory and pilot data acquisition. Subsequently, the process was subjected to a techno-economic evaluation to assess the performances of the HySWEET® process for the typical application cases and compare them with those of conventional amine processes. The results have been confirmed in industrial operations at the Lacq sour gas plant, where half the gas production is now treated with the HySWEET® solvent.

2. Development of the HySWEET® process

Analysis of existing technologies led Total to consider the research and development of a new hybrid solvent targeting high performance for mercaptan and acid gases removal and low affinity for hydrocarbons.

The R&D program began by screening potential co-solvents for a water-amine solvent. One key parameter here was the solubility of the gaseous impurities to be absorbed simultaneously: CO_2, H_2S, COS and mercaptans. Besides this, different technical and economic criteria, designed to ensure the process would be competitive compared to other existing processes, were also considered: industrial property situation, stability, volatility, commercial availability, cost. Based on these different criteria, ThioDiGlycol (TDG) was selected as the best candidate (Capdeville et al., 2006).

Different formulations of the new hybrid solvent were explored, by mixing TDG with various types of amines. The HySWEET®-DEA (i.e. water + DEA + TDG) solvent was selected for total deacidification of a natural gas, i.e. the complete elimination of all the acid gases. The HySWEET®-MDEA solvent (i.e. water + MDEA + TDG) was retained for the selective absorption of sulphur compounds over CO_2.

The HySWEET® solvent, studied first at laboratory scale, offers a greater capacity to absorb mercaptans in the presence of acid gases than the reference solvents of the AdvAmine™ technology. N-paraffins and aromatic hydrocarbons (benzene, toluene and xylene) have higher solubilities than in the reference solvent. However, these higher hydrocarbon solubilities fairly remain within an acceptable range.

Once the new hybrid solvent formulations had been identified, a data acquisition program was set up to acquire thermodynamic, kinetic and physicochemical data in a range of temperatures, pressures and solvent compositions spanning the typical operating conditions of the process. The aim is to adapt all the proprietary models specifically used for amine solvents to enable them to represent a three-component solvent. The rigorous models under development will be integrated in a proprietary amine plant simulator, enabling Total to propose optimal gas sweetening unit designs with HySWEET® solvent.

This entailed verifying the capability of the solvent to reach the low mercaptan concentrations in a real absorber column, where, besides equilibrium and reaction kinetics, the physical properties of the solvent such as its viscosity, and contactor efficiency also play an important role. The impact of reboiler heat duty on the quality of the regenerated solvent was also an important parameter to be investigated. An existing solvent pilot unit (Lallemand et al., 2008a) was used to pursue these studies. The pilot experiments showed that a stringent mercaptan specification, as low as 4 ppmv, could be reached under certain operating conditions. This confirmed the ability of the hybrid solvent to reach a low total sulphur specification, provided that the number of transfer stages is sufficient. The regeneration heat duty of the new hybrid solvent was also measured on the pilot plant, in comparison with that of the reference solvent for the same rich solvent CO_2-loading. For example, the reference solvent requires 130 kg of steam/ton of solvent to achieve 0.03 mol/mol lean CO_2-loading, while the hybrid solvent needs only 110 kg of steam/ton of solvent to achieve the same CO_2-loading. Reboiler heat duty values measured during these tests are specific to the pilot plant configurations and operating conditions, and should therefore not be used directly in process design. However, these results give a clear indication that the energy consumption of an industrial amine unit could be reduced by 10-15% if the process is switched from **Hi**Load**DEA** to HySWEET®.

In parallel to the pilot tests, corrosion, foaming, stability and degradation issues were studied (Lallemand et al., 2008a), and confirmation was obtained that the process would not be impaired by operational problems arising from the use of the new hybrid solvent.

3. Techno-economic evaluation

The laboratory data and models obtained during the process development were used to evaluate the performances of the HySWEET® process against those of the conventional processes of the Advamine™ series. Different formulations of the solvents were investigated, in pairs, according to the sweetening performance required: HySWEET®-DEA and **Hi**Load**DEA** for total deacidification, HySWEET®-MDEA and **MDEA**max for selective deacidification. The main operating parameters of the processes were evaluated and compared, namely the solvent flow rate, the duty required to regenerate the solvent, the mercaptan removal efficiency, and the hydrocarbon losses in the acid gas.

Various cases of highly sour gas with high mercaptan content were considered. In all of them, the deacidification process requires a high solvent flow rate to remove large amounts of acid gases, in order to reach stringent specifications for both H_2S and mercaptans, and possibly CO_2 too. For these application cases, the HySWEET® process removes more than 95% of the mercaptans, whereas the mercaptan-elimination capability of a water-amine solvent is limited to 55-60% (Figure 2). The hydrocarbon content in the separated acid gas is greater with the new hybrid solvent but remains below 1 vol. %, which is considered as an acceptable level of concentration, deemed not to impair the performances of a downstream Claus unit. Moreover, the HySWEET® process consumes less energy in solvent regeneration than the classical amine processes.

Several cases of sour gas with low acid-gas contents were also examined, some of them with no stringent CO_2 specification for the treated gas, allowing the use of MDEA-based selective solvents. For these low acid-gas content cases, the HySWEET® solvent flow rate required to remove mercaptans down to the specification would be much higher than that required to achieve solely the H_2S/CO_2 specification. Thus, the economic optimum calls for the integration of an additional mercaptan removal step, as in the classical amine process, but the unit will be much smaller because of the lower mercaptan content of the gas downstream of the HySWEET® unit.

The HySWEET® process, its efficiency and suitability compared to chemical solvent processes need to be evaluated on a case-by-case basis to establish an optimum process scheme. Generally speaking, for highly sour gases, it will simultaneously eliminate both acid gases and mercaptans to the level specified for the treated gas (Figure 2), rendering any additional mercaptan removal step (e.g. molecular sieves) unnecessary. It therefore represents considerable savings in the capital and operating costs compared to AdvAmine™ technology.

On the other hand, for the treatment of gases with low acid-gas content, HySWEET® will in most cases not, economically, reach a severe mercaptan specification without the installation of a polishing unit (Figure 2). This however, will be smaller than that necessary after an amine, because of the significantly higher mercaptan-removal efficiency attained by the HySWEET® process. And, as with highly sour gases, the energy required to regenerate the hybrid solvent remains less than for regeneration of the amine solvent.

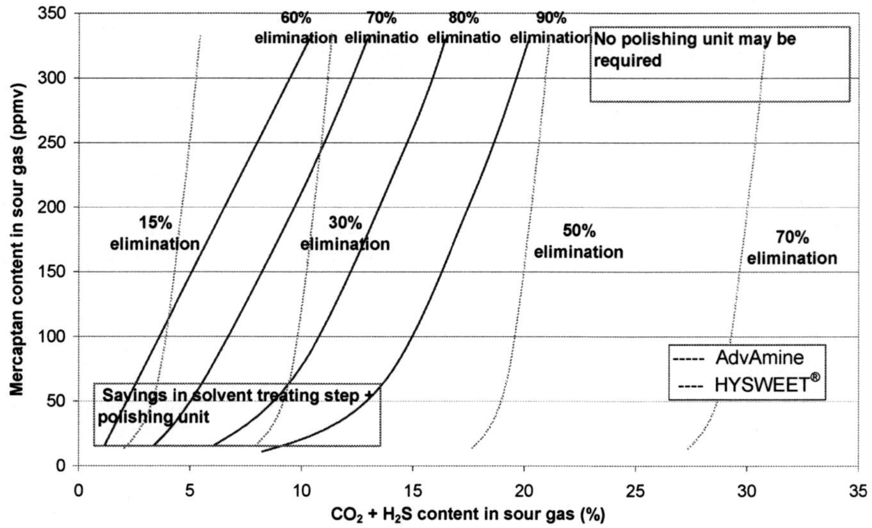

Figure 2: Mercaptan elimination vs. CO_2+H_2S content in the raw gas

4. Industrial operation of the HySWEET®-DEA process at the Lacq plant

Following the positive results of the techno-economic evaluation, an opportunity arose in 2007 to carry out an industrial demonstration of the HySWEET®-DEA process in the T9 unit of the Lacq plant (Lallemand et al., 2008b), originally designed for DEA. The original DEA was replaced by the HySWEET®-DEA formulation to verify the performances of the new process in industrial operation, with the perspective of converting the two other amine sweetening units of the Lacq plant to the new process should the demonstration be successful.

A reference test using the **Hi**Load**DEA** solvent was first carried out, after which the final composition of the HySWEET®-DEA solvent was achieved by successive additions of TDG. By analysing the different parameters defining the technical and economic efficacy of the new hybrid solvent, and their evolution versus the composition of the solvent (Figure 3), it was then possible to identify an optimum, specific to the tests conducted on the T9 unit. At this optimum concentration of TDG (Figure 3), the treated gas leaving the absorber column contained on average 35 ppmv of mercaptans, corresponding to 95% mercaptan removal, compared to the 60% achieved during the reference test.

The test performed on the T9 unit also confirmed that implementation of the new solvent does not degrade the operability of the sweetening unit compared to that of the well-established **Hi**Load**DEA** process. The quality of the separated acid gas was confirmed, and no adverse consequences were observed on the operation and performance of the sulphur recovery units. Reboiler energy savings were also measured, and found to be as expected.

After this successful industrial demonstration, it was decided to implement the HySWEET®-DEA process in the T12 unit of the Lacq plant in 2008 (Cadours et al., 2009). In contrast to the test on T9 unit, which was an R&D test at industrial scale, implementation of the HySWEET®-DEA process in T12 is a routine commercial operation. Half the plant's gas production is now treated with the hybrid solvent.

The same mercaptan removal efficiency observed during the industrial demonstration on T9 was also measured in routine operation of the T12 unit. Moreover, with the swap from **Hi**Load**DEA** to HySWEET®-DEA, the Lacq plant has reduced reboiling energy consumption by 15% and the regeneration gas flow rate of the molecular sieves by 12%.

Co-absorption of hydrocarbons was measured on T12: for the same solvent circulation rate, the co-absorption of light hydrocarbons and BTX with HySWEET®-DEA process was found to be respectively 39% and 62% higher than with the **Hi**Load**DEA** process. However, the total hydrocarbon content in the acid-gas stream does not exceed 0.6 vol. %, with no direct impact on routine operation of the Sulphur Recovery Units.

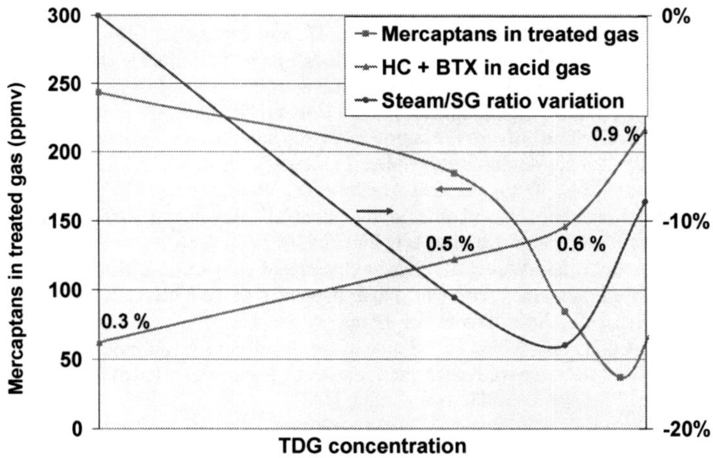

Figure 3: Performances of the HySWEET®-DEA process on T9 unit

It must be pointed out that the performances of the HySWEET®-DEA process observed on units T9 and T12 do not represent the maximum performance that can be obtained with this process, as the units were not modified from their original design based on the DEA solvent: better performances would be achieved with a unit specifically designed for the HySWEET® process.

5. Conclusion

Total is at the final stage of development of the HySWEET® process, based on a hybrid solvent containing an amine solution and thiodiglycol, which will significantly reduce the total sulphur content in the treated gas, with a lower reboiler heat load while keeping co-absorption of hydrocarbons at an acceptable level.

The techno-economic study carried out during the process development has shown the efficiency of the HySWEET®-DEA and HySWEET®-MDEA processes for total and selective deacidification of gases, respectively. Two years of operation with the HySWEET®-DEA process have confirmed a number of the conclusions of the techno-economic study, chiefly the ability of the new solvent to improve the mercaptan separation and significantly reduce solvent regeneration duty.

The HySWEET® process represents an alternative to the present gas sweetening solutions for the development of new sour gas fields, as well as a possibility of retrofitting existing plants to comply with more stringent total sulphur specifications.

References

Bonis, M., Ballaguet, J.P., and Rigaill, C., "A critical look at amines: a practical review of corrosion experience over four decades", paper presented at the 83rd Annual GPA Convention, New Orleans, Louisiana, USA, March 14-17, 2004.

Cadours, R., Magné-Drisch, J., Normand, L., Roquet, D., and Perdu, G., "COS removal from natural gases by absorption in alkanolamine solutions", paper presented at the 85th Annual GPA Convention, Grapevine, Texas, USA, March 5-8, 2006.

Cadours, R., Shah, V., Weiss, C., Lallemand, F., and Roquet, D., "Industrial operation of HySWEET®, a new hybrid solvent for improved mercaptan removal", paper presented at the GPA European Annual Conference, September 23, 2009, Venice, Italy.

Capdeville, S., Peytavy, J.L., Frémy, G., and Anglerot, D., "Process for purifying gaseous mixtures containing mercaptans and other acidic gases", French Patent 06 00448 January 18, 2006, International Publication Number WO 2007/083012 A1.

Lallemand, F., Roquet, D., and Weiss, C., "Development and industrial validation of a new hybrid solvent for improved mercaptan removal", paper presented at the 87th Annual GPA Convention, March 2-5, 2008a, Grapevine, Texas.

Lallemand, F., Roquet, D., and Weiss, C., "Status of the development and industrial validation of a new hybrid solvent for improved mercaptan removal", paper presented at the SOGAT conference, April 27- May 1, 2008b, Abu-Dhabi, UAE.

Streicher, C., Lallemand, F., Minkkinen, A., and Lecomte, F., "A new look at amines: Extending the gas industry "workhorse" to new limits", paper presented at the 83rd Annual GPA Convention, New Orleans, Louisiana, USA, March 14-17, 2004.

Proceedings of the 2nd Annual Gas Processing Symposium
Farid Benyahia and Fadwa T. Eljack (Editors)

FLEXSORB® SE A Proven Reliable Acid Gas Enrichment Solvent

L. E. Parks, D. Perry , R. Fedich

ExxonMobil Research and Engineering Company, 3225 Gallows Road, Fairfax, Va 22037, USA

Keywords: amine; sulfur; gas absorption; hydrogen sulfide; selective removal

1. AGE Background

To bring natural gas to market, either via pipeline to nearby consumers or via liquefaction to distant markets, various impurities in the produced gas must be removed. Chief among these are H_2S and CO_2. These acid gas components are removed from the sales gas using traditional amine treating processes also known as Acid Gas Removal (AGR) units. Typically, the AGR absorbers are run at high pressure 35-83 barg (500-1200 psig) to minimize recompression costs.

Typical specifications for H_2S in sales gas are <4 vppm. For pipeline gas, the residual CO_2 is usually determined by the overall heating value of the gas stream, but generally ranges between 2-5 mol%. For Liquefied Natural Gas (LNG), the H_2S and CO_2 specifications are significantly reduced to <2 and <50 vppm respectively to avoid corrosion and freezing problems in downstream processing.

As a result of the sales gas specifications, AGR's remove essentially all of the H_2S and CO_2 from the produced natural gas. AGRs can utilize a variety of solvents, including generic and formulated MDEA. In some cases, controlled CO_2 removal can be used using a selective solvent such as FLEXSORB or MDEA, to leave some of the CO_2 in the sales gas. For gas fields with high CO_2 to H_2S ratios, the acid gas stream from the AGR regenerator will have an unfavorably high CO_2 to H_2S ratio resulting in a poor quality feed to a Claus plant.

As the amount of sulfur removed from the sales gas exceeds a threshold limit between 1 and 10 MT/D, depending on local regulations, it is generally required to recover the sulfur. The preferred recovery technology is the modified Claus process. One of the keys to good Claus plant performance is to maintain a stable flame and high temperatures in the main reaction furnace (>926°C, 1700°F). Higher temperatures increase the conversion of H_2S to elemental sulfur. CO_2 in the acid gas feed to the Claus unit acts as a diluent, reducing reaction furnace temperatures. To achieve high temperatures in the Claus furnace an acid gas feed stream typically requires at least 50 mole percent H_2S. If the H_2S concentration in the acid gas from the AGR is lower than 50%, several Claus plant design options are available. The most common is the split-flow design where a portion of the dilute acid gas is bypassed around the front of the main reaction furnace. All of the combustion air and a fraction of the acid gas are fed to the main burner. Bypassing some of the acid gas to a second furnace chamber raises the temperature in the combustion zone but lowers the sulfur recovery efficiency. If the acid gas H_2S concentration is below 20%, a split flow design cannot achieve the required temperature in the main reaction furnace. In these cases, Claus plant designers can choose to fire a supplemental fuel, or utilize oxygen enrichment of the combustion air, or preheat the air and/or acid gas feed to the Claus furnace to maintain the proper temperatures. None of these options are preferred.

In the last two decades, a new option to process dilute acid gas streams called Acid Gas Enrichment (AGE) has become an increasingly economic option. As the name implies, acid gas enrichment concentrates the H_2S from the AGR system by further gas treatment in a second amine unit utilizing a selective amine solvent. Except for the use of the selective amine solvent, an AGE unit is similar to other traditional amine treating units. Figure 1 shows a simplified flow diagram of an acid gas enrichment unit. The AGR acid gas is fed to the base of an absorber column equipped with either trays or packing where the H_2S is absorbed via counter-current contacting with the descending amine solvent. The AGE absorber typically operates at low pressure (~0.5 barg, 7 psig), compatible with the operating pressure of the upstream AGR regenerator overhead system. The selective amine in the AGE preferentially absorbs the H_2S and allows the CO_2 to remain in the treated gas (also known as "CO_2 slip"). The rich amine from the bottom of the absorber is then pumped through a rich-lean heat exchanger and on to a regenerator tower. The regenerator produces an enriched acid gas product overhead and the lean amine bottoms product to be recycled to the absorber.

To achieve the twin goals of low H_2S in the treated gas and low CO_2 in the enriched acid gas, the AGE amine solvent must maximize the selectivity for absorbing H_2S. AGE process simulations by Weiland (2008) have demonstrated that the CO_2 and H_2S partial pressure driving forces in the AGE absorber work against achieving these goals simultaneously. As the gas moves up the absorber tower, the H_2S partial pressure is decreasing reducing the mass transfer driving force. At the same time, the CO_2 partial pressure is increasing, making CO_2 pick-up more difficult to avoid.

The severe steric hindrance of the FLEXSORB SE molecule is specifically tailored to give high H_2S absorption capacity in the presence of CO_2. This allows FLEXSORB SE and SE PLUS solvents to achieve high H_2S cleanup, high solvent capacity and high H_2S

selectivity at low solvent circulation rates. The FLEXSORB SE PLUS solvent is especially formulated to facilitate lean solvent regeneration, thus allowing treatment of

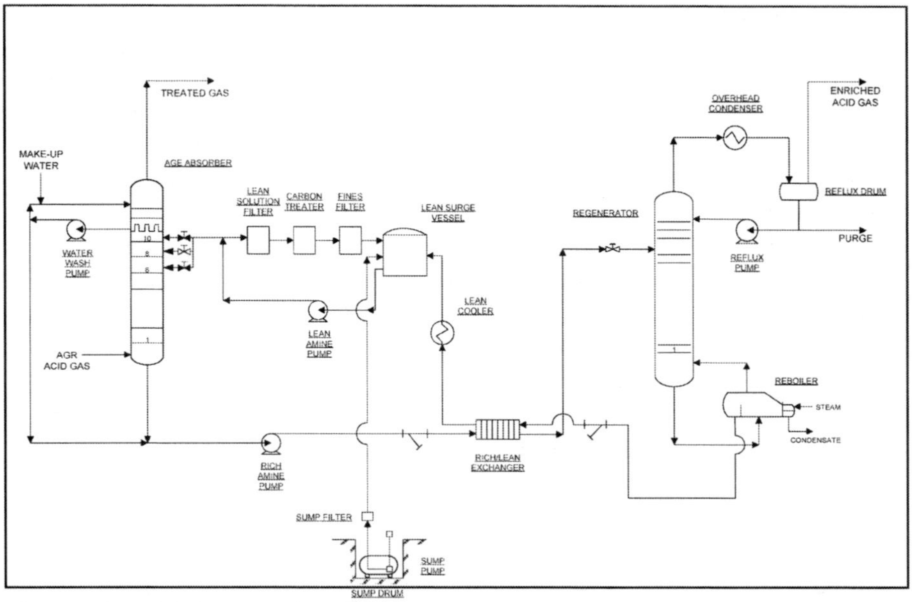

Figure 1: Standalone AGE Flow Diagram

gases to less than 10 vppm H_2S. No solvent reclaiming is required in AGE or TGTU services because the FLEXSORB SE amine molecule is chemically very stable.

Chludzinski and Iyengar (1995) describe the application of these sterically hindered amines to AGE units as well as outline the operating conditions and some of the possible unit configurations.

2. What levels of enrichment are possible?

The level of H_2S enrichment depends on several factors including the H_2S concentration in the feed gas, the H_2S specification in the treated gas, process operating temperatures, the design of the AGE absorber tower internals, and of course, the selective solvent selection. Higher processing temperatures may require increased lean amine circulation rates to ensure that the H_2S specification in the treated gas is met. Increased lean amine circulation rate generally reduces selectivity. Similarly, a tight specification for residual

H_2S in the treated gas (CO_2 rich) stream may also require higher lean amine circulation rates or greater mass transfer contacting area, both of which will directionally reduce selectivity.

The design of the tower internals influences the mass transfer rates and H_2S selectivity as highlighted by Weiland. ExxonMobil has found that tray columns generally provide better selectivity than packed columns in AGE service. One of the reasons is that in a tray column, the mass transfer parameters can be finely tuned by adjusting the number of liquid passes, weir heights and open area on the trays. It is also easier to provide multiple lean amine inlets in a tray column for optimizing the mass transfer zone, thereby providing operators with additional flexibility to meet changing gas feed compositions.

To successfully achieve the AGE treating goals of a low H_2S content in the treated gas and maximum CO_2 slip requires a solid base of supporting data and careful engineering. Table 1 below, illustrates the range of H_2S enrichment that can be achieved using FLEXSORB SE or SE PLUS solvent in a single enrichment step.

Table 1: Examples of H2S enrichment

Project	A	B	C	D
Inlet H_2S, mol% (wet)	4.0	14.7	21.9	32.1
Enriched H_2S, mol% (wet)	41.5	51.4	59.5	75.3

At very low levels of H_2S in the AGR acid gas, two consecutive stages of enrichment can be used. Alternatively, a recycle of AGE enriched acid gas back to the AGE absorber inlet can be used to achieve sufficiently high levels of H_2S in the Claus feed.

3. Commercial Experience

Several published case studies have documented the engineering details, project economics and operating results of FLEXSORB AGE units. Collectively, these case studies illustrate the wide variety of process configurations that can be utilized with the FLEXSORB SE or SE PLUS solvent. The optimum configuration for each new project must be determined by the individual project constraints including local environmental regulations, gas compositions, and economics.

Table 2 provides a concise summary of the range of experience for FLEXSORB SE and SE PLUS AGE units.

Table 2: FLEXSORB AGE Experience Bands

	Minimum	Maximum
Acid Gas Feed Conditions:		
Rate, kNm^3/h (MSCFD)	2.0 (1.8)	76.2 (68.3)
Pressure, barg (psig)	0.3 (4.5)	0.8 (11)
Temperature, °C (°F)	27 (80)	49 (120)
Inlet H_2S, mol%	1.4	39

Royan et al. (1992) described the performance of a FLEXSORB AGE with a very high CO_2 to H_2S ratio. A single stage FLEXSORB AGE is used to produce an enriched acid gas of about 20% H_2S which was then processed in a split flow Claus plant. The FLEXSORB also produced less than 10 vppm H_2S in the treated gas which was vented to atmosphere.

ExxonMobil started gas production at the Mobile Bay plant in March 1994. Natural gas is produced offshore where associated liquids are separated. The gas is then treated for hydrocarbons, dehydrated and piped onshore for further treating and sulfur recovery. (True, 1994) The gas is produced from separate formations where the H_2S concentration can vary from 50 vppm to 1.5 mol%. The CO_2 concentration was relatively constant at 3-4 mol%.

Because of the possible wide variability in the inlet H_2S concentration, the onshore gas treating facilities were designed to be very flexible. The sales gas is first treated to produce a pipeline gas containing less than 4 vppm H_2S and 2 mol% CO_2. The raw acid gas from the AGR regenerator contains from 1% to 25 mol% H_2S. This raw acid gas is treated in an AGE absorber using FLEXSORB SE PLUS solvent utilizing a lean / semi-lean configuration. The lean FLEXSORB SE PLUS is fed to the top absorber section and the partially-loaded rich solvent from the two tail gas absorbers is fed along with additional lean solvent to the lower section of the absorber. The lean plus semi-lean solvent configuration in the AGE reduces the total solvent circulation rate and thereby reduces the size of the regenerator and its auxiliary systems. The lean plus semi-lean combined AGE / TGT units have demonstrated total solvent circulation savings of 20% or more compared to the solvent required for two independent absorbers. The rich solvent from the AGE is sent to the FLEXSORB SE regenerator to produce an enriched acid gas feed to the two Claus plants. Typically, the enriched acid gas contains about 70 to 80 mol% H_2S and essentially no hydrocarbons.

Garrison et al. (2002) describes a retrofit application where the AGE was added to an existing gas plant to help meet new, higher sulfur recovery requirements. A detailed engineering study compared the performance of FLEXSORB SE PLUS to a proprietary MDEA solvent. FLEXSORB was chosen because it offered higher H_2S in the enriched acid gas, lower H_2S in the treated gas, lower circulation rate and lower capital cost. Due to the enrichment capabilities of the FLEXSORB SE PLUS solvent, part of the primary acid gas was bypassed around the AGE while still being able to meet the minimum H_2S concentration required for the Claus plant feed. This approach helped minimize the capital cost of the new facilities. The FLEXSORB AGE consistently enriched the acid gas from about 30 mol% to 65 mol% H_2S.

Jones et al. (2004) describes the integration of an AGE followed by acid gas injection. The AGE unit utilizes FLEXSORB SE solvent to reduce the volume of acid gas to be compressed and re-injected by an order of magnitude from 15.6 kNm^3/h to 1.45 kNm^3/h (14 to 1.3 MSCFD). This project also illustrated the operating flexibility advantages of designing the AGE absorber with alternate lean amine feed locations. Operating experience demonstrated that using the lowest lean amine inlet maximized CO_2 slip and still met H_2S specification in the treated gas using only four actual trays.

Hanna and Ilyasak (2009) describe the newly commissioned Qatargas LNG trains 4&5 (also known as the QatarGas II Project or QGII) which use FLEXSORB SE PLUS solvent in both the AGE and TGT units. The QGII trains are the world's largest LNG trains rated for 7.8 million tons of LNG per year. The AGE absorber feed gas H_2S concentrations ranged from 16-34 mol% for the design cases studied. The CO_2-rich treated gas from the AGE absorber can be sent to the TGT absorber or directly to the incinerator. The treated gas from the TGT absorber is sent to the incinerator. Each treated gas stream from the AGE or TGT contains less than 150 vppm H_2S with lean amine temperatures as high as 45°C (113°F).

The AGE tower is fitted with 12 valve trays with 3 alternate lean amine inlet locations while the TGT absorber and regenerator towers are designed using 50mm random packing. Each absorber also includes a small water wash section to minimize solvent losses. The recirculating water wash has an intermittent purge that is sent with the rich amine stream to recover the FLEXSORB SE PLUS.

Qatargas III and IV (trains 6 and 7) will share essentially the same design as QGII.

4. Conclusion

ExxonMobil's FLEXSORB SE or SE PLUS solvents are in use in a wide variety of acid gas enrichment unit designs around the world. These units include both ExxonMobil affiliates and licensees. These plants range in size from 3 to more than 1000 tons per day sulfur capacity. CO_2 slip ranges from 73-94%.

Multiple independent project analyses have repeatedly demonstrated the advantages of FLEXSORB SE and SE PLUS over competing MDEA-based solvents. FLEXSORB SE has advantages in better H_2S selectivity and lower circulation rates. This property allows FLEXSORB SE to achieve higher H_2S cleanup in the AGE treated gas and higher H_2S concentration in the enriched acid gas. These advantages result in substantial investment savings for new units, debottleneck capacity for retrofit applications, and operating cost savings for either type.

References

Chludzinski, G. R., and J. N. Iyengar, 1993, "Lean Acid Gas Enrichment with Selective Hindered Amines," World Intellectual Property Organization, Patent Number WO 93/10883, June 10, 1993.

Garrison, J., S. Loucks, B. K. Moore, P. S. Sheth, and C. S. Tanzi, 2002, "Keyspan Energy Canada Rimbey Acid Gas Enrichment with FLEXSORB SE PLUS Technology," Proceedings of the 2002 Laurance Reid Gas Conditioning Conference, Norman, Oklahoma, USA.

Hanna, J. M. and M. Ilyasak, 2009, "Qatargas Expansion Projects: Why Change the Gas Treating Concept from Sulfinol-D?," SOGAT 2009, Proceedings of the 5th International Converence, March 31-April 1, 2009, Abu Dhabi, UAE.

Jones, S. G., D. R. Rosa, and J. E. Johnson, 2004, "Design, Cost & Operation of an Acid Gas Enrichment and Injection Facility," Proceedings of the 2004 Laurance Reid Gas Conditioning Conference, Norman, Oklahoma, USA.

Royan, T., R. Warchola, and M. Clarkson, 1992, "Acid Gas Enrichment Using FLEXSORB®," Proceedings of the 1992 Laurance Reid Gas Conditioning Conference, Norman, Oklahoma, USA.

True, W.R. (1994), "New Mobile Bay complex exploits major sour gas reserve," Oil & Gas Journal, v 92, n 21, pp 49-51, May 23, 1994.

Weiland, R. H., 2008, "Acid Gas Enrichment - Maximizing Selectivity," Proceedings of the 2008 Laurance Reid Gas Conditioning Conference, Norman, Oklahoma, USA.

Proceedings of the 2nd Annual Gas Processing Symposium
Farid Benyahia and Fadwa T. Eljack (Editors)

Computational Study of a Lifted Turbulent Jet Flame in a Cross-flow: Flame Length and Emissions

M.S. Lawal,[a] M. Fairweather,[b] D.B. Ingham,[a] L. Ma,[a] M. Pourkashanian,[b] and A. Williams[a]

[a] *Centre for Computational Fluid Dynamics* ,[b]*School of Process, Enviromental and Materials Engineering, University of Leeds, Leeds, LS2 9JT, United Kingdom*
**Communicating author: mnmsl@leeds.ac.uk*

Abstract

A lifted turbulent non-premixed jet flame of methane in a cross-flow of air has been investigated computationally. The purpose of the study is to predict the effect of variable flow conditions on the flame length and its greenhouse gas emission characteristics (NO_x, CO), information relevant in routine and emergency process flaring. Turbulence and combustion were modelled using the two equation k - ε turbulence model and the Eulerian particle flamelet model, respectively. Emission indices of NO_x and CO were calculated in a post-processing stage. Radiation was modelled using the discrete ordinates method based on the weighted-sum-of-grey-gases-model. Comparison of predicted results with experimental data shows good agreement for the flame length, radiant heat flux and the species emission indices for the range of jet-to-cross-flow momentum ratios (10 to 800) investigated. The results show that in the high momentum ratio regime, the far-field radiative heat flux and emission index of NO_x is lower, whereas the emission index of CO is relatively higher.

Keywords: flame length, radiation, flamelets, soot, flares.

1. Introduction

The UK government's commitment to sustainable energy operations, by reducing greenhouse gas emissions by 8% below the 1990 level by the year 2020, requires an accurate estimate of emissions. This involves several measurements; some of which are difficult to obtain in complex situations such as flares. The problem can be partially resolved through the numerical study of the jet flame in cross-flow (JFICF), to predict the flame structure, radiation field and the emissions characteristics of flares. The study is also relevant in optimising combustion processes in burners and for fuel-mixing enhancement in gas turbines.

The JFICF can be grouped into two main flow regimes – lifted and non-lifted. The possibility of the flame being lifted off from the burner depends on the magnitude of the cross-flow velocity (u_{cf}) (Kalghatgi, 1981; Huang and Chang, 1994). When this velocity is low, and below a certain critical value, which depends on the fuel type, increasing the fuel jet velocity (u_j) may cause the lift-off of the flame above the burner exit. However, higher u_{cf} produce a non-lifted flame significantly deflected in the direction of the cross-flow. This study is concerned with the lifted JFICF typically found in emergency and well-test flares.

Extensive measurements of the structure, stability, temperature and radiation field of the lifted JFICF have been reported (Gollahali et al., 1975; Kalghatgi, 1981; Birch et al., 1989; Kodata et al., 1990). Bandaru and Turns (2000) experimentally studied the emission characteristics. The study demonstrated the important influence of flow variables on emissions from the lifted JFICF. Most of the previous numerical studies of the JFICF have used the laminar flamelet combustion model as a convenient approach of accounting for finite-rate kinetic effects at a reasonable computational cost. Fairweather et al. (1991, 1992) employed this approach to calculate the flame length and radiation field. The study reported that increasing the u_{cf} produces an initial decrease, and then an increase, in the flame length. However, it is yet to be determined if changes in the fuel exit velocity would have an analogous or different effects on the flame length and the emission pattern of the JFICF. This aspect is investigated in the present study.

A detailed review of the open literature shows that only a few numerical studies have examined the emission characteristics of the JFICF, except the recent study by Lawal et al. (2009). This study investigated the effect of changes in u_j on the flame structure and emissions in the high jet-to-cross-flow momentum ratio M (100 to 800). The flow field was calculated using the steady, Reynolds averaged Navier Stokes (RANS) equations, incorporating the realizable k - ε turbulence closure for the Reynolds stresses. The combustion process and the flame lift-off were modelled using the laminar flamelet equations and the diffusion flamelet quenching criterion (Peters, 1984), respectively. The transport equation of the mixture fraction and its variance were solved, to calculate a multiple flamelet profile, based on a presumed-shape probability density function of the mixture fraction and the scalar dissipation rate χ (0.01 to 39 s^{-1}). Reacting species with long residence times, i.e. NO and CO, were computed in a post-processing mode using the Eulerian particle flamelet model (Barths et al., 1998). The study predicted the temperature, the flame length and the emission index of NO_x ($EINO_x$) in agreement with the experimental data of Bandaru and Turns (2000).

In the present study, the model described above is modified by including smaller values of χ in the flamelet profile. The aim is to account for the effect of low χ on the trace species. The model is then used to investigate the lifted JFICF in both the high and the low momentum ratio M (10 to 800). In the subsequent sections, a brief description of the numerical models used and computational details are provided. Then, the results obtained are presented and discussed, and finally, the main conclusions are highlighted.

2. Numerical models

As outlined above, the numerical model used in the present study incorporates smaller values of the scalar dissipation rate in the flamelet profile, from 0.0001 to 39 s^{-1}. The flamelet library consists of a set of 20 flamelets calculated with steps in the scalar dissipation of 2.5 s^{-1}. The scalar dissipation rate represents the departure of the flame from chemical equilibrium, and is represented in the flamelet equation (Peters, 1984):

$$\rho \frac{\partial \phi_i}{\partial t} + \rho \frac{\partial \chi}{\partial Z} \frac{\partial \phi_i}{\partial Z} = \frac{\rho}{Le_i} \frac{\chi}{2} \frac{\partial^2 \varphi_i}{\partial Z^2} + \omega_i \qquad (1)$$

$$\rho \frac{\partial T}{\partial t} = \frac{\rho \chi}{2} \frac{\partial^2 T}{\partial Z^2} - \frac{1}{c_p} \sum_{i=1}^{N} H_i \omega_i + \frac{1}{2c_p} \rho \chi \left[\frac{\partial c_p}{\partial Z} + \sum_{i=1}^{N} c_{p_i} \frac{\partial \phi_i}{\partial Z} \right] \frac{\partial T}{\partial Z} \qquad (2)$$

where H_i, C_{pi}, ρ and ω_i are, the specific enthalpy, the specific heat, density, and the net production rate of species *i,* respectively. Reacting scalars, such as the species mass fraction and temperature, are represented by ϕ_i, while Z denotes the conserved scalar, the mixture fraction. The temperature, the mixture averaged specific heat and the species Lewis number are represented by T, C_p and Le_i, respectively.

The scalar dissipation rate may also be interpreted as the inverse of the characteristics diffusion time. Therefore, smaller values of χ reduce the diffusive term in Eqs. (1) and (2), resulting in a slower diffusive transport of species and enthalpy. This has an effect similar to increasing residence times, and has been reported to influence slow reacting species, such as NO (Heyl and Bockhorn, 2001). The effect of soot on the flame radiation was also investigated using the Brookes and Moss (1999) soot model. This model solves a transport equation for the soot mass fraction and the particle number density of soot nuclei, with acetylene as the precursor species. The model accounts for soot particle nucleation, coagulation, surface growth and oxidation.

3. Computational details

The flame investigated in this study is based on the JFICF experiments by Bandaru and Turns (2000). The flame configuration consisted of a vertical air jet with a horizontal concentric burner enclosed within a screen, as schematically represented in Figure 1. A detailed description of the set up is given in Bandaru and Turns (2000). In order to investigate flames at a low M, three additional M (10 to 75) were included in the simulations, as indicated in Table1. As in the experiment, the cross-flow velocity was kept constant at 2.3 ms^{-1}. The range of the Reynolds numbers of the fuel jet is 3,000 to 27,300 and that for the cross-flow is 103,500.

The computational domain consists of a Cartesian coordinate system, with the origin at the centre of the burner inlet. The calculation was simplified by assuming symmetry. Therefore, only half of the domain, represented by a rectangular box of $1.67 \times 0.68 \times 0.34$ m^3 was modelled. The burner is represented by a cylindrical tube of length 0.1 m, with internal and external diameters of 4.12 mm and 5.54 mm, respectively. The distance of the cross-flow inlet from the burner is 0.17 m, and the burner exit projects 40 mm above the lower floor of the box. The grids consist of unstructured hexahedral elements. This provides the flexibility of varying the mesh sizes, from 0.025d to 10d (where d is the pipe inner diameter), with the fine elements placed in the pipe region.

The boundary conditions prescribed included symmetry on the x-y plane at z = 0, and a uniform cross-flow velocity inlet with 3% turbulence intensity (within the range used in the experiment) up-stream of the y-z plane.

Table 1. Summary of the investigated flow velocities and momentum ratios.

Run	1	2	3	4	5	6	7	8	9
$u_j (ms^{-1})$	10	20	27	34	42	56	67	76	87
M^a	10	41	75	117	193	332	455	600	770

$^a M = \rho_j u_j^2 / \rho_{cf} u_{cf}^2$, where ρ_j and ρ_{cf} are the respective densities of the jet and the cross-flow.

Zero normal gradient of all flow variables, except pressure, were specified at the domain outlet. At the inflow of the pipe, a constant mean velocity with 10% turbulence intensity was specified. The sensitivity of the results to these turbulence intensities was tested and found to be insignificant. On the x-z plane and x-y planes, corresponding to the lower and top boundaries, no-slip wall boundary condition were specified. The GRI 3.0 mechanism was used for the chemical kinetic modelling, and the SIMPLE algorithm of Pantakar (1980) for the pressure-velocity coupling. All equations were discretised using the second-order upwind scheme as implemented in the Fluent 6.3.26 code.

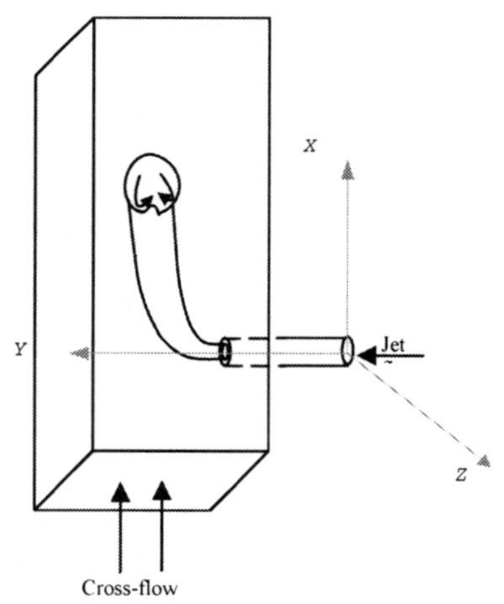

Figure 1: Schematic of a turbulent jet flame in a cross-flow.

4. Results and discussion

4.1. Radiative heat flux

As noted in Section 2, the effect of soot on flame radiation and chemical species was evaluated through two separate calculations, without and with the soot model. The calculation with soot incorporates the Brookes and Moss (1999) soot model for the case:

u_j = 76 ms^{-1} and u_{cf} = 5 ms^{-1} (M = 125). The radiative heat flux predictions were compared with the experimental data of Birch et al. (1989) due to unavailability of similar data from the study of Bandaru and Turns (2000). Predicted peak radiant heat flux with the soot model was found to be higher by only 2.2%, relative to the result without the model. This small difference in the predicted radiant heat flux at atmospheric pressure shows that the soot level in the methane-air jet flame investigated is low, as might be anticipated. Consequently, the effect of soot was not considered in the remaining simulations.

Figure 2 (a) shows a comparison of the predicted and measured radiative heat flux on the plane of symmetry. Good agreement with measurements can be seen, particularly in the near-field of the flame. However, the data are slightly under-predicted further downstream, beyond the flame tip. This is likely due to the over-prediction of the strength of the counter-rotating vortices, present in JFICFs by the gradient diffusion turbulence model (Fairweather et al., 1991). The radiative heat flux predicted at different momentum ratios is shown in Figure 2 (b), which reveals an overall decrease in the flame radiation at higher momentum ratios. This trend correlates with the increase in flame length (discussed below), and the reduction in peak flame temperature at higher momentum ratio reported previously (Lawal et al., 2009). Cook et al. (1987) also measured a lower fraction of heat radiated from large scale flares at high exit velocities.

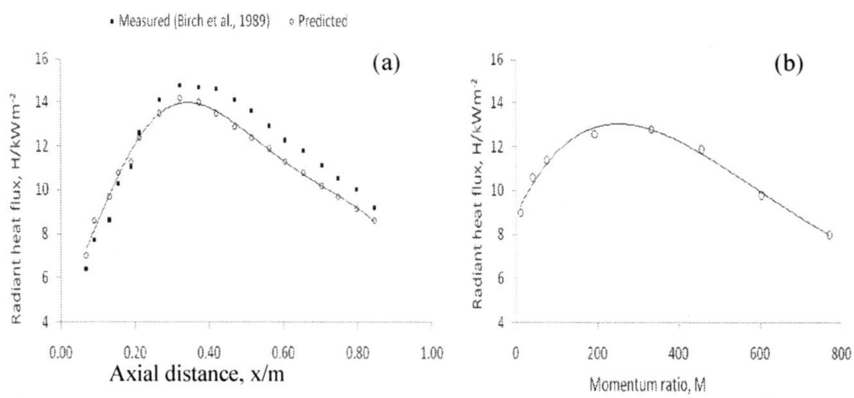

Figure 2: (a) Predicted and measured radiative heat flux on the plane of symmetry, and (b) predicted radiative heat flux at different momentum ratios.

4.2. Flame length

The flame length has an important influence on the flame radiative heat flux, and is defined as the curvilinear distance along the jet axis from the burner pipe exit to the flame tip. The location of the flame tip was computed based on the contour of a time-averaged mean temperature of 1200 K, which matches the location of the mean stoichiometric mixture fraction. In the absence of experimental data to validate the calculations in the low momentum ratio regime considered in this study, predictions were made across the full range of momentum ratios (10 to 800). Interestingly, the fit to the present predictions, shown in Figure 3 agrees with the trend reported by Gollahali et al. (1975), which was also reproduced by the predictions of Fairweather et al. (1991).

Figure 3 shows that at low jet exit velocities, or low momentum ratios, the flame is relatively longer. This is due to considerable bending of the flame by the wind, resulting in lower air entrainment and small velocity difference between the jet and the cross-flow. As the momentum ratio is increased, the increasing relative velocity enhances turbulent fluctuation and mixing, resulting in a decrease in the flame length. At some point, however, a reversal of this trend occurs, where the flame length reaches a minimum. Beyond this point, further increase in the momentum ratio cause the flame length to increase. The point where the initial trend reverses represents a limiting condition, which occurs in this case at $M \approx 336$. This correspond with velocity ratio ($u_{cf} / u_j = 2.3/56$) of 0.04, in agreement with the value reported by Fairweather et al. (1991). Therefore, this result suggests that for the lifted JFICF, increasing the fuel jet exit velocity for a fixed u_{cf}, produces an effect on the flame length similar to an increase in the cross-flow velocity for a fixed u_j, and vice versa. This trend of the flame length with M has an important effect on the flame emission characteristics, discussed below.

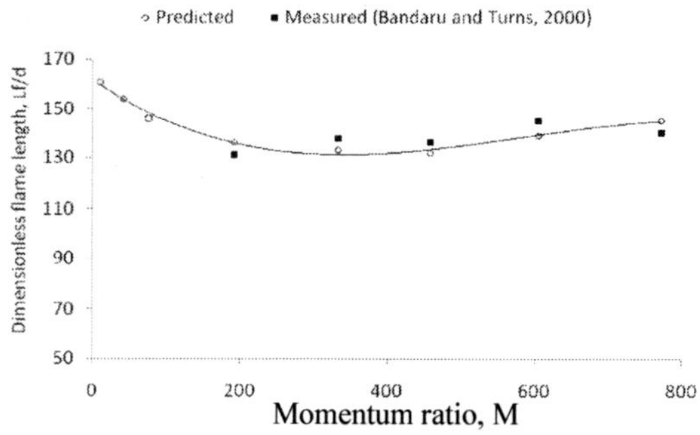

Figure 3: Predicted and measured flame lengths at different momentum ratios.

4.3. Emissions index

The concentration of the pollutant species calculated at the exit of the solution domain is reported in terms of the emission index. This is defined as the quantity in grams of the species per kilogramme of fuel burned (Al-Fawaz et al., 1994). Figure 4 presents a comparison of the measured and predicted emission indices of CO and NO at different M. It reveals an increase in the predicted concentrations of the trace species in the present calculations, where smaller χ (0.0001 to 0.01) were included in the flamelet profile. This increase is associated with the slower rate of species diffusion from the reaction zone resulting from the inclusion of small scalar dissipation flamelets. This is akin to increasing residence times, and hence more accurately accommodating species that are sensitive to residence times and temperature.

Although the inclusion of smaller scalar dissipation flamelets provides some improvement in the predicted slow reacting species, slight discrepancies still exist between the predictions and the measurements, particularly for CO concentration, where there is little over-prediction of the EICO, see Figure 4. This over-prediction may be connected with the inability of the RANS turbulence models to properly capture the effect of coherent structures on mixing (Coats, 1996) and fluctuations in the scalar dissipation rate (Pitsch, 2002). Accounting for these phenomena may require a more advanced turbulence simulation techniques, such as large eddy simulation which resolves the large energy-containing eddies.

Figure 4 also indicates that in the low momentum ratio regime, the emission indices of NO_x and CO are, respectively, higher and lower. This flow regime corresponds to where lower peak flame temperatures (< 2100 K) were observed. A possible explanation for the higher $EINO_x$ may be the dominance of the prompt NO mechanism at lower temperatures (Turns and Myhr, 1991). On the other hand, at higher momentum ratios, an increase in EICO and a decrease in $EINO_x$ were observed. The decrease in $EINO_x$ could be linked to the increase in flame length and volume leading to reduction in peak flame temperature, and hence lower thermal NO_x emissions at higher momentum ratio. In contrast, the increase in EICO may be attributed to an increase in unburned hydrocarbon, due to rapid quenching of the flame by the cross-flow. Consequently, subsequent partial oxidation of the unburned fuel species may results in an increase in CO emission at higher lift-off height and momentum ratio.

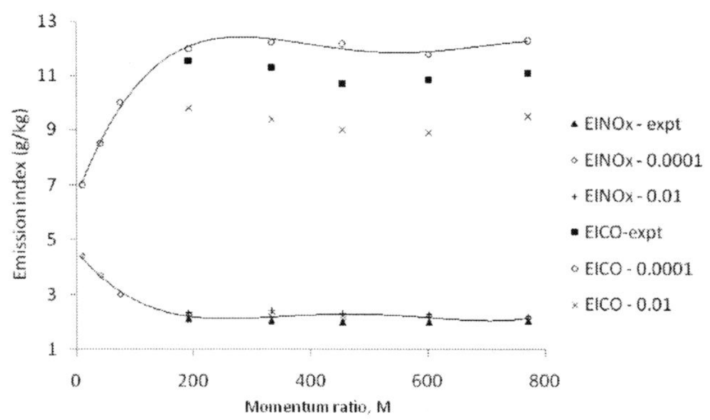

Figure 4: Predicted and measured (Bandaru and Turns, 2000) emission indices of NO and CO at different momentum ratios.

In the light of results discussed above, it is clear that consideration of the low momentum ratio regime is vital when investigating lifted JFICFs, because of the markedly different profiles of the radiative heat flux and emission species. Furthermore, the lower far-field radiative heat flux and $EINO_x$, as well as the higher EICO which does not change significantly at higher momentum ratios, suggest that flare operation in this flow regime is expected to be more environmentally acceptable.

5. Conclusions

A computational study of a lifted methane-air jet flame in cross-flow has been performed. The method used in an earlier study was modified, by incorporating smaller scalar dissipation rate in the flamelets profile used in the combustion model. Reasonable agreement of predicted results with experimental data of Bandaru and Turns (2000), and Birch et al. (1989), have been obtained.

The main conclusions from this study are that consideration of the low momentum ratio regimes is important in evaluating the characteristics of the lifted JFICF. Increasing the jet exit velocity, for a fixed cross-flow velocity, has also been shown to have an effect on the flame length, similar to that observed when increasing the cross-flow velocity for a fixed jet velocity. This produces a trend of increasing flame length and flame volume at higher momentum ratio, translating into a lower radiative far-field heat flux and $EINO_x$. Although, the EICO is relatively higher, its increase with M at higher momentum ratios is not significant. Therefore, flare operation in the high momentum ratio regime is expected to be more environmentally friendly.

Acknowledgments

M.S. Lawal acknowledges the financial support of the Nigerian Government provided through the Petroleum Technology Development Fund PhD Scholarship Scheme.

References

Al-Fawaz, A.D., L.M. Dearden, J.T. Hedley, M. Missaghi, M. Pourkashanian, A. Williams and L.T. Yap, 1994. Nox formation in geometrically scaled gas-fired indusrial burners, Proc. Combust. Inst., 25, 1027-1034.

Bandaru, R. V. and S.R.Turns, 2000. Turbulent jet flames in a crossflow: effects of some jet, crossflow, and pilot-flame parameters on emissions, Combust. Flame, 121: 137-151.

Barths, H., N. Peters, N. Brehm, A. Mack, M. Pfitzner and V. Smiljanovski 1998. Simulation of pollutant formation in a gas-turbine combustor, Proc. Combust. Inst. 27, 1841-1847.

Birch, A.D., D.R. Brown, M. Fairweather, and G.K. Hargrave, 1989. An experimental study of a turbulent natural gas jet in a cross-flow, Combustion Science Technology, 66: 217- 232.

Brookes, S.J. and J.B. Moss, 1999. Predictions of soot and thermal radiation properties in confined turbulent jet diffusion flames, Combust. Flame, 116, 4: 486-503.

Coats, C.M., 1996, Coherent structures in combustion, Prog. Energy Combust. Sci., 22: 427-509.

Cook, D.K., M. Fairweather, J. Hammonds and D.J Hughes, 1987. Size and radiative characteristics of natural gas flares, Chem. Eng. Des., 65, 318-325.

Fairweather, M., W.P.Jones, R.P. Lindstedt, and A.J. Marquis, 1991. Predictions of a turbulent reacting jet in a crossflow, Combust. Flame, 84: 361-375.

Fairweather, M., W.P.Jones, R.P. Lindstedt, 1992. Predictions of radiative transfer from a turbulent reacting jet in a cross-wind, Combust. Flame, 89: 45-63.

Gollahali, S.R., T.A., Brzustowski and H.F. Sullivan, 1975. Characteristics of turbulent propane diffusion flame in a cross-wind, Trans. Canadian Soc. Mech. Engrg. 3:205-214.

Heyl, A. and H. Bockhorn, 2001. Flamelet modelling of NO formation in laminar and turbulent diffusion flames, Chemosphere, 42: 449-462.
Huang, R. F. and J. M., Chang, 1994. The stability and visualized flame and flow structures of a combusting jet in cross-flow, Combust. Flame, 98:267-272.
Kadota, T., J.X. Wang, and T. Kawaoka, 1990. Structure of a propane jet diffusion flame in a cross flow, Japan Society of Mechanical Engineers, Series II, 33: 575-581.
Kalghatgi, G.T., 1981, Blow-out stability of gaseous jet diffusion flames part II: effect of cross-wind, Combust. Sci. Tech., 26: 241-244.
Lawal, M.S., L. Ma, M. Pourkashanian, A. Williams and D.B. Ingham, 2009. High momentum jet flame in crossflow: flame structure and emission characteristics of flares, 10th Intl. Conf. on Energy for a Clean Environment, 7-9 July, Paper 11.3.
Pantakar, S.V., 1980, Numerical heat transfer and fluid flow, McGraw Hill, New York.
Peters, N., 1984, Laminar diffusion flamelet models in non-premixed turbulent combustion, Prog. Energy Combust. Sci., 10: 319-339.
Pitsch, H., 2002. Improved pollutant predictions in large-eddy simulations of turbulent non-premixed combustion by considering scalar dissipation rate fluctuations, Proc.Combust. Inst., 29, 1971-1978.
Smith, G.P., D.M. Golden, M. Frenklach, W.N. Moriarty, B. Eisteneer, M. Goldenberg, C.T. Bowman, R.K. Hanson, S. Song, W.C. Gardiner, V.V. Lissianski, and Z. Qin http://www.me.berkeley.edu/gri_mech/.
Turns, S.R. and F.H.Myhr, , 1991. Oxides of nitrogen emissions from turbulent jet flames:
Part 1 - fuel effect and flame radiation, Combust. Flame, 87: 319-335.

Proceedings of the 2nd Annual Gas Processing Symposium
Farid Benyahia and Fadwa T. Eljack (Editors)

Towards Novel Hybrid Biomass and Coal Processes for Satisfying Transportation Fuel Demands

Richard C. Baliban, Josephine A. Elia, & Christodoulos A. Floudas*

Department of Chemical Engineering, Princeton University, Princeton, NJ, 08544, USA

Abstract

Novel biomass and coal flowsheets are developed to satisfy United States transportation sector fuel needs using the principles of the H_2Car process. A fuel-carbon conversion near 100% is achieved for all systems using a renewable source of hydrogen and the reverse water gas shift reaction. Each flowsheet is fully heat and power integrated using a heat exchanger and heat engine network to sequentially minimize the annualized utility cost and the annualized heat exchanger cost. A detailed economic analysis reveals the crude oil price for which the processes become competitive with current petrochemical systems.

Keywords: gasification, Fischer-Tropsch, hybrid biomass/coal, heat/power integration

1. Introduction and Background

The increased energy use of industrialized nations coupled with the rapid growth of China and India has led to concern that world oil production will not be able to satisfy global demand (NRC, 2008). Additionally, the United States only holds 2.4% of the total proven world oil reserve but requires 22.5% of the world oil yearly consumption (BP, 2009). This will have a marked effect on the sustainability of the United States transportation sector as 66.3% of consumption is associated with this sector (EIA, 2007). Though extensive research has been devoted to the use of alternative fuels such as hydrogen and electricity, the technical and economic challenges have so far prevented their widespread use (MacLean and Lave, 2003; NRC, 2004). It is therefore desirable to produce hydrocarbon fuels using more abundant domestic resources such as biomass or coal. One such biomass to liquid (BTL) or coal to liquid (CTL) technology focuses on gasification of the feedstock to produce a synthesis gas followed by a subsequent Fischer-Tropsch (FT) reaction to generate a range of hydrocarbons (Bechtel, 1998; Kreutz et al., 2008; Hamelinck et al., 2004; Vliet et al., 2009).

Despite the large amount of focus on FT fuels, issues still remain with the unreacted CO_2 produced by the process. A strong focus on reducing greenhouse gas emissions has motivated carbon capture and sequestration (CCS) where the CO_2 is stored in oil/gas reservoirs or saline aquifers. However, the large quantities of CO_2 that must be sequestered may have severe consequences on soil chemistry and aquatic life (Service, 2004) and will require a significant investment in sequestration and monitoring systems

*Corresponding author; Tel: (609) 258-4595; Fax (609) 258-0211; E-mail: floudas@titan.princeton.edu

(Agrawal et al., 2007). Recently, the novel H_2Car process (Agrawal et al., 2007) has shown that it is possible to deliver transportation fuels without producing *any* CO_2 by using input H_2 generated from a renewable source of energy. The H_2 will react with the CO_2 via the reverse water-gas shift (RGS) reaction to generate CO which can be used in the FT unit.

The purpose of this paper is to expand upon the methodology of the H_2Car process by producing gasoline, diesel, and kerosene in mass ratios consistent with the United States consumption of transportation fuels (EIA, 2008). Using herbaceous biomass and Illinois #6 Coal (Kreutz et al., 2008), six process flowsheets were developed using Aspen Plus (2007) using (a) a pure biomass feedstock (BTL), (b) a pure coal feedstock (CTL), and (c) a hybrid feedstock (CBTL) consisting of equal mass ratios of biomass and coal along with either (1) low-temperature (LT) or (2) high-temperature (HT) FT conditions. Hydrogen and electricity will be generated using either (i) solar, (ii) wind, or (iii) nuclear energy. In addition to providing near 100% carbon conversion of the feedstock, the novel flowsheets serve as a means for the development and analysis of centralized, distributed hydrogen production technology (NRC, 2004). For each flowsheet, a fully integrated heat exchanger and heat engine network (Floudas, 1995) is developed to sequentially minimize the total utility usage and the total capital heat exchanger cost. Finally, a detailed cost analysis is provided that determines the crude oil price for which the H_2Car process is competitive with current petroleum based processes.

2. Process Flowsheets

2.1. Syngas Generation

Syngas is generated from biomass and coal using distinct gasification trains (Figure 1). Each gasifier is modeled stoichiometrically where the extent of reaction of a linearly independent set of chemical reactions is calculated based on the (i) input oxygen amount, (ii) input steam amount, (iii) operating temperature, and (iv) feedstock composition. The functionality of the reaction extents was obtained using a non-linear optimization (NLP) model where the effluent of multiple case studies was correlated to the four inputs listed above for both biomass and coal. Additional information for each train is described below.

2.1.1. Biomass Feedstock

Herbaceous biomass is lockhopped with CO_2 to the circulating gasifier (T = 900 °C, P = 30 bar, Kreutz et al., 2008) designed with distinct chambers for gasification and combustion. The feedstock is fed to the gasification chamber along with oxygen and steam to facilitate the gasification of char as well as cracking of residual tars and oils. The effluent of the gasification chamber is predicted using the NLP model and is passed

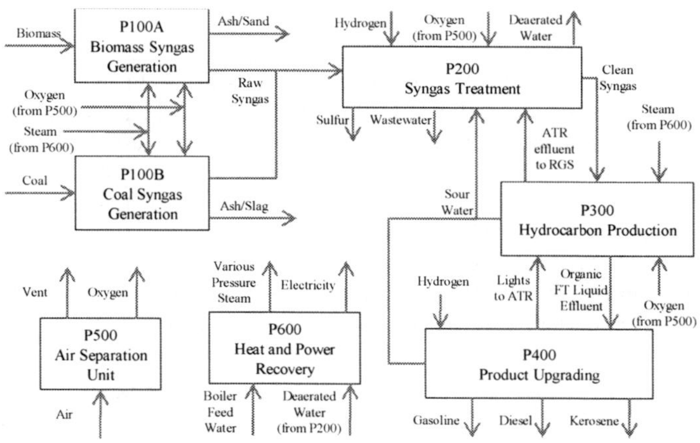

Figure 1: Overall flowsheet design for the production of transportation fuels.

through a series of cyclones to remove particulates and then sent to a tar cracker to further decompose tars and oils. Residual ash, char, and sand (olivine) are sent to a combustion chamber (T = 1300 °C) where the char is completely combusted. The solids are recycled back to the gasification chamber after a purge to prevent ash build-up.

2.1.2. Coal Feedstock

Illinois #6 coal is dried to 2% moisture (Bechtel, 1992) and fed to the gasifier (T = 1437 °C, P = 42 bar; NETL, 2007) with CO_2 carrier gas, where it reacts with oxygen and steam. Molten slag is removed, depressurized with lockhoppers, and cooled for disposal. The high temperature syngas is cooled below the ash melting point to 891 °C (NETL, 2007) to separate out the ash that is carried out in the gas. It is further cooled to 200 °C and is passed through another separator unit to remove any flyslag and remaining particulates in the syngas. Finally, the syngas is heated to 800 °C before it enters the reverse water gas shift reactor.

2.2. Syngas Treatment

The raw syngas (Figure 1) is sent immediately to a reverse water-gas shift (RGS) reactor where hydrogen is added to maintain a 2/1 H_2/CO ratio in the effluent stream and oxygen is added to control the temperature (T = 800 °C). The RGS effluent is cooled to 185 °C and is hydrolyzed to convert HCN to NH_3 and COS to H_2S (NETL, 2007). The gas is then cooled, passed through a flash unit to remove water, and sent to a two-stage Rectisol unit to separately remove H_2S and CO_2 (Kreutz et al., 2008). The H_2S is sent to a Claus recovery system (NETL, 2007) and the CO_2 is compressed and recycled back to the RGS reactor. All knock-out water is sent to a sour stripper (SS) where the sour gas is removed and recycled back to the Claus system and the recovered water is sent to a deaerator to remove residual vapors.

2.3. Hydrocarbon Production

The clean syngas from the Rectisol unit enters the Fischer-Tropsch (FT) unit at either a high (320 oC) or low temperature (240 oC). To represent the formation of oxygenated compounds, pseudocomponents were developed for vapor, organic, and aqueous phase compounds. It is assumed that the amount of total converted carbon present in these pseudocomponents is 0.1%, 1.0%, and 0.4%, respectively (Bechtel, 1998). All remaining converted carbon will follow an Anderson-Schulz-Flory distribution that is slightly modified to account for increased formation of light hydrocarbons. The high-temperature process has a lower chain growth probability (α = 0.65) and favors the formation of gasoline-length hydrocarbons while the low-temperature chain growth probability (α = 0.71) favors heavier hydrocarbons and waxes (Dry, 2002).

The FT process produces both paraffin and olefin (one double bond) compounds. The paraffin percentage is 20% for C2-C4, 25% for C5-C6, and 30% for C7-C20. C21-C29 compounds are represented by pseudocomponents that have properties derived from 30% paraffin and 70% olefin and all C30+ compounds are represented by a generic wax pseudocomponent (Bechtel, 1998). For the C4-C6 compounds, a portion of the hydrocarbons (5% for C4, 10% for C5-C6) is represented as a branched form while the balance is represented by the straight chain form (Bechtel, 1998). The products are cooled and any residual vapor is sent to an auto-thermal reactor (ATR; T = 800 oC) before being recycled back to the RGS unit (Figure 1). Any knock-out liquid water is sent to the SS and all organic liquid is sent to the hydrocarbon upgrading section.

2.4. Hydrocarbon Upgrading

The input hydrocarbons are first sent to a distillation unit where they are separated into vapor, naphtha, kerosene, distillate, wax, and wastewater (Bechtel, 1998). Kerosene production was incorporated into the model by assuming that a cut will be taken from the hydrocarbon distillation unit between the liquid naphtha and the distillate. The vapor is sent to the ATR and the wastewater is sent to the SS (Figure 1). The naphtha, kerosene, and distillate pass through hydrotreaters to remove sour water and form product kerosene and diesel (Bechtel, 1998). Operation of the kerosene hydrotreater is assumed to be the same as the distillate hydrotreater. The gasoline output is sent through a naphtha reformer and a C_5/C_6 isomerizer to upgrade the quality of the fuel (Bechtel, 1998). The overall yields of gasoline, diesel, and kerosene were set to be in proportion with the current United States mass demands by taking a fraction of the naphtha, kerosene, or distillate from the distillation unit outlet and sending it to the ATR.

3. Heat and Power Recovery Network

The waste heat streams can either provide process steam or generate electricity using a heat exchange and power recovery network (HEPN). To provide the maximum possible heat and power integration, we will include all heat that is removed from or absorbed by individual units at a specified temperature (i.e., FT, upgrading units, Claus reactor). We can define the optimal HEPN as the system that minimizes the annualized cost of heat exchange for the process using any necessary heat or power utilities (Floudas, 1995).

The formulation of this model can be prohibitively large, especially when heat capacity is temperature dependent and power integration must be considered. Thus, we decompose the problem to first find the minimum hot/cold/power utility cost and then find the heat exchanger topology with the lowest annualized cost that uses the given minimum utilities.

3.1. Minimum Utility Cost

A model for the minimum hot/cold/power utility cost was developed recently (Holiastos and Manousiouthakis, 2002) using heat engines and pumps to provide the electricity to be generated by a hot and cold process stream. However, this model is only capable of giving *target* utility usage since the heat engines are assumed to have Carnot efficiency. We address this issue by postulating a set of heat engines that will interact with the process streams to produce a given amount of electricity. Since the conditions of the turbine and pump are known *a priori* for a given heat engine in the set, we may directly calculate the electricity delivered for that particular engine. Furthermore, we know the conditions of the working fluid at the pump and turbine inlet/outlet, so we can consider the streams within the engine as process streams of unknown flow rate.

To prevent the excessive use of heat engines, the maximum number of heat engines is set to three. We then construct a Mixed-Integer Linear Optimization (MILP) model to determine (i) the amount of cooling water (CW) required, (ii) the amount of hot steam generated (through deaerator output water or purchased boiler feed water

Table 1: Utility amounts requiring the minimum annual cost (May 2009 $).

	CW	BFW	Steam (bar, kg/s)						Elec.	Ut. Cost
	(kg/s)	(kg/s)	5	25	35	45	75	125	(MW)	($MM/yr)
BTL,LT	576	148	-2.33	-72.1	0	0	0	-72.3	59.9	-$1.341
BTL,HT	445	327	-4.44	0	0	-145	0	-176	87.2	-$3.919
CTL,LT	1828	145	0	-89.4	0	0	0	-89.8	206	-$0.337
CTL,HT	1833	457	-8.89	0	0	0	-23.8	-330	354	-$1.914
CBTL,LT	416	302	0	-175	0	0	0	-118	88.4	-$2.913
CBTL,HT	26.2	659	-0.22	0	0	0	-272	-260	120	-$6.602
Cost[1]	$30.6	$918	$15.8	$24.5	$26.3	$27.6	$30.2	$33.1	$20.7	

1. Costs are in $/$10^6$ kg for cooling water (CW) and boiler feed water (BFW) , $/1000 kg for steam, and $/GJ for electricity (Elec).

(BFW)), (iii) the amount of electricity (Elec) generated from the heat engines, (iv) the working conditions of the heat engines, and (v) the pinch point temperatures for the system. The MILP is a restricted model (Papoulias and Grossmann, 1983) that prevents inappropriate matches between (a) distinct point sources (operating units), (b) point sources and process streams, and (c) hot utilities with the heat engines. Hot utility interaction with the heat engines may cause an unbounded minimum in the annualized utility cost (Holiastos and Manousiouthakis, 2002). The results are given in Table 1 along with the utility costs. Negative values for utility cost indicate that the particular utilities can be sold at that price. Note that the high temperature systems consistently have a lower minimum utility cost than the corresponding low temperature systems for each feedstock. Additionally, the hybrid systems have a lower utility cost than either of the pure feedstock systems for both low and high temperature processes.

3.2. Minimum Heat Exchanger Network Annualized Cost

Given the operating conditions of the heat engines and the flow rates of the hot/cold utilities, we now seek to find the heat exchanger topology that yields the minimum capital cost. This is accomplished by first determining the minimum number of heat exchanger matches followed by finding the optimal topology which corresponds to the set of matches (Floudas, 1995). The minimum matches MILP model yields a large number of distinct sets of heat exchanger matches that give the same minimum value. We distinguish among them by using vertical heat transfer (Gundersen and Grossmann, 1990) along with weighting the matches according to their relative proximity in the plant (Papoulias and Grossmann, 1983). That is, we find the set(s) of matches which minimize the vertical heat transfer and then choose from among them the set of matches that corresponds to the minimum weight (closest relative proximity). The complexity of the MILP is reduced by decomposing the HEPN into distinct subnetworks based on the pinch point(s) determined in the minimum utility model (Floudas, 1995).

Given the optimal set of matches, we determine the optimal heat exchanger topology by developing the superstructure (Floudas et al., 1986) and minimizing the annualized cost to construct the HEPN. A mixed-integer non-linear optimization (MINLP) was formulated to account for the fact that the stream heat capacity was a function of the temperature intervals established in the minimum utility model. Two hundred starting points (i.e., initial flow rates, temperatures for the superstructure) were chosen to help find a local minimum that can approximate the global optimum.

4. Economic Assumptions

4.1. Capital Cost Assumptions

The direct permanent investment (DPI) of all pumps, compressors, turbines and flash units was calculated using ASPEN Icarus (ASPEN IPE, 2007) while the DPI of the remaining process units was calculated using estimates from several data sources (Bechtel, 1998; Kreutz et al., 2008; NETL, 2007; NRC, 2004; Spath et al., 2005) using the cost parameters in Table 2 and Equation 1:

$$DPI = (1 + BOP) \cdot C_o \cdot (S / S_o)^{sf} \cdot n^{0.9} \qquad (1)$$

where C_o is the base cost, S_o is the base capacity, S is the actual capacity, n is the total number of trains, sf is the cost scaling factor, and BOP is the balance of plant (BOP) percentage (site preparation, utility plants, etc.). BOP was calculated for the FT unit, the hydrocarbon recovery unit, and all upgrading units as a function of the feedstock higher heating value (Eqn. 4 of Kreutz et al., 2008). We have expressed all results in May 2009 dollars and used the Chemical Engineering Plant Cost Index (Chemical Eng. Magazine, 2009) to convert the original price when applicable.

Table 2: Process flowsheet reference capacities, costs (May 2009 \$), and scaling factors.

Unit Name	C_o (MM\$)	S_o	S_{max}	Units	*sf*	*BOP*	Ref.
Biomass H&D[1]	\$25.86	2,000	N/A	tonne/day	0.67	Included	a
Biomass Gasifier	\$144.87	815	568	MW LHV	0.67	15.50%	b
Coal H&D[1]	\$75.99	2,464	2,616	tonne/day	0.6	Included	c
Coal Gasifier	\$126.73	2,464	2,722	tonne/day	0.6	Included	c
RWGS	\$2.97	2,566	2,600	tonne/day	0.65	15.50%	c
COS Hydrolysis	\$2.84	4,975	7,500	tonne/day	0.67	15.50%	c
Acid Gas Recovery	\$50.32	200,000	700,000	Nm^3/hr	0.63	15.50%	b
Fischer Tropsch	\$37.88	226,669	228,029	Nm^3/hr	0.75	25.69%	d
Hydrocarbon Rec.	\$0.69	152.32	2,176	tonne/day	0.7	25.69%	d
Wax Hydrocracker	\$8.94	97.92	6,256	tonne/day	0.55	25.69%	d
Dist. Hydrotreater	\$2.40	31.55	7,072	tonne/day	0.6	25.69%	d
Nap. Hydrotreater	\$0.72	22.85	7,072	tonne/day	0.65	25.69%	d
Ker. Hydrotreater	\$2.40	31.55	7,072	tonne/day	0.6	25.69%	d
Nap. Reformer	\$4.99	36.99	8,160	tonne/day	0.6	25.69%	d
C5/C6 Isomerizer	\$0.92	13.06	2,720	tonne/day	0.62	25.69%	d
Gas Blender	\$8.46	2,740	N/A	tonne/day	0.6	25.69%	e
ATR	\$3.04	430,639	9,438,667	Nm^3/hr	0.67	Included	b
Claus Plant	\$22.40	135	N/A	tonne/day	0.67	15.50%	c

1. H&D = Handling and Drying; a) Spath et al., 2005; b) Kreutz et al., 2008; c) NETL, 2007; d) Bechtel, 1998; e) Bechtel, 1992

4.2. Feedstock, Product, and Utility Assumptions

The price ("as-received", delivered to plant gate, May 2009 \$) of herbaceous biomass and Illinois #6 Coal is \$4.79/GJ_{HHV} (Kreutz et al., 2008) and \$44.5/tonne (EIA, 2007), respectively. Disposal costs of wastewater and sand/ash are included in the fixed operating costs of the process producing those wastes. Utility costs are given in Table 1 for steam, water, and electricity. Steam utility costs were extrapolated assuming that the price varies logarithmically with the pressure (ASPEN IPE, 2007). Due to the variable marketability of sulfur, no credit was taken for this byproduct (NETL, 2007).

Table 3: Additional economic assumptions.

G&A Cap. Overhead (% of DPI)	3%	G&A Op. Expenses (% of SOC)	8%
Contract Fees (% of DPI)	3%	Yearly Operating Capacity	8000 hrs
Capital Contingencies (% of TDI)	18%	Tax Rate	40%
Working Capital (% of TPI)	5%	Desired Rate of Return	15%
Construction Time	2.5 years	Salvage Value (% of TPI)	20%
Startup Time	0.5 years	Project Capital Escalation	5%/year
Book life and economic life of investment	30 years	Products Escalation	5%/year
Maintenance Costs (% of TPI)	1%	Raw Material Escalation	3.5%/year
Labor Costs (\$/hr; operators, supervisor)	\$300, \$50	OL&M Escalation	3%/year
Operating Charges (% of Labor Costs)	25%	Utilities Escalation	3%/year
Plant Overhead (% of OL&M Costs)	50%		

The resale cost of the transportation fuels is based on the current crude oil price and the refiner's margin for each product. The refiner's margin is the difference between the sale price of petroleum products and the purchase price of crude oil and is estimated as the 1992-2003 average (EIA, 2009) after adjustment with the U.S. gross domestic index (Kreutz et al., 2008). The margin for gasoline, diesel, and kerosene is $0.312, $0.250, and $0.204, respectively. The margin for diesel is $0.05 higher than the average due to the estimated additional cost for production of low-sulfur diesel (Kreutz et al., 2008).

4.3. Additional Economic Assumptions

Table 3 lists the additional economic assumptions used in the paper. The total depreciable capital (TDC) is the sum of the DPI plus G&A capital overhead and contract fees. The total permanent investment (TPI) is the sum of the TDC plus the capital contingencies. The distribution of the TPI over the three year construction/startup period is ¼ in the first year, ½ in the second, and ¼ in the third. The subtotal operating costs (SOC) are calculated by adding the cost of the raw materials and utilities to the operating labor and maintenance (OL&M) costs, operating charges, and plant overhead.

5. Results and Discussion

5.1. Break Even Crude Oil Price

The economic value of a plant can be measured using the price of crude oil that would make net present value (NPV) of the plant equal to zero at the end of the economic lifespan. The break even crude oil price (BEOP) as a function of hydrogen cost is shown in Table 4. It is clear that the price of hydrogen determines the choice of low or high temperature FT conditions. When comparing the biomass and coal processes, the BEOP of the biomass feedstock is higher for the low temperature FT process and lower for the high temperature FT process. We also point out that the low temperature coal process has the lowest BEOP compared to the four BTL and CTL technologies for almost all hydrogen values, with the high temperature biomass process becoming competitive at $1.10/kg.

When comparing all low temperature processes, we see that the BEOP of the hybrid process is lower than the biomass process but higher than the coal process for all hydrogen prices considered (Tab. 4). Alternatively, the high temperature hybrid process BEOP is lower than both the high temperature biomass and coal processes for all hydrogen prices. Furthermore, the high temperature hybrid process has the lowest BEOP of all six processes for hydrogen prices below $1.83/kg. At the current H_2 steam reforming of methane selling price of $2.04/kg (May 2009 $; Mathure, 2007), the low-temperature coal process has the lowest BEOP of $86/bbl while the high temperature hybrid process is slightly higher at $93/bbl.

5.2. Land Use

It is important to comment on the total amount of biomass land area that is required to satisfy the *entire* United States transportation demand. The requirements based on 2007 demand (13,752 thousand barrels/day; EIA, 2009) are 1.11E+06 km^2 for the pure biomass feedstock and 5.01E+05 km^2 for the hybrid feedstock (10 tonne/hectare-year;

Tab. 4; NRC, 2004). The land requirement is 12.1% of the U.S. total (9.2E+6 km^2; Agrawal et al., 2007) for a pure biomass feedstock and 5.43% of the total for the hybrid processes.

Table 4: Economic results (May 2009 $) and land use requirements (km^2) for each flowsheet.

	BTL,LT	BTL,HT	CTL,LT	CTL,HT	CBTL,LT	CBTL,HT
BEOP ($/barrel)						
$2.50/kg H_2	$118	$130	$101	$135	$111	$117
$2.25/kg H_2	$109	$116	$93	$122	$101	$103
$2.00/kg H_2	$91	$102	$84	$108	$91	$89
$1.75/kg H_2	$90	$89	$76	$95	$81	$74
$1.50/kg H_2	$80	$75	$67	$82	$71	$60
$1.25/kg H_2	$71	$62	$59	$68	$61	$46
$1.00/kg H_2	$61	$48	$50	$55	$51	$31
Land Use (km^2)	1.11E+06 (12.1%[1])		0		5.01E+05 (5.43%[1])	

1. Percent of U.S. total land area

References

Agrawal R., N. Singh, F. Ribeiro and W. Delgass, 2007, Sustainable fuel for the transportation sector, PNAS, 104, 12, 4828.

ASPEN Plus Version 2006.5, 2007, Aspen Plus User Guide, Aspen Technology, Cambridge, MA.

ASPEN Icarus Process Evaluator Version 2006.5, 2007, Aspen IPE User Guide, Aspen Technology, Cambridge, MA.

Bechtel, 1992, Baseline Design/Economics for Advanced Tischer-Tropsch Technology, Contract No. DE-AC22-91PC90027.

Bechtel, 1998, Aspen Process Flowsheet Simulation Model of a Battelle Biomass-Based Gasification, Fischer-Tropsch Liquefaction and Combined-Cycle Power Plant, U.S. Department of Energy Topical Report, Project Number DE-AC22-93PC91029.

BP, 2009, BP Statistical Review of World Energy.Chemical Engineering Magazine, 2009, http://www.che.com/pci/. Energy Information Administration, Refiner Petroleum Product Prices by Sales Type, www.eia.doe.gov.

Department of Energy, 2008, 20% Wind Energy by 2030: Increasing Wind Energy's Contribution to U. S. Electricity Supply, Contract No. DOE/GO-102008-2587.

Dry M., 2002, The Fischer-Tropsch process: 1950-2000, 71, 3-4, 227.

Energy Information Administration, 2008, Product Supplied, www.eia.doe.gov.

Energy Information Administration, 2009, Annual Energy Review 2008, DOE/EIA0384 (2008).

Floudas C.A., Ciric A. and I.E. Grossmann, 1986, Automatic Synthesis of Optimum Heat Exchanger Network Configurations, AIChE J., 32, 2, 276.

Floudas, C.A., 1995, Nonlinear and Mixed-Integer Optimization: Fundamentals and Applications, New York, Oxford University Press.

Gundersen T. and I.E. Grossmann, 1990, Improved optimization strategies for automated heat exchanger network synthesis through physical insights, Comp. & Chem. Eng., 12, 6, 503.

Hamelinck C., A. Faaij, H. Uil and H. Boerrigter, 2004, Production of FT transportation fuels from biomass; technical options, process analysis and optimisation, and development potential, Energy, 29, 1743.

Kreutz T., E. Larson, G. Liu and R. Williams, 2008, Fischer-Tropsch Fuels from Coal and Biomass, 25th Annual International Pittsburg Coal Conference.

MacLean H. and L. Lave, 2003, Evaluating automobile fuel/propulsion system technologies, Progr. Energy Combustion Sci., 29, 1.

Mathure P., A. Patwardhan, and R. Saha, 2007, Steam Reforming of Methane for Bulk and Small Scale Production of Hydrogen, Indian Chemical Engineer, 49, 4, 480.

National Energy Technology Laboratory, 2007, Cost and Performance Baseline for Fossil Energy Plants: Volume 1: Bituminous Coal and Natural Gas to Electricity, Report Number DOE/NETL-2007/1281.

National Research Council, 2008, America's Energy Future: Summary of a Meeting, Washington, D.C., The National Academies Press.

National Research Council and National Academy of Engineering, 2004, The Hydrogen Economy: Opportunities, Costs, Barriers, and R&D Needs, Washington, D.C., The National Academy Press.

Papoulias S. and I.E. Grossmann, 1983, A Structural Optimization Approach In Process Synthesis – II. Heat Recovery Networks, Comp. & Chem. Eng., 7, 6, 707.

Spath P., A. Aden, T. Eggeman, M. Ringer, B. Wallace, and J. Jechura, 2005, Biomass to Hydrogen Production Detailed Design and Economics Utilizing the Battelle Columbus Laboratory Indirectly-Heated Gasifier, NREL Techincal Report TP-510-37408.

Vliet O., A. Faaij and W. Turkenburg, 2009, Fischer-Tropsch diesel production in a well-to-whell perspective: A carbon, energy flow and cost analysis, En. Con. Man., 50, 855.

Proceedings of the 2nd Annual Gas Processing Symposium
Farid Benyahia and Fadwa T. Eljack (Editors)

The Impact of PHEV Adoption on Natural Gas Demand in Electricity Generation

Shisheng Huang, Bri-Mathias S. Hodge, Joseph F. Pekny, Gintaras V. Reklaitis

School of Chemical Engineering, Purdue University, 480 Stadium Mall Drive, West Lafayette, IN 47907, USA

Abstract

Natural gas is an important energy resource that, because of its versatility, exerts an influence over many disparate sectors of the energy system. One such area is the electricity grid where natural gas serves as an important fuel for most electricity produced during peak periods. The widespread adoption of plug-in hybrid electric vehicles (PHEVs) could significantly change the current electricity load profile and hence, the demand for natural gas from the electricity sector. An integrated electricity supply and demand model is developed to help quantify the changes to the energy system that are anticipated due to the adoption of PHEVs. The Californian electricity market is used as a case study to determine the effects of PHEVs on natural gas demand for electricity generation

Keywords: PHEVs, Electricity Generation Modeling, Electricity Demand Modeling

1. Introduction

The demand for natural gas and petroleum combines to account for almost 60% of energy consumption in the United States (EIA, 2009a). Traditionally, uses for these two resources are disparate. Petroleum is predominantly used in the transportation sector and natural gas use spans a large variety of sectors, but with minimal penetration in the transportation industry. Recent concerns regarding high oil prices, oil dependency, energy security and climate change have driven both public and private sector entities to seek alternatives to petroleum based energy sources for the transportation sector. Automotive companies have recently revealed advanced technologies that allow for vehicles to utilize electricity instead of gasoline as their main source of energy. The Chevrolet Volt and Nissan Leaf are examples of such electric cars that are projected to be commercially available within the next few years.

The Chevrolet Volt in particular is a Plug-in Hybrid Electric Vehicle (PHEV), which differs from traditional all-electric vehicles by including a secondary source of energy. Such a vehicle can use the electricity stored in a battery as the primary means of propulsion or utilize a backup energy source, such as gasoline, when the electricity source is exhausted. PHEVs have the potential to be commercially viable without the need for a major infrastructure overhaul because they rely greatly on the existing electricity grid, requiring only a normal electrical socket and a small charging unit in order to dispense their fuel. Because of this great reliance on existing infrastructure the wide-spread adoption of these vehicles could have far ranging implications. Firstly, since their primary energy source is non-petroleum based in nature, they could significantly reduce oil consumption. This also presents a possible secondary environmental benefit of enormous magnitude if the electricity used comes from a source that is renewable. Secondly, since the primary energy source used to propel the vehicles would be electricity, it would represent a significant paradigm shift in the energy system where the transportation sector would become closely linked to the electricity market.

The greatest impact on the electricity market would be through increased residential electricity demand. Consumers would, in theory, exhaust the battery of the PHEV during the day and charge the vehicles through the night. If the penetration of the PHEVs is high enough, it could dramatically alter the electricity demand profile in the electricity market. This could in turn affect the mix of energy sources that provide electricity to the grid. In regions like California where natural gas constitutes a major part of the electricity generation profile, it could greatly affect the demand for natural gas in the region.

It is therefore important to be able to quantify and visualize the impact of the adoption of PHEVs on the electricity market. In this study, the impact of PHEVs is modeled from the bottom-up, being coupled together with residential electricity demand that is determined at the household level. This data is then merged with industrial and commercial electricity demand. The combined demand is then passed on to the electricity supply model which schedules electricity generating units to match the anticipated demand, taking into consideration the cost and availability of each individual generating plant. California is used in this study as a case study to demonstrate the implementation of the model framework although it could easily be adapted for other regions where natural gas is also a major energy source.

2. Electricity Demand Modeling

A discrete event modeling framework is used as the basis for the generation of electricity demand load profiles in the residential sector, with the consumption decision making disaggregated down to the household level. The household is modeled as a collection of appliances, with each appliance having a probability of being turned on at any particular time. The probability of this event happening is dependent on the hour of the day, season, daily frequency of the event happening and whether the appliance allows for multiple instances of the same object being used at the same time. The framework for the model was developed for Finnish households (Paatero and Lund, 2006) and has been modified to fit the California market.

In order to accommodate the impact of PHEVs, the model has been nested within an agent based modeling structure to allow communication between the different modeling paradigms. The discrete event model is used to forecast the household electricity demand while the individual PHEV users are modeled as agents and may develop their own charging patterns for their batteries based on certain heuristics. One such heuristic involves the pricing scheme used for electricity in the region. If the price of electricity is constant throughout the day, the PHEVs may charge at anytime they are at home and, therefore, are not being used. Otherwise, if the price of electricity changes according to a peak or off-peak pricing scheme, the PHEV user may decide to charge only if the price of electricity drops below a certain threshold. Another benefit of framing the PHEV user behavior as an agent is the ability to more realistically examine the travel patterns of individual PHEV users. This in turn allows more realistic battery charging characteristics to emerge from the model. This level of autonomy has not been considered for PHEVs in the published literature and could be valuable in helping to further hone in on the expected impact of PHEV use on the electricity market.

In this study, the impact of PHEVs on the commercial and industrial electricity demand is assumed to be minimal. Both industrial and commercial electricity demands are broken down into individual components as described in literature (Pfannenstiel et al., 2007). It is assumed that daily usage patterns for both these demand sectors do not change with the adoption of PHEVs in the model. Average daily usage data is obtained from relevant sources and together with component data obtained previously, an average daily load profile for both the industrial and commercial sector is obtained. A daily random factor and seasonality factor is obtained via regression of historical data and used to simulate daily fluctuations in demand for the electricity demand for both the industrial and commercial sectors.

3. Electricity Generation Modeling

An agent-based modeling framework has been used to represent the numerous different generators that encompass the electricity supply and the market in which the supply interacts with demand. The agent-based approach has been applied to many computational economic problems and a survey of their application to electricity market models may be found in a recent summary (Weidlich and Veit, 2008).

Every individual electricity generating entity is modeled as an agent. Different electric utility agent types are differentiated primarily by their fuel source and subsequently by their unit availability percentage and production cost. The utility agents interact through a scheduling agent that assigns utilities, based upon the lowest cost unit first, as needed in order to fulfill the demands provided by the integrated demand model discussed in the previous section. The scheduling agent acts as the clearing house for two distinct markets: the day-ahead market and the spot market. The scheduling agent forecasts the electricity load at the hourly timeframe for the following day and then assigns utilities that have reported available to fulfill the anticipated demand in the specified periods. As the actual demands are passed into the supply side of the model from the demand side the scheduling agent once again assigns available units to fulfill unanticipated demand or to account for an unplanned shutdown of a plant already assigned via the spot market.

This generation assignment model thus ensures that electricity supply always meets or exceeds the electricity demand for the time period under consideration. The resulting schedule of generation can be decomposed in order to examine the amount of production from each energy source.

4. California Case Study

Due to its large population, California is the second largest electricity market in the United States, trailing only Texas in total retail sales (EIA, 2009b). As California already produces almost 55% of its electrical power needs from natural gas (EIA, 2009b) any additional electricity demand from PHEVs could lead to significant increases in the consumption of natural gas within the state. The goal of this study is therefore to quantify the additional natural gas that could be expected to be consumed by the California electricity sector with the wide-spread adoption of PHEVs by consumers. In order to do so a multi-paradigm modeling framework (Hodge et al., 2009) has been used to allow communication between a discrete event electricity demand model and an agent-based electricity supply model. Both the demand and supply side of the model have been encapsulated in an agent based model on the highest abstraction, with communication of required information through ports.

California, being the most populous state in the United States, also has the most number of registered automobiles. As discussed previously, since PHEVs directly impact the electrical demand of a residential household, the significant adoption of PHEVs in California would have an impact on neighboring states electricity generation and subsequently on the entire energy system of the United States. California has roughly 12.2 million households (Census, 2009) with a registered automobile population of 19.8 million (BTS, 2008). In this study we are interested in determining the effect of one PHEV vehicle per household, in other words, a penetration rate of about 62% for PHEVs. With some of the highest average electricity prices in the nation, it is not surprising to find that Californian households are some of the lowest consumers of electricity per capita in the United States (EIA, 2009d). With their lower than average per capita electricity consumption, any increase in electricity consumption due to the PHEV would have a relatively large impact on a Californian household. When all of the households are considered together the impacts could be far-reaching.

In order to determine the residential electricity demand for California, a "bottom-up" approach to electricity profile modeling was adopted. The approach is to model an individual household based on a set of appliance that the household owns. The appliances are then randomly switched on following usage probabilities that vary throughout the day. The model has been localized for households in California and the details can be found in a separate publication (Huang et al., 2009). In this study, however, the PHEV has been redefined as an agent with guiding heuristics that help determine its charging patterns. The PHEV used is modeled after the anticipated characteristics of the Chevrolet Volt, due for commercial release in 2011. The PHEV is assumed to follow a typical usage pattern of a working individual during the week (BLS, 2009). The length of each trip that the vehicle makes is determined from a distribution of historical data (FHA, 1990). Vehicles are assumed to charge only when they are at their residence, and are plugged in whenever they are at home, regardless of the time of the day. The relevant data for the industrial and commercial sectors are determined from a combination of various sources. The usage pattern and regression for random and seasonality factors are determined from CAISO (CAISO, 2009) and the literature (Brown and Koomey, 2003), while average daily use figures are obtained from the California Energy Commission (Pfannenstiel et al., 2007). The results of the simulation's breakdown of electricity demand for the state of California for an average summer weekday can be seen in Figure 1. A 62% penetration rate for PHEVs would account for about 12% of daily electricity consumption in California. The average daily electricity demand for the state would then rise to roughly 800 GWh day^{-1}.

In order to fulfill the enormous amounts of demand associated with its large population, and significant industrial and agricultural sectors, California also has a correspondingly large generating capacity. Once again, the state trails only Texas in the total amount of generating capacity with a summer peak capacity of 63,813 MW. According to the EIA there were 1,492 generating units active in California in 2007, with an additional 55 units on standby. In this analysis we have only considered those units that were operating in 2007 and did not include the reactivation of units on standby or of retired plants. Of the operational generating units many refer to multiple units on the same company site. For this analysis we have combined all operating units of a similar type and on one site into a single agent. Thus there are 716 electricity generating agents included in the supply model, each assumed to be running at full capacity when scheduled.

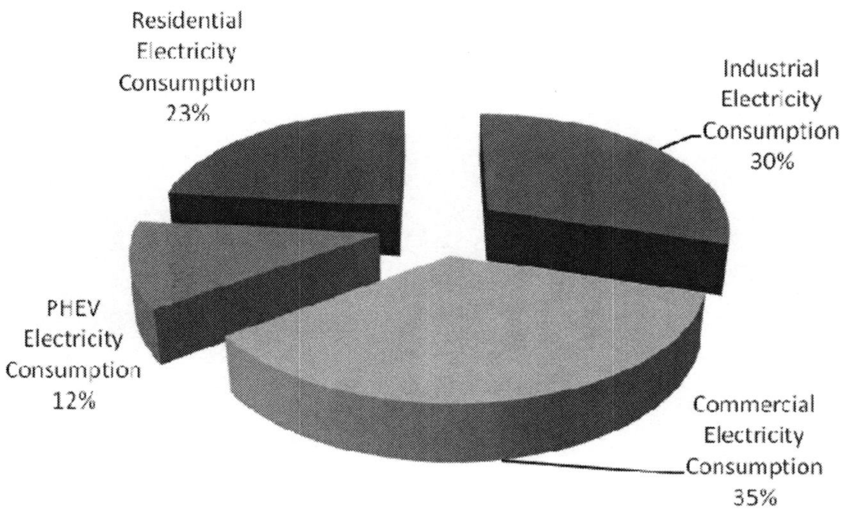

Figure 1: Electricity consumption by sector for the state of California on a summer weekday with PHEV adoption.

Each generating unit is assigned a pair of parameters based upon its energy source: unit marginal cost and unit availability. The specific cost for the unit is taken from a beta distribution with shape parameter values of three and the upper and lower bounds for the cost specified. The upper and lower bounds of marginal generating costs are those used by the IEA (IEA, 2005). Average availability data has been taken from the North American Electricity Reliability Corporation (NERC, 2009). Where only bounds for the availability data were available, such as with wind and solar electricity units, the same beta distribution described above for costs was used. The individual unit availability figures does not take into account restrictions on usage due to plants exceeding their allowed emissions of pollutants, such as NO_x, an aspect not considered in the model, but the relevance of which was realized during the 2000 California electricity crisis. Another factor that figured in the crisis, the decreased availability of hydroelectric power due to a dry summer season, is also not considered within the model.

Both the electricity demand and supply models are encapsulated in a broader agent-based modeling framework. In this study, the electricity demand model simulates the electricity demand profile based on minute steps and at the end of the hour synchronizes with the supply model through a data port and allows for the demand model to allocate electricity supply on the hourly time scale. The simulated electricity generation mixture for the state of California can be seen in Figure 2. When compared to actual data from the EIA (EIA, 2009c), it can be seen that the model produces a reasonable fit to the actual data. Looking at the top 4 sources, California obtains approximately 56% of the electricity produced in state from natural gas, 13% from conventional hydroelectric sources, 18% from nuclear sources and 6% from geothermal sources; figures which are reasonably replicated by the simulated data.

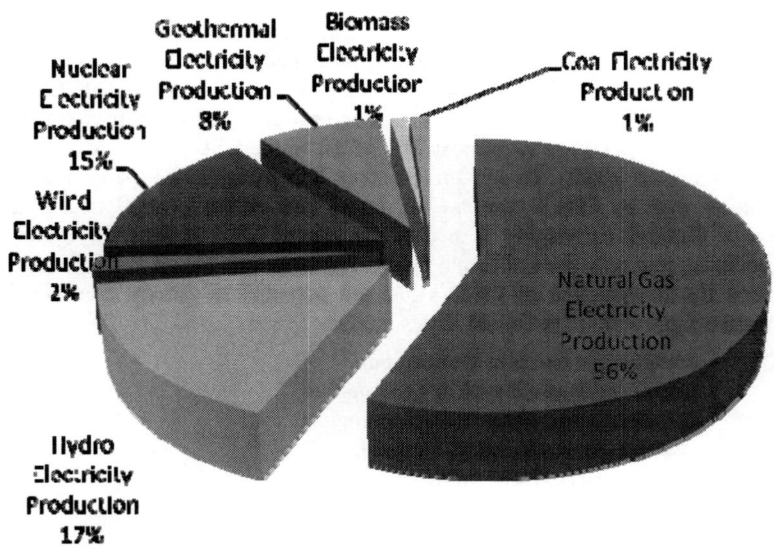

Figure 2: Simulated electricity generation mix for California without PHEV electricity demand

If every household in California replaces one conventional automobile with a PHEV, it can be seen from Figure 3 that the end result is the emergence of a peak at 19:00 hours when a majority of vehicle owners reach home and plug their vehicles in to be charged. It can also be seen that the majority of this increase in electricity demand is fulfilled by an increase in usage of natural gas for electricity generation. The proportion of electricity supplied by natural gas would increase from 56% to 59% according to the results of the simulation.

A linear regression of natural gas consumed for electricity production versus the amount of electricity produced by natural gas was performed with EIA data in order to translate the number of megawatt hours of electricity produced by a natural gas generation unit into millions of cubic feet of natural gas consumed. The data used was for the entire United States electricity market and consisted of 29 months of both reported natural gas usage and electricity output from January 2007 until May 2009. As would be expected, the fit was very good with a R^2 value of over 0.9966. On an average summer weekday in California the total electricity consumption from natural gas was found to be approximately 395.8 GWh. With the addition of PHEVs at a 62% penetration rate, and assuming an unregulated charging pattern, the state-wide natural gas fueled electricity consumption increases to approximately 472 GWh per day. This represents 3,337 million cubic feet of natural gas per day without the addition of PHEVs and 3,990 million cubic feet per day with PHEVs. These figures indicate that the introduction and widespread adoption of PHEVs could cause an increase in natural gas consumption of roughly 19.5% in the California electricity generating sector. This corresponds to a state-wide increase in natural gas consumption of approximately 6.5% across all sectors. Given that the size of the California natural gas market is over 2.3 trillion cubic feet per year, even this seemingly small increase corresponds to an increase of approximately 150 billion cubic feet per year or almost 1% of all United States dry natural gas production for the year 2007. To put this number into perspective: the increase in California alone due to PHEV use would equal the entire yearly natural gas consumption of Finland, or roughly 13.6% of the annual LNG exports for Qatar in 2006. Considering that only the California market has been considered in this analysis, it is clear that the introduction of PHEVs has the potential to greatly increase the demand for natural gas within the United States market.

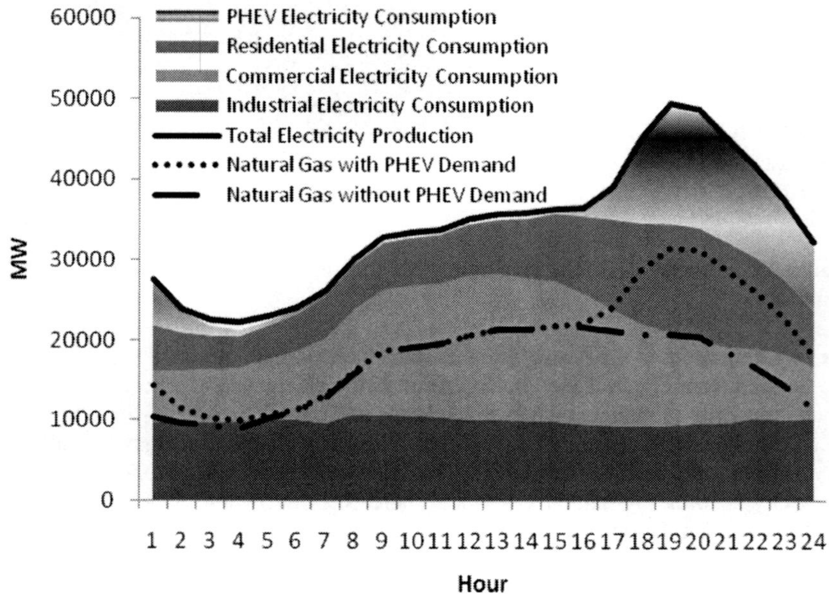

Figure 3: Daily Californian electricity load profile for a summer weekday. The stacked areas in the figure represent electricity demand with top to bottom order corresponding to the top four items in the legend. The lines represent the electricity demand. The change in natural gas electricity generation can be seen from the difference in the dotted lines.

5. Conclusion

While PHEVs are an attractive option to reduce the consumption of gasoline in the transportation sector and hence the United States' total petroleum imports, the shift in energy demand from petroleum to electricity can create unforeseen problems. Before relying on the existing infrastructure, namely the electricity grid, to handle the shifted burden currently held up by petroleum products, a thorough study of the system's limitations must be undertaken in order to anticipate the ripple effects of this switch. An integrated electricity demand and supply model has been developed to help quantify this anticipated paradigm shift and a case study of the Californian electricity market has been conducted. For a 62% penetration rate of PHEVs in California, it is expected that the total natural gas demand from the state would increase by 6.5%. If the adoption of PHEVs becomes widespread throughout the country, the aggregated demand increase would be significant. It is therefore important to be careful when planning for this transition from petroleum based transportation to an electricity based system in order to anticipate these and other foreseeable effects of the transition.

References

Bls. 2009. American Time Use Survey [Online]. Bureau of Labor Statistics. Available: http://www.bls.gov/tus/home.htm [Accessed 08/29 2009].

Brown, R. E. & Koomey, J. G. 2003. Electricity Use in California: Past Trends and Present Usage Patterns. Energy Policy, 31, 849-864.

Bts 2008. State Transportation Statistics 2008. U.S Department of Transportation.

Caiso. 2009. System Load Reports [Online]. California ISO. Available: http://oasishis.caiso.com/ [Accessed 08/29 2009].

Census. 2009. State Housing Unit Estimates: 2000 to 2008 [Online]. U.S. Census Bureau. Available: www.census.gov/popest/housing/HU-EST2008.html [Accessed 08/29 2009].

Eia 2009a. Annual Energy Review 2008. Energy Information Administration.

Eia 2009b. Electric Power Annual 2007. Energy Information Administration.

Eia 2009c. Electric Power Monthly August 2009. Energy Information Administration.

Eia. 2009d. Table 5. U.S. Average Monthly Bill by Sector, Census Division, and State 2007 [Online]. Available: http://www.eia.doe.gov/cneaf/electricity/esr/table5.html [Accessed 08/29 2009].

Fha 1990. Nationwide Personal Transportation Survey - Urban Travel Patterns. U.S Department of Transportation.

Hodge, B.-M. S., Pekny, J. F. & Reklaitis, G. V. 2009. A Multi-Paradigm Energy Model for Liquid Natural Gas Analysis. In: Alfadala, H., Reklaitis, G. V. R. & El-Halwagi, M. M. (eds.) 1st Annual Gas Processing Symposium. Qatar: Elsevier.

Huang, S., Hodge, B.-M., Taheripour, F., Pekny, J., Reklaitis, G. V. & Tyner, W. 2009. The Effects of Electricity Pricing Structure on the Competitiveness of Plug-in Hybrid Electric Vehicles in the California Market. Purdue University. (Submitted Manuscript)

Iea 2005. Projected Costs of Generating Electricity 2005 Update. Nuclear Energy Agency International Energy Agency, Organisation for Economic Co-Operation and Development.

Nerc. 2009. Generating Availability Data System (Gads): Reports [Online]. North American Electric Reliability Corporation. Available: http://www.nerc.com/page.php?cid=4|43|47 [Accessed 08/29 2009].

Paatero, J. V. & Lund, P. D. 2006. A Model for Generating Household Electricity Load Profiles. International Journal of Energy Research, 30, 273-290.

Pfannenstiel, J., Gessman, J. L., Blevins, B. B., White, M. L., Garfield, S., Walker, C. & Nelson, Y. 2007. 2007 Integrated Energy Policy Report.

Weidlich, A. & Veit, D. 2008. A Critical Survey of Agent-Based Wholesale Electricity Market Models. Energy Economics, 30, 1728-1759.

Proceedings of the 2nd Annual Gas Processing Symposium
Farid Benyahia and Fadwa T. Eljack (Editors)

Design of Sustainable Processes: Systematic Generation & Evaluation of Alternatives

Ana Carvalho [a,b], Rafiqul Gani [b], Henrique A. Matos [a]

[a]*CPQ- Dep of Chem. & Bio. Eng, Ins. Sup. Técnico, Av Rovisco Pais, 1049-001 Lisboa, Portugal*

[b] *CAPEC- Dep of Chem Eng, Tech Univ of Denmark, DK-2800 Lyngby, Denmark*

Abstract

In order to address the sustainability of chemical processes various methodologies have been proposed in the past decade. The objective of this paper is to describe a systematic methodology for sustainable chemical process design and, a software tool, called *SustainPro*, which employs this a methodology to improve the sustainability of existing processes as well as new processes. The developed methods and tools are able to generate, screen and then identify sustainable alternatives in any chemical process by locating the operational, environmental, economical and safety related bottlenecks inherent in the process. It can be applied to a wide range of processes such as refineries, gas processing, pharmaceutical processes, etc, operating in batch as well as continuous modes. The application of the methodology and software is highlighted with a case study related to a gas processing application.

Keywords: Process design, Sustainability metrics, Mass and Energy Indicators, Safety indices, Software

1. Introduction

Chemical industry plays a very important role worldwide. It is responsible for the turnover of millions of Euros per year, it employs a large number of people and it produces the goods required for the day life of the modern society. However, chemical industrial activity is also responsible for the depletion of raw materials, the enormous consumption of energy and the huge production of wastes/residues. The sustainability of the processes has therefore become a major issue the sustainability of the processes. Many methodologies have been proposed to address and to solve these issues. Some methodologies were introduced using heuristic rules for the generation of new design alternatives (Rapoport *et al.*, 1994). Other methodologies based on mathematical programming concepts such as, mixed integer nonlinear programming (MINLP) have also been proposed (Jackson and Grossmann, 2002). Another approach for continuous processes has been the development of methodologies based on the resynthesis of the entire process by incorporating operating units with enhanced performance (Liu *et al.*, 2006). In the area of batch operations, several methodologies for retrofit design have been proposed. For example, Halim and Srinivasan (2008) proposed an intelligent

simulation–optimization framework for identifying comprehensive sustainable alternatives for batch processes. Simon *et al.* (2008), on the other hand, presented an indicator, heuristics, and process model based decision support framework for retrofitting chemical batch processes. Their framework considered the identification of improvement opportunities in a batch plant by considering first the product market situation. Recently, Carvalho *et al.*, (2008) and Carvalho *et al.*, (2009) presented a systematic and generic methodology to generate new sustainable design alternatives for a wide range of chemical processes. This work is based in a set of mass and energy indicators and a series of sensitivity analysis, which lead to a more sustainable alternative, which is evaluated through a set of sustainability metrics and safety indices.
The objective of this paper is to provide a general overview of the new combined methodology together and introduce the features of *SustainPro*. The ammonia production process, which processes natural gas, is used as a case study to illustrate the new features of the methodology and the software and consequently its applicability to gas processes.

2. Methodology

The main steps of the methodology are briefly described below in Figure 1. For continuous processes the main features of the methodology can be found in Carvalho *et al.* (2008) and the extended methodology for batch processes is described in Carvalho *et al.* (2009). Figure 1 presents the flow-diagram of the indicator-based methodology.

<u>Step 1: Collect data</u>
The plant and/or the model generated results (mass and/or energy balance) are the input for the methodology and consequently for the software. In continuous processes, steady state data related to the mass and the energy balance is needed. In the batch processes the information required is the time of each operation, the equipment volume, the initial and the final mass for each compound in each operation and the energy used in each step. The purchase and sale prices for each chemical are also needed for both cases.

<u>Step 1A: Transform equipment flowsheet in operational flow-diagram</u>
This is an intermediate step that is only applied to batch processes. In this step the equipment based-flowsheet is transformed to an operation-based flowdiagram (Carvalho *et al.* 2009). This flowdiagram allow

<u>Step 2: Decompose Flowsheet</u>
In this step, a flowsheet decomposition is performed to identify all open- (OP) and closed-paths (CP) for each compound in the continuous process flowsheet as well as in the batch operations flow-diagram. For batch operation flow-diagram, a path related to the accumulation of mass and energy has also been introduced. This path is called accumulation-path (AP) and corresponds to the accumulation in a given operation.
SustainPro automatically generates a list of all the paths.

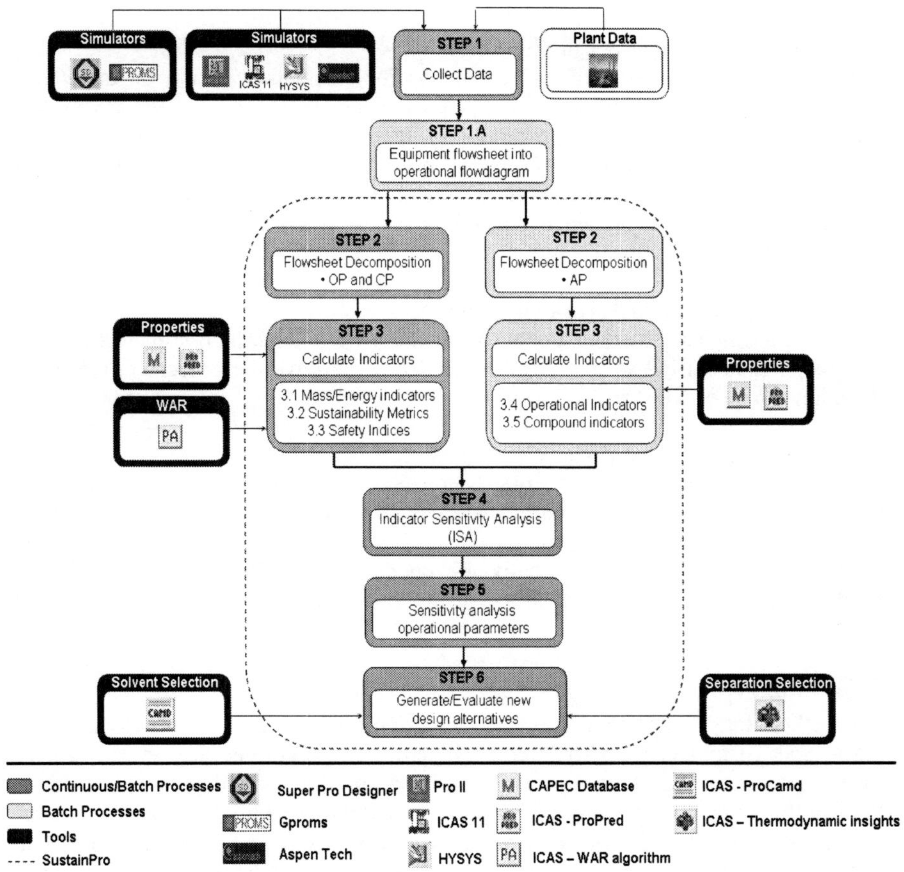

Figure 1: Flow-diagram of the indicator-based methodology in *SustainPro*

Step 3: Calculate indicators, sustainability and safety parameters

The mass and energy indicators are calculated for all the OP and CP determined in step 2 for batch and continuous processes. Two sets of indicators for batch processes have been developed: The Operation Indicators (compare the operation performance) and the Compound Indicators (indicate for each operation which compound is most likely to cause operational problems). Through the values obtained for the full-set of indicators it is possible to identify the locations within the process where the mass/energy "paths" face "barriers" with respect to costs, benefits, or accumulation problems. These critical points present high potential for process improvements. The analysis of the process impact is determined through the calculation of a set of sustainability metrics. The safety of the process is measured by the use of a set of safety indices.

Step 4: Perform Indicator Sensitivity Analysis

The objective of this step is to identify the indicators that should be targeted for improvement using an algorithm for sensitivity analysis (Carvalho *et al.* 2008), Here, the indicators having the highest potential for improvements are identified first. Then an objective function such as gross-profit or process total cost is specified. The sensitivity analysis is performed to determine the indicators that cause the largest improvement in the objective function. The final result is a list of target indicators and the desired changes in their values.

Step 5:Perform operational sensitivity analysis

Through this analysis the operational (parameters) variables, which influence the target indicators, is identified so that changes needed to improve the process in the desired direction can be established.

Step 6: Generation of new design alternatives

New alternatives are generated using a collection of methods and tools for process synthesis. To find new separation processes the thermodynamic insights based method of Jaksland *et al.* (1995), is used. New solvents are selected applying the algorithm of Harper and Gani, (2000) and finally the operational conditions for separation and reactional processes are defined based in the reverse approach method developed by D'Anterroches and Gani, (2005).

3. *SustainPro*

The methodology presented in the previous section has been implemented, together with other tools (database, property prediction, processes synthesis, etc) into a software called *SustainPro* has been developed to perform the methodology. The software follows the work-flow described in Figure 1. *SustainPro* is an *Excel* based software, divided in 21 different *Excel* sheets, where two of the *Excel* sheets are Principal Menus and the remaining sheets represent different steps and sub-steps of the methodology presented in Figure 1. The needed data on chemical properties is retrieved through tools such as the CAPEC data-base, property prediction package (ProPred) and the WAR Algorithm, which is available in ICAS (Carvalho *et al.* 2008).

4. Ammonia- Case Study

Over 100 million tons of ammonia are produced each year. The rewards for reducing costs, increasing efficiency and improving the profitability of ammonia plants are enormous. A PRO/II simulation model of ammonia synthesis plant is presented here. The entire plant is modelled, from the reforming of the hydrocarbon feedstream to synthesis gas through its purification to its conversion to ammonia in a synthesis reactor. The Ammonia process flowsheet (Casebook files, (1992)) can be divided into four sections as shown in Figure 2.

The ammonia process is a single train process and is divided into four stages, which operate sequentially:

- Catalytic Reforming: the natural gas undergoes catalytic reforming to produce hydrogen from methane and steam. In this stage nitrogen required is introduced.
- Shift and Methanation: The resulting syngas is purified by the removal of carbon monoxide and carbon dioxide.
- Compression: This stage consists of compression of the syngas up to the pressure required in order to remove water.
- Conversion: The compressed synthesis gas is dried, mixed with a recycle stream and introduced into the synthesis loop.

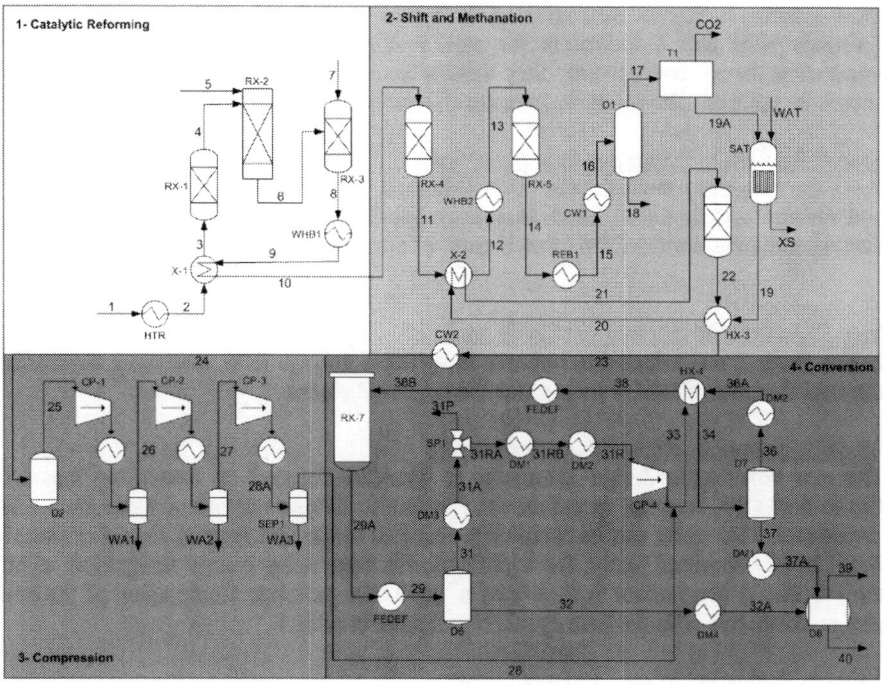

Figure 2: Flowsheet of ammonia production process

Step 1: Collect data

The required detailed process data for the ammonia process plant was taken from a simulation model available on ProII software package (Casebook files, (1992)).

Step 1.A: Transform equipment flowsheet in operational flow-diagram
This is a continuous process and consequently this step does not need to be performed.

Step 2: Decompose Flowsheet
The process involves 57 streams, 42 units and 13 compounds. For this case study, the operational flow-diagram decomposition generated 13 closed-paths, 109 open-paths with flowrate different from 0.

Step 3: Calculate indicators, sustainability and safety parameters
For the entire set of flow-paths, the full-set of indicators were calculated and only the most sensitive indicators were selected for further study. In this case study 13 indicators for open-paths and 2 indicators for closed-paths were showing high potential for improvements and consequently they were selected. The water metrics were high and should be reduced. The safety index were also calculated and its value was 44.

Step 4: Perform Indicator Sensitivity Analysis
From the sensitivity analysis, the MVA (Material Value Added) and the EWC (Energy and Waste Cost) for OP 6, which is an open-path for water, and the AF (Accumulation Factor) and EWC for the CP 6 (closed-path) of nitrogen were identified as targets.

Step 5: Perform operational sensitivity analysis
In this study it was selected to improve the process through OP6, that is, by decreasing the total flowrate of the OP to improve the target indicators.

Step 6: Generation of new design alternatives
The new sustainable design alternative is therefore focused on decreasing the total flowrate of OP6. In order to reduce the flowrate in OP6, a recycle of water should be considered. This water can be completely recycled to the first reactor. However since it needs to be vaporized firstly, the vaporization is done using energy integration within the process. A compressor is then used to adjust the pressure. Comparison of the new design alternative with the base case is highlighted in table 1.

Table 1: Evaluation of new design alternative

	Base Case	**Improved Alternative**
Revenues ($/year)	23306380	23306380
Variable Costs ($/year)	7383605	6995780
Raw Material Costs ($/year)	7120174	2068059
Waste Treatment Costs ($/year)	173186	173186
Profit ($/year)	8632877	14072817

With this new alternative the target indicator was improved by 100% (MVA_{Base} = -259 x 10^3; $MVA_{Improved}$ = 0). The sustainability metrics and the safety index are listed for the base case and for the new design alternative in table 2.

Table 2: Sustainability Metrics and safety index

Metrics	Base	New	Deviation
Total Net Primary Energy Usage rate (GJ/y)	862693.44	876315.84	-1.58%
% Total Net Primary Energy sourced from renewables	0.99	0.98	1.55%
Total Net Primary Energy Usage per Kg product (kJ/kg)	9044.94	9187.77	-1.58%
Total Net Primary Energy Usage per unit value added (kJ/$)	6.73	5.20	22.78%
Total raw materials used per kg product (kg/kg)	1.62	1.62	0%
Total raw materials used per unit value added	0.0012	0.0012	0%
Net water consumed per unit mass of product (kg/kg)	41.38	33.90	18.07%
Net water consumed per unit value added	0.03	0.02	37.71%
WAR per kg product (impact/kg)	852354	852354	0%
Safety Index	44	44	0%
Profit ($/yr)	8632877	14072817	63.01%

From table 2, it is possible to see that the water metrics decreased by 18% and 38% respectively. The profit increased by 63% since the costs were reduced considerably. All the remaining performance criteria remained unchanged except the energy metrics. The energy metric per kg of final product increased by 1.6% due to the energy consumed in the new compressor, however, the energy metric per value added improved 23% due to the increase in the profit. The process with the alternative of recycled water can therefore be considered more sustainable.

5. Conclusions

The development of a systematic and generic indicator-based methodology for continuous and batch process analysis and for generating sustainable process improvements has been presented together with a corresponding software (SustainPro). Through the software it has been shown that more sustainable process designs can be achieved without any trade-offs. As shown through the ammonia case study, the size of the problem is not an issue since the objective is to locate the most sensitive indicators and their related variables, which are usually not that many. It can be concluded that this methodology has a wide range of applicability.

6. Acknowledgments

The authors gratefully acknowledge financial support from Fundação para a Ciência e a Tecnologia (under Grant No. SFRH/BD/24470/2005).

References

Carvalho, A., Matos, H. A., Gani. R., 2008, Design of Sustainable Chemical Processes: Systematic Retrofit Analysis Generation and evaluation of alternatives, *Process Safety and Environmental Protection*, 86: 328–346

Carvalho, A., Matos, H. A., Gani. R., 2009, Design of batch operations: Systematic methodology for generation and analysis of sustainable alternatives, *Comput. Chem. Eng*, doi:10.1016/j.compchemeng.2009.06.015

Casebook files- ProII, 1992

D'Anterroches, L. and Gani, R., 2005, Group contribution based flowsheet synthesis, design and modelling. Fluid Phase Equilibria, 228/229: 141–146.

Halim, I. and Srinivasan, R., 2008, Designing sustainable alternatives for batch operations using an intelligent simulation–optimization framework, *Chemical engineering research and design*, 86:809–822

Harper, P.M. Gani, R., 2000, A Multi-Step and Multi-Level Approach for Computer Aided Molecular Design, *Comput. Chem. Eng*, 24, 2-7, 677-683.

Jackson, J. R.; Grossmann, I. E., 2002, High-level optimization model for the retrofit planning of process networks, *Ind. Eng. Chem. Res.*, 41: 3762-3770.

Jaksland, C., Gani, R. and Lien, K., 1995, Separation Process Design and Synthesis Based on Thermodynamic Insights, *Chemical Engineering Science,* Vol. 50., No. 3, pp 511-530

Liu, J.; Fan L. T.; Seib, P.; Friedler, F.; Bertok, B., 2006, Holistic Approach to Process Retrofitting: Application to Downstream Process for Biochemical Production of Organics, *Ind. Eng. Chem. Res.,* 45: 4200-4207.

Rapoport, H., Lavie, R., Kehat, E, 1994, Retrofit Design of New Units into an Existing Plant: Case Study: Adding New Units to an Aromatics Plant, *Comput. Chem. Eng.*, 18: 743-753.

Simon L. L., Osterwalder N., Fischer U., Hungerbuhler K., Systematic Retrofit Method for Chemical Batch Processes Using Indicators, Heuristics, and Process Models, *Ind. Eng. Chem. Res.*, 47:66-80

Proceedings of the 2nd Annual Gas Processing Symposium
Farid Benyahia and Fadwa T. Eljack (Editors)

Steady State Simulation for Optimal Design and Operation of a GTL Process

Mehdi Panahi, Sigurd Skogestad, Ramprasad Yelchuru

Chemical Engineering Department, Norwegian University of Science and Technology (NTNU), 7491, Trondheim, Norway

Abstract

This study of gas to liquid (GTL) production uses an authothermal reformer (ATR) for synthesis gas production and a slurry bubble column reactor (SBCR) with Cobalt catalyst for the Fisher-Tropsch (FT) conversion to liquid fuels. The well-known Satterfield kinetics are used for the FT reactor; the remaining reactions are simulated assuming thermodynamic equilibrium. The process also includes high-pressure CO_2 removal. Important process parameters, which are subject to optimization, include feed rates of water and oxygen, inlet ATR temperature and recycle flows. In addition to determining the process and equipment design, optimization is used to find the controlled variables. The aim is to identify "self-optimizing variables" that indirectly give close-to-optimal operation with constant setpoints, in spite of disturbances.

Keywords: Gas to liquid process, Optimization, Plantwide control, Self-optimizing control, Process control,

1. Introduction

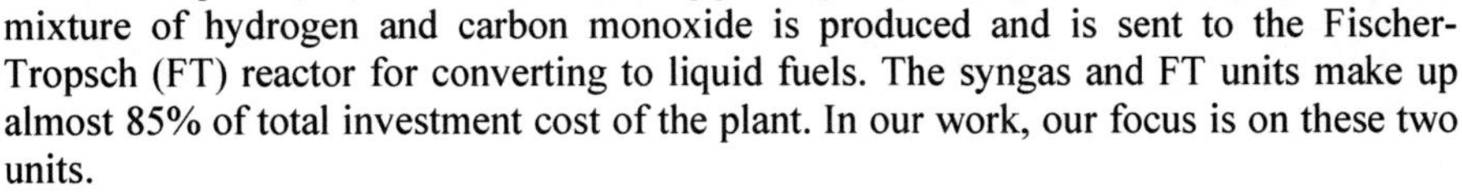

Figure 1: Main parts of GTL process

The GTL process in Fig.1 has 3 main steps (i) production of synthesis gas (syngas), (ii) Fischer-Tropsch (FT) reactor and (iii) upgrading units. (Fig.1). In the syngas unit, a mixture of hydrogen and carbon monoxide is produced and is sent to the Fischer-Tropsch (FT) reactor for converting to liquid fuels. The syngas and FT units make up almost 85% of total investment cost of the plant. In our work, our focus is on these two units.

This paper focuses on the optimal operation of such plants so that we can capture in practice the benefits in terms of energy savings and productivity that are predicted when designing the process. Skogestad (2004) has proposed a systematic method for designing the control structure of a complete chemical plant. The first step of this procedure is the selection of controlled variables (self-optimizing control) which is the focus of this paper.

2. Conceptual design of a GTL process

2.1. Feed data

Natural gas feed is assumed at the following conditions:

Flow rate: 8195 $\frac{\text{kmol}}{\text{hr}}$, Pressure: 73.5 bar, Temperature: 40 °C

Composition: $CH_4(95.5\%), C_2H_6(3\%), C_3H_8(0.5\%), nC_4H_{10}(0.4\%), N_2(0.6\%)$

Nominal feed rates of water and oxygen (see Fig. 1): $\frac{H_2O}{\text{Natural Gas}} = 0.53$, $\frac{O_2}{\text{Natural Gas}} = 0.67$

2.2. Syngas unit

In GTL processes, there are different routes for syngas production from natural gas, including Auto Thermal Reforming (ATR), steam reforming, partial oxidation of methane and CO_2 reforming. ATR is a combination of steam reforming and oxidation of methane. It is claimed to be the best route for syngas production (Bakkerud, 2005) and has been selected for this study.

To avoid the potential problem that the ATR works as a steam cracker, producing olefins from higher hydrocarbons in the feed, an adiabatic pre-reformer is introduced. Here, the temperature is 350-550°C and all higher hydrocarbons are converted (Aasberg-Petersen et al, 2001) according to the following reactions:

Complete conversion reactions (endothermic, so energy is needed):

$$C_2H_6 + 2H_2O \rightarrow 5H_2 + 2CO \qquad \Delta H = +350\frac{kJ}{mol} \qquad \Delta G = +201.86\frac{kJ}{mol} \tag{1}$$

$$C_3H_8 + 3H_2O \rightarrow 4H_2 + 3CO \qquad \Delta H = +500\frac{kJ}{mol} \qquad \Delta G = +302.78\frac{kJ}{mol} \tag{2}$$

$$C_4H_{10} + 4H_2O \rightarrow 9H_2 + 4CO \qquad \Delta H = +650\frac{kJ}{mol} \qquad \Delta G = +403.71\frac{kJ}{mol} \tag{3}$$

Equilibrium reactions:

$$CO + 3H_2 \Leftrightarrow CH_4 + H_2O \qquad \Delta H = -210\frac{kJ}{mol} \qquad \Delta G = -151.65\frac{kJ}{mol} \tag{4}$$

$$CO + H_2O \Leftrightarrow CO_2 + H_2 \qquad \Delta H = -41\frac{kJ}{mol} \qquad \Delta G = -19.09\frac{kJ}{mol} \tag{5}$$

In our case, natural gas and water are preheated to 455°C and fed to the adiabatic pre-reformer. In spite of the exothermic equilibrium reactions the overall reactions are endothermic and the outlet temperature is about 416°C.

The pre-reformer outlet stream is mixed with recycled flue gas from the FT unit and heated in a fired heater to 675 °C before entering the adiabatic authothermal reformer

(ATR). Oxygen is preheated to 200 °C and is also fed to the ATR. The ATR is the main reactor in producing the synthesis gas and the following three main reactions take place:

Oxidation of methane (exothermic):

$$CH_4+\frac{3}{2}O_2 \Leftrightarrow CO+2H_2O \qquad \Delta H=-520\frac{kJ}{mol} \qquad \Delta G=-562.65\frac{kJ}{mol} \tag{6}$$

Steam reforming of methane (endothermic):

$$CH_4+H_2O+210\frac{kj}{mol} \Leftrightarrow CO+3H_2 \qquad \Delta H=+210\frac{kJ}{mol} \qquad \Delta G=+151.65\frac{kJ}{mol} \tag{7}$$

Shift Reaction (exothermic):

$$CO+H_2O \Leftrightarrow CO_2+H_2 \qquad \Delta H=-41\frac{kJ}{mol} \qquad \Delta G=-19.09\frac{kJ}{mol} \tag{8}$$

Because of the large heat generated by combustion in reaction (6) the net reactions are exothermic, and the outlet temperature of the ATR is about 1000 °C.

A high-pressure CO_2 capturing process with MDEA as an absorbent is used to remove most of the CO_2 from the syngas. A mixture of hydrogen and carbon monoxide is the main product of ATR and a ratio of $\frac{H_2}{CO}$ around 2-2.3 is desired to have maximum conversion to liquid fuels in the subsequent FT reactor (see Fig. 2). To set this ratio, a small amount of CO_2 is recycled to the ATR. This avoids that too much CO is shifted to CO_2 according to reaction (8).

2.3. FT reactor

The syngas is sent to the FT reactor. The FT synthesis with cobalt catalyst produces mostly n-alkanes and 1-alkenes. The FT reaction and kinetics using cobalt catalyst was taken from Satterfield and Yates (Yates and Satterfield, 1991). The main reaction is

$$nCO+2nH_2 \rightarrow (\text{-}CH_2\text{-})_n+nH_2O \tag{9}$$

$$-R_{CO}=\frac{aP_{CO}P_{H_2}}{(1+bP_{CO})^2} \tag{10}$$

The a and b are temperature-dependent constants; a representing a kinetic parameter and b an adsorption coefficient (Table1). The amount of cobalt catalyst is 392.2 $kg/m^3_{reactor}$ (based on data from Sehabiague et al, 2008).

Table 1: Kinetic parameter for FT reaction

Reactor Temperature (°C)	a $\left(\frac{\text{mmol}}{\text{min.g of catalyst. Mpa}^2}\right)$	b$\left(\frac{1}{\text{MPa}}\right)$
240	75.76	11.61
220	53.11	22.26

$$a = 8.01368e^{-\frac{37326}{8.31*T}}(\frac{kmol}{kg_{cat.}.\sec.MPa^2}),\ b = 1.248*10^{-6}e^{\frac{68402}{8.31*T}}(\frac{1}{MPa})$$

The product distribution of the hydrocarbons "(–CH2-)$_n$" are assumed to follow an Anderson-Schulz-Flory (ASF) distribution to methane, alkanes and alkenes, which can be expressed as:

$$\frac{W_n}{n} = (1-\alpha)^2\alpha^{n-1} \tag{11}$$

$$R_{CH_4} = -(1-\alpha)^2 R_{CO} \tag{12}$$

$$R_{C_nH_{2n+2}} = \frac{\alpha^{n-1}}{1+\gamma} R_{CH_4} \tag{13}$$

$$R_{C_nH_{2n}} = \frac{\gamma}{1+\gamma}\alpha^{n-1} R_{CH_4} \tag{14}$$

Here, W_n is the weight fraction of hydrocarbon molecules containing n carbon atoms. α is the chain growth probability, which is determined by the catalyst and the process conditions. α for C_1 and C_{2+} were set to 0.45 and 0.85, respectively (Fox III and Tam 1995). γ is the selectivity to olefins which is approximately 4 (Ahón et al 2005). Cobalt is not very active for the water gas shift reaction so the shift reaction is negligible. In total, 59 reactions (to CH_4 and to alkanes and alkenes with number of carbons from 2 to 30) were modelled in a CSTR (SBCR) reactor.

The FT reactions are highly exothermic and the generated heat in the reactor is removed by boiling water at high pressure to generate steam. The FT products at 227 °C are then cooled to 50 °C and separated in a 3-phase separator (water, liquid fuels and flue gas). The flue gas, mostly unreacted H_2 and CO, is recycled (90%) to the FT reactor, some (9%) is recycled to the ATR reactor, while about 1% is purged.

The effect of reactor volume on liquid fuels production is shown in Fig. 3. From this figure, it can be observed that a volume greater than about 2200 m^3 does not significantly increase liquid fuel production, so the reactor volume was selected to be 2200 m^3. The final process flowsheet is shown in Fig. 4. All units were modelled using the Unisim flowsheet simulator.

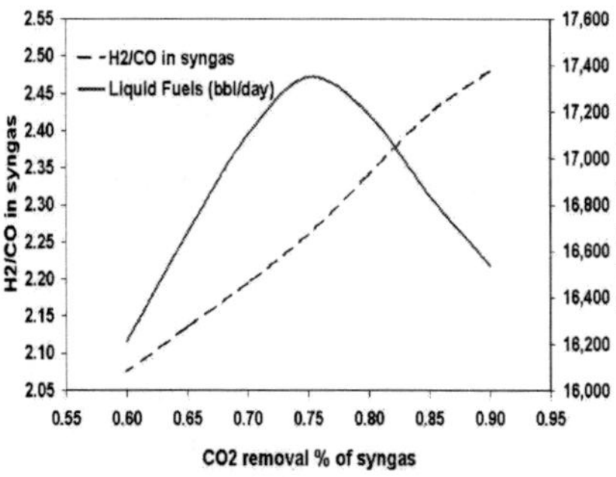

Figure 2: Effect of %CO2 removal on (1) H2/CO in syngas and (2) production of liquid fuels

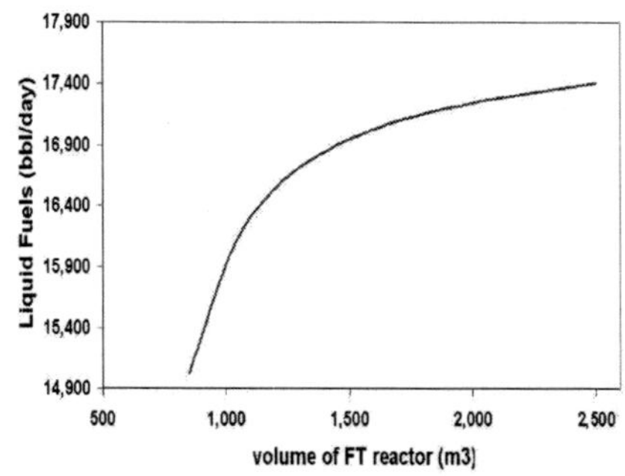

Figure 3: Liquid fuels production as function of volume of FT reactor

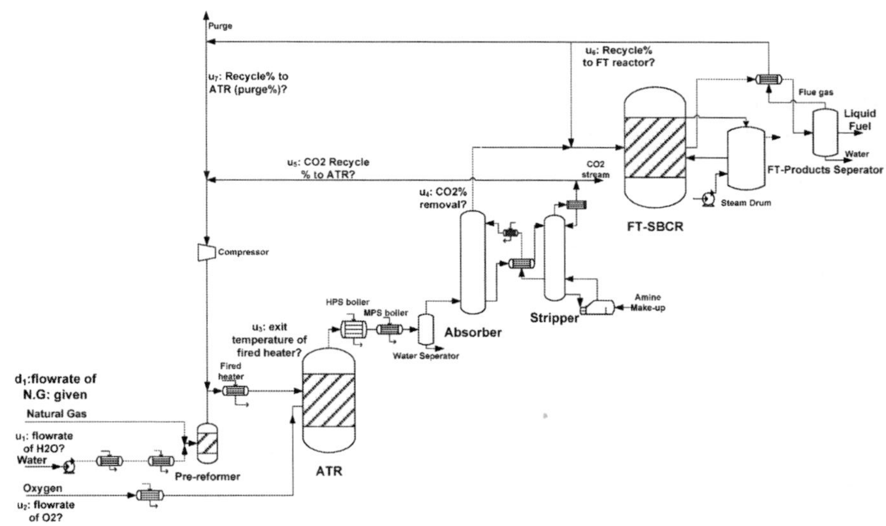

Figure 4: Process flowsheet for GTL process

3. Optimal operation and Self-optimizing control

With a given process design the objective is now to study operation and we use the procedure of Skogestad (2004). One important step is to perform optimization to find optimal conditions of the process for various disturbances.

Step 1: Define the objective function

We consider the case where the product price of the liquid product (fuel) is high so it is optimal to maximize the yield. Hence, the objective function is defined as:

Maximize the liquid fuels production with a given feed rate of natural gas.

The following operational constraints should be satisfied:

(a) The $\frac{H_2O}{\text{Natural Gas}}$ ratio should be larger than 0.3 to avoid soot formation in the ATR.

(b) The fired heater exit temperature should not exceed 675 °C due to limitations on construction material. (Bakkerud, 2009)

3.1. Step 2: Degrees of freedom (DOFs)

With a given natural gas feed rate we identify seven degrees of freedom that can be used for optimizing the objective. These were here selected to be:

1- Water feed rate (u_1)
2- Oxygen feed rate (u_2)
3- Exit temperature of fired heater (u_3)
4- Fraction CO_2 removed in the CO_2 capturing plant (u_4)
5- Amount of CO_2 recycled to the ATR (u_5)
6- Amount of flue gas recycled to FT reactor (u_6)
7- Amount of flue gas recycled to the ATR (u_7)

Optimization with respect to these 7 degrees of freedom was performed using the SQP algorithm in UniSim to find the optimal nominal operating point.

Optimal objective function (liquid production): 17,315 bbl/day

There is 1 active constraint (exit temperature of fired heater); leaving 6 unconstrained degrees of freedom at the optimum. The optimal values for the 7 degrees of freedom are as follows

$u_1(H_2O) = 6025$ kmol/hr ($\frac{H_2O}{\text{Natural Gas}} = 0.735$)

$u_2 (O_2) = 5803$ kmol/hr ($\frac{O_2}{\text{Natural Gas}} = 0.708$)

u_3 (exit temperature of fired heater) = 675 °C (active constraint)

u_4 (CO_2 removal %) = 77.8

u_5 (recycled % of removed CO_2) = 1.34

u_6 (recycled % of flue gas to FT reactor) = 88.69

u_7 (recycled % of flue gas to ATR) = 91.53

The active constraint (exit temperature of fired heater) should always be controlled, but it is not clear how we should use the remaining 6 unconstrained degrees of freedom. Preferably, we would like to find 6 variables that give close-to-optimal operation with constant setpoints ("self-optimizing" control). To look for potential self-optimizing variables, we may reoptimize the process for different disturbances and look for variables with a small optimal variation. Based on this we may suggest potential CVs, and we can then test alternatives by considering the economic loss when they are kept constant.

3.2. Step 3: Disturbances

We consider 4 major disturbances for the GTL process:

1- flowrate of natural gas (±10%)
2- composition of hydrocarbons in natural gas (-10%)
3- change in FT kinetics (±10% in kinetic parameter a),
4- change in exit temperature of fired heater (±25 °C)

Reoptimizing the process for different disturbances show that the constraint on the exit temperature of the fired heater always remains active. This means that 6 CVs associated with the 6 unconstrained DOFs need to be identified.

To keep control simple, we first try whether we can keep some of the original DOFs (u_1-u_7) constant at their nominal optimal value and achieve near-optimal operation with small loss in the presence of different disturbances and also implementation errors (which is when we are not able to keep the DOFs constant). The ideal case is when the loss is small so there is no need to reoptimize any of DOFs in presence of disturbances or implementation error.

3.2.1. First disturbance: Change in flowrate of natural gas

The flowrate of natural gas is allowed to change ±10%. Fig.5 shows the value of reoptimized objective function when we let to all the six unconstrained DOFs vary (solid line), The dotted line shows the value if we keep all DOFs constant at their optimal nominal points. We see that if the flowrate of natural gas increases by 10%, then the simple policy with constant DOFs gives a 20.6% loss in liquid fuel production, which is not acceptable. We also considered the optimal variation in the six DOFs, and found that the largest variation was in the oxygen flowrate (Fig. 6). This suggests that we may attempt to let the other DOFs be constant. Indeed, we see from the dashed line in Fig. 5 that if we keep all DOFs (u_i) at their optimal nominal points, except for the oxygen flowrate, then we have an acceptable loss (1.3%) in the worst case. However, there are other disturbances, so it is too early to conclude that we can keep all these DOFs constant.

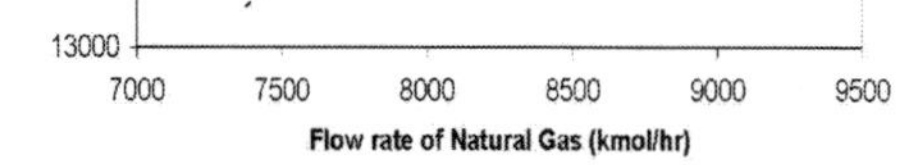

Figure 5: Effect on operation of changes in flowrate of natural gas feed.

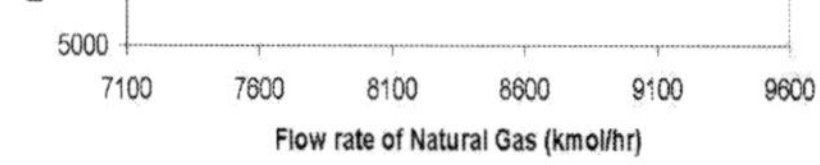

Figure 6: Relation between reoptimized oxygen flowrate and natural gas flowrate

3.2.2. Second disturbance: Change in composition of hydrocarbons in feed

The same method was applied to the second disturbance, which is an increase in the N_2 contents in the range from 0.6% to 10%. The results in Fig.7 show that the effect of this disturbance is quite small. In the worst case, the fuel production loss is 3.3%. If we allow for reoptimization of the O_2 feed rate (Fig. 8) then the loss is almost zero (Fig. 7).

3.2.3. Third disturbance: Model mismatch in kinetic parameter of FT reactions

This disturbance is not important. Fig.9 shows that a ±10% change in the kinetic parameter a gives a loss of less than 1% even with all DOFs kept constant.

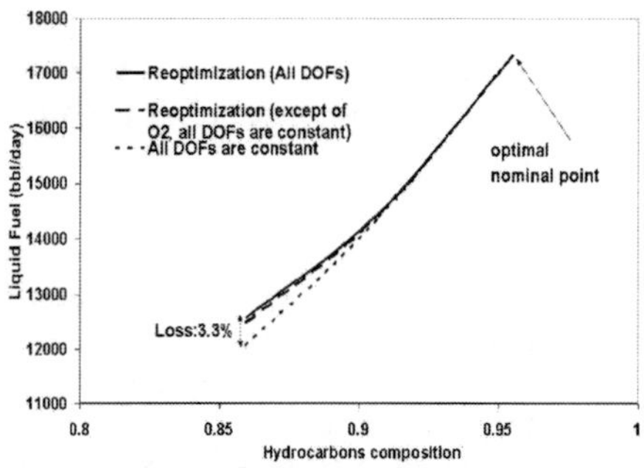

Figure 7: Effect on operation of changes in feed composition

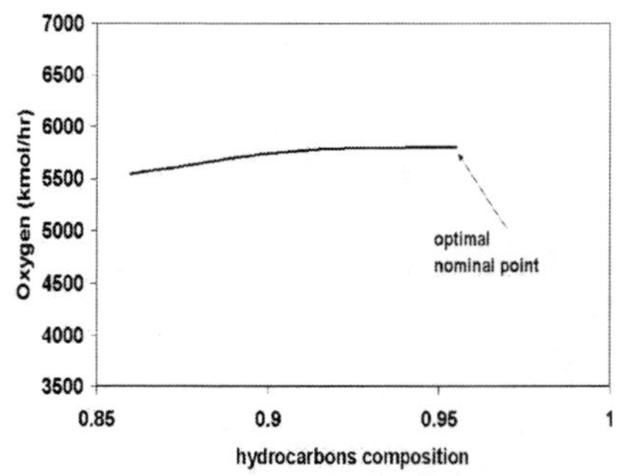

Figure 8: Relation between re-optimized oxygen flowrate and composition of hydrocarbons in feed

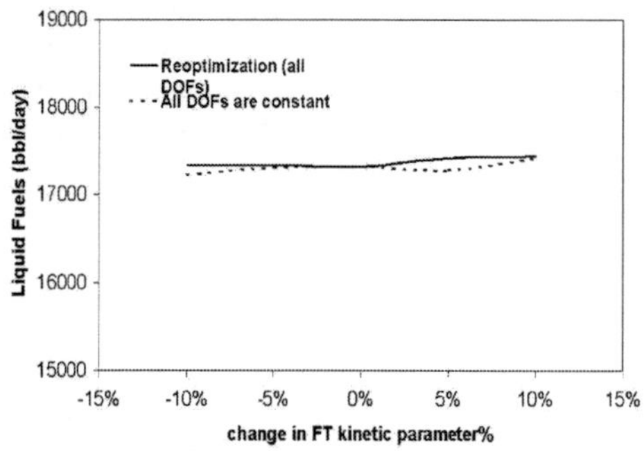

Figure 9: Effect on operation of change in FT kinetic parameter a

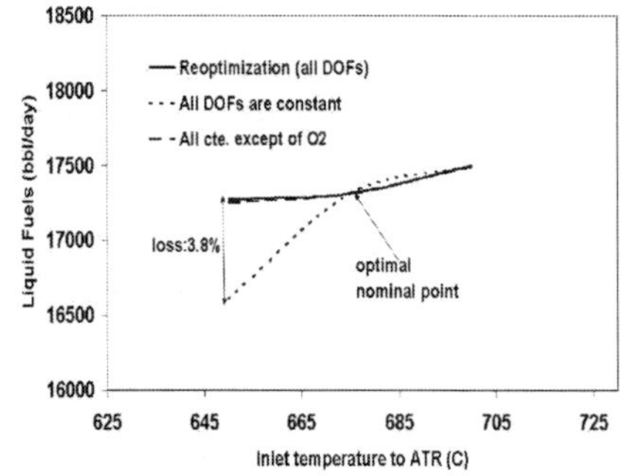

Figure 10: Effect on optimal operation of changes in change to exit fired heater temperature to ATR

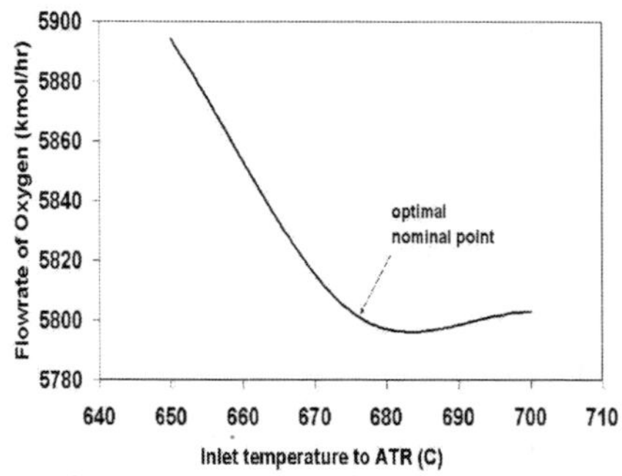

Figure 11: Relation between re-optimized flowrate of oxygen and exit fired heater temperature

3.2.4. Fourth disturbance: Change in fired heater exit temperature

Fig. 10 shows that if the fired heater exit temperature decreases by 25 °C, without reoptimizing, the fuel production loss is 3.8%. If we reoptimize the oxygen feedrate, the loss is almost zero. The reoptimized flowrate of oxygen is shown in Fig. 11.

Table 2: Summary of losses for various disturbances

no.	disturbance	worst case of each disturbance		Loss (%), if all DOFs are constant except of O_2 flowrate
		Change from nominal point	Loss (%)	
1	Flowrate natural gas	+10%	20.6	1.3
2	Inlet temperature to ATR	-25 °C	3.8	0.1
3	Hydrocarbons in the feed	-10%	3.3	0.8
4	Kinetic FT parameter a	+5%	0.8	-

Table 3: Effect of CV implementation error on objective function (loss)

CVs	Implementation error	Loss (%)
Recycled flue gas % to ATR (u7)	-15%	13.55
H_2O feedrate (u1)	-10%	1.9
Recycled flue gas % to FT (u6)	-15%	0.51
CO_2 removal% (u4)	-5%	0.21
Recycled CO_2% to ATR (u5)	-15%	0.02

The results for all four disturbances are summarized in Table 2. It seems that we can always keep all the DOFs constant, except of the O_2 feedrate. However, in addition to disturbances, we should also examine the effect on the objective function of the implementation errors in the selected controlled variables, for example caused by measurement errors. From Table 3 we see that objective function is sensitive to implementation errors in recycled flue gas to ATR (u_7) and H_2O feedrate (u_1), while its sensitivity to the other CVs (u_4, u_5, u_6) is negligible. Note that the implementation error in the active constraint (u_3) has already been considered as a disturbance, while we have already decided that the oxygen feedrate (u_2) should not be kept constant.

In summary, we still have 3 unconstrained DOFs (flowrate of oxygen u_2, flowrate of water u_1 and recycled flue gas to ATR, u_7) for which we have not identified good (self-optimizing) controlled variables. Some potential candidates are: H_2/CO ratio at ATR outlet, temperature at ATR outlet, methane contents in flue gas, oxygen/hydrocarbon feed ratio and water feedrate. For the last two "open-loop" policies one may need good flow measurements to reduce the implementation error. The selection of the best variables will be the subject of future work.

4. Conclusion

The UniSim process simulator was used to simulate and optimize a GTL plant. A rule of thumb for GTL is that for every 10,000 SCF (=282.6 Sm^3 =11.953 kmol) of natural gas, approximately 1 bbl (=159 liter) of fuel is produced (Patel, 2005). In our case study, we produce 17,315 bbl/day from 8195 kmol/hr natural gas, which is 5.23% higher than the rule of thumb. In the synthesis gas part, the optimal ratios of oxygen and water to methane at the nominal operating point were $\frac{O_2}{CH_4}=0.708$, $\frac{H_2O}{CH_4}=0.735$ and a good volume of the FT reactor was found to be 2200 m^3.

There are 7 operational degrees of freedom (DOFs): Feed rates of oxygen and water, exit temperature of fired heater, flue gas recycle to ATR and FT, CO_2 removal rate and recycled CO_2 to ATR). By optimizing the operation for various disturbances, we found that 1 of the DOFs is always needed to satisfy an active constraint at the fired heater exit temperature. Furthermore, we found that 3 of the original DOFs can be kept constant at their optimal nominal values (flue gas recycle to FT, CO_2 removal rate, recycled CO_2 to ATR) when there are disturbances, so these are good self-optimizing variables. Further work is needed to find good controlled variables for the remaining 3 unconstrained DOFs ($\frac{O_2}{\text{Natural Gas}}$, $\frac{H_2O}{\text{Natural Gas}}$ and recycled flue gas to ATR).

References

Aasberg-Petersen, K., J.-H Bak Hansen, T. S. Christensen, I. Dybkjaer, P. Seier Christensen, C. Stud Nielsen, S. E. L.Winter Madsen and J. R. Rostrup-Nielsen, 2001, Technologies for large-scale gas conversion, Applied Catalysis A: General 221, 379-387.

Aasberg-Petersen, K., T. S. Christensen, C. Stud Nielsen and I. Dybkjær, 2003, Recent developments in authothermal reforming and pre-reforming for synthesis gas production in GTL applications, Fuel Processing Technology, 83, 253-261.

Ahón,V. R. , Jr. E. F. Costa, J. E.P. Monteagudo, C.E. Fontes, Jr. E. C. Biscaia, P. L. C. Lage, 2005, A comprehensive mathematical model for the Fischer-Tropsch synthesis in well-mixed slurry reactors, Chemical Engineering Science, 60, 677-694.

Bakkerud, P. K., 2009, Personal commiunication.
Bakkerud, P. K., 2005, Update on synthesis gas production for GTL, Catalysis Today, 106, 30-33.
Fox III , J. M. and S. S. Tam, 1995, Correlation of slurry reactor Fischer-Tropsch yield data, Topics in Catalysis, 2, 285-300.
Patel, B., 2005, Gas Monetization: A Techno-Economic comparison of Gas-To-Liquid and LNG, 7th World Congress of Chemical Engineering, Glasgow.
Sehabiague, L., R. Lemonie, A. Behkish, Y. J. Heintz, M. Sanoja, R. Oukaci, B. I. Morsi, 2008, Modeling and optimization of a large-scale slurry bubble column reactor for producing 10,000 bbl/day of Fischer-Tropsch liquid hydrocarbons, Journal of the Chinese Institute of Chemical Engineers, 39, 169-179.
Skogestad, S., 2004, Control Structure Design for Complete Chemical Plants, Computers and Chemical Engineering, 28, 219-234.
Yates, I. C. and C. N. Satterfield,1991, Intrinsic Kinetics of the Fischer-Tropsch Synthesis on a Cobalt Catalyst, Energy & Fuels, 5, 168-173.

Proceedings of the 2nd Annual Gas Processing Symposium
Farid Benyahia and Fadwa T. Eljack (Editors)

Techno-Economic Analysis of Gas-To-Liquid process

Buping Bao[a] and Mahmoud M. El-Halwagi[a], Nimir O. Elbashir[b]

[a] *The Artie McFerrin Department of Chemical Engineering, Texas A&M University, College Station, TX 77843-3122, U.S.A*

[b] *Department of Chemical Engineering, Texas A&M University at Qatar, P.O. Box 23874, Doha, Qatar*

Abstract

Gas-to-liquid (GTL) process, with the core technology of Fischer-Tropsch (FT), is a way to produce fuels as well as other hydrocarbon products. There is a growing market of GTL processes in face of the dwindling oil reserves. The process basically involves the following steps: syngas reforming from natural gas followed by syncrude production from syngas and finally syncrude upgrading and separation into various products from LPG to wax. This work involves the simulation and integration of a GTL process. A framework is provided in the work for evaluating and improving the performance of GTL plants. First, a base-case flowsheet is synthesized. The developed process is simulated using computer-aided process simulation tool ASPEN Plus for a process of **128,000 barrels per day (BPD) capacity**.Then, the process performance is determined along with key design and operating variables and specifications. Next, mass and energy integration analyses are undertaken focusing on the following items: 1. thermal analysis for heating and cooling utilities, 2. combined heat and power (process cogeneration), 3. handling of process water recycle, 4. allocation targeting of tail-gas, and 5. performance optimization for recovery of catalyst-supporting hydrocarbon medium. After that, the economic impact of the various options is estimated. Finally, a break-even-point analysis is carried out along with a sensitivity analysis.

Keywords: GTL, process integration, techno-economic analysis.

1. Introduction to Gas-To-Liquid Technology

The key concept of gas-to-liquid (GTL) process is to chemically convert natural gas into hydrocarbon products ranging from gasoline to middle distillates up to heavy hydrocarbons (wax). These products are environmentally attractive because of their low sulfur content and lower CO, NO_x, hydrocarbons, and particulate matters upon combustion. GTL fuel fractions have also high cetane number allowing superior performance for engine design. Because of these environmental benefits, GTL products could be blended with refinery stock to gain environmental advantages. One key advantage of GTL is that liquid fuels are much easier to transport relative to the natural gases.

The main parts of the process can be broken into three major steps, including: reforming of natural gas to synthesis gas or syngas (CO and H_2 mixture), then, Fischer-Tropsch (F-

T) reaction, which is the heart of GTL technology where conversion of syngas into chains of hydrocarbon products take place on a catalyst surface, and the final stage is the upgrading of syncrude products through cracking and hydro-processing units to produce a variety of fuel fractions and value added-chemicals (schematic of this process is shown in Figure 1). The F-T reaction equation can be simplified as the following equation (Suehiro, et al., 2004)

$$(2n+1)H_2 + nCO \rightarrow C_nH_{(2n+2)} + nH_2O \quad -\Delta H_{298}=-167 \text{ kJ/mol/CO} \qquad (1)$$

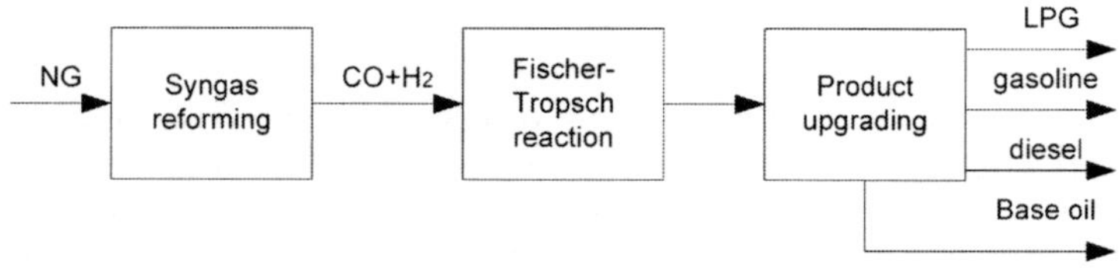

Figure 1: The main step of GTL process

One of the most known models used to determine the chain growth process in F-T is the Anderson-Schulz-Flory (ASF) model (Kuipers, et al., 1996). The main equation of this model is as follows:

$$W_n/n = (1-\alpha)^2\alpha^{n-1} \qquad (2)$$

where W_n is the mass fraction of the hydrocarbons with n carbon numbers, and α is the chain growth probability. The α-value, which represents the chain growth probability, can then be determined from the slope of the line ln(Wn/n) vs. n that is shown in Figure 2.

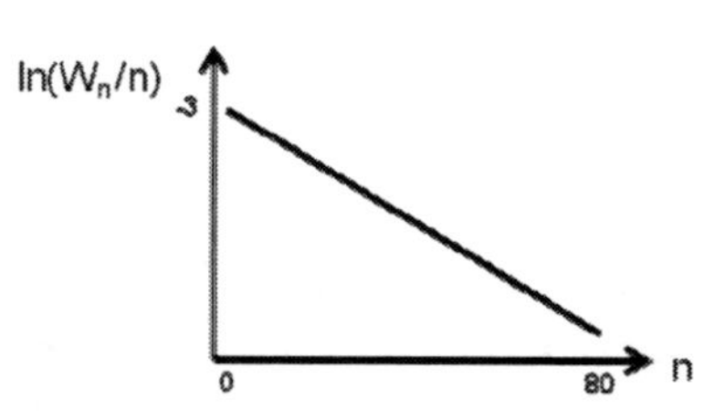

Figure 2: ASF hydrocarbon product distribution of GTL products with log scheme

Typical advanced commercial F-T reactor technology is part of Shell's GTL plant in Malaysia that started operation in 1993 (Steynberg and Dry, 2004). Shell's F-T reactor technology is known as the middle distillate synthesis (SMDS) technology. Similar F-T reactor technology will be used in Shell's new GTL plant in Qatar (the Pearl project), which is considered as the largest GTL project worldwide of capacity reaching 140,000 bbl/d. Another big project is Oryx GTL plant using slurry phase distillate process (SPD). Although GTL process is already in industrial application, it's still challenged by many design and scale-up problems, including the high operation cost of large capacity plants. Since the reaction is highly exothermic, the system needs lots of heat exchange equipments, and efficient processes to control mass and heat transfer inside the reactor as well. Furthermore, the reactor should stand high pressure required for the F-T reaction. The major objective of our study is to find out the optimal performance for a GTL process through simulation and integration of the units involved in the process on an economic basis. Given a typical process with certain units and design specifications, a techno-economic analysis for the process is addressed, targeting for cost reducing and energy efficiency improvement. Development of simulation model in this field has not yet been considered as established literature because of the complex nature of the process and the limited information in open literature. This can be implemented to the fact that the main contributors in this field are from industry who normally competes in technology development. The development of simulation models in this field may assist in development of reactor technology in F-T field where only "three" commercial reactor technologies found their way to commercialization since 1926. The methodology explained in this paper is based on conventional techniques in addition to a modern approach to simultaneously optimize the process while considering its techno-economic feasibility.

2. Methodology and Approach

The process synthesis and analysis used in this study were aimed to deal with elements incorporation and structure generation following the method described in El-Halwagi (2006). A detailed description of the sequential step for analyzing the process is as follows: firstly, a base case process is designed with parameters and operation conditions obtained from literature and real plant. Secondly, this flowsheet is simulated with computer-aided simulation package (ASPEN Plus) to evaluate the specific characteristics of the main equipments and streams entering and leaving units. Thirdly, sensitivity analysis is developed to determine alternative ways for the initial design. After mass and heat balance calculation, mass and energy integration is performed through pinch analysis. A mixed integer linear program (MILP) is carried out and solved by the optimization software LINGO to get the target. Fourthly, to do the cost evaluation, a software (ICARUS) is employed to assess the cost and to provide the economic feasibility of the process. Finally, the break-even point is determined for a process capacity.

3. Case Study

The base case for the GTL process was assumed at 1.16 standard cubic feet (SCF)/day natural gas as a feedstock. The flowsheet (illustrated in Figure 3) can be divided into three sections. The first section before the F-T reactor (the second section) is the reformer that is responsible for the production of synthesis gas ($CO+H_2$) at specified ratio, and the section after the F-T reactor recycles tailgas and at the same time treats the products from the F-T reactor (known as synthesis crude or syncrude). The production is estimated to be 128,000 barrel/day. The thermodynamic property method used in modeling the process is NRTL-RK. In order to represent a typical range of products in the F-T stream we have selected number of hydrocarbons compounds to keep a proper description for the ASF distribution as well. In ASPEN Plus © Requil model is chosen to simulate the autothermal reforming reactor. The pressure is set at 18 bars and temperature at 1300 K. The F-T reactor follows low temperature slurry phase reactor (typical commercial technology is Sasol's Slurry Middle Distillate technology in their Oryx plant in Qatar), with cobalt based catalyst that operates at 30 bars and 510 K. To properly represent the products from F-T reactor, syngas usage ratio (H_2/CO) is set around 2 to maintain an excellent conversion rate, and the extents of the reactions in the F-T reactor are adjusted to keep the distribution along the desired ASF line. The distillation curve (or boiling point fraction assay) of the products is represented in Figure 4.

4. Results and Analysis

First, material and energy balance calculations are carried out to get the information necessary for mass and energy integration.

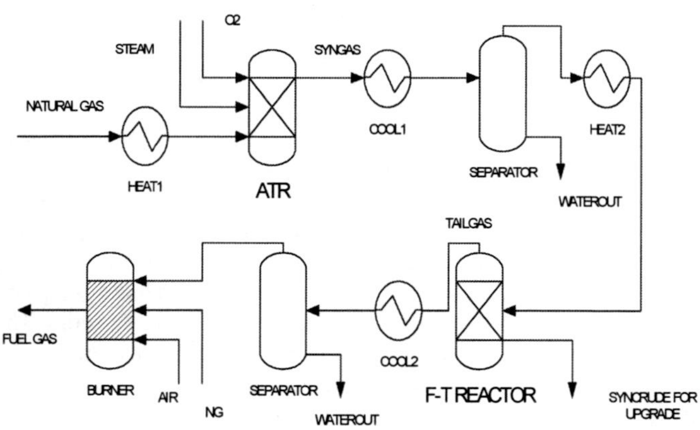

Figure 3: Schematic representation of the base-case GTL flowsheet

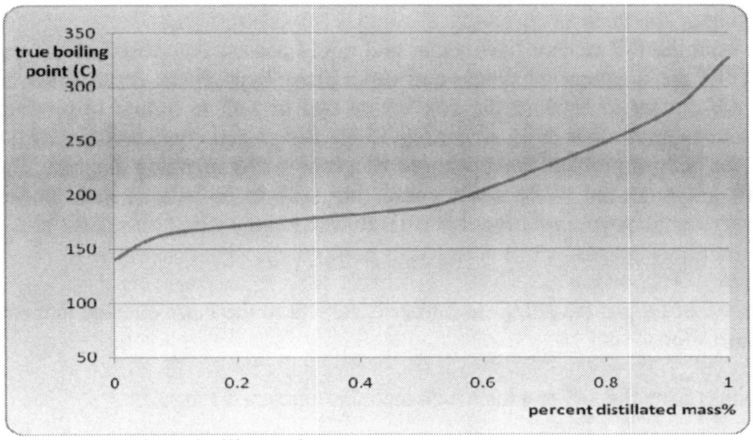

Figure 4: Products boiling point curve

4.1. Heat Integration and Targeting

To evaluate potential for exchanging heat between hot and cold streams, a steady state ASPEN Plus © simulation is used to calculate the heat duty for each unit. In the second step we conducted thermal pinch analysis with a minimum approach temperature of 10 K between the hot and the cold streams. After examining the most negative value and adding equivalent external heating utility from the top to keep the heat flow downward, the minimum utility requirement will be obtained as listed in Table 1. There are numerous high pressure (HP) and low pressure (LP) steams needed in the process. It is possible to let down a portion of the HP steam through a turbine to supply part of LP steam, and at the same time generate power from turbine for other units consumption. The cogeneration potential can be read by plotting extractable power versus the steam flow rate. From plotting, 224,000 hp power is generated from the turbine, equaling to 93,627,000 \$/yr, and LP steam saved from this step has a value of 153 MM \$/yr. These together contribute to 246 MM \$/yr savings from cogeneration measurements.

4.2. Mass Integration

Mass integration has been focused on four categories: including tail gas utilization, catalyst-supporting medium recovery, water management, and CO_2 separation.

Table 1: Heating and cooling utilities savings

	Heating utilities (Btu/hr)	**Cooling utilities (Btu/hr)**
before integration	4.2E+09	10.6E+09
after integration	0	6.16E+09
Savings	100%	41.8%
(MM\$/yr)	671	213

4.2.1. Utilization of Tail Gas

The products from the F-T reactor have vapor and liquid phases, comprised of tail gas and syncrude. Tail gas is unreacted syngas and vapor phase byproducts. It can either be redirected to the reactor to enhance the conversion rate or sent to burner to produce energy needed to support other units. According to the design and characteristics of the GTL process, the tail gas problem is represented as a source sink mapping diagram. The objective is to minimize the waste while complying with mass balance and species equilibrium, varying the source splitting that will allocate fractions into different sinks.

Minimize tailgas $CO+H_2$ waste

The minimization of *tailgas* ($CO+H_2$) to burner is subject to the mass balance that can be calculated as follows:

Source (the tailgas from the F-T reactor which includes unreacted syngas) =
F-T + ATR + burner (3)

Source* Source fraction=
F-T*F-T fraction+ ATR*ATR fraction + burner* burner fraction (4)

as well as to the following constraints:

(1) The maximum inlet mass fraction for F-T reactor should be kept below 0.87 so as to get the usage ratio around 2. Dropping the value will negatively influence the products yield. The ratio cannot be 1 to the ATR, otherwise the fractions of components in the reactor will affect the syngas ratio.
(2) After simulation and calculation, the ratio for the tail gas to the ATR is calculated to be 0.25 based on the abovementioned analysis. This will keep a high and feasible yield for the syncrude production.

4.2.2. Management of Water

Reverse osmosis can be used to clean water up which costs \$0.2 /m^3. In our case study of the GTL process there is 2.9 MM kg/hr water produced, where 600,000 kg/hr is separated with syncrude at a volume ratio of 0.7. There are different venues of the recovered water using as heating/cooling water for the process or for other uses in industrial or irrigation purpose. From treatment management based on the above data, savings of process water can be calculated up to 22.6 MM \$/yr.

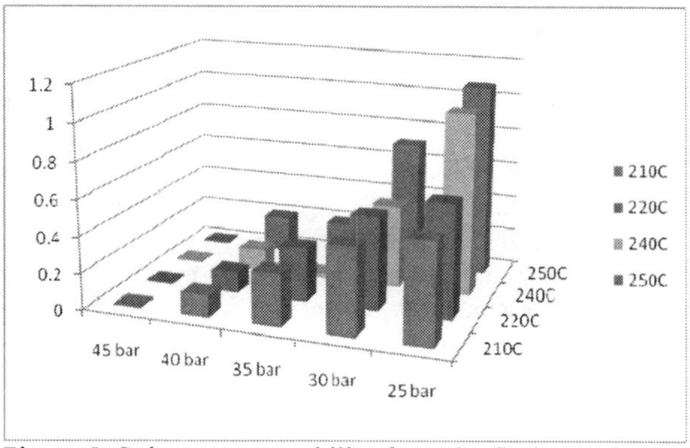

Figure 5: Solvent recoverability from the flash as a function of temperature and pressrue

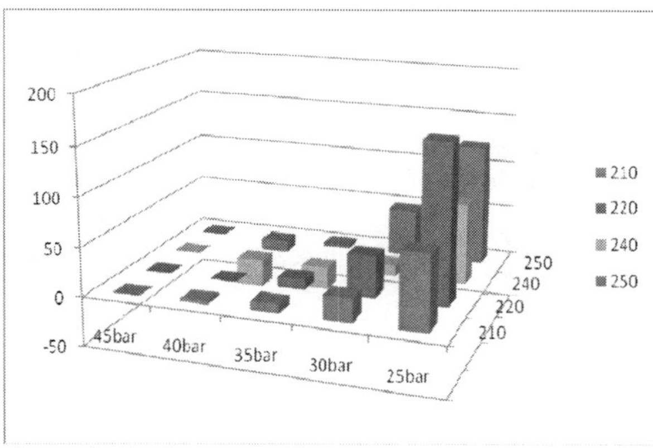

Figure 6: C_{10+} increase with temperature and pressure change in the flash

4.2.3. Recovery of the Catalyst Supporting Medium

Recently, several studies reported advantages of utilizing supercritical fluids (SCF) in F-T reaction to provide a medium that combines the advantages of the liquid-phase or slurry reactors (density and heat capacity to provide excellent temperature distribution) and the gas-phase or fixed-bed-reactor (diffusivity and easiness of operation to provide higher reaction rates) (see Elbashir and Roberts, 2005; Elbashir, et al., 2005). In our simulation model for supercritical phase F-T we have selected C_5-C_7 as the supercritical fluid since they are portion of the hydrocarbon products distribution of F-T. Nevertheless, success in utilizing SCF in F-T requires the design of proper capabilities to extract and recycle the solvent. In a preliminary investigation in this field we ran a simulation to compare between two major separation units which can be used for this purpose. These units are flash distillation column and are modeled on ASPEN Plus using the RADFRAC distillation column which is a rigorous method for simulating multicomponent separation. Our simulation data showed that the flash distillation column is much better on economical basis while it is still maintaining high recoverability of the solvent. The simulated conditions of the F-T reactor are syngas:solvent molar ratio of 1:1, total pressure of 45 bars while temperature varied from 210-250°C. In order to evaluate the solvent recoverability without affecting the heavy hydrocarbon products (represented by C_{10+} fractions), ASPEN simulation is conducted whereby both temperature and pressure were varied while looking on composition of upper stream and lower stream of the distillation column. The result of this simulation as shown in Figures 5 and 6 implies that as temperature increases, solvent recoverability increases while dropping pressure also assists in recovering the solvent. On the other hand, C_{10+} hydrocarbon products are not much influenced by changes in temperatures while as pressure decreases the fraction is found to increase as shown in Figure 6. In general, these results indicate that simultaneous drop of temperature and pressure at the reactor outlet would help in solvent recoverability and recycling without affecting heavier hydrocarbons.

4.2.4. CO_2 Separation

Because of the growing concern about increasing CO_2 concentrations in the atmosphere, it is important to endeavor to reduce the carbon footprint of the process. In the unit of F-T reactor, CO_2 is produced as a main byproduct from water gas shift reaction. In order to cut down CO_2 emissions, separation step is undertaken to reduce the amount. The cost and normalizing calculation are listed below (Table 2).

$$\frac{operating_A}{operating_B} = \frac{capacity_A}{capacity_B} \tag{5}$$

$$\frac{FCI_A}{FCI_B} = \left(\frac{capacity_A}{capacity_B}\right)^{0.6} \tag{6}$$

FCI stands for fixed capital investment and the subscripts A and B stand for processes with different capacities A and B, respectively. From literature data for CO_2 separation cost of certain capacity, the cost for CO_2 separation in this process is calculated from the above equations.

Table 2: Estimation of CO_2 separation cost

	Total capital investment (\$)	**Total annual operating cost (\$/yr)**	**Capacity (Mt/yr)**
Cost from literature (Singhal, 2000)	467,200,000	30,980,000	4.6
Cost for the case study	79,428,000	1,616,000	0.24

4.3. Total Cost for GTL Plant

With resort to process integrations taken in the above sections, the total cost for GTL plant for this case is evaluated and reduced tremendously. Based on reported capital investment data of $12-18 billion for Pearl plant with a 140,000 bbl/d capacity, the fixed cost for this study of 128,000 bbl/d capacity can be estimated to be 9.67 billion dollars, assuming the fixed capital investment is 85% of the total capital investment And the total capital investment is calculated to be 11.3 billion dollars accordingly. The fixed and operating capital costs are determined by ASPEN ICARUS. The plant size makes the fixed cost not so accurate since it goes beyond the normal size Aspen can hold. Reported data for fixed cost is taken. Operating cost is from ICIS and EIA. Setting working capital investment at 15% of total capital investment, the total capital investment for GTL process before and after integration are shown in Table 3.

4.3.1. ROI Calculation

$$ROI = \frac{profit}{TCI} \times 100\% = (5.4\text{-}4.82)/11.3 = 5.1\%$$

This return of investment (ROI) is low. It's worth noting a reduction in the cost of feedstock like natural gas will make the process more profitable, since some companies own their own gas wells and can get special discounts from the host country. Let's assume the natural gas price set to 5 $/thousand SCF, the cost of raw material goes down to 3.1 billion $/yr. This renders the ROI to be (5.4-3.576)/11.3 = 16.2% which goes significantly up.

Table 3: TAC calculation for GTL plant

Items	value
capacity	128,365 BPD
useful life period	20 years
FCI ($)	9.67 billion
TCI ($)	11.3 billion
total operating cost before integration($/yr)	5,543,727,000
total operating cost after integration($/yr)	4,390,022,000
total production income ($/yr)	5,415,921,000
salvage value ($)	0.967 billion
Depreciation/annualized fixed cost ($/yr)	0.43 billion
total annualized cost (TAC) ($/yr)	4.82 billion

4.3.2. Break Even Point Calculation

A plant production should exceed a break even point to make profit. Capacities below this point will subject a plant to financial loss. As such, break even point analysis is carried out. At relatively small capacity of 4,300 BPD, the fixed and operating cost are calculated and plotted on the figure versus this capacity. From the two capacities of 128,000 and 4,300 BPD, production line and total investment line is generated. The cross point shows the break even point of 68,000 BPD production rates. Therefore, for the case in this process, the higher the capacity the more profitable the plant can be, as long as the capacity is above 68,000 BPD.

5. Conclusions

This study gives a framework for developing and analyzing the GTL process performance. It deals with following work: Synthesizing a typical GTL flowsheet, optimizing the design and operating conditions for the plant, integrating heat and cogenerating power for the process, conducting mass integration measures touching upon tail gas recycle, catalyst medium recover, process water management, and CO_2 separation, and finally evaluating performance and cost for the process in the way of ROI and breakeven point. These measurements provide a way to optimize a typical GTL process and evaluate it from an economic view.

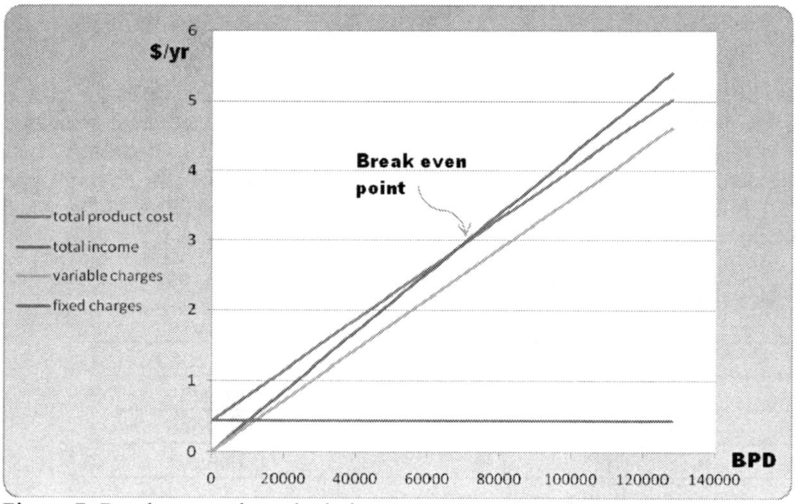

Figure 7: Breakeven point calculation

A case study with capacity of 128,000 BPD production is analyzed to show the improvement made with the abovementioned efforts. Simulation is done by ASPEN Plus with these results: A reduction of 884 MM $/yr from heat integration and 246 MM$/yr from cogeneration step, savings of 22 MM $/yr from water integration. Process ROI increases from 5% to 16% with the natural gas price coming down from $8 to 5/1000 SCF. Breakeven point analysis shows that plant production should go over 68,000 BPD to make profit under the conditions described in this study. With reduction in the cost of feedstock and with larger plant size, the process can be more profitable.

Reference

Elbashir N.O., Dutta, P., Manivannan, A., Seehra, M.S., Roberts, C.B., 2005. Impact of Cobalt-based Catalyst Characteristics on the Performance of Conventional Gas-phase and Supercritical-phase Fischer–Tropsch Synthesis Applied Catalysis A: General. 285, 169-180.

Elbashir, N.O., Roberts, C.B., 2005. Enhanced Incorporation of α-Olefins in the Fischer-Tropsch Synthesis Chain-Growth Process over an Alumina-Supported Cobalt Catalyst in Near-Critical and Supercritical Hexane Media. Industrial Engineering Chemistry Research. 44, 505-521.

El-Halwagi, M. M., 2006. Process Integration. Process Systems Engineering Series, Vol 7, Academic Press, San Diego, CA.

Kuipers, E.W., Scheper, C., Wilson, J.H., Winkenburg, I.H., Oosterbeek, H., 1996. Non-ASF Product Distributions due to Secondary Reactions during Fischer-Tropsch Synthesis. Journal of Catalysis. 158, 288-300.

Singhal, R.K., Singhal, A.K.M., 2000. Environmental Issues and Management of Waste in Energy and Mineral Production. Taylor and Francis, Alberta, Canada.

Steynberg, A.P., Dry, M.E., 2004. Fischer-Tropsch Technology. Elsevier. Amsterdam, the Netherlands.

Suehiro, Y., Nakamura, A., Sacomota, A., 2004. New GTL Process- Best Candidate for Reduction of CO_2 in Natural Gas Utilization. SPE Asia Pacific Oil and Gas Conference and Exhibition, Perth, Australia, 1423-1429.

Proceedings of the 2nd Annual Gas Processing Symposium
Farid Benyahia and Fadwa T. Eljack (Editors)

The Role of Molecular Thermodynamics and Simulation in Natural Gas Sustainable Processes

Ioannis G. Economou[1,2], and Zoi A. Makrodimitri[1]

[1]*National Center for Scientific Research "Demokritos", Institute of Physical Chemistry, Molecular Thermodynamics and Modeling of Materials Laboratory, GR-15310 Aghia Paraskevi Attikis, Greece*
economou@chem.demokritos.gr, www.mtmml.gr
[2]*The Petroleum Institute, Chemical Engineering Program, PO Box 2533, Abu Dhabi, United Arab Emirates*

Abstract

Advances in applied statistical mechanics and molecular thermodynamics together with the unprecedented increase of computing power over the last two decades have made molecular simulation a powerful tool for the elucidation of molecular mechanisms and microscopic phenomena and subsequent prediction of macroscopic physical properties of complex chemical systems of interest to oil and gas industry. A realistic atomistic representation of the intra- and intermolecular interactions in these systems results in predictions that are of comparable accuracy with experimental measurements.

This work focuses on molecular dynamics simulation for modeling novel elastomer polymers with promising properties as membranes for natural gas mixture separation. A systematic investigation is presented regarding the chemical structure of silicon containing polymers and its effect on solubility, diffusivity and permeability of gases and liquids in these polymers. For such purposes, a detailed atomistic model was developed that accounts explicitly for intra- and intermolecular interactions. Bond stretching, bond angle bending, torsional angle displacement, repulsive and attractive dispersion interactions are accounted using models parameterized to quantum mechanics calculations and experimental data. The agreement between experimental data and molecular simulation predictions for the properties calculated ranges from good to excellent.

Keywords: Silicon polymers, Solubility, Diffusion, Selectivity, Molecular dynamics

1. Introduction

The accurate knowledge of the transport properties of polymeric materials based on their chemical structure is crucial for technological applications, since these properties govern the behavior of the materials at end-use conditions. For example, the accurate evaluation of the diffusion of *n*-alkane molecules dissolved in a polymer influences the quality of the material to be used in a membrane for separation of natural gas mixtures. Typically, these properties are measured experimentally. Often though, experimental measurements are cumbersome and / or expensive. This is true for the measurement of the diffusion coefficients of penetrants in polymers in the low concentration limit and at high temperatures and pressures. In such cases, theoretical models or accurate computational approaches are preferred.

Considerable work has been devoted to the prediction of transport properties through theoretical models or detailed computer simulation studies. From the theoretical point of view, a model that describes accurately the diffusion of small molecules in polymers is the free volume theory of Vrentas and Duda (1977). This theory is based on the fundamental assumption of Cohen and Turnbull (1959) that free volume redistribution is responsible for the molecular diffusion. Although widely used, for the accurate correlation of experimental data a number of system specific properties need to be fitted to the data.

On the other hand, molecular dynamics (MD) simulations have been used successfully in the past for the investigation of the molecular structure of membrane materials based on amorphous polymers and of the sorption and diffusion of low molecular weight penetrants in these materials (Müller-Plathe, 1991, Sok et al. 1992). In most of these studies, the diffusion coefficient of small gaseous penetrants, such as He, O_2, N_2, CH_4 and CO_2, in polymers was calculated. The duration of these MD simulations was limited to several hundred or few thousand ps.

The main objective of this work is to extend diffusion and solubility studies to higher molecular weight penetrants in silicon containing polymers and perform substantially long MD runs, on the order of 100 ns, in order to obtain reliable results. A united atom (UA) force field is developed for the calculation of microscopic structure, thermodynamic and conformational properties of pure poly(dimethylsilamethylene) (PDMSM), poly(dimethylsilatrimethylene) (PDMSTM), alternating poly(DMSM-alt-DMSTM) copolymer and poly(dimethylsiloxane) (PDMS) melts over a wide temperature and pressure range. Detailed *ab initio* quantum mechanics calculations on corresponding monomers that appeared recently in the literature are used for the parameterization of the bonded and nonbonded local intramolecular force field (Raptis and Melissas, 2006). For the intermolecular and nonbonded nonlocal intramolecular interactions, parameters are obtained from accurate force fields proposed in the literature together with fitting to experimental thermodynamic data at ambient conditions.

Subsequently, the model is used for the prediction of solubility and diffusivity of noble gases, *n*-alkanes and *n*-fluoroalkanes, again at different conditions. Both single and mixed solutes are examined. In this way, the selectivity of the various polymer structures to mixtures of components is obtained. Simulation predictions are compared to literature experimental data resulting in excellent agreement in all cases.

2. Force Filed Development

All of the polymers here examined are flexible polymers consisting of methyl, methylene, silicon and oxygen (in the case of PDMS) atoms or groups. Realistic atomistic force fields for macromolecules account explicitly for the molecular geometric characteristics as well as for the non-bonded intra- and inter-molecular interactions. In this work, force fields based on the united atom (UA) representation were used, thus reducing the number of "atoms" significantly (Makrodimitri et al., 2006).

The potential energy function is written as the sum of contributions due to bond stretching, bond angle bending, dihedral angle torsion and non-bonded interactions between UAs in the same or different chains. In PDMSM, PDMSTM and the copolymer, non-bonded interactions consist of short range van der Waals repulsive and dispersive interactions, while in PDMS long range electrostatic (Coulombic) interactions also exist (Makrodimitri et al., 2007).

MD simulations were performed in the isobaric – isothermal (*NPT*) ensemble using Nosé and Klein extended method and in the canonical (*NVT*) ensemble. In *NPT* simulations, an integration time step of 0.5 fs and the fifth order Gear predictor-corrector scheme were used to integrate the equations of motion in Cartesian coordinates. *NVT* simulations were performed for the estimation of the diffusion coefficient of light gases and *n*-alkanes in PDMS matrix. In these simulations, an increased time step of 2 fs was used and the equations of motion were solved using the velocity Verlet algorithm.

For the evaluation of the chemical potential of light gases and *n*-alkanes in polymers, *NPT* simulation runs of 6 ns were performed; the first nanosecond corresponded to the "equilibration" stage and the remaining 5 ns to the "production" stage where instantaneous properties are averaged. In order to improve statistics in the calculated solubility properties, for each state point four runs were performed starting from completely independent different initial configurations. In each simulation run, 5,000 configurations were recorded at equal time intervals. These configurations were used for the calculation of the excess chemical potential of light gases, *n*-alkanes and *n*-perfluoroalkanes in polymers using the Widom test particle insertion method. According to Widom's method, the chemical potential is calculated from the ratio of the partition function of a system containing $N+1$ molecules divided by the partition function of a system containing N molecules, where both systems are under the same temperature and volume or pressure conditions. The "ghost" molecule is inserted randomly into the simulated system. The interaction energy of the "ghost" molecule with the remaining molecules (U_{ghost}) is evaluated and subsequently used for the calculation of the excess chemical potential (μ^{ex}) of the component:

$$\mu^{ex} = \mu - \mu^{ig} = -\frac{1}{\beta}\ln\left[\frac{1}{\langle V\rangle_{NPT}}\left\langle V\left\langle \exp\left(-\beta U_{\text{ghost}}^{\text{intra}} - \beta U_{\text{ghost}}^{\text{inter}}\right)\right\rangle_{Widom}\right\rangle_{NPT}\right]$$
$$+\frac{1}{\beta}\ln\left\langle \exp\left(-\beta U_{\text{ghost}}^{\text{intra}}\right)\right\rangle_{\text{ideal gas}} \quad (1)$$

where $\beta = 1/kT$, $U_{\text{ghost}}^{\text{inter}}$ is the intermolecular energy and $U_{\text{ghost}}^{\text{intra}}$ is the intramolecular energy of the "ghost" molecule. The latter is calculated independently in a single chain simulation. V is the instantaneous volume of the system and the brackets denote ensemble averaging over all configurations and spatial averaging over all "ghost" molecule positions.

From the excess chemical potential, one may calculate the infinite dilution solubility coefficient S_o based on the expression:

$$S_o = \frac{22400\ cm^3\,(STP)/mol}{RT}\lim_{x_{solute}\to 0}\exp\left(-\beta\mu_{solute}^{ex}\right) \quad (2)$$

NVT simulations were also performed for systems consisting of 3 PDMS chains of 80 monomer units each and a number of gas molecules (10 He, Ne or Ar, or 5 O_2 or N_2 molecules) or *n*-alkane molecules (10 CH_4 or 5 C_2H_6 up to *n*-C_6H_{14} molecules). Simulations were performed in the temperature range 300 – 450 K. Long simulation runs, corresponding to real time of 50 – 150 ns were executed in order to obtain reliable results. The self-diffusion coefficient was calculated from the linear part of the mean square displacement of the centers of mass of the solute molecules at the beginning of time and after time t, denoted as $r_i(0)$ and $r_i(t)$ respectively, according to the Einstein equation:

$$D = \lim_{t\to\infty}\frac{1}{6}\frac{d}{dt}\left\langle \left|r_i(t) - r_i(0)\right|^2\right\rangle \quad (3)$$

In Figure 1 a representative snapshot for a well equilibrated PDMS – *n*-butane mixture is presented.

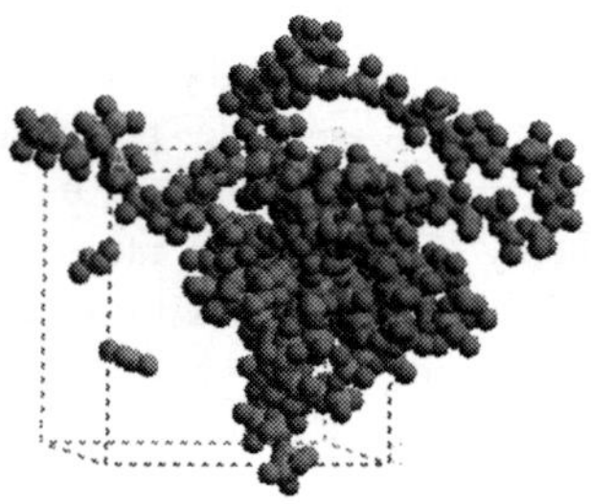

Figure 1: A typical snapshot of PDMS (long chains) – *n*-butane (green tetramers) mixture

3. Results and Discussion

3.1. Model validation

Initially, the developed force field was tested for the prediction of pure polymer melts at various temperatures and pressures. For PDMSM, PDMSTM and the copolymer MD simulations were performed from 300 to 400 K and from 0.1 to 160 MPa. For PDMS, MD simulations were executed at 300, 375 and 450 K and for each temperature at 0.1, 10 and 40 MPa (Makrodimitri et al., 2007).

In Figure 2a, the melt density of PDMSM, PDMSTM and the copolymer at 0.1 MPa is shown as a function of temperature. Experimental data for PDMSM and copolymer are also shown (open symbols). The agreement between experimental data and simulation is very good in all cases. For PDMSM in particular, the melt density at 300 K has been measured by several authors. The force field developed here predicts a density value which agrees within experimental uncertainty with some of these data. In Figure 2b, experimental and simulated melt density of PDMS is shown as a function of temperature. MD predictions are in excellent agreement with either of the experimental data sets at 0.1 MPa. Some deviations are observed at higher pressures.

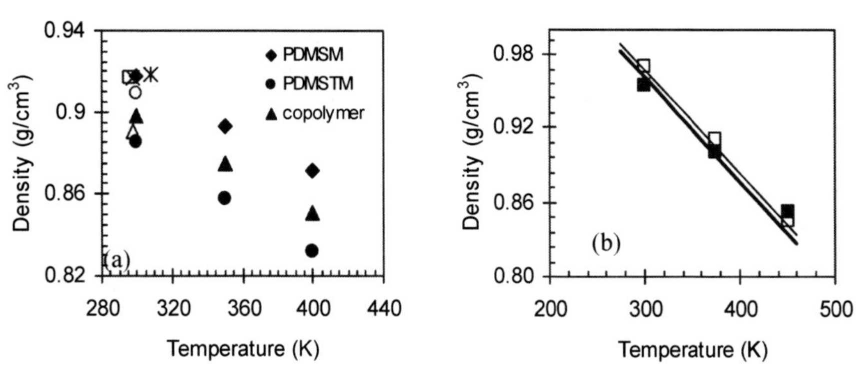

Figure 2: Melt density of (a) PDMSM, PDMSTM and copolymer and (b) PDMS at 0.1MPa. Open symbols and lines correspond to experimental data and closed symbols to MD predictions.

3.2. Solubility Properties

Polymer melt configurations generated by MD were used for the calculation of solubility of various components in the polymers. The Widom test particle insertion method was used to calculate the excess chemical potential of *n*-alkanes, *n*-fluoroalkanes, and noble and light gases in PDMSM, PDMSTM, PDMS and the copolymer at various temperature and pressure values. Representative results for S_o of *n*-alkanes in PDMSM at 300 K and PDMS at 300 K and 450 K and 0.1 MPa are shown in Figure 3. In all cases, solubility increases with *n*-alkane carbon number. Simulation results are in very good agreement with experiment, when available. Furthermore for a given *n*-alkane, S_o values are very similar for the various polymers. In other words, the chemical structure for these polymers has very little effect on the solubility of the solutes (Economou et al., 2007).

3.3. Transport Properties

Long *NVT* MD simulations on the order of 100 ns were performed for the estimation of the diffusion coefficient of solutes in PDMS. The solute diffusion coefficient, *D*, was calculated from the mean square displacement (msd) of solute molecules with time, according to the Einstein equation (Eq. 3). In Figure 4a, experimental data and MD simulation results for the diffusion coefficients of *n*-alkanes in PDMS at 300 K are presented. MD predictions are lower than experimental values which, in turn, exhibit significant scatter which is also a measure of the uncertainty associated with these measurements (Makrodimitri and Economou, 2008).

The effect of penetrant concentration on the diffusion coefficient value was also examined. MD simulations for PDMS – CH_4 and PDMS – *n*-C_4H_{10} mixtures at 350 K and various compositions were performed. Simulations were performed in the *NVT* ensemble and mass density was kept constant for a given mixture, independent of composition. In Figure 4b, *D* values for CH_4 and *n*-C_4H_{10} are presented as a function of the penetrant weight fraction. In both cases, diffusion coefficient increases with penetrant concentration. This behavior is consistent with literature experimental data.

Separation of a binary mixture of components *i* and *j* (where *i* is typically the most permeable of the two components) by a given polymer membrane is characterized by the ideal separation factor, which is the ratio of permeabilities for components *i* and *j* according to the expression:

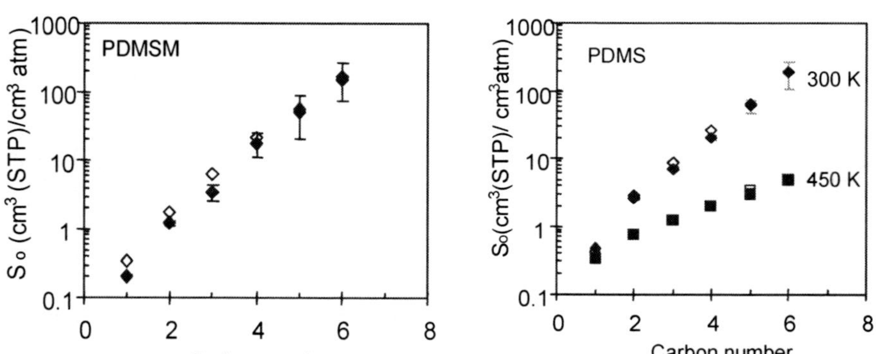

Figure 3: Experimental data (open symbols) and MD predictions (closed symbols) for the infinite dilution solubility coefficient of *n*-alkanes in PDMSM and PDMS.

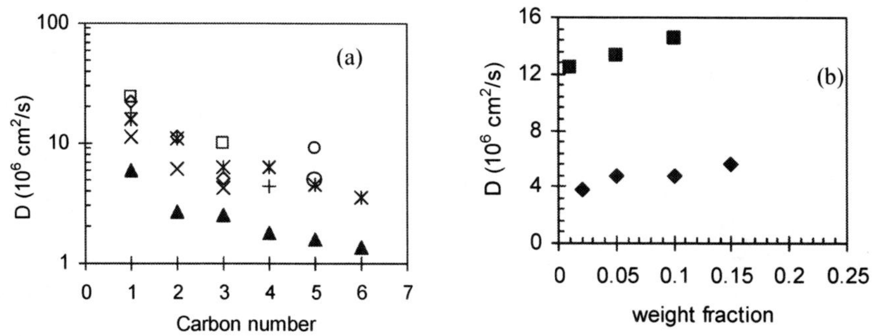

Figure 4: (a) Diffusion coefficient of *n*-alkanes in PDMS. Open symbols are experimental data and closed symbols are MD predictions and (b) Diffusion coefficient of CH_4 (squares) and n-C_4H_{10} (diamonds) in PDMS at 350 K as a function of solute weight fraction calculated by MD simulations

$$\alpha_{ij}^{id} = \frac{P_i}{P_j} = \left(\frac{S_i}{S_j}\right)\left(\frac{D_i}{D_j}\right) \tag{4}$$

where the ratios $\alpha_{ij}^{S} = S_i/S_j$ and $\alpha_{ij}^{D} = D_i/D_j$ represent the solubility selectivity and the diffusivity selectivity, respectively. In rubbery polymers α_{ij}^{D} is less than unity, while $\alpha_{ij}^{S} >> 1$, so ideal separation factor is governed by selectivity of sorption. PDMS is a widely used polymer membrane and so ideal separation factors for various binary gas and liquid mixtures have been measured. The separation factor for n-C_4H_{10} / CH_4 mixture is used widely as a benchmark for hydrocarbon mixture separation capability of a given membrane material. In Figure 5, pure gas n-C_4H_{10} / CH_4 solubility, diffusivity and permeability selectivities in the range 273 – 400 K together with experimental data from Raharjo et al. (2007) are shown. Solubility selectivity decreases significantly as temperature increases. MD predictions are in excellent agreement with experiments over the entire temperature range.

3.4. Simulation Results for Mixtures of Penetrants

For the accurate design of a polymer membrane for the separation of a real mixture, mixture permeability data are needed. It is often assumed that in rubbery polymers penetrants permeate independently of one another. However, this behavior needs to be confirmed for a given system. Significant experimental and simulation work has been devoted to the permeability and diffusivity of mixed gases in polymers. Recent experimental data for the n-C_4H_{10} – CH_4 mixture in PDMS showed an increase in CH_4 solubility in the presence of n-C_4H_{10} in the polymer. On the other hand, only a weak influence of CH_4 on n-C_4H_{10} solubility was reported.

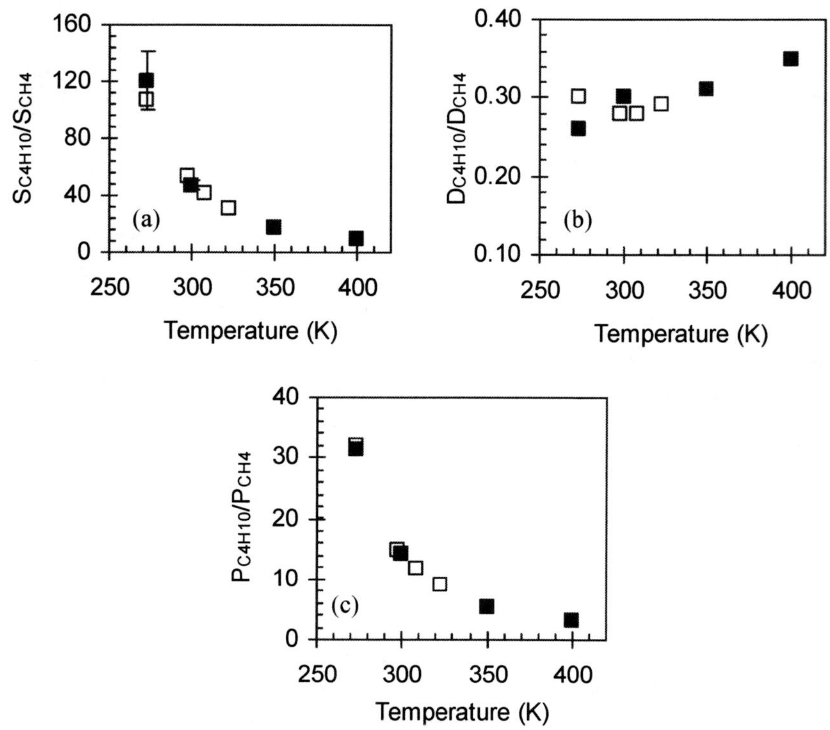

Figure 5: Effect of temperature on pure n-C_4H_{10} – CH_4 (a) solubility, (b) diffusivity and (c) permeability selectivity in PDMS. Open symbols are experimental data and closed symbols are MD predictions.

In this work, molecular simulation sorption calculations were performed for this ternary mixture. More specifically, the infinite dilution solubility coefficient of CH_4 in a PDMS – n-C_4H_{10} mixture was examined. Three different mixtures were simulated consisting of 3 PDMS chains of 80 monomer units each with a n-C_4H_{10} weight fraction equal to 0.10, 0.13 and 0.20, respectively. *NPT* simulations of PDMS – n-C_4H_{10} mixtures at 300 and 450 K were performed lasting 6 ns each and the resulting configurations were used for CH_4 molecule insertions in order to calculate the infinite dilution solubility coefficient of the latter with the Widom particle insertion method. Results depicted in Figure 6 indicate that at both temperatures, CH_4 solubility increases as n-C_4H_{10} concentration in the polymer increases. Clearly, the presence of n-C_4H_{10} molecules in the polymer creates a more favorable environment for CH_4 sorption, which enhances CH_4 solubility. This behavior is consistent with the recent experimental data of Raharjo et al. (2007) at 298 K shown in Figure 6.

The diffusion coefficients of a mixture of CH_4 and n-C_4H_{10} in PDMS at ambient conditions were also calculated and compared to pure gas diffusion calculations. Two systems were examined containing (a) 3 PDMS chains of 80 monomer units each + 1 wt % CH_4 (10 CH_4 molecules) + 2 wt % n-C_4H_{10} (5 C_4H_{10} molecules) and (b) 3 PDMS chains of 80 monomer units each + 1 wt % CH_4 (10 CH_4 molecules) + 10 wt % n-C_4H_{10} (27 C_4H_{10} molecules), respectively. *NPT* simulations of 50 – 100 ns were performed at 300 K and 0.1 MPa. In Figure 7, mixed gas *D* predictions together with pure gas *D* predictions are displayed. Clearly, CH_4 molecules move faster in the presence of n-C_4H_{10} molecules in PDMS matrix than in pure polymer. The same behavior is observed for n-C_4H_{10} in the presence of CH_4 molecules. The presence of a second penetrant species swells the polymer matrix resulting in an increase in the diffusion coefficient of the first penetrant. For example, pure CH_4 diffusion coefficient of $6x10^{-6}$ cm^2/s corresponds to density of 0.942 g/cm^3. The presence of 10 wt % n-C_4H_{10} in the polymer decreases density to 0.899 g/cm^3 and approximately doubles CH_4 diffusion coefficient value. On the other hand, the presence of 1 wt % CH_4 in polymer caused a slightly decrease in system density from 0.942 g/cm^3 to 0.928 g/cm^3 and an increase in n-C_4H_{10} diffusion coefficient from $1.8x10^{-6}$ cm^2/s to $2.6x10^{-6}$ cm^2/s.
The swelling behavior of PDMS in the presence of mixed gases and the consequent increase in diffusivity and permeability coefficients of the corresponding gases has also been reported experimentally by other investigators (Pinnau et al. 2004, Raharjo et al. 2007).

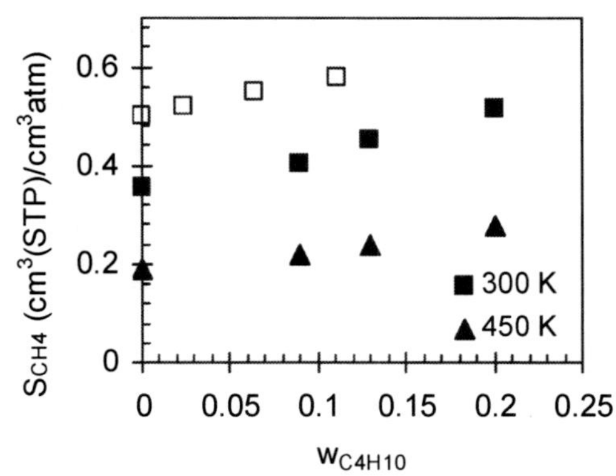

Figure 6: Mixed gas CH_4 solubility in PDMS at 300 and 450 K as a function of n-C_4H_{10} weight fraction in PDMS predicted by MD. Experimental data (open points) and MD predictions (closed points).

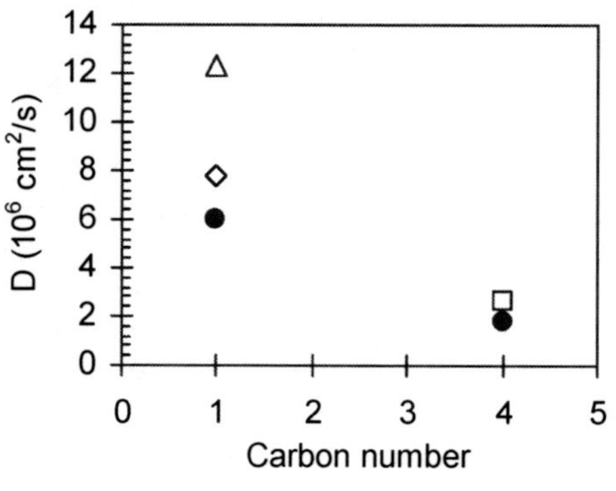

Figure 7: Diffusion coefficient of pure and mixed *n*-alkanes in PDMS at ambient conditions. Solid circles correspond to pure CH_4 and *n*-C_4H_{10} diffusion coefficient in PDMS. Open symbols correspond to *n*-alkanes in mixture: (◇) CH_4 mixed with 2% *n*-C_4H_{10} in PDMS, (△) CH_4 mixed with 10 % *n*-C_4H_{10} in PDMS, and (□) *n*-C_4H_{10} mixed with 1% CH_4 in PDMS.

4. Conclusions

The thermodynamic properties and microscopic structure of silicon polymers were examined using MD. Model predictions for pure polymers and experimental data were in good agreement in all cases. Subsequently, the solubility of *n*-alkanes, *n*-fluoroalkanes and noble and light gases was also calculated at a wide temperature range. Results agreed very well with limited experimental data available from the literature. Moreover, permeability coefficients and separation factors (selectivities) were calculated for various *n*-alkane and binary *n*-alkane mixtures in PDMS, respectively. Predictions concerning the former are lower than experimental data, due to lower diffusion coefficient predictions. However, the developed model for PDMS results in excellent prediction of selectivities over a wide temperature range.

Finally, molecular simulations were performed for binary penetrant mixtures with PDMS. The presence of a second penetrant increases both the solubility coefficient and the diffusion coefficient of the first penetrant species, in agreement with limited experimental data available (Pinnau et al. 2004, Raharjo et al. 2007). In summary, molecular simulation using detailed atomistic force fields is a powerful tool for the elucidation of structure and prediction of permeability properties of rubbery polymers, in most cases in good agreement with experimental data.

Acknowledgment

Financial support of this project through the European Union – European Social Fund, the Greek Secretariat of Research and Technology and Bayer Technology Services GmbH is gratefully acknowledged.

References

Cohen, M. H. and Turnbull, D. 1959, "Molecular Transport in Liquids and Glasses", J. Chem. Phys. 31, pp. 1164.

Economou, I.G., Makrodimitri, Z.A., Kontogeorgis, G.M. and Tihic, A. 2007, "Solubility of gases and solvents in silicon polymers: molecular simulation and equation of state modeling", Molec. Simul. 33, pp. 851.

Makrodimitri, Z.A., Raptis, V.E. and Economou, I.G., 2006, "Molecular Dynamics Simulation of Structure, Thermodynamic, and Dynamic Properties of Poly(dimethylsilamethylene), Poly(dimethylsilatrimethylene) and Their Alternating Copolymer", J. Phys. Chem. B 110, pp. 16047.

Makrodimitri, Z.A., Dohrn, R. and Economou, I.G., 2007, "Atomistic Simulation of Poly(dimethylsiloxane): Force Field Development, Structure and Thermodynamic Properties of Polymer Melt and Solubility of n-Alkanes, n-Perfluoroalkanes, Noble and Light Gases", Macromolecules 40, pp. 1720.

Makrodimitri, Z.A. and Economou I.G., 2008, "Atomistic Simulation of Poly(dimethylsiloxane) Permeability Properties to Gases and n-Alkanes", Macromolecules 41, pp. 5899.

Müller-Plathe, F., 1991, "Diffusion of penetrants in amorphous polymers: A molecular dynamics study", J. Chem. Phys. 94, pp. 3192.

Pinnau, J., He, Z., 2004, "Pure- and mixed-gas permeation properties of polydimethylsiloxane for hydrocarbon/methane and hydrocarbon/hydrogen separation" J. Membr. Sci. 244, pp. 227.

Raharjo, R. D., Freeman, B. D. and Sanders, E. S., 2007, "Pure and mixed gas CH_4 and n-C_4H_{10} sorption and dilation in poly(dimethylsiloxane)", J. Membr. Sci. 292, pp. 45.

Raptis, V.E. and Melissas, V.S., 2006, "Force Field Development for Poly(dimethyl-silylenemethylene) with the Aid of ab Initio Calculations", J. Phys. Chem. B 110, pp. 14929.

Sok, R.M, Berendsen, H.J.C., van Gunsteren, W.F. 1992, "Molecular dynamics simulation of the transport of small molecules across a polymer membrane", J. Chem. Phys. 96, pp. 4699.

Vrentas, J. S. and Duda, J. L., 1977, "Diffusion in polymer - solvent systems. I. Reexamination of the Free-Volume Theory", J. Polym. Sci. Pol. Phys. 15, pp. 403.

Proceedings of the 2nd Annual Gas Processing Symposium
Farid Benyahia and Fadwa T. Eljack (Editors)

Efficient Design, Operating and Control Strategies for LNG Plants

Francois Chantant[a], Matthew Swatton[a]

[a]*Shell Global Solutions International BV, Carel van Bylandtlaan 23, 2501CM The Hague, the Netherlands*

Abstract

This paper presents concrete examples of how to increase the LNG plant efficiency for existing assets through improved operation and for new projects with new design approaches.

In today's environment, a strong focus exists in the gas industry to maximise the output of an LNG plant. Environmental constraints, high construction costs, difficult feed gases and remote locations are all current challenges that plant designers and operators face. To overcome these challenges, there needs to be a continued effort to optimise the operation of existing LNG plants and implement lessons learned in any new design.

Operational feedback from LNG sites all over the world, including tropical environments and sub-arctic locations, enables Shell to select the most appropriate technology for any new development. The operational experience gained from the different technologies e.g. water-cooled or air-cooled, powered by steam turbine or gas turbine, contributes to increased awareness of the importance of integration and how to best implement it in new projects.

For existing LNG plants, the potential increase in energy efficiency can often be limited by the installed equipment. However, though the application of modern control systems and optimising technologies a lot of opportunities exist to further enhance the plant efficiency, reduce green house gas emissions and flaring, whilst at the same time increasing production. Modern computer tools enable accurate modelling and real time prediction of what is happening in a plant. Transient behaviours of new control systems can be modelled to assess their potential benefits and drawbacks.

Development of new technologies and key equipment plays an important role towards achieving a more energy efficient design. Examples of this include heat integration and the use of high performance gas turbines. Alternative process design can also lead to an increase in efficiency, for example the ability to accommodate the variation in ambient temperatures while keeping the different refrigeration loops at their optimal working point.

Keywords: LNG, CO_2, liquefaction, operation, operating window

1. Introduction

Energy challenges are becoming tougher every day. The world's energy demand is expected to increase rapidly after the current economic downturn, the environmental regulations (NOx, SOx, etc) are becoming stricter, and the focus on reducing CO_2 emissions is growing. As natural gas is the cleanest-burning fossil fuel, it is an attractive solution for providing an abundant and clean energy source. But the natural gas reserves are often remote from markets and the use of liquefied natural gas (LNG) is becoming increasingly important as an alternative to gas pipeline transport. It also provides more flexibility for the importing countries since they can import LNG from several locations and are less dependent on a single producer using pipelines.

In this context, technologies with lower environmental footprints such as a combined cycle LNG plant are needed to help this source of energy to grow.

Since the 1960's Shell has been involved in the development and operation of LNG projects. The experience gained from existing operations is continuously used in new designs to give better plant performance through innovative technical solutions. To minimize the CO_2 emissions across the LNG value chain it is important to design and operate efficiently, saving fuel gas and maximizing LNG output. This paper presents concrete examples on how to increase the efficiency of an existing LNG plant through improved operation, and for new projects by a new design approach.

2. Background

From its early involvement in the LNG plant in Brunei until the recent start up in Sakhalin, Shell has been developing and operating a wide range of technologies for very different environments. Brunei LNG (start up in 1972) is a steam driven Propane/MR plant with a closed cooling water system located near the equator. Sakhalin (start up in 2009) is a gas turbine driven Dual Mixed Refrigerant (DMR) plant with waste heat recovery, using air cooling in a sub arctic environment. Throughout the years, new technologies have been developed to provide efficient solutions to unlock natural gas reserves.

There is also an ongoing effort to modify existing plants to best suit the current operating conditions. Examples from Brunei LNG are an illustration of this:

- Integration of a new combined cycle power plant and a new feed gas compression station with the LNG plant
- Waste heat recovery from the gas turbine exhaust in these new facilities provides the majority of the steam required for BLNG
- Replacement of the Main Cryogenic Heat Exchangers (MCHE's) to increase the plant capacity and reliability
- Solvent swap in the acid gas removal facilities (Ahmad et al., 2004) to provide more flexibility for higher CO_2 content in the feedgas and higher efficiency of the unit

3. Excellence in operation: how to get the most out of an existing facility

There are several challenges in the day-to-day life of an LNG asset manager / operator: the integrity and (process) safety of the asset, delivery of the production and reliability of the equipment are some of these. An increasingly important one is to get the most out of an existing facility and to do this with minimal costs and impact on the environment.

Fuel gas is one of the largest contributors to the total operating cost of an LNG complex. Depending on the pricing structure in some cases it can be as much as 55% (Wilson and Van der Wal, 2008). At Oman LNG a program has been developed and implemented to monitor and increase the plant thermal efficiency. As a result, the CO_2 emissions and fuel gas usage are reduced, especially when operating under part load. Modern process control technologies have been used such as Advanced Process Control (APC) to keep the plant operation at its optimum. APC can reduce the CO_2 emissions from an LNG plant by up to 3 percent with very limited hardware changes.

3.1. Thorough understanding of an existing process: case of a C3-MR process

In a C3-MR liquefaction system, the natural gas is precooled in a propane system using kettles where the propane is evaporated against the natural gas (NG) and the mixed refrigerant (MR). The propane vapors are then compressed through a centrifugal compressor before being condensed against (sea) water or air.

The precooled NG is then liquefied in the MR system, where MR is evaporated in the MCHE to provide the cooling duty. The vapors are then also compressed and sent through ambient cooler and propane kettles to be partly liquefied. The vapor and the liquid MR phase are separated to provide two refrigerant cooling streams. Both refrigerant streams, Heavy Mixed Refrigerant (HMR) and Light Mixed Refrigerant (LMR), are used in the MCHE to liquefy the NG. Refer to Figure 1.

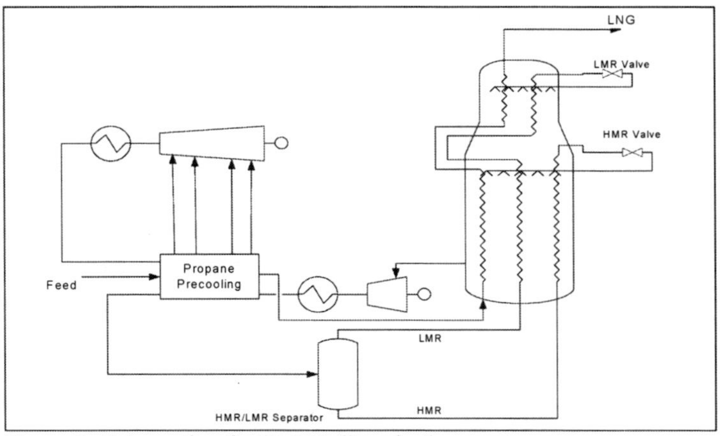

Figure 1: Schematic of a C3-MR liquefaction process

The plant's operating window is defined by the limits of the installed equipment. For a liquefaction unit, this window is limited by a set of constraints that are change according to seasonal variations. Shell has developed dynamic simulation models of generic C3-MR and Dual Mixed Refrigerant (DMR) liquefaction processes to optimize their operation and produce guidelines for this.

These advanced simulation techniques increase understanding of phenomena that can't easily be observed directly in a normal plant operation, for example the distribution of refrigerant hold up in a liquefaction loop. The hold up of mixed refrigerant vaporizing inside the MCHE is normally not known. However, the vaporization of this refrigerant provides the cooling needed to liquefy the natural gas. A good understanding of the vaporization process can help to predict the pro-active actions that can be taken in order to operate at the optimal working point.

In a C3-MR system, the operating envelope of the MR refrigeration loop is limited by certain constraints:

- Power available from the drivers
- Compressor surge and stonewall
- Lack or excess of total mass of refrigerant
- High and low level of the HMR/LMR separator

These constraints define the operating envelope of a MR loop, as shown in reference (Opstal, 2009). However, some of these constraints are continuously changing as both the ambient conditions the plant throughput change. For example, if the MR system is driven by a gas turbine, the power available is a function of the ambient temperature. Temperature variations between day and night cause the power available to change as a function of the inlet air temperature of the gas turbine as shown in Figure 2.

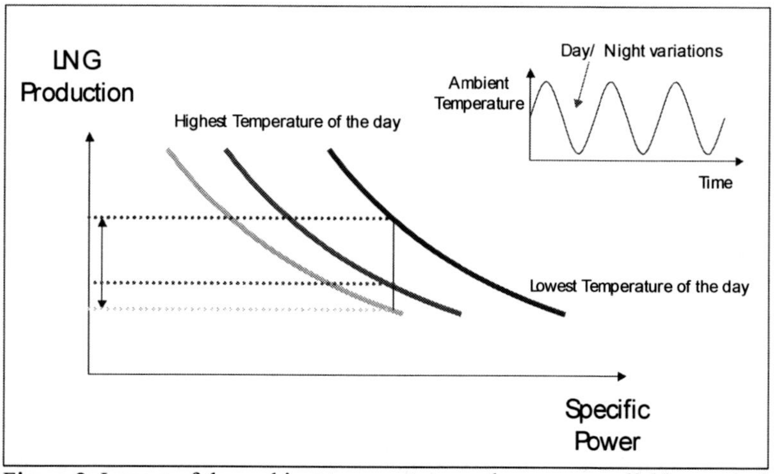

Figure 2: Impact of the ambient temperature on the power available constraint

Also during low load on the gas turbine (like turndown operation), other limits like high NOx emissions or very low gas turbine efficiency will define an area where operation should be avoided.

Other limits are independent of the ambient conditions, like compressor surge and stonewall. These are a characteristic of the equipment. For any given compressor, there is a limit, which is defined by a pressure ratio and volumetric flow map beyond which the compressor operates in a surge region and has the potential to damage the machine. At the other end of its operating curve, during stonewall the MR flow is very large and the gas velocity at the tip of the blades may exceed the speed of sound. The pressure ratio over the compressor is minimum, and thus it is not possible to increase the MR flow rate any further.

The compressor operating point must remain within these boundaries under all operating modes: turndown, transients (start-up, shutdown, etc.) or high ambient temperature. Some margin is necessary for some of these constraints to avoid accidentally transgressing into an unfavourable region. These margins are plant dependent: the surge control line can be set between 7% and 10% to the actual surge, for instance.

Constraints can interact with each other. For example, depending on the composition of the MR, the limits on level in the MR separator and the ability to provide the desired MR rate are all linked. A refrigerant composition that is too rich can overfill the MR separator and does not have sufficient cooling capability at low temperatures.

The refrigerant inventory of the loop is therefore critical to optimize the process conditions. Vjacheslav et al (2001) have shown the relation between of the refrigeration efficiency and the inventory for a theoretical system. Figure 3 reproduces a similar effect at an operating LNG plant. The charge of refrigerant and its quality can be varied to optimize the process conditions.

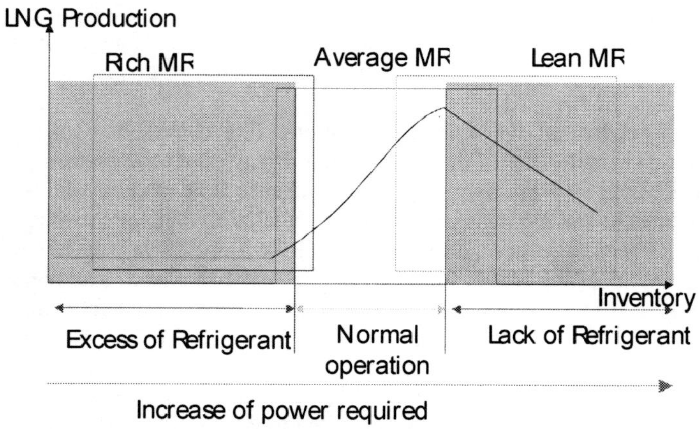

Figure 3: Impact of the refrigerant inventory on the LNG production

Since the constraints are continuously changing, an operator cannot be continuously making small adjustments to the control setpoints. Modern process control technology can be applied to overcome this challenge.

APC is a tool which can drive the plant towards the operator target within its operating window (Opstal, 2009). If the operator wants to have the most efficient operation of the plant, APC can be used to determine the optimum set of conditions to minimize fuel usage, in terms of valve opening or quality of the refrigerant for example. As a consequence, the CO_2 emissions will be reduced and the LNG production will be increased. This control strategy has been successfully applied for instance in Oman LNG with an efficiency gain of 3%.

3.2. Accommodating efficiently large changes in ambient environment: case of a DMR process

Running an LNG plant at its optimum point is an even bigger challenge in a subarctic environment where ambient temperature changes between winter (-25°C) and summer (+25°C) are large. The operation of Sakhalin LNG is another example where the design accommodates this temperature variation and is suited for efficient operation all year round.

Operating the Sakhalin plant efficiently throughout the year while maximizing production and efficiency has been a key driver to the selection of the Double Mixed Refrigerant (DMR) process. In this liquefaction process, the propane loop is replaced by a mixed refrigerant pre-cooling (PMR) system where coiled wound heat exchangers are also replacing kettles. Refer to Figure 4.

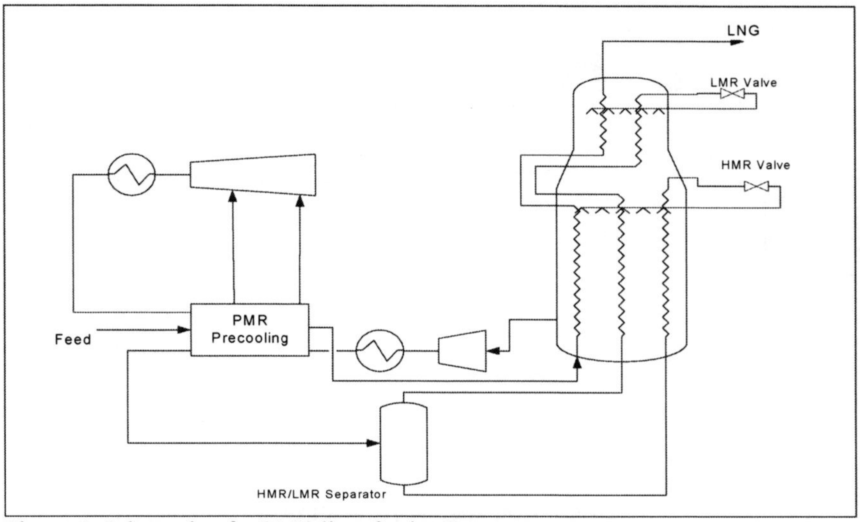

Figure 4: Schematic of a DMR liquefaction Process

The key feature is the capability of adapting the pre-cooling refrigerant composition to best suit the ambient temperature. The colder the ambient temperature, the leaner the pre-cooling refrigerant should be. This unique design feature enables the liquefaction process to keep the precooling gas turbine fully loaded during winter. Also the volumetric flow of the refrigerant going through the compressor can be adjusted to the optimum conditions.

As the gas turbine and the compressor both operate at their optimum design conditions, the CO_2 emissions are also reduced and the LNG production increases. Figure 5 illustrates the flexibility of the DMR process versus ambient temperature.

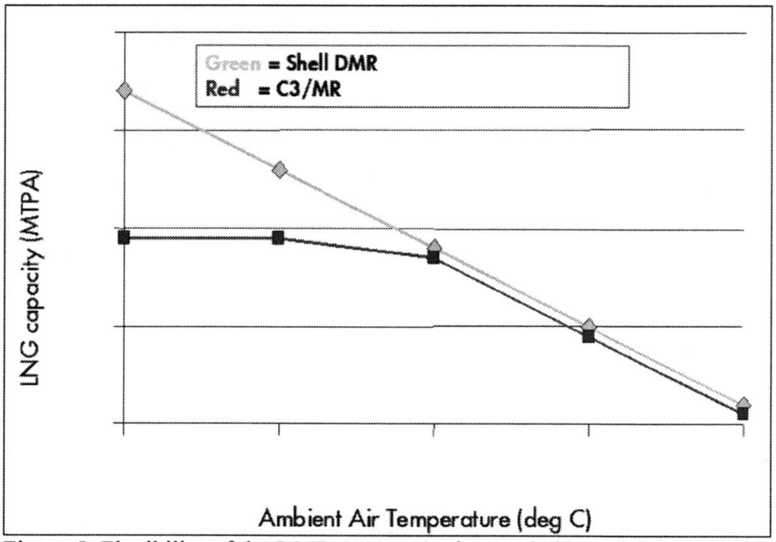

Figure 5: Flexibility of the DMR process against variations of the ambient temperature

The know-how obtained from optimizing the operation of existing plants enables Shell to propose new designs that could incorporate these learnings. Therefore, the new proposed line-ups are more robust against environment changes and are more fuel-efficient.

New designs take into consideration the process efficiency during changing ambient conditions. Processes that can accommodate the changes in ambient conditions (or minimize them) have been proposed as the new solutions for LNG plant.

4. New liquefaction line ups: including the lessons learned for a more efficient operation

During the design phase of a project, key choices have to be made. Utilizing the exhaust heat from the refrigerant gas turbines exhaust to generate steam and use this steam to drive another refrigerant compressor is an example of an energy efficient design.

One advantage of these designs is the reduced sensitivity of the propane driver (steam driven) to the ambient temperature variations. The power available from the driver is not linked to the ambient conditions. Therefore, the fluctuations of the propane process conditions are smaller, and the process operation is easier.

This idea has been developed to fully integrated design and presented by van Loon et al. (2008).

The key results are higher plant fuel efficiency with only 6 % auto consumption, enabled by employing such a "Combined cycle LNG plant". Increased availability of 5 streamdays and low emissions (NOx, SOx, CO_2) are other key advantages. This concept can achieve a 30% reduction of CO_2 emissions compared to the traditional LNG plant line-ups. The plant wide integration of heat and power is a step-out in terms of improved efficiency.

Another way to increase the plant efficiency is by using high efficiency gas turbines. Aero-derative gas turbines can also be an option like presented by Meher et al. (2009). These drivers achieve very high efficiency whilst being simpler than a combined cycle system. However, aero-derivative gas turbines are sensitive to high seasonal temperature variations and the delivered power is less than large heavy duty gas turbines.

These new liquefaction technologies can enable more efficient designs that are easier to operate, helping the plant operator to make the most out of the incoming gas.

5. Conclusions

This paper has presented a number of options for optimizing plant operation and plant design. Better understanding of process behavior helps the existing operating sites to maximize their efficiencies and stay at the best operating point. Additionally, feedback from the operating sites helps designers to develop improved design for future projects, where energy efficiency is becoming a key feature.

References

Ahmad, S., Wilson, J.; Thee, H.K.; Bradley, A and Klinkenbijl, J (2004). "Accelerated amine technology carbon dioxide removal at Brunei LNG", LNG Journal, Sep/Oct page 28-29

Meher-Homji, C.; Messersmith, D.; Masani, K. and Weyermann H. (2009). "The application of Aeroderivative Engines for LNG liquefaction- Higher plant thermal efficiency, Lower CO_2 footprint, and Modularization Capability", Gastech.

van Loon, M.; Nagelvoort, R. K.; Bosma, P. and Kaart, S. (2008). "Green Novel and Large LNG Trains", 6th Doha Natural Gas Conference.

van Opstal, J.; Sturm, W. ; Parra-Calvache, M. and Chantant, F. (2009). "Unlocking the potential of Moderna Control and Optimization strategies in LNG production", Proceedings of the 1st Annual Gas Processing Symposium, Elsevier

Vjacheslav, N. ; Rozhentsev, A. and Wang, C. (2001). "Rationally based model for evaluating the optimal refrigerant mass charge in refrigerating machines" Energy conversion and management, (42), page 2083-2095.

Wilson, J. ; and Van der Wal, G. (2008). "Oman LNG boosts efficiency and performance at baseload plant", LNG Journal, October, page 17-21

Proceedings of the 2nd Annual Gas Processing Symposium
Farid Benyahia and Fadwa T. Eljack (Editors)

Thermodynamic Analysis on Post Combustion CO_2 Capture of Natural Gas Fired Power Plant

Zeinab Amrollahi,[1] Ivar S. Ertesvåg, Olav Bolland
Department of Energy and Process Engineering, Norwegian University of Science and Technology, NO-7491 Trondheim, Norway

Abstract

A chemical absorption, post-combustion CO2 capture unit is simulated and an exergy analysis was conducted, including irreversibity calculations for all process units. With pinpointing major irreversibilities, new proposals for efficient energy integrated chemical absorption process were suggested. Moving further to the whole natural gas combined cycle plant with a CO_2 capture unit, it has been analyzed on an exergetic basis. By defining exergy balances and black-box models for plant components, investigation has been made to determine effect of each component on overall exergy efficiency. Simulation of chemical absorption plant was done using UniSim Design software with Amine Property Package which maintains thermodynamic data. For overall power plant design, GT PRO software (Thermoflow, Inc.) was used for simulation of a natural gas combined cycle. For exergy calculations, spreadsheets were created with Microsoft Excel by importing data from UniSim and GT PRO. Results show that for current chemical absorption plant, the exergetic efficiency compared to the reversible separation work lies between 9% to 21%.

Keywords: CO_2 capture, Absorption, Exergy analysis

1. Introduction

For a natural gas-capture from flue gases using chemical absorption with aqueous monoethanolamine (MEA), is one of the most near-term technologies. Flue gas containing CO_2 is flowing through absorber while contacting with MEA solvent counter-currently. Meanwhile reaction is happening between MEA solvent and CO_2 forming a water soluble salt. A rich MEA stream which contains the chemically bound CO_2, preheated in a heat exchanger is entered to a stripper column to reverse the reaction by means of heat maintained by a reboiler and lose CO_2 content as a stream leaving at the top of the column. The lean MEA is recycled back to the absorption column while the CO_2 stream is going to compression section.

[1] Corresponding author. Email: zeinab.amrollahi@ntnu.no

Although it is a well-established separation method, the energy consumption and the costs of CO_2 separation are substantially high and lead to consumption of more fossil fuel for the same power generation. In order to increase the energy efficiency and prevent forced extra costs and energy consumption, it is necessary to optimize the process and evaluate the performance of the whole system by means exergy analysis which identifies the energy consumption potential improvements and thermodynamic irreversibility amounts.

2. Exergy analysis

The exergy method of evaluating energy-intensive systems integrates the first and second laws of thermodynamics at the state of particular environmental conditions. Exergy analysis with its own certain methods of process evaluation has proven to be an efficient method to define the second law efficiency of processes. It combines the principles of conservation of mass and conservation of energy together with the second law of thermodynamics to characterize the thermodynamic losses of each component of a system through the whole design and it enables to make possible improvements of work and energy consumption. This is an advantageous method to approach the goal of more efficient energy-resource use, since it specifies the locations, types, and real magnitudes of irreversibilities either to be recovered or inevitably lost.

In absence of potential and kinetic energy, exergy of stream is divided into physical exergy and chemical exergy. Physical exergy equals to maximum amount of work obtainable when the stream of substance is brought from its actual state to the environmental state defined by P_0 and T_0 (Szargut et al., 1988) by physical processes involving only thermal interaction with the environment. It is depicted as:

$$\varepsilon_{ph}=(h-h_0)-T_0(s-s_0) \tag{1}$$

Where h and s are the specific enthalpy and entropy and $h_0=h\ (T_0, P_0)$ and $s_0=s\ (T_0, P_0)$ for the flowing matter.

The chemical exergy of a substance is the minimum work requirement to deliver it in the environmental state from the environmental substances by means of processes involving heat transfer and exchange of substances only with the environment. There are tables of calculated standard chemical exergy of various substances in literature (Kotas, 1995). molar chemical exergy of an ideal mixture is expressed as

$$\tilde{\varepsilon}_{0M} = \sum_i x_i \tilde{\varepsilon}_{0i} + \tilde{R}T_0 \sum_i x_i \ln x_i \tag{2}$$

Exergy loss of each individual unit can be calculated by finding the difference between the exergy of input and output streams of a unit operation. To pinpoint irreversible losses in each unit operation, the exergy balance for steady state steady flow is used;

$$\sum_{in} \dot{m}_j \varepsilon_j + \sum_{l} \dot{Q}_l (1 - \frac{T_0}{T_l}) = \sum_{out} \dot{m}_k \varepsilon_k + \dot{W} + \dot{I} \quad (3)$$

Flow exergy into system — Heat exchange — Flow exergy out of system — Work — Irreversibility

Exergy analysis can be done when composition and thermodynamic properties of all streams involving in capture process are available. For this purpose, specific simulation software model is used to simulate the whole CO_2 capture process. By transferring stream physical properties and compositions to excel spreadsheets, exergy calculations are performed and reported.

To calculate the chemical exergy of each stream containing MEA component there is a need of chemical exergy of the MEA molecule in the liquid phase. The value which is used in these calculations is not found directly from literature but estimated. The value is $1.274 \cdot 10^6$ kJ/kmol.

3. Methodology

The plant subsystems that are analyzed include gas turbine, heat recovery steam generator (HRSG), steam turbine and condenser, CO_2 absorption column, main heat exchanger of CO_2 capture plant, stripping section, compression section etc. Each of the processes consists of exergy inputs and exergy outputs with some exergy losses.

Particularly plant has been divided into control volumes with exergy inputs and outputs from each representing the different process flows. The processes were approximated to steady or quasi steady-state flow conditions. Relevant thermodynamic data is taken from UniSim software which is process modeling software and transferring them into Excel spread sheet. Chemical and physical exergy of all streams is functioned in excel spreadsheets. The relevant amount of exergy losses and irreversibility is calculated in each part of process considering the exergy balance in each control volume.

Furthermore, exergy analysis calculations for the designed power plant are derived from GTPro Thermoflow software calculation which is going to be depicted later. The reference environment will be the local environment of the place where the natural gas fired power plant is located which it is assumed in here with ambient temperature T_0= 298.15 K and pressure P_0=101.325 kPa.

As mentioned before, this study is limited to the analysis of the physical exergy and chemical exergy. Other forms of exergy as kinetic and potential are insignificant in these processes so they are ignored. The degradation and consumption of the MEA solvent was neglected in CO_2 capture unit.

4. Base case model

As a base case, the CO_2 separation with MEA absorption model shown in figure1 is designed according to the capture rate that is set to 93%. This capture rate for the base case was attained by MEA wt% of 30%, solvent circulation rate of 2358 t/h and reboiler duty of $6.12 \cdot 10^8$ kJ/h. Reboiler energy consumption is 4.36 (MJ/kg of separated CO_2) which is produced by the steam flow of 76.45 kg/s .Total mechanical work needed for the capture and compression unit is mentioned in Table 1. CO_2 compression was done in 3 stages with adiabatic efficiencies of 85%, 85% and 80% respectively with intermediate cooling after each stage. A pump further raised the pressure from 79.7 bara to 110 bara. The pump adiabatic efficiency was set to 75%.

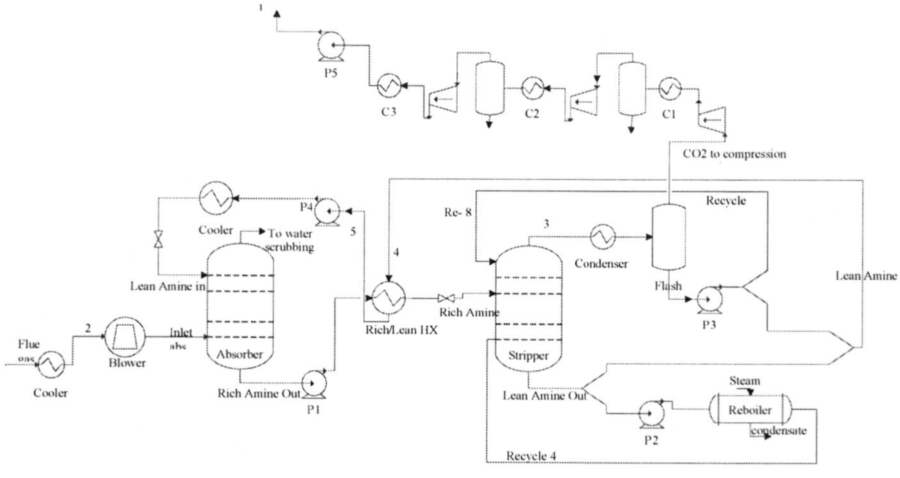

Figure 1. Flow sheet of CO_2 capture and compression units designed by UniSim Design

The virtual power plant that is connected to the CO_2 capture process provides mechanical work to cover the demand of the CO_2 capture unit as well as the steam demand of the regeneration reboiler. A complete schema of the designed combined cycle power plant is shown in Figure 2 with key stream information. The plant key data are shown in table 2.

The fuel was considered as natural gas without H_2S with 722087 kW thermal as lower heating value and flow of stack gas is 686.4 kg/s with molar composition of 3.82% CO_2, 12.54% O_2, 8.24 % H_2O, 75.4% N_2 and temperature of 412.5 K which is going to be cooled in capture unit.

Table 1: Total mechanical work demand for post combustion CO_2 capture plant

Work demand	MJ/kg CO_2 separated
Power production penalty	1.00
compression work	0.30
Auxiliary power	0.16
Total	1.46

Table 2: Power plant summary

	Power Output kW		LHV Heat Rate kJ/kWh		Elect. Eff. LHV%	
	gross	net	gross	net	gross	net
Gas Turbine	291600		8915		40.38	
Steam Turbine	89322					
Plant Total	380922	373761	6824	6955	52.75	51.76

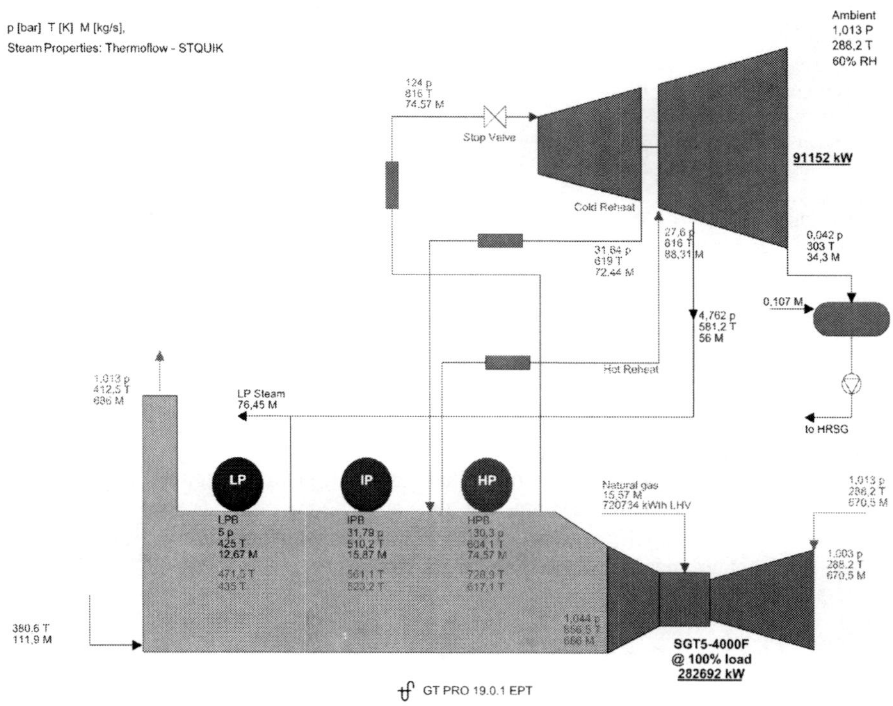

Figure 2. Flowsheet of the designed power plant

5. Results

5.1. Natural gas fired power plant

The results of exergy calculation for specified natural gas fired power plant designed by GTPro Thermoflow software are shown in table 3.

Table 3: Plant exergy analysis

	kW	MJ/kg CO_2 separated
Exergy In	741527	18.88
Fuel exergy	729153	18.56
Ambient air exergy	115.6	0.00
Condenser cooling water in	6228	0.16
Process condensate return	6029	0.15
Makeup water	1.852	0.00
Exergy Out	443391	11.29
Net electric output	373760	9.52
Process steam/water exergy @ delivery	55318	1.41
Condenser cooling water out	1362.8	0.03
Stack gas exergy	12950	0.33
Exergy Loss	298136	7.59
GT exergy loss	250133	6.37
HRSG exergy loss	20218	0.51
Steam turbine exergy loss	10173	18.88
Condenser exergy loss	6062	18.56
Non-heat balance related auxiliaries	2520.7	0.00
Transformer loss	1904.6	0.16
Miscellaneous exergy loss*	1868.6	0.15
Unaccounted exergy loss**	5257	0.0

* Includes piping loss, ST leakage to external sink, fuel compressor loss, condensate pump loss

** Includes losses from desuperheating, mixing, and throttling, small water streams, misc. aux. and heat rejection

5.2. CO_2 capture plant

The detailed description of CO_2 capture plant has been shown in Figure1. Figures 3a-3c shows a simplified block schema of the absorption unit, desorption unit and compression unit which presents the streams transferring exergy through this subsystems that are material streams and work streams.

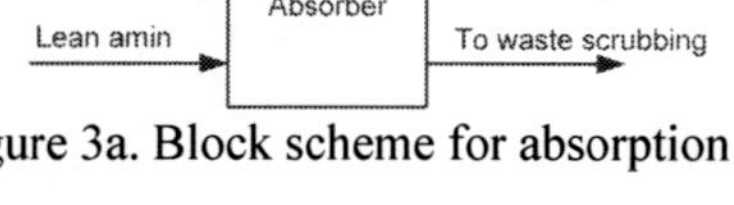

Figure 3a. Block scheme for absorption section

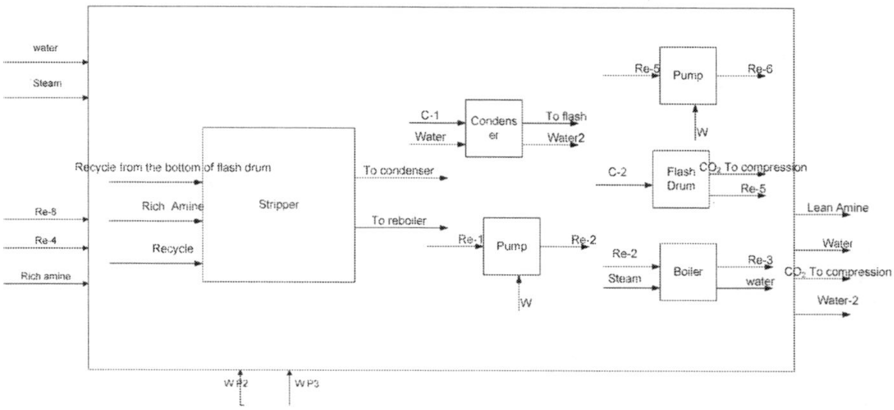

Figure 3b. Block scheme for stripping section

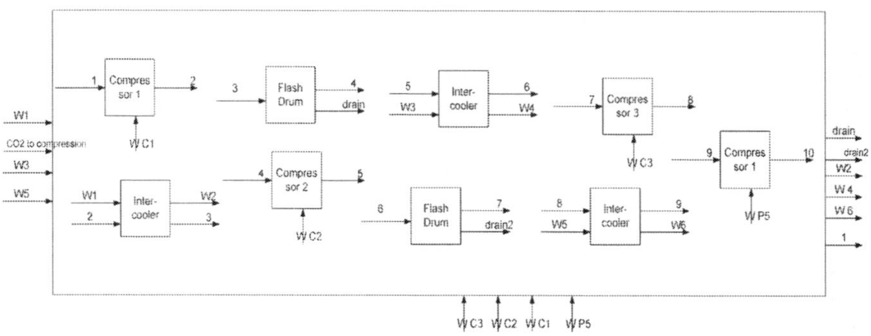

Figure 3c. Block scheme for CO2 compression section

In Table 4, physical stream's characteristics and relevant calculated exergy which is used to find irreversibility amounts according to exergy balance formula are shown; it should be noted here that for simulation of streams containing amine component, UniSim Design software (Honeywell 2008) developed a specific property package which predicts behavior of systems containing MEA solvent. For simulation of other streams Peng-Robinson equation of state is used.

Table 4: Themodynamical data and exergy of streams

Stream	Temperature (°C)	Pressure (kPa)	Mass Flow (kg/s)	Exergy (MJ/kg CO_2 separated)
1	24.66	11000.00	39.3	0.49
CO2 to compression	28.00	167.20	39.6	0.49
2	60.00	101.30	686.4	0.39
inlet abs	67.11	107.30	686.4	0.50
Flue gas	100.00	101.30	715.8	0.50
to water scrubbing	46.91	101.30	651.0	0.40
Re-4	150.54	194.00	91.0	16.11
Re-8	28.15	649.97	28.2	0.50
3	103.26	172.37	71.0	1.04
4	119.71	186.20	676.3	115.20
5	57.83	146.20	676.3	112.49
Lean Amine Out	119.74	186.20	764.0	130.69
Steam	176.85	400.00	76.4	2.66
condensate	142.87	392.00	76.4	1.43
Lean Amine in	20.96	107.00	680.5	111.32
Rich Amine Out	49.14	106.30	715.8	112.36
Rich Amine	105.12	180.00	715.8	114.88

Table 5: irreversibility amounts by unit sections

	Irreversibility MJ/kg CO_2
Flue gas cooler	0.14
Blower	0.03
Absorption section	0.42
Rich /lean heat exchanger	0.01
Stripping section:	0.71
reboiler	0.20
Flasher(condenser+ flash)	0.69
compression section	0.31
Total	1.62

6. Concluding remarks

Although the exergy loss in CO_2 capture and compression units are rather small comparing to those lost in Gas turbine, HRSG and steam turbine, there are some points of potential improvements. Results show that for current chemical absorption plant, minimum reversible separation work which is calculated based on the approach presented by Cengel et al. (2006) is 0.194 MJ/kgCO_2 and consequently the relevant exergy efficiency is 19.36%. Additionally, current study -as in Table 5- shows that flasher, absorber and reboiler have the most irreversibility amounts and exergy losses. Use of other solvents with lower binding energy is suggested to decrease the exergy loss of reboiler section. In order to minimize the exergy loss, it is important to have uniform exergy degradation along equipments, which can be an optimization idea for the regeneration column, flasher and reboiler. Furthermore in order to divide exergy losses through the absorption column and stripping column, process configuration changes such as stream splitting can be performed (Adisorn, 2005).

References

Aroonwilas, Adisorn (2005). Evaluation of split-flow scheme for CO2 absorption process using mechanistic mass-transfer and hydrodynamic model; Greenhouse Gas Control Technologies 7, Pages 1841-1844.

Cengel, Yunus A., Boles, MA. (2006). Thermodynamics: An Engineering Approach (6th edition): McGraw-Hill; 6 edition.

Kotas, TJ. (1995). The exergy method of thermal plant analysis. Malabar (FL): Krieger Publishing Company.

Szargut J, Morris DR, Steward FR. (1988). Exergy analysis of thermal, chemical and metallurgical processes. New York: Hemisphere Publishing Corp.

UniSim design user guide; Honeywell; March 2008 R380 Release.

Proceedings of the 2nd Annual Gas Processing Symposium
Farid Benyahia and Fadwa T. Eljack (Editors)

Self-Optimizing and Control Structure Design for a CO_2 Capturing Plant

Mehdi Panahi, Mehdi Karimi, Sigurd Skogestad, Magne Hillestad, Hallvard F. Svendsen

Department of Chemical Engineering, Norwegian University of Science and Technology(NTNU), 7491 Trondheim, Norway

Abstract

Capturing and storing the greenhouse gas carbon dioxide (CO_2) produced by power plants could play a major role in minimizing climate change. In this study a post-combustion CO_2 capture plant using MEA is designed, simulated, and optimized using the UniSim process simulator. The focus of this work is the subsequent optimal operation and control of the plant with the aim of staying close to the optimal operating conditions. The cost function to minimize is the energy demand of the plant. It is important to identify good controlled variables (CVs) and the first step is to find the active constraints, which should be controlled to operate the plant optimally. Next, for the remaining unconstrained variables, we look for self-optimizing variables which are controlled variables that indirectly give close-to-optimal operation when held at constant setpoints, in spite of changes in the disturbances. For the absorption/stripping process, a good self-optimizing variable was found to be a temperature close to the top (tray no.4) of the stripper. To validate the proposed structure, dynamic simulation was done and performance of the control structure was tested.

Keywords: Process control, Optimal operation, Plantwide control, Controlled variable selection

1. Introduction

Aqueous absorption/stripping with aqueous solvents such as MEA has been used effectively for removing acid gases (CO_2 and H_2S) from natural gas, oil refineries, power plant flue gas and the production of ammonia and synthesis gas. Fig. 1 shows a typical flow diagram of the process for a simple reboiled stripper. The system consists of two columns: the absorber, in which the CO_2 is absorbed into an amine solution via a fast chemical reaction, and the stripper, where the amine is regenerated and then sent back to the absorber for further absorption. Prior to CO_2 removal, particulates, sulfur dioxide, and NO_x are removed from the flue gas. The flue gas from the power plant is typically cooled before the absorber from 150 to 55 °C (its adiabatic saturation temperature) or to 40 °C if cooling water is used.

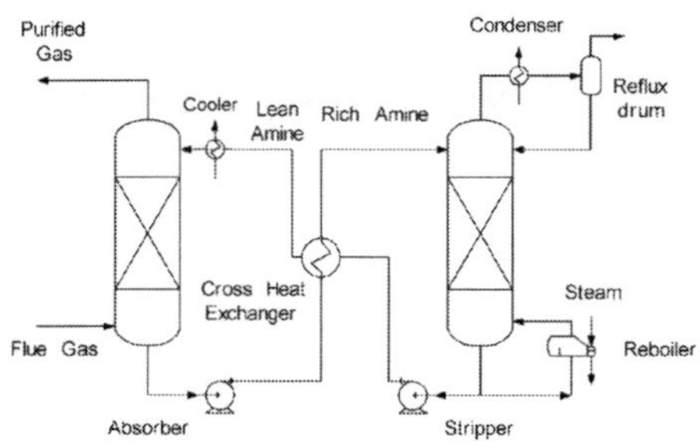

Figure 1. Typical absorber/stripper process for CO_2 capture

One problem with using MEA as a solvent is the high cost of operation. This is simply due to the excessive energy requirement for solvent regeneration, which contributes about 70 per cent of the process utility cost. In fact, the energy consumption in the CO_2 capturing plant is estimated to be 15-30% of the net power production of a coal-fired power plant. A lot of work have been done to reduce energy consumption of CO_2 units, but little has been done on studying how this can be implemented in practice when the process is subjected to disturbances. This is the aim of the present study where we focus on selecting good controlled variables which can be kept at constant setpoints without the need to re-optimize when disturbances occur. To select the controlled variables we look for self-optimizing control, one may use the stepwise procedure of Skogestad (Skogestad, 2004). The plant was modelled using the UniSim flowsheet simulator from Honeywell using the amine package for the thermodynamic calculations.

2. Self-optimizing control of a CO_2 capturing plant

2.1. Step 1: Define objective function and constraints

In the CO_2 plant there are operational costs related to the two utilities: Steam (heat) for the reboiler of stripper and electricity (power) for driving the pumps. To avoid using prices we convert the heat to equivalent thermodynamic work (power). We assume that the temperature of steam in reboiler (T_H) is 10°C higher than reboiler temperature and steam condenses at 40°C in the turbine (T_C). The total equivalent work for the plant (the objective function) is then

$$W_{eq} = Q_r\left(1-\frac{T_C}{T_H}\right)\times\eta + W_{Pumps} \qquad W_{eq}\left(\frac{kJ}{kg\,CO_2}\right) \qquad (1)$$

Where $T_H = T_C + 10$ [K] and $T_C = 313$ K . The efficiency η of the imagined Carnot cycle (heat pump) that generates heat from power is assumed to be 75% .

The constraints are:

1. Environmental requirement: Capture 90% of CO_2.
2. Temperature of lean solution to the absorber is 51°C (to get a good operation of the absorber).
3. Because of the MEA degradation problem, pressure should be less than 2 bar. Stripper top pressure is therefore kept at 1.8 bar.

4. The stripper condenser temperature should be as low as possible and is here assumed to be at 30°C.

2.2. Step 2. Determine DOFs for optimization

We have 9 valves (Fig.2) which give 9 dynamic degrees of freedom. However, there are 4 levels (2 in stripper, 1 in absorber, 1 surge tank) that need to be controlled and since these levels have no steady state effect, the number of degrees of freedom (DOFs) for steady-state optimization is 5.

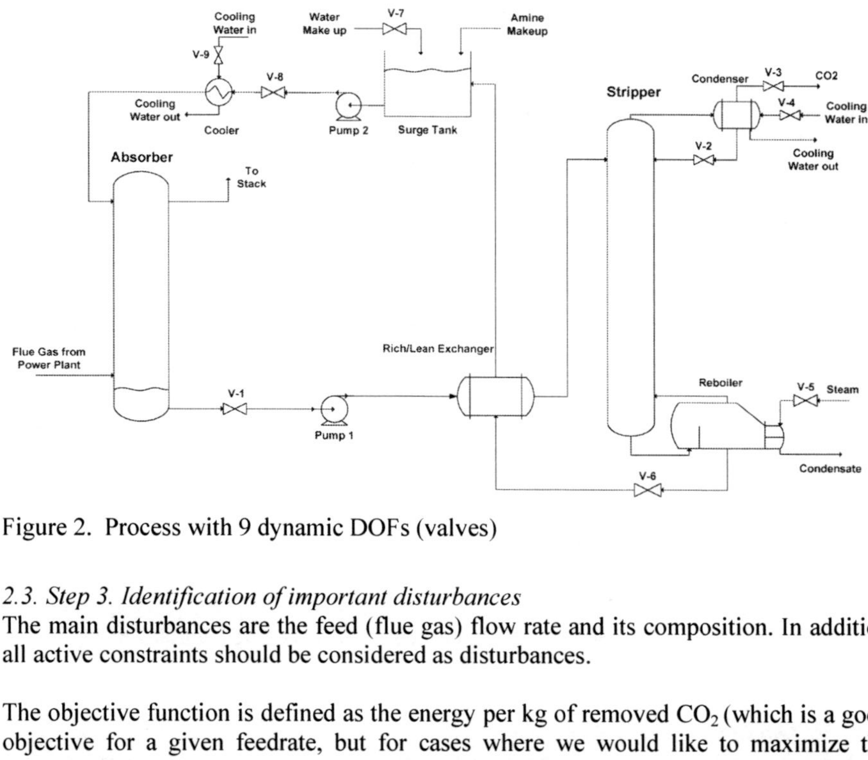

Figure 2. Process with 9 dynamic DOFs (valves)

2.3. Step 3. Identification of important disturbances

The main disturbances are the feed (flue gas) flow rate and its composition. In addition all active constraints should be considered as disturbances.

The objective function is defined as the energy per kg of removed CO_2 (which is a good objective for a given feedrate, but for cases where we would like to maximize the amount of treated gas), so small variations in the CO_2 recovery constraint have a small influence on the objective function. In practice, the inlet temperature of lean solution is around 51°C and even if it changes in the range 40-60°C has no effect on the energy consumption. The only equality constraint that may have significant affect on the objective function is change in pressure of the stripper.

Finally, we consider three main disturbances. (table 1)

Table 1- Main disturbances

Disturbance		Nominal	Change
d1	Gas flowrate	219.3 kmol/hr	±5%
d2	Gas composition	CO_2: 0.1176, N_2: 0.7237,O_2: 0.0502, H_2O: 0.1085	±5%
d3	Stripper pressure	Top: 180 kPa, Bottom: 200 kPa	+10 kPa

2.4. Step 4. Optimization (nominally and with disturbances),

To control the 4 equality constraints we need 4 DOFs and we need 4 DOFs to control 4 levels then we have one degree of freedom left for optimization, $N_{opt.free} = 9 - 4 - 4 = 1$.

Objective function: min. W_{eq}

Subject to: The four constraints in section 2.1 and:

5. $0.005 \leq CO_2$ fraction in bottom of stripper ≤ 0.05.

At the nominal operating (no disturbances) point we get:

Optimal objective function: $W_{eq} = 640.5 \dfrac{kJ}{kg\,CO_2}$

CO_2 composition in the bottom of stripper = 0.0227 (so the optimum is unconstrained).

2.5. Step 5. Identification of candidate controlled variables.

The remaining unconstrained DOF could for example be selected as the reboiler duty. However, rather than keeping it constant, it may be better to use it to control some other variables (CV), and we consider two alternatives:

1. Tray temperature at some stage in the stripper column.
2. CO_2 composition in the bottom of the stripper.

2.6. Step 6. Evaluation of loss

For evaluation and initial screening of the candidate controlled variables we use the maximum scaled gain rule (Skogestad and Postlethwaite, 2005).

2.7. Procedure for scalar case:

1. Make a small perturbation in each disturbances *di* and re-optimize the operation to find the optimal disturbance sensitivity for each candidate CV, $\dfrac{\partial \Delta y_{opt}}{\partial d_i}$,where Δd_i denotes the expected magnitude of disturbance i. From this we compute the overall optimal variation (here we choose the 2-norm):

$$\Delta y_{opt} = \sqrt{\sum_i \left(\frac{\partial \Delta y_{opt}}{\partial d_i} . \Delta d_i \right)^2} \qquad (2)$$

2. Identify the expected implementation error n for each candidate controlled variable y (measurement).
3. Make a perturbation in the independent variables u (in our case u is reboiler duty) to find the unscaled gain (G),

$$G = \frac{\Delta y}{\Delta u} \quad (3)$$

4. Scale the gain with the optimal span where n is implementation error to obtain for each candidate output variable y:

$$Span\ y = \Delta y_{opt} + n \quad (4)$$

The scaled gain is then:

$$|G'| = \frac{|G|}{Span\ y} \quad (5)$$

The worst-case loss $L_{wc} = J(u,d) - J\text{opt}(u,d)$ (the difference between the cost with a constant setpoint and re-optimized operation) is then for the scalar case:

$$L_{wc} = \frac{|J_{uu}|}{2} \frac{1}{|G'|^2} \quad (6)$$

Where $J_{uu} = \frac{\partial^2 J}{\partial u^2}$ is the Hessian of the cost function J. In our case $J = W_{eq}$.

Note that J_{uu} does not matter for selecting CVs in the scalar case.

By using a Matlab script interfaced with Unisim, the scaled gains were found for different candidate CVs. The results are shown in table 2.

Table 2. Scaled gain for different candidate CVs

Candidate CV	**Scaled gain**	**Candidate CV**	**Scaled gain**
CO_2 composition	0.2463	Temp. Gain Tray 11	0.1358
Temp. Gain Tray 1	0.0203	Temp. Gain Tray 12	0.151
Temp. Gain Tray 2	0.056	Temp. Gain Tray 13	0.108
Temp. Gain Tray 3	0.1276	Temp. Gain Tray 14	0.0788
Temp. Gain Tray 4	**0.2845**	Temp. Gain Tray 15	0.0614
Temp. Gain Tray 5	0.2693	Temp. Gain Tray 16	0.0499
Temp. Gain Tray 6	0.2279	Temp. Gain Tray 17	0.0409
Temp. Gain Tray 7	0.1913	Temp. Gain Tray 18	0.0334
Temp. Gain Tray 8	0.1632	Temp. Gain Tray 19	0.0264
Temp. Gain Tray 9	0.1446	Temp. Gain Tray 20	0.0200
Temp. Gain Tray 10	0.1332		

From table 2, the best candidate CV is found to be the temperature on tray no. 4. The CO_2 composition has also a good (high) scaled gain but it is still ranked 3rd after temperature of tray no.5. To validate the proposed controlled variable, dynamic simulation were performed in the next step.

3. Dynamic simulation

To switch to the dynamic mode in the UniSim simulator, sizing of the equipments and pressure flow specification was done. There is some discrepancy between the steady-state and dynamic models which seems to be caused by a difference in the thermodynamic models used by UniSim for two modes. The main effect is that recycle amine flow between the columns is smaller, which results in a smaller objective function (W_{eq}) in the dynamic case. For our purposes this does not matter very much and the relative order of the control structures remains the same.

All control loops were implemented and tuned individually using the SIMC method. (Skogestad, 2003) The final control structure with 9 feedback loops is shown in Fig.3 for the proposed case where the CV is stripper tray temperature no.4. The paring of the loops is quite obvious in this case and is based on minimizing the effective time delay from inputs to outputs. The reboiler duty is used as the MV to control tray temperature no. 4.

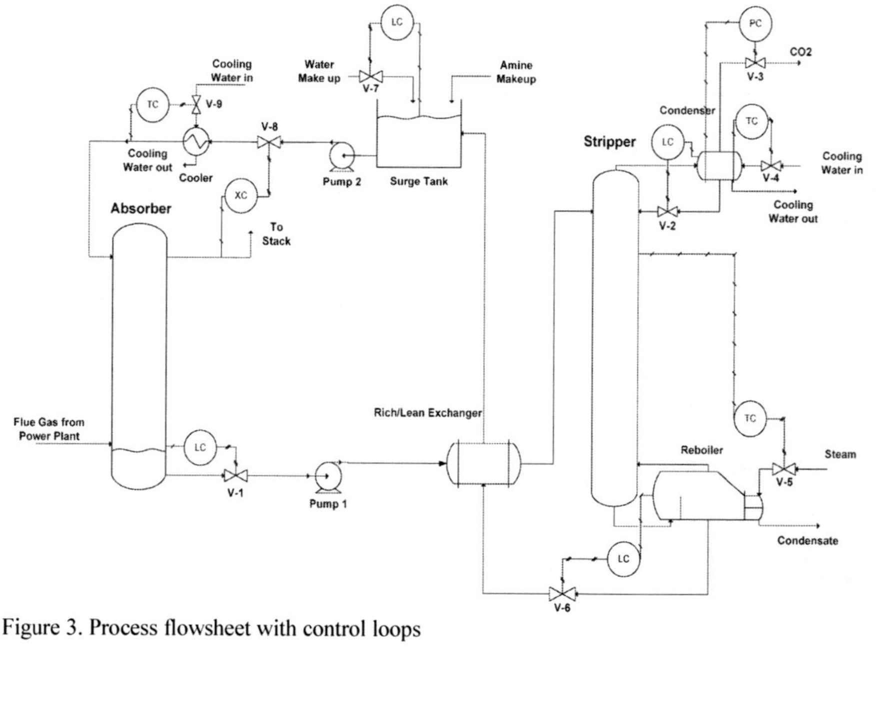

Figure 3. Process flowsheet with control loops

Figure 4a shows the performance of the proposed structure. This can be compared to the case where bottom temperature (tray no.20) is controlled (Fig. 4b) which results in larger losses, especially at steady-state for the pressure disturbance (disturbance 6).

As expected the losses are also small if we control the CO_2 composition in the bottom of stripper. (Figure 4c). However, temperature control is much easier, faster and cheaper than composition. Therefore, control temperature of tray no.4 that we found by self-optimizing concept is the best controlled variable.

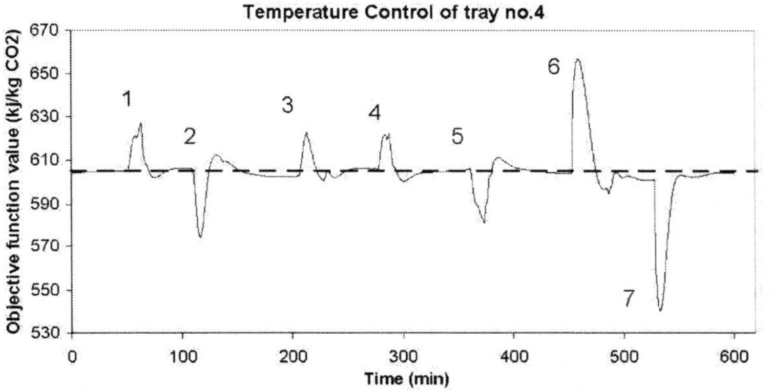

(a)

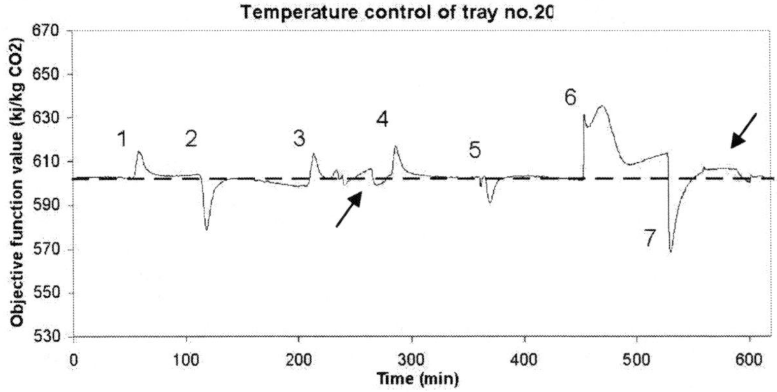

(b)

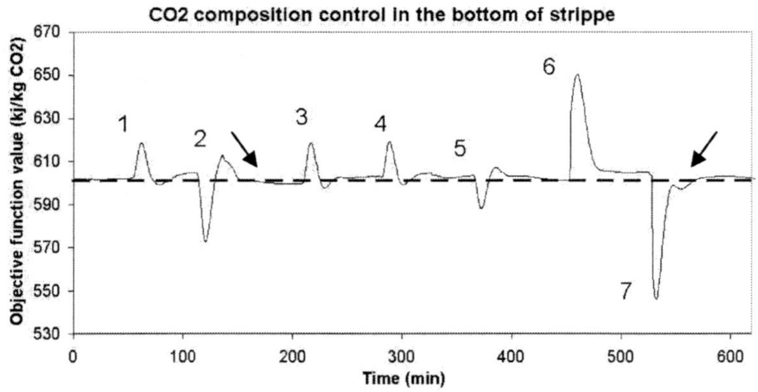

(c)

Figure 6. Objective function (W_{eq}) in presence of disturbances 1) d1:+5% change from base case, 2) d1:-10%, 3) back to base case 4) d2:+5% change from base case 5) back to base case, 6) d3:+10 kPa, 7) back to base case. Arrows indicate cases with large steady-state losses.

4. Conclusion

A self-optimizing concept control structure was designed for a post-combustion CO_2 capturing plant. The losses are small which means that it is not necessary to re-optimize the process when different disturbances occur. The plant has 9 dynamic degrees of freedom; 4 of them were used to control equality constraints and 4 of them were used for level control. We found the temperature close to the top (tray no. 4) of the stripper to be a good CV for the remaining unconstrained degree of freedom.

References

Grainger, D ., M-B. Hägg, 2008, Techno-economic evaluation of a PVAm CO_2-selective membrane in an IGCC power plant with CO_2 capture. Fuel, 87, 1, 14-24

Jassim, M. S., G. T. Rochelle, 2006, Innovative Absorber/Desorber Configurations for CO_2 Capture by Aqueous Monoethanolamine, Ind. Eng. Chem. Res., 45, 2465-2472

Jensen, J. B., S. Skogestad, 2008, Optimal Operation of Refrigeration Cycles, PhD thesis

Skogestad, S., 2004, Control Structure Design for Complete Chemical Plants, Computers and Chemical Engineering, 28, 219-234

Skogestad, S., 2000, Plantwide control: The search for the self-optimizing control structure, Process Control, 10, 487-507

Skogestad, S., I. Postlethwaite, 2005, Multivariable Feedback Control Analysis and Design, 2nd edition

Skogestad, S., 2003, Simple analytic rules for model reduction and PID controller tuning, Process Control, 13, 291-309

Proceedings of the 2nd Annual Gas Processing Symposium
Farid Benyahia and Fadwa T. Eljack (Editors)

Pilot Plant Study of 3-(methylamino)Propylamine Sarcosine for Post-combustion CO_2 Capture

Ugochukwu E. Aronu[a], Hallvard F. Svendsen[a1], Karl Anders Hoff[b], Hanna Knuutila[b]

[a] *Department of Chemical Engineering, Norwegian University of Science and Technology, N-7491 Trondheim, Norway*
[b] *SINTEF Materials and Chemistry, N-7465 Trondheim, Norway*

Abstract

A laboratory pilot plant that processes 150Nm^3/h of gas with capture capacity of about 10 kgCO_2/h was used to study the carbon dioxide absorption, desorption, and corrosion properties of the mixture of 3-(methylamino)propylamine and sarcosine (SARMAPA). Results compared to monoethanolamine (MEA) show that the absorption rates of SARMAPA, estimated by its mass transfer coefficient, Kg; were slightly lower than that of MEA. It was found that SARMAPA potentially has a lower desorption heat requirement than MEA. The stripping steam requirement for the SARMAPA system was found to be half of that of MEA mainly due to countering of its stripping steam requirement by reduction in sensible heat duty against the case for MEA where higher specific heat duty was required. SARMAPA had no corrosive effect on process lines during tests.

Keywords: Sarcosine; 3-(methylamino)propylamine; amino acid salt; CO_2 absorption; desorption; pilot plant.

1. Introduction

To meet the global energy demand and at the same time reduce global warming, scientist are faced with challenges of finding new and more efficient ways of increasing the energy production while reducing the emissions of the major green house gas, CO_2 to the atmosphere, the major source for energy being fossil fuel combustion (IPCC, 2007). One method of mitigating CO_2 emissions is post combustion CO_2 capture from large point sources such as power plants, oil refineries, petrochemical facilities, fertilizer and gas-processing plants, steel works and pulp and paper mills and its further utilization in many technologies such as coal conversion, organic synthesis, destructive oxidation of hazardous wastes, enhanced oil recovery and activated carbon regeneration (Lee et al., 2008). One of the most attractive methods of CO_2 separation from such point sources is absorption with chemical reaction using amine solutions. Amines such as monoethanolamine (MEA), diethanolamine (DEA), di-isopropanolamine and

[1] Corresponding author. Tel.: +47-7359-4100; Fax: +47-7359-4080.
E-mail address: hallvard.svendsen@chemeng.ntnu.no

methyldiethanolamine (MDEA) have been used industrially over many years (Kohl and Nielsen, 1997), however some of these amines have been found to have adverse effects on the environment (Eide-Haugmo et al., 2009). In searching for solutions to the CO_2 emissions reduction by chemical absorption, it is important that efforts should be directed at using solvents that are environmentally benign.

Amino acids salts have been mentioned as attractive alternatives to amines as solvents for carbon dioxide absorption (Kumar et al., 2003 and Hamborg et al. 2007). They have pKa values favourable for carbon dioxide absorption and by using a solvent solution in salt form, the ionic nature of the absorbent makes it less volatile (Goan, 1960), thus avoiding significant solvent vaporization losses. Amino acids have the added advantage that some are naturally occurring and have no environmental issue. Previous researchers, Jensen and Faurholt (1952), Penny and Ritter (1983), Kumar et al, (2003) and Hamborg et al., (2007) investigated Amino Acids Salts (AAS) from neutralization of the amino acid with an inorganic base. However, one problem with these systems is the formation of carbonate salt precipitates at high loadings (Hook, 1997). Aronu et al., (2009) reported that Amine Amino Acid Salts (AAAS) from the reaction between organic base (an amine) and amino acids do not form precipitates at high CO_2 loadings and may have a better CO_2 absorption potential than AAS.
This work has further tested these solutions and presents results of a pilot plant experimental study of the 3-(methylamino)propylamine salt of sarcosine (SARMAPA), resulting from the reaction between sarcosine and the organic base, 3-(methylamino)propylamine (MAPA). Sarcosine and MAPA have been reported by Eide-Haugmo et al., (2009) to be biodegradable and have low ecotoxicity. The pilot plant test results are compared with results from MEA tests from Tobiesen et al., (2007 and 2008).

2. Laboratory Pilot Plant Description

A simplified diagram of the laboratory pilot rig is shown in figure 1. The pilot plant consists of a 0.15 m ID absorber equipped with 4.19m of Mellapak 250.Y packing and a 0.1 m ID desorber with 3.89 m of the same packing material. As described by Tobiesen et al., (2007) it is a closed loop pilot plant where both gas and liquid are recycled. The plant capacity is about 150Nm^3/h of gas and the capture capacity is about 10kg CO_2/h. The CO_2 captured in the absorber is desorbed in the stripper and returned to the absorber to complete the cycle. The system is fully automated and files of logged data were recorded continuously during operation. The plant uses a feed gas mixture of N_2 or air and CO_2 at the required partial pressures to represent the desired process exhaust gas. Gas from the absorber bottom is contacted counter currently with the lean solvent typically at 40-50^oC from top of the absorber. CO_2 in the gas stream is absorbed by the solvent through chemical reaction between the weak acid gas and weak base solvent resulting to formation of water soluble carbamate and bicarbonate. These reactions are reversible and temperature dependent. The temperature in the absorber varies typically between 40 to 65^oC as the absorption reaction proceeds at about 100kPa. The CO_2 cleaned gas exits from the top of the absorption column.

CO_2 rich solution leaves the bottom of the absorber and is pumped to the stripper through the heat exchangers where it gains sensible heat to about 95-110°C. At a pressure of about 200kPa in the stripper the CO_2 rich solution is contacted with steam from the reboiler thus stripping off CO_2 from the solution. A mixture of steam and stripped CO_2 exits the top of the stripper through the condenser where it is cooled to 15-25°C thus separating CO_2 from water which is returned directly to the reboiler in our pilot rig.

The amount of CO_2 from the stripper is measured in kg/h by passing through a mass flow meter between the stripper and absorber before the CO_2 is recycled and mixed with effluent cleaned exhaust.

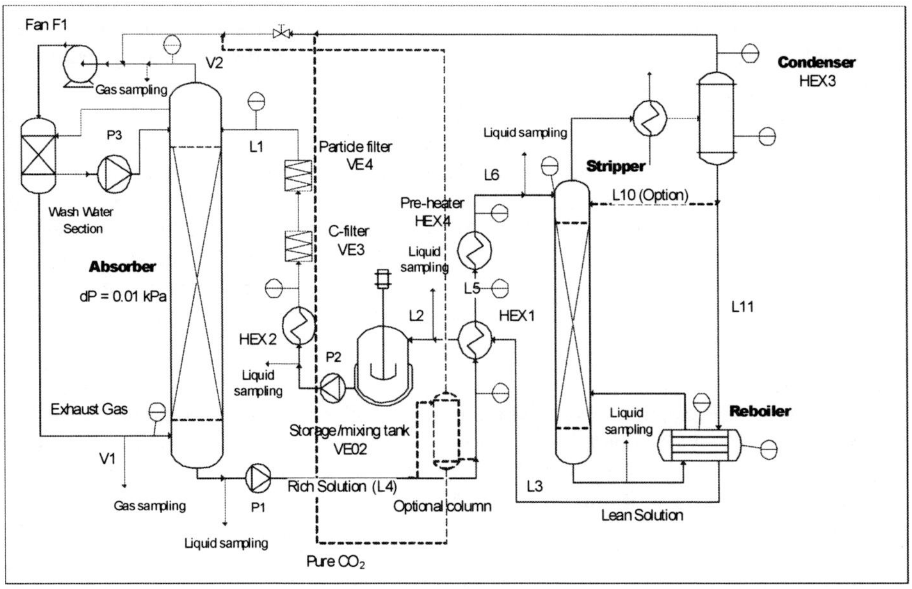

Figure 1: Simplified Process Flow Sheet of Laboratory Pilot Plant

This gas mixture is then returned to the bottom of the absorber as absorber inlet gas after passing through a water wash section. The regenerated lean solution from the stripper at about 120°C is then cooled in the lean/rich heat exchanger with rich solvent from the absorber and stored in the mixing tank before returning to the absorber.

3. Experimental

3.1. Tests and Materials

N-methyl-glycine (sarcosine), purity 99% and 3-(methylamino)-propylamine (MAPA), purity 99% obtained from Carl Roth GmbH + Co. KG and Acros Organics respectively were used without further purification. The 3-(methylamino)propylamine salt of sarcosine was prepared with deionized water by blending equinormal amounts of sarcosine with MAPA. The resulting solution was allowed to cool to room temperature and its concentration was cross-checked by titration with standard H_2SO_4 solution. The CO_2 (purity > 99.99mol %) and N_2 (purity > 99.99 mol %) gases used were obtained from AGA Gas GmbH. Three different tests, absorber, desorber and corrosion tests were conducted on the solvent system in the laboratory pilot.

3.2. Absorber tests

The absorber tests enabled determination of the mass transfer properties of the solvent. For this system the absorber was tested for a coal fired exhaust gas case. This implies that the inlet gas CO_2 content was between 12 and 15kPa on dry basis. The absorber packing height does not allow the whole operating range of an industrial contactor in a single experimental run but rather a section of it. However, tests were run such that the whole operating range of a normal absorber could be studied Experiments were conducted by first studying the low loading region of the absorber by loading the system to achieve approximately 1.2vol% CO_2 out from absorber top. The inlet gas CO_2 content was in this case about 3% and this represents the top absorber section industrially. The operation of the absorber was further divided into 3 other regions; 3-6%, 6-9% and 9-12% representing the concentrations of CO_2 in and out of parts of an industrial column. This was achieved by loading CO_2 into the absorber system after reaching a steady state, thus the CO_2 loading was built up from low (top absorber) to high loading representing the bottom sections of an industrial absorber. Tests were run at 6 l/min liquid rate and 100m^3/h gas rate. A total of 15 temperature measurements were made for each experimental run. 8 of these temperature measurements were along the absorber column.

3.3. Desorber tests

The desorber tests enabled determination of the reboiler energy requirement. These tests involved closed loop operation of the whole plant. The flow rates used were respectively 3.0 l/min for liquid and 100 to 120 m^3/h for gas and the heat input ranged between 4.0 to 12.0 kW in this case. The operation was continued until stable values of temperatures, pressures, condensate, and CO_2 partial pressures in and out of absorber as well as stripper were achieved. 11 temperature measurements of which 5 were along the stripper column were taken during each experimental run.

Each time the operating conditions in the pilot were changed for new measurements it took about 5 to 6 hrs to attain a new steady state. Samples were then taken by collecting 100ml liquid samples from 4 different sampling points as shown in figure 1 after withdrawing about 100 to 150ml liquid to ensure representation of the true state of the process. Liquid samples were analyzed for CO_2 and amine concentration by the precipitation and titration method as described by Ma'mun et al., (2006). Pressure measurements from the top, bottom, inside the absorber, and in the atmosphere were also recorded in addition to the online continuous operating data logging which were averaged over a 30min interval during liquid sampling. The IR-CO_2 analyzer (Rosemount Binos 100) was calibrated on a daily basis with at least 5 points covering the CO_2 concentrations region of interest.

3.4. Corrosion monitoring

100ml liquid samples were taken at different time intervals from the boiler outlet sampling point for corrosion test. The pilot plant is made of 316 stainless steel material. The amount of iron in the solution in mg/kg solution was measured with an atomic absorption spectrometer which was calibrated each time before use. The results were calculated based on 3 parallel analyses.

4. Results and Discussion

4.1. CO_2 Absorption Kinetics

The absorber test measurement results describing the mass transfer rates in the SARMAPA system are shown in figure 2, where the mass transfer coefficients, Kg were calculated disregarding the equilibrium partial pressure of CO_2 over the solvent. It shows that the SARMAPA absorption rate with CO_2 is somewhat lower than that of MEA at the same mean CO_2 loading.

4.2. Desorption Energy

CO_2 desorption energy requirement plots for SARMAPA and MEA are shown in figure 3. They clearly show that the specific heat requirement decreases with increase in rich loading, as expected, and in agreement with results from Sakwattanapong et al. (2005). The figure indicates that SARMAPA could have a lower desorption energy requirement than MEA.

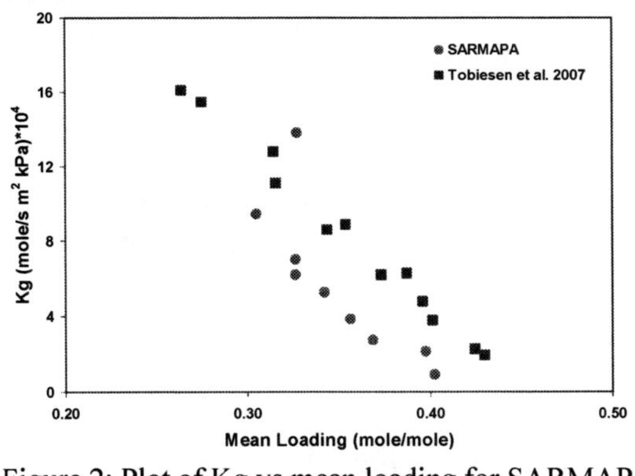

Figure 2: Plot of Kg vs mean loading for SARMAPA and MEA.

Tobiesen et al. (2008) noted that the triangle 'Δ' points are uncertain caused by the liquid sampling of the rich amine. The energetic advantage of SARMAPA is further underlined in figure 4 where we see that less liquid is heated per mole CO_2 released in SARMAPA, evidenced by big changes in loading (Δ loading) leading to a reduction in the sensible heat requirement compared to the very small changes in loading in the MEA tests still requiring higher specific heat input.

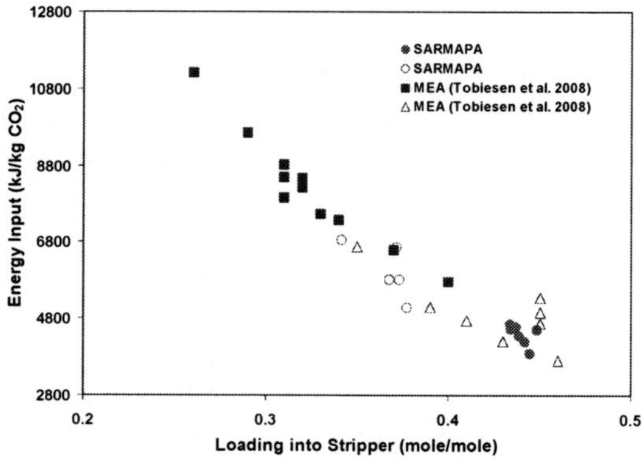

Figure 3: Energy input as a function of loading into stripper for SARMAPA and MEA.'○' lower loading points SARMAPA; '●' higher loading points SARMAPA; '■', certain points MEA; 'Δ', uncertain points MEA.

The MEA data are all for a very narrow of delta in loading (rich minus lean loading) as seen in figure 4. This, combined with the trend in heat requirement with loading level shown in figure 3, clearly shows that the heat requirement decreases with increased rich CO_2 loading. Figure 5 shows that this is due mainly to the reduction in required stripping steam. The ratio of partial pressure of water to that of CO_2 (shown as reflux ratio in kg H_2O/kg CO_2) decreases with an increase in loading.

In figure 6 the filled points for SARMAPA show that the stripping steam consumption increases with an increase in delta in loading between stripper and absorber at an almost constant rich loading. Comparing the SARMAPA filled points in Figure 4 and 6 shows that in figure 4 the increase in heat requirement with increasing delta in loading, seen in figure 6, does not penetrate to the total heat requirement. The reason for this is that

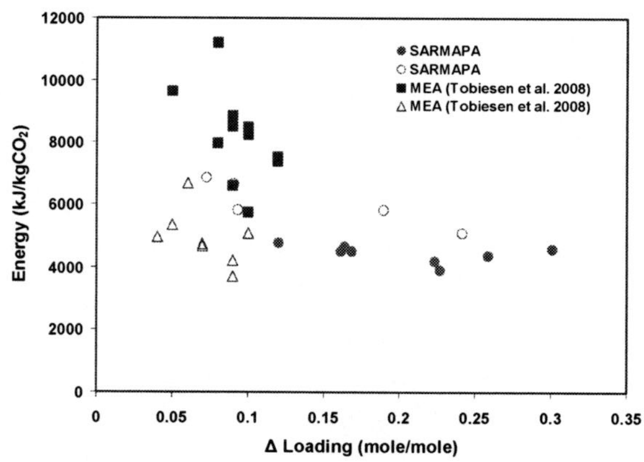

Figure 4: Energy input as a function of change in loading between absorber and stripper for SARMAPA and MEA.

increased delta in loading implies more carrying capacity and thereby a reduction in the sensible heat requirement. These two factors partly counteract each other.

Now comparing MEA and SARMAPA, figure 6 further shows the high stripping steam requirement in MEA for the very small changes in loading even at high loadings. The points with zero reflux ratios must be disregarded as no measurements were made. This figure clearly shows that the stripping steam requirement for SARMAPA is of the order half of that of MEA. For example at loadings of 0.34 and 0.37, reflux ratios for SARMAPA are respectively, 0.46 and 0.71 while MEA is 1.05 and 1.29 This is also shown by figure 5 taking into account the uncertainty of the open triangles for the MEA system. Taking this into account also in Figure 3 and also the fact that all the MEA points are at very low delta in loading, it is clear that the SARMAPA system is significantly better than 30 wt% MEA energetically. However, the absolute value of this improvement can not be determined accurately from these pilot runs and must be part of a further study.

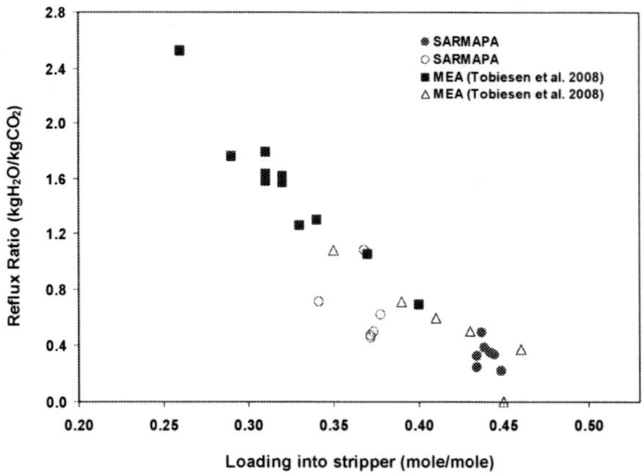

Figure 5: Reflux ratio as function of loading into stripper for SARMAPA and MEA systems.

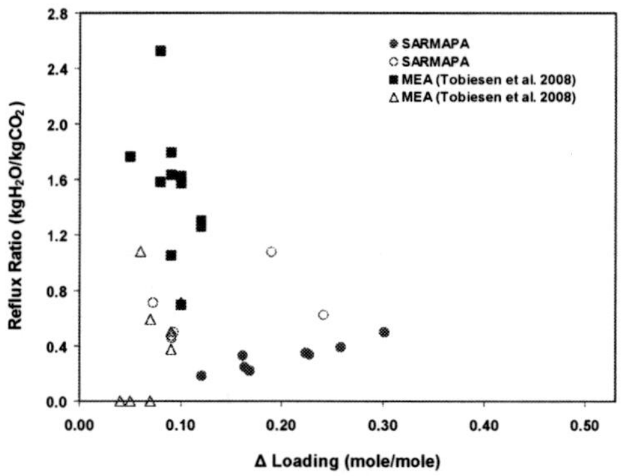

Figure 6: Reflux ratio as function of change in loading between absorber and stripper for SARMAPA and MEA systems.

4.3. Process Corrosion

Kohl and Nielsen, 1997 noted that the most serious operating problem encountered with alkanolamine gas-purification plants is corrosion. This also goes for MEA solutions. Corrosion monitoring of pilot process lines by solvent during testing was thus conducted. Corrosion was estimated by samples analyses for iron. The results in figure 7 show that the corrosion rates in the SARMAPA system were so small that they had practically no effect on process lines during testing.

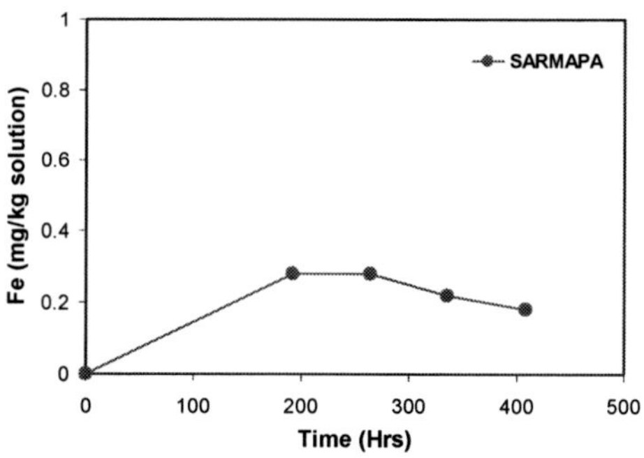

Figure 7: Process corrosion test result for SARMAPA system.

5. Conclusions

Carbon dioxide absorption and desorption charactistics of 3-(methylamino)propylamine salt of sarcosine (SARMAPA) from the reaction between amino acid, sarcosine and organic base, 3-(methylamino)propylamine were determined successfully in a laboratory pilot rig. Absorber test results showed that SARMAPA has slightly lower absorption rates than MEA. Desorber test results, however, showed that SARMAPA can significantly lower the reboiler heat requirement compared to MEA. Lower reboiler heat duty in the SARMAPA system was mainly due to a reduction in the stripping steam consumption. Stripping steam consumption estimated from reflux ratios in kg H_2O required per kg CO_2 removed in SARMAPA was found to be about half that required for MEA. Corrosion results showed that SARMAPA had no corrosive effect on the pilot process lines during testing.

Acknowledgement

This publication forms a part of the BIGCO2 project, performed under the strategic Norwegian research program Climit. The authors acknowledge the partners: StatoilHydro, GE Global Research, Statkraft, Aker Kværner, Shell, TOTAL, ConocoPhillips, ALSTOM, the Research Council of Norway (178004/I30 and 176059/I30) and Gassnova (182070) for their support.

References

Aronu, Ugochukwu E., Svendsen, Hallvard F., Karl Anders Hoff, Knuutila, Hannah, Kim Inna and Jonassen, Øystein 2009, Amine Amino Acid Salt for Carbon Dioxide Absorption. Conference Presention; The 5th Trondheim Conference on CO_2 Capture, Transport and Storage 16-17 June 2009.

Eide-Haugmo, Ingvild; Brakstad, Odd Gunnar; Karl Anders Hoff, Sørheim, Kristin Rist; da Silva, Eirik Falck; Svendsen, Hallvard F. ,2009, Environmental Impact of Amines. Energy Procedia, vol 1, issue 1.,1297-1304.

Hamborg, Espen S.; Niederer, John P. M.; Versteeg, Geert F., 2007, Dissociation Constants and Thermodynamic Properties of Amino Acids Used in CO2 Absorption from (293 to 353) K. Journal of Chemical & Engineering Data, 52(6), 2491-2502.

Hook, Robert J.,1997, An Investigation of Some Sterically Hindered Amines as Potential Carbon Dioxide Scrubbing Compounds. Industrial & Engineering Chemistry Research, 36(5), 1779-1790.

Intergovernmental Panel on Climate Change (IPCC), 2007, Fourth Assessment Report; Climate Change 2007: Synthesis Report.

Jensen, Arne; Faurholt, Carl., 1952, Carbamates. V. The carbamates of α-alanine and β-alanine. Acta Chemica Scandinavica, 6 385-94.

Kohl, A. L. and Nielsen, R., 1997, Gas Purification, fifth edition. Gulf publishing company, Houston, Texas.

Kumar, P. S.; Hogendoorn, J. A.; Versteeg, G. F.; Feron, P. H. M., 2003, Kinetics of the reaction of CO2 with aqueous potassium salt of taurine and glycine. AIChE Journal , 49(1), 203-213.

Lee, Seungmoon; Maken, Sanjeev; Park, Jin-Won; Song, Ho-Jun; Park, Jong Jin; Shim, Jae-Goo; Kim, Jun-Han; Eum, Hee-Moon., 2008, A study on the carbon dioxide recovery from 2 ton - CO2 / day pilot plant at LNG based power plant. Fuel, 87(8-9), 1734-1739.

Ma'mun, Sholeh; Jakobsen, Jana P.; Svendsen, Hallvard F.; Juliussen, Olav., 2006, Experimental and Modeling Study of the Solubility of Carbon Dioxide in Aqueous 30 Mass % 2-((2-Aminoethyl)amino)ethanol Solution. Industrial & Engineering Chemistry Research, 45(8), 2505-2512.

Penny, David E.; Ritter, Terence J., 1983, Kinetic study of the reaction between carbon dioxide and primary amines. Journal of the Chemical Society, Faraday Transactions 1: Physical Chemistry in Condensed Phases, 79(9), 2103-9.

Sakwattanapong, Roongrat; Aroonwilas, Adisorn; Veawab, Amornvadee, 2005; Behavior of Reboiler Heat Duty for CO2 Capture Plants Using Regenerable Single and Blended Alkanolamines. Industrial & Engineering Chemistry Research , 44(12), 4465-4473.

Tobiesen, Finn Andrew; Juliussen, Olav; Svendsen, Hallvard F., 2008, Experimental validation of a rigorous desorber model for CO2 post-combustion capture. Chemical Engineering Science, 63(10), 2641-2656.

Tobiesen, Finn Andrew; Svendsen, Hallvard F.; Juliussen, Olav., 2007, Experimental validation of a rigorous absorber model for CO2 postcombustion capture. AIChE Journal, 53(4), 846-865.

Proceedings of the 2nd Annual Gas Processing Symposium
Farid Benyahia and Fadwa T. Eljack (Editors)

Qatar, LNG, Spill Experiments and Process Safety

Ruifeng Qi[a], Mohammed AlBaker[b], Omar Basha[b], Raid Hassiba[b], Sam Mannan[a], Tomasz Olewski[b,c], and Simon Waldram[b,*]

[a]*Mary Kay O'Connor Process Safety Center, Artie McFerrin Department of Chemical Engineering, Texas A & M University, College Station, TX 77843, USA*
[b]*Texas A&M University at Qatar, PO Box 23874, Education City, Doha, Qatar*
[c]*On leave from Technical University of Lodz, Poland*

Abstract

The Liquefied Natural Gas (LNG) industry is developing rapidly throughout the world. The State of Qatar is the major player in this process. The need for internationally accepted design standards and risk assessment procedures for land based LNG facilities is stressed. Risk assessments necessitate modeling worst credible scenarios using the best available techniques. The Health and Safety Laboratory (HSL) and National Fire Protection Association (NFPA) have summarized LNG field trial data that can be used for validation of models and software. Our medium scale experiments have been conducted at Brayton Fire Training Field (BFTF) and Computational Fluid Dynamics (CFD) modeling of these experiments using the CFX software is discussed.

Keywords: Natural gas, LNG, Spills, Computational fluid dynamics, CFD modeling.

1. World Natural Gas (NG) Reserves

The largest proven NG reserves are in Russia but reliability of pipeline deliveries has recently been poor. Iran and Qatar, with 15.7% and 14.4% of total reserves, are ranked second and third in terms of proven gas fields. Qatar's reserves to production (RP) ratio is in excess of 100 years (BP, 2009). Small and politically stable, Qatar has been hugely ambitious in terms of LNG investment and is bringing new facilities on line with perfect timing for supply of LNG to Europe and the USA. It will be the world's largest exporter of LNG by the end of the decade with a projected production of 77 mtpa. As a consequence many new facilities around the world are being developed for import, storage and re-gasification of LNG. These will be serviced by, amongst others, Qatar's purpose built fleet of 43 new LNG supertankers.

* *Corresponding author:* simon.waldram@qatar.tamu.edu

2. The Global Impact of an LNG Incident

The LNG industry has an enviable safety record: this includes offshore exploration and production, onshore processing, storage, tanker loading, transportation and offloading, and storage and re-gasification. But because of the international aspect of LNG trading a single major incident would probably have world-wide ramifications. Operations in all parts of the LNG industry must be subjected to the most rigorous risk assessments and facilities designed so that they remain adequately safe for both operators and the public even during worst credible maloperations. Professional concerns about the possibility and consequences of a major incident remain, so research and field trials continue (GAO, 2007; Luketa-Hanlin et al., 2008; Havens and Spicer, 2007).

3. Design Standards for LNG Facilities

From section 2 it should be clear that standards and guidance to ensure best practice when designing LNG facilities are required. In this paper we concentrate on onshore facilities. For these there is already a good standard in the form of the latest edition of NFPA 59A (NFPA, 2009). This applies to facilities that liquefy, store, vaporize, transfer or handle LNG as well as to the training of all personnel involved with LNG and the design, location, construction, maintenance, and operation of LNG facilities. High quality risk assessments of credible incidents necessitate an excellent ability to model them, to predict consequences and to have confidence in the validity of these predictions. Recently the Health and Safety Laboratory (HSL), the research agency of the United Kingdom Health & Safety Executive (HSE), under contract to the Fire Protection Research Foundation of the NFPA has prepared a report on "Evaluating Vapor Dispersion Models for Safety Analysis of LNG Facilities" (Ivings et al., 2007). A Model Evaluation Protocol (MEP) has been developed which can be used as a guideline to choose a suitable software model to predict LNG dispersion over land. A summary of the models considered is provided. As an example, three of these (DEGADIS, FEM3A and DOE-NETL LNG Dispersion Module for FLUENT) have been reviewed using the proposed MEP (Ivings et al., 2007).

These models are categorized by Ivings and co-workers into four groups: Workbooks or Correlations, and Integral, Shallow layer and Computational Fluid Dynamics (CFD) models. The most popular are integral and CFD models: some analysis of their advantages and disadvantages is reported (Coldrick et al., 2009). One critical conclusion is that no model should be used without validation against experimental data: it is for this reason that high quality experimental data derived from medium and large scale tests has to be publically available. In Feb. 2009 the research group from HSL reviewed such data and prepared a database (hereafter called the HSL database) of experiments that they recommend are suitable for validating LNG vapor dispersion models. These data meet the quality standards necessary in a software validation context. The HSL database contains 33 data sets obtained during 8 test series. Five of these series involve field experiments, whereas three were performed under controlled, and therefore more repeatable, conditions in wind tunnels. Only 4 of the field trials involve releases of LNG. The others involve dense gases such as cold nitrogen, carbon dioxide, sulfur hexafluoride or Freon 12. Full details are provided (Coldrick et al., 2009).

4. Brayton Fire Training Field LNG Spill Tests

Between 2005 and 2009 the Mary Kay O'Connor Process Safety Center (MKOPSC) at Texas A&M University, together with Texas Engineering Extension Service, has performed several medium scale experiments at Brayton Fire Training Field (BFTF) in College Station, TX. All of these tests were with LNG containing 98%-99.8% methane: the experiments were highly instrumented with thermocouples, gas detectors, water and LNG flow meters, anemometers, and radiation sensors. These were situated both in the immediate area of the experiment as well as in the downwind direction. Meteorological data were recorded by Davis weather stations at different elevations. These experiments were part of research sponsored by BP Global Gas SPU. The focus of the tests was in the following three areas:

1) LNG vapor dispersion: studying key parameters for CFD modeling of LNG vapor dispersion and obtaining a relevant database for model validation.
2) Water curtain application: examining the effectiveness of water curtains for mitigating horizontal dispersion of LNG vapor and studying the major physical mechanisms of gas-liquid interaction in two specific types of water curtain.
3) Application of high expansion foam: studying the effectiveness of foam for vapor control and pool fire mitigation.

In each day of testing approximately 11,000 US gallons (41 m^3) of LNG was spilled into purpose built concrete pits: in some cases these contained water. Some analysis of the experiments has been reported by Cormier (2008)

4.1. BFTF LNG Test Facilities

BFTF provides education and training for firefighters and emergency responders so that they can cope with various hazards including those associated with LNG spills. These include the cryogenic hazard, flammable vapor cloud dispersion, and pool fires. LNG props were therefore developed with the sponsorship of BP to create four scenarios for trainees. The props are composed of three concrete pits and one L-shape trench: these are illustrated in the photograph shown in Figure 1.

4.2. 2009 BFTF Test Setup and Summary

Our most recent tests were performed on March 5th 2009 and have not been previously reported. A total of 41m^3 of LNG was released onto water or into concrete pits with the supply rate varying from 80 to 90 US gpm (0.3 to 0.34 m^3/min). Pit II with dimensions of 10.1m × 6.4m × 1.2m, was used to perform the vapor dispersion tests. The pit was filled with water to the brim: the large volume of water below the LNG promotes rapid vaporization and ensures that the LNG evaporation rate is essentially equal to its discharge rate. Both 1.2 and 1.8m solid, wooden, vapor fences were erected in turn around the pit to assess their mitigation effects on vapor dispersion. LNG was delivered from a tanker via an insulated 60-m fixed line and released onto water via an L-shaped, downward facing pipe. In a separate experiment pit I was also filled with water and used

1. Pit I: 10ft × 10ft × 4ft
2. Pit II: 33ft × 21ft × 4ft
3. L-shape Trench
4. Pit III: 22ft × 22ft × 8ft

Figure 1: Brayton Fire Training Field (BFTF) LNG training props.

to conduct water curtain application tests. LNG was also released onto the water surface but in this case was constrained to a 1.5m × 1.5m area by a wooden frame on the surface of the water. The performance of full cone and 2-D water sprays was studied. Finally, pit II was used for a high expansion foam application test. LNG was fed onto the dry concrete bottom of the pit until there was a 20 cm deep pool. Two foam generators were installed 1.2 m above the pit so as to be able to apply high expansion foam to a pool fire or in case of any emergency. Several pictures are shown in Figure 2.

Figure 2: On-site pictures of tests. From left to right: Pit II with an LNG vapor cloud during pit filling, vapor fences shown and foam generator at the left: Pit I with LNG delivery line and wooden frame to limit the LNG pool dimensions: Application of High expansion foam to burning LNG in pit II.

Important findings resulted from the detailed analysis of the test data: some of these are listed below:

- A humidity gradient exists at the boundary of the LNG vapor cloud: within this the humidity changes from the free atmosphere value to the saturated state.
- The turbulence above the LNG pool results from evaporation, buoyancy changes, wind turbulence and the recirculation of the vapor cloud within the impounded area. This parameter has a strong influence on the shape of the vapor cloud near to the LNG source and therefore affects the downwind hazard.
  impact area. The intensity of the turbulence was measured using both ultrasonic and mechanical anemometers and ranged between 10% and 100%.
- The full cone fan spray generates more gas phase turbulence than the flat two dimensional spray and dilutes the LNG vapor concentration mainly by augmenting turbulent mixing with air. By way of contrast the two dimensional fan spray creates a partial barrier in the path of the LNG cloud and alters the free gas dispersion of LNG by imparting upward momentum to it as well as transferring heat.
- High expansion foam applied to burning LNG reduced the downwind back-radiation heat flux from the fire by up to 91%. Likewise, under the conditions of the BFTF tests, the crosswind distance to the location at which the radiation is 5 kWm^{-2} was reduced by approximately 72%: this was estimated based on the recordings from 2 fixed radiometers and 2 portable radiometers located at specific distances crosswind from the pool fire.
- Details of data from these tests are being prepared in the standard format advocated by HSL, see Coldrick et al, 2009.

5. Numerical Simulation of LNG Vapor Dispersion Using the CFX Code

The simulation of LNG vapor dispersion with ANSYS CFX is described for one March 5th 2009 BFTF test - the dispersion of vapor from pit II following LNG release onto water with a 1.2 m high surrounding vapor fence. Data were used to set up key aspects of the CFD simulation and to validate its results. Essential inputs to the simulation include atmospheric conditions, the LNG pool area and shape, the vapor temperature and the turbulence intensity in the source term. A computation domain of 60m × 85m × 15m was created to define the boundary of the dispersion region. Within this the governing equations for fluid flow and heat transfer were solved. Geometric features, including the vapor fences and foam generators, as shown in Figure 3(a), were constructed inside the domain which was then discretized into control volumes using the CFX-mesh generating software with structured grids near walls and unstructured grids in the bulk of the domain. This resulted in 1.09×10^6 control volumes with almost 99% being tetrahedra and the balance wedges. The geometry construction and meshing details for this test scenario are illustrated in Figure 3(b).

The boundary condition at the extremity of the computational grid was set as "open" thus allowing flow into, or out of, the domain. Wind velocity, temperature and turbulence profiles were developed with Monin-Obukhov similarity theory and weather measurements at two elevations so as to represent the key meteorological data. The LNG pool area and shape, evaporation rate, vapor temperature and turbulence in the source term were the key inputs when setting up the LNG pool boundary. A semi-circular area at the base of the computational grid surrounded by the fence was specified

Figure 3: (a) Test setup and (b) the CFD grid representation of the geometry.

as the inlet boundary at which the LNG vapor flows into the computational domain. This shape resulted from the effects of wind and pit edge and its dimensions were determined from the readings of thermocouple very close to the water surface. The vapor temperature at this boundary was taken as 111 K, the atmospheric boiling temperature of LNG. Due to the large thermal mass of the water, the evaporation rate of LNG is essentially equal to its discharge rate. The "ground" was set to be a no-slip boundary, where obstacles (vapor fence and foam generators) are represented as a feature of the "ground." The test data used to characterise the aforementioned boundary conditions are summarized in Table 1.

Table 1: Summary of key test data

Parameter	Value
LNG flow rate, m^3/s	0.00536 ± 0.00032
LNG pool area, m^2	28.8[a]
Average wind speed & direction @ 2.4m, m/s	3.3±1.15 from the South
Average wind speed & direction @ 10m, m/s	5.1 ± 0.88 from the South
Air temperature @ 2.4m and 10m, K	299.95 and 299.35
Absolute air pressure, Pa	102258.2
Free air relative humidity, %	46
Atmospheric stability class	C
Monin-Obukhov length, m	-116.5
Roughness height, m	0.2

[a] estimated from temperature recordings of thermocouples installed at the water surface

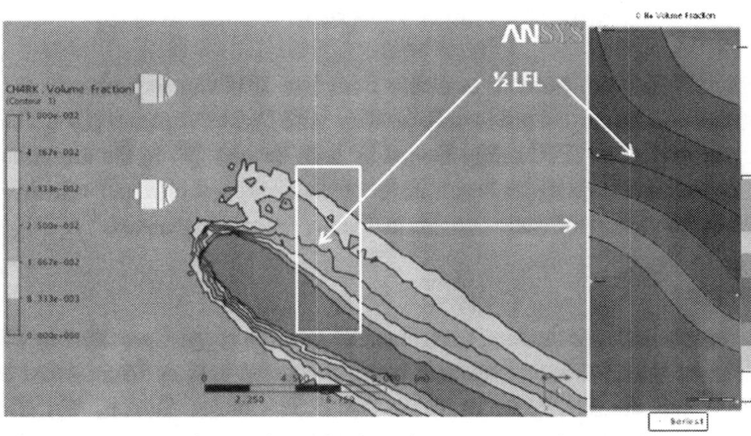

Figure 4: Comparison of 10 ft elevation concentration profiles between simulation output (left) and the test data (right).

The simulation was set up with the specific inputs pertinent to the 5th March 2009 test using the approach described in previous work (Cormier et al., 2009). The simulation was for the steady state and the convergence criteria set as root mean square (RMS)

values of all equation residues of less than 10^{-4}. The results for gas concentrations are compared with the test data in Figure 4. The test data were extracted from the database over a 30-s time period, within which there was little variation in wind speed or direction. The gas detectors were oriented downwind at the start of the test but as it proceeded the wind direction shifted resulting in approximately half of the vapor cloud being outside the detector array. However, from the available half contours, the distance to the simulated ½ Lower Flammable Limit (LFL) concentrations agrees well with the experimental observations.

6. Experiments at Ras Laffan Emergency and Safety College (RLESC)

Gas processing in Qatar is centred in Ras Laffan Industrial City (RLIC) some 80km North of Doha where construction of the RLESC commenced in the second quarter of 2009. The College will provide specialized training for emergency responders within RLIC as well as those who will deal with civil emergencies. Many of the amenities are modeled on those at BFTF and props for LNG spills and fires are included. These will allow trainees to hone fire fighting skills in realistic situations. Whilst the primary purpose of these is for training, some additional features have been incorporated into their design to enhance their value for TAMUQ's LNG research. These include embedded thermocouples and heat flux sensors in and around the concrete LNG pits, local weather stations, and marshalling junction boxes for interfacing the many sensors with our data acquisition system. Hydrocarbon video cameras will be available.

As soon as RLESC facilities become available from late 2010 our experiments will move to Qatar and will be carried out in collaboration with Qatar Petroleum (QP). This is part of a 5 year project on LNG safety funded by both BP and QP. In the meantime smaller scale measurements on major heat transfer mechanisms and on vapor velocities and trajectories in the vicinity of water curtains and vapor fences are planned.

Research Support

We would like to acknowledge the long term financial research support provided by BP Global Gas SPU for the LNG safety research being conducted both at Texas A&M at Qatar (TAMUQ) and the MKOPSC at Texas A&M in College Station. We also acknowledge the support of QP in the form of the facilities to be used for the medium scale experiments to be conducted at RLESC and the supply of the LNG for these tests.

References

BP, June 2009, BP Statistical Review of World Energy, available online at www.bp.com/statisticalreview.

S. Coldrick, C.J. Lea and M.J. Ivings, 2009, Validation database for evaluating vapor dispersion models for safety analysis of LNG facilities, Guide to the LNG model validation database, Health and Safety Laboratory report for the Fire Protection Research Foundation.

B.R. Cormier, R. Qi, G. Yun, Y. Zhang and M.S. Mannan, 2009, Application of Computational Fluid Dynamics for LNG Vapor Dispersion Modeling: A Key Parameters Study, Journal of Loss Prevention in the Process Industries, 22, 332-352.

B.R. Cormier, 2008, Computational Fluid Dynamics for LNG Vapor Dispersion Modeling: A Key Parameters Study, PhD dissertation, Texas A&M University.

J.A. Havens, T.O. Spicer, 2007, United States regulations for siting LNG terminals: problems and potential, Journal of Hazardous Materials, 140, 439 – 443.

M.J. Ivings, S.F. Jagger, C.J. Lea and D.M. Webber, 2007, Evaluating vapor dispersion models for safety analysis of LNG facilities, Health and Safety Laboratory report for the Fire Protection Research Foundation.

A. Luketa-Hanlin, M. Hightower and S. Attaway, 2008, Breach and safety analysis of spills over water from large liquefied natural gas carriers, Sandia report SAND2008-3153.

National Fire Protection Association (NFPA), 2009, NFPA 59A: Standard for the production, storage, and handling of liquefied natural gas (LNG), http://www.nfpa.org.

United States Government Accountability Office (GAO), 2007, Maritime security reports: Public safety consequences of a liquefied natural gas spill need clarification, report 07-633T and Public safety consequences of a terrorist attack on a tanker carrying liquefied natural gas need clarification, report 07-316.

Proceedings of the 2nd Annual Gas Processing Symposium
Farid Benyahia and Fadwa T. Eljack (Editors)

Optimization of Flare Header Platform Design in a Liquefied Natural Gas Plant

E. Mahdi, K. Nasser and M. Gharbia

Qatar University,P.O. Box. 2713 Doha-Qatar

Abstract

The high demand of energy in the world made natural gas producers construct mass production gas liquefaction plants. This has resulted in new challenges in the design of these oil/gas plants. Flare headers are one of the substantial elements in oil/ gas plant where unneeded gases and light liquids are burned into the atmosphere. This paper is devoted to investigate a case of flare header misalignment which has been classified as a serious problem in LNG plants. It is well-known that failure in platform of flare headers will pose a risk to the production, employees and the environmental impact. To this end linear and nonlinear finite element analysis models have been developed to study the effect of platform support type and geometry on the flare header misalignment. Results showed that under non-cyclic operating pressure the flare header behaves elastically and the observed misalignment could be due to cyclic operating pressure. It is determined that the misalignment in flare header platform could be due to the fact that the pipeline is subjected to cyclic pressure rather than static operating pressure. The later can generates different of failure modes.

Keywords: nonlinear Finite Element Analysis, flare header, optimization, pipe support, stresses analysis.

1. Introduction

Flares are considered as the last defence line in plants that produces flammable products. One of its major duties is burn harmful, unneeded gases and light liquids into the atmosphere. A typical flare is simply a set of equipments that safely combusts waste hydrocarbons at a pressure drop which does not compromise plant relief systems or can not be utilized. In a perfect operating condition, flare would be eliminated as the waste hydrocarbons could become products (American Petroleum Institute, 1996; FEAL internal Report 'Plane Strain', 1986). However, facilities to recover large amounts of released hydrocarbons under emergency conditions are currently not feasible. The typical flaring system handles the discharge of all reliefs inside a designated plant or number of plants. A flaring system generally consists of the following major components:

1. A knock-out drums to recover liquid hydrocarbon from the gas stream
2. A flare header to the site
3. An anti-surge system (liquid seal)
4. A flare stack with flare tip

5. A fuel gas system for pilots together with igniters

A Liquefied Natural Gas company has faced a serious problem when a flare header is started to move away of its normal track in a new mass production plant. Figure 1 shows a misalignment in flare header. This case was classified as a serious problem due to the criticality of liquefied natural gas plants and the consequences of such misalignment may lead to catastrophic failure. After performing an extensive investigation, it was proven that the stress analysis and pipe supports design need to be reassessed. The main features of the flare headers are listed in Table 1:

Table 1: Flare header properties

Item	Value	Remarks
Service Fluid	Dry Gas	
Material	SS 304L	SS: Stainless Steel
Max Operating Pressure	2.5 bar	Design pressure: 3 bar (never reached)
Outer Diameter	64"	Assuming uniform diameter
Pipeline Thickness	9.5 mm	Assuming uniform thickness
Length in X axis	157 m	Full length of the pipeline =183.8 m
Length in Z axis	26.8 m	
Elevation of the pipeline	15.3 m	As per Isometric (slopes neglected)
Number of supports	24 Supports	Located at different distances
No. of Anchor Supports	2	Located at first and last supports

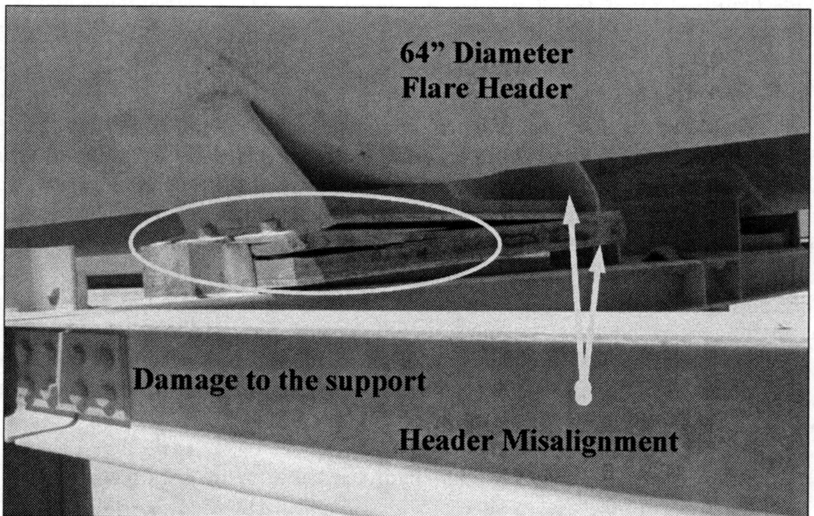

Figure 1: Misaligned flare header

2. Finite Element Model Development

2.1. Finite Element Analysis

Finite element analysis is a fairly recent discipline crossing the boundaries of mathematics, physics, and engineering. This method which is based on numerical analysis has wide application and enjoys extensive utilization of the structural, thermal and fluid analysis areas. The finite element method is comprised of three major phases (Guide for Pressure-Relieving and Depressuring Systems, 1997). These are: pre-processing phase, in which the analyst develops a finite element mesh to divide the subject geometry into sub-domains for mathematical analysis, and applies material properties and boundary conditions, solution phase, during which the program derives the governing matrix equations from the model and solves for the primary quantities and post-processing, in which the analyst checks the validity of the solution, examines the values of primary quantities (such as displacements and stresses), and derives and examines additional quantities (such as specialized stresses and error indicators).

2.2. Model development

The finite element simulation was designed to predict the deformation and stresses throughout the flare header complete and semi-circular curved compound systems. Detailed three-dimensional finite element models of a full expansion loop of the flare header was developed using LUSAS finite element package (LUSAS Modular,). Typical meshes generated and it was found that the model consists of 54642 nodes and 5754 elements as shown in Figure 2. Figure 3 shows the flow chart of finite element analysis procedures.

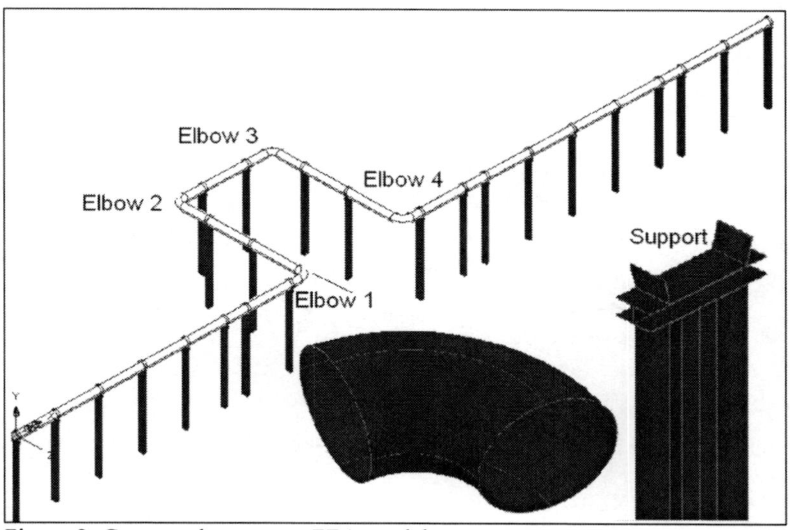

Figure 2: Generated geometry FEA model

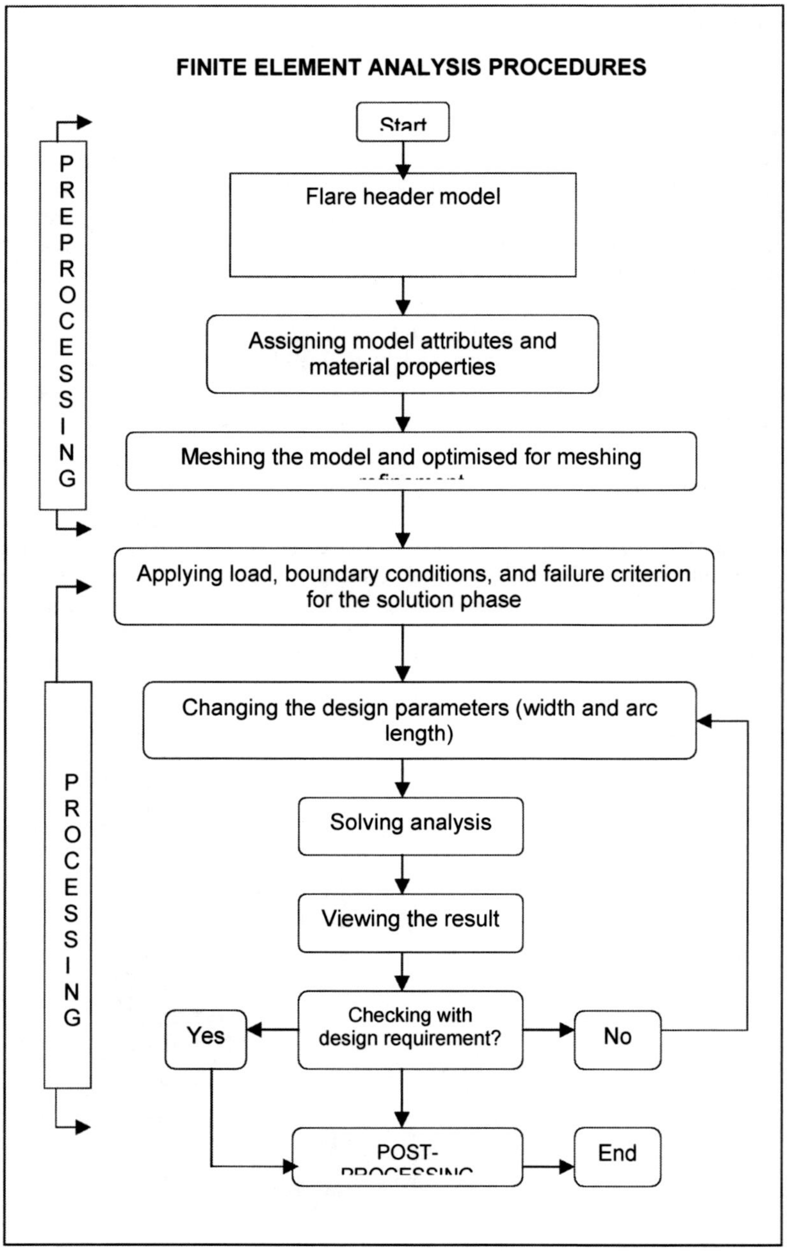

Figure 3: The Flow Chart of Finite Element Analysis Procedures

Hexahedral element (HX20) with twenty nodes was used since it was expected to give accurate stress and strain results. This type of element also includes out of plane stresses effects. Hence, HX20 element was chosen for modelling the flare header system, for static non-linear analysis prediction. Each node has three degrees of freedom, which include three displacements, u_x, u_y, u_z. With out of plane stress effects, this element is expected to be more accurate compared with the other stress element. Fixing degrees of freedom at two ends of anchor supports and applying the internal pressure load throughout header by using a face load simulated the operating conditions.

3. Results and Discussion

3.1. Linear Analysis

The main objective of conducting this analysis was to optimize the supports' width and arc length... This study was conducted by building and simulating several models which have different supports widths. After finishing the design simulation process, the displacement results were recorded and represented as viewed in Figure 4 which shows the variation of displacement with changing the (support width over total pipeline length) ratio. The results shows that the optimum support width for this particular case was in the range of 40 to 50 cm. the existing supports were conforming to these results.

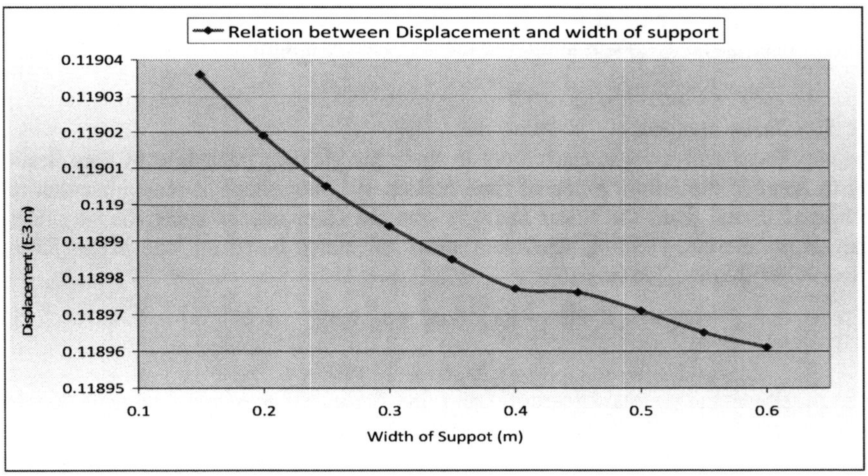

Figure 4: The effect of varying the support width with the Lateral Displacement DZ

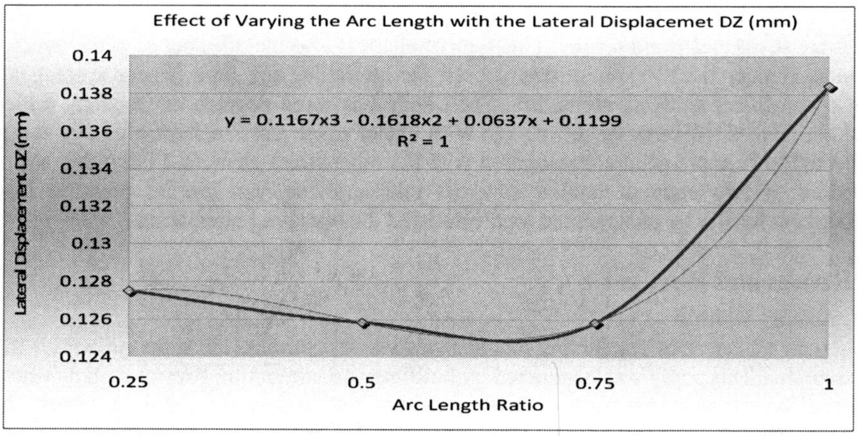

Figure 5: The effect of varying the support arc length with the lateral displacement DZ

Figure 5 shows the variation between the arc length ratio and lateral displacement. The optimum value was around 0.65 of a circle or 235 degrees. It is very important to mention that this analysis was similar to the one used in optimizing the support width.

3.2. Non-linear Analysis

The non-linear analysis was carried out to study the yielding behaviour of flare header and to identify the failure mode of flare header. It is important to remember that the optimized model from the linear analysis was the candidate to study the non-linear analysis on. Herein, yielding equivalent stress calculated based on Von Mises failure criterion as (Shigley, 2004):

$$\sigma_y^2 = (\sigma_1 + \sigma_2)^2 + (\sigma_1 + \sigma_3)^2 + (\sigma_2 + \sigma_3)^2 + \sigma_1^2 + \sigma_2^2 + \sigma_3^2 \quad (1)$$

Figure 6 shows the equivalent stress. From this figure, the maximum equivalent stress found to be 317 MPa, at which the header will be no more withstand any pressure and is expected to experience burst failure. It is interesting to note that the header start to yield at equivalent stress (Von Mises stress) of 173 MPa, where the header experienced very low plastic deformation.

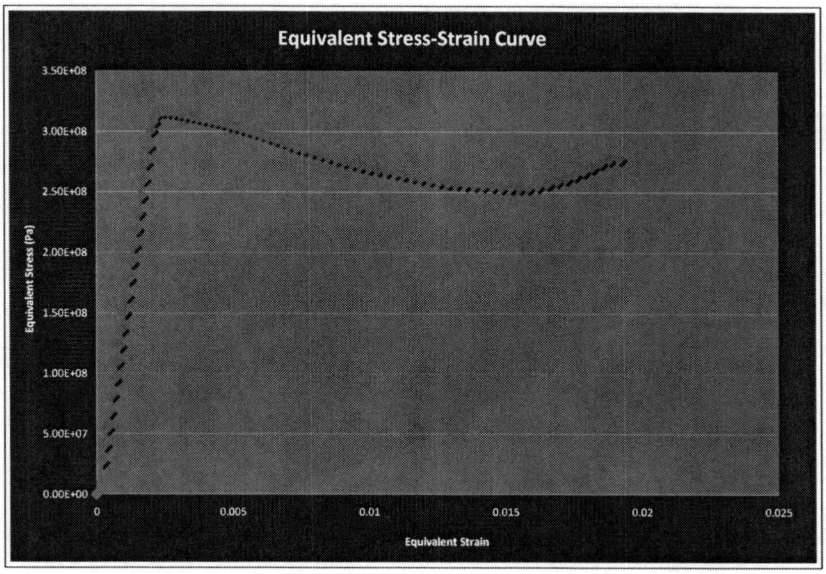

Figure 6: Equivalent stress-strain curve

3.2.1. Yielding behaviour

It is computed that the structure start to yield at load case 12, where the pressure calculated to be 11.25 bar, while the flare header maximum operating pressure is designed to be 2.5 bar. This indicates that the flare header misalignment was not due to operating pressure, since it is far away from yielding pressure. This result leads to a strong probability that the misalignment is due to impact hammer, which is normally caused by the variation in natural gas flow rate. This might cause high vibration levels on the flare tip caused the platform to vibrate, posing a risk to personnel and production. It is worth to mention that in design of pipeline, elbows are suspicious to be stress concentrating areas. Therefore, it will be very much interesting to investigate the equivalent stress concentration throughout the elbows. Figure 7 shows the equivalent stress contours at elbow 1 as representative. It can be seen that very high equivalent stresses were found to be at the joint between the elbow and the pipe. The maximum equivalent stress value computed to be 72.514 MPa at node 465 of elbow1. The recorded maximum equivalent stress at elbow is still lower than the yielding stress (173MPa). This is supporting the finding that the static operating pressure is not the cause of plastic permanent deformation (i.e. misalignment).

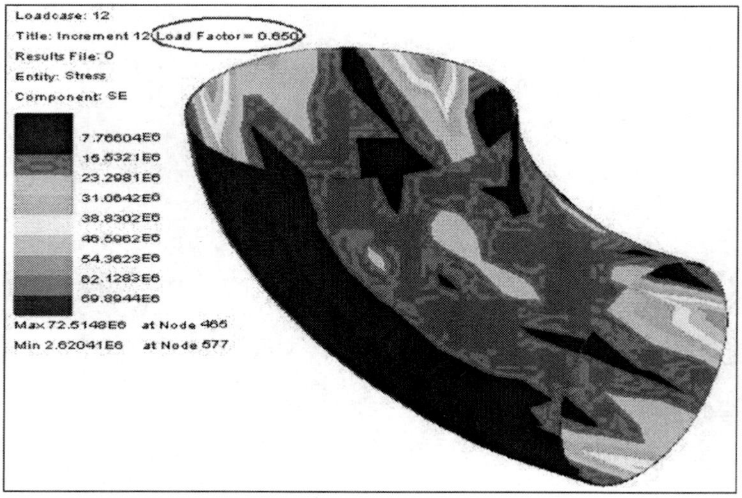

Figure 7. Equivalent stress countor at Elbow 1

3.2.2. Lateral Displacement

Figure 8 shows the variation of lateral displacement (i.e. translation displacement along z-axis). It can be seen that the lateral displacement reached its maximum value of 2.34mm at elbows 1 and 4. The abrupt changing in pipeline structures lead to high pressure in elbows 1 and 4, which in turns lead to high lateral displacement in elbows 1 and 4.

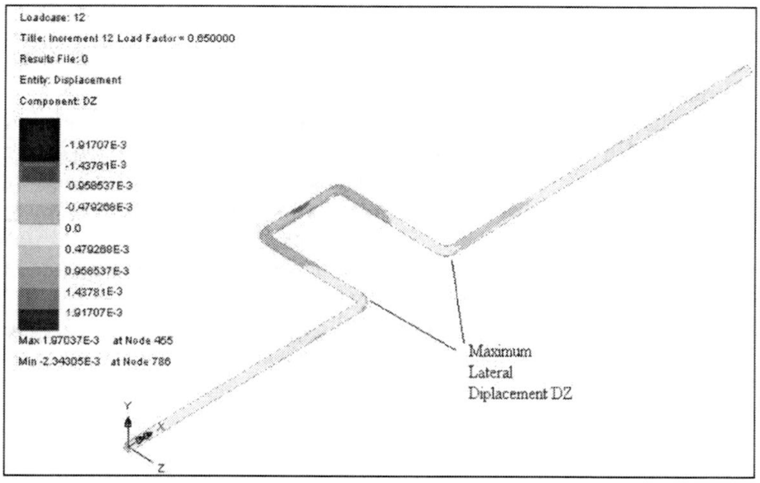

Figure 8: Lateral displacement contour through the flare header

4. Conclusion

Linear and nonlinear finite element analyses were carried out to investigate misalignment in flare header plat form. It is determined that the misalignment in form could be due to the fact that the pipeline is subjected to cyclic pressure, which in turns generates different of failure modes. Modal finite element analysis has to be carried out to identify the vibration mode shapes to determine stresses in the system for a fatigue assessment.

References

American Petroleum Institute, 1996, Introduction to Oil and Gas Production, Information Handling Services, fifth edition, USA

FEAL internal Report 'Plane Strain', 1986, FEAL708, Finite Element Analysis Ltd. Forge House, 66 High Street, Kingston Upon Thames, Surrey KT1 1HN.

Guide for Pressure-Relieving and Depressuring Systems, 1997, API Recommended Practice 521, 4th Ed., American Petroleum Institute, March.

LUSAS Modular, FEA Software Version 14.0 -3, FEA Ltd.

Shigley, Joseph E., 2004, Standard Handbook of Machine Design, 3rd Edition, McGraw-Hill.

Proceedings of the 2nd Annual Gas Processing Symposium
Farid Benyahia and Fadwa T. Eljack (Editors)

A Method to Design an Advanced Gas-to-Liquid Technology Reactor for Fischer-Tropsch Synthesis

Nimir O. Elbashir[a], Fadwa T. Eljack[b]

[a]*Chemical Engineering Program, Texas A&M University at Qatar, P.O. Box 23874, Doha, Qatar*

[b]*Gas Processing Center, Qatar University, P.O Box 2713, Doha, Qatar*

Abstract

This paper describes a novel approach to the design of an advanced Fischer Tropsch synthesis (FTS) reactor technology that consists of several units including reactions as well as separation processes and where streams are recycled between different units. This advanced reactor is designed to facilitate conducting FTS reaction in both conventional media (gas-phase FTS and liquid-phase FTS (e.g. trickle bed)) and non-conventional media (near critical and supercritical-phase FTS). The suggested design approach is composed of several phases that incorporate fundamentals studies on phase behavior of FTS reaction mixture, development of reaction pathway, kinetics and modeling for the non conventional FTS media, and advanced process integration and process control studies for the complex dynamic system, which is non-trivial to control. The outcome of the synthesis and simulation models will be used to build the reactor bed whereby all experimental investigation will be conducted. The ultimate goal of this work is to design an advanced FTS reactor technology to operate in both conventional and non-conventional fluid media.

Keywords: Gas-to-Liquid technology, Fischer-Tropsch synthesis, reactor design, process integration and optimization, and process control

1. Introduction

During the past twenty years, tremendous investments from the major players in the energy market have been directed towards Gas-to-Liquid (GTL) technology. This has resulted from a significant increase in global demand of crude oil. The advantages of GTL technology in energy production are that it furnishes a broad range of environmentally clean fuels, additives and value-added chemicals that can be obtained from abundant natural resources such as natural gas, coal, or biomass. The primary commercial technology of GTL for the production of liquid hydrocarbons and value added chemicals involves the Fisher-Tropsch synthesis (FTS) process. The typical steps in the GTL process include the gasification or reforming process to obtain synthesis gas or syngas (consisting primarily of a carbon monoxide and hydrogen mixture), followed by the synthesis of hydrocarbons and oxygenates from syngas via FTS technology, and finally the upgrading of liquids hydrocarbons (known as syncrude) via hydrocracking, isomerisation, and/or hydro-isomerisation processes to produce clean kerosene, naphtha, gasoline, diesel, base oil, and value added chemicals. FTS technology is the heart of the GTL process and was developed in early 1923 by Franz Fischer (1877-1947) and Hans Tropsch (1889-1935) (Fischer and Tropsch, 1926). FTS is a classical heterogeneous reaction whereby CO and H_2 molecules react on the surface of an active metal (Co, Fe, Ru, Ni, etc.) to produce chains of higher gaseous, liquid, and solid hydrocarbons composed of paraffins, olefins, and oxygenated compounds. This invention was soon upgraded to an industrial plant by Ruhrchemie A.G. (Oberhausen, Germany) in 1935. Within a short period of time, eight industrial plants using Fischer-Tropsch technology for fuel production were operating in Germany (Pichler, 1952).

Diesel fractions from FTS have unique characteristics that include very low sulfur and aromatic content, high cetane index and exceptional clean burning in compression-ignition engine. Compared to crude oil derived fuel fractions, FTS fuels such as diesel and jet fuels have shown to reduce the emissions of carbon monoxide, nitrogen oxides, hydrocarbons and other particulates. The first generation of FTS reactors was the fixed-bed-reactors (FBR). The latest commercial version of an FBR reactor is the multitubular reactor of Shell's Middle Distillate Synthesis (SMDS^{TM}) technology. This is part of Shell's Bintulu plant in Malaysia (an upgraded version of this reactor will be used in Shell PearlGTL plant, which is the world's largest GTL plant located in Ras Laffan, Qatar) (Editor, 2003). There is a better opportunity to control catalyst size and shape in FBR to facilitate high diffusivity of reactant gases into the catalyst pores, thereby resulting in higher reaction rates and thus higher syngas conversions per pass. In addition, the fixed bed reactor configurations are straightforward to operate over wide ranges of temperatures and pressures. Nevertheless, major challenges face FBR and multitubular FBR reactors that involve the efficient removal of reaction heat from these highly exothermic reactions and hence affect the ability to control a reasonably uniform temperature profile inside reactor bed(Lang et. al, 1995). Therefore, these reactors are equipped with high load heat exchangers to cool down the catalyst bed. The second generation of FTS reactors was the fluidized-bed reactor (either circulated fluidized-bed or fixed-fluidized bed), which are the FTS technology that Sasol and its partners played major role in its invention and commercialization. Sasol then introduced in 1999 the fixed-fluidized bed, referred to as the advanced synthol (SAS), and reported several advantages of this technology over the circulated fluidized bed reactors (Duvenhage and Shingles, 2002). The latest FTS technology is the slurry phase reactors that provide an excellent temperature distribution for this highly exothermic reaction than the fixed-bed-reactor. This is considered a major advantage of slurry reactors over other FTS reactor technologies; furthermore, they operate at moderate temperatures favoring the production of middle distillate hydrocarbons (specifically ultra-clean diesel fuel fraction). Sasol Oryx GTL plant in Qatar is based on slurry reactor (Slurry Phase Distillate or SPDTM technology), and ExxonMobil has also invested in the design of slurry FTS reactor as part of their advanced GTL technology known as AGC21TM (Advanced Gas Conversion 21st Century).

In a recent study, Elbashir et al. (2009) have reported technical comparison between the advantages and disadvantages of each FTS commercial reactor technologies. It is clear from the brief review above of commercial FTS technologies that the selection of a proper reactor technology for a given GTL plant was always a non-optimal compromise between the advantages and disadvantages of each existing reactor technology (de Deug et. al, 2003). Industry urgently needs the development of an advanced FTS technology to overcome the limitations of the existing, conventional FTS technology platforms as was well described in the presentation of Dr. Ben Jager at the 2003 spring meeting of the American Institute of Chemical Engineering in New Orleans. Dr. Jager is a distinguished scientist in the field of GTL technology and a leading researcher from Sasol who won numerous awards for his contribution in designing Sasol's slurry bubble column reactor for low temperature FTS, and for the development of an innovative concept in the use of advanced autothermal reforming. Dr. Jager concluded his presentation by making the following statements (1) GTL technology is at early stage of development; (2) there is an incentive for improved FTS technology; and (3) new FTS reactors are early on the learning curve (Jager,2003).

This paper briefly introduces a new approach to the design of advanced FTS technology utilizing non conventional reaction media. The background of this approach has been described in details elsewhere (Elbashir et. al, 2009).

2. Novel Approach to the Design of Advanced Fischer-Tropsch Reactor Technology

The challenges facing the development of new GTL technology can be attributed to many fundamental reasons including the complex nature of the synthesis process and reactions involved as well as certain technical difficulties in controlling the thermophysical characteristics of the reaction in conventional reaction media (i.e. gas-phase or liquid-phase). While the GTL market continues to grow, there are still many hurdles to overcome in order to optimize the use of this important technology – all of which are related to technology limitations. Because of complexity in reactor configuration and process control, commercial GTL technologies requires tremendous financial and operational support, which is available only to large corporations, while small companies are still seeking the development of profitable small scale GTL plants with fewer technical complications.

On the other hand, the classical development route for a reactor technology in industry normally starts with a lab-scale reactor to gather basic information on kinetics, catalyst stability, and product distribution and then moves to a pilot-plant scale whereby the reactor configuration and typical operating conditions for long time-on-stream (TOS) can be evaluated prior to building a commercial scale plant. The final product (commercial reactor) goes through a series of screening processes utilizing different approaches that mainly depend on technology development methods designed by the research team (composed of chemists, engineers and technicians); these techniques are normally based on team members expertise in FTS technology development and prior accumulated experiences in upgrading reactor technology. Unfortunately, very little information about these experiences finds their way into the open literature because of obvious reasons, such as protection of trade secrets from competition.

The proposed process here for FTS consists of several units that include reactions as well as separation processes and where streams are recycled between different units. Such a design has to be considered as a complex dynamic system which is non-trivial to control (see Figure 1). These units are designed to facilitate running FTS reaction in both conventional media (gas-phase FTS and liquid-phase FTS (trickle bed)) as well as for non-conventional media (near-critical and supercritical FTS phase). The advantages of conducting FTS in the non conventional media have been explained in details

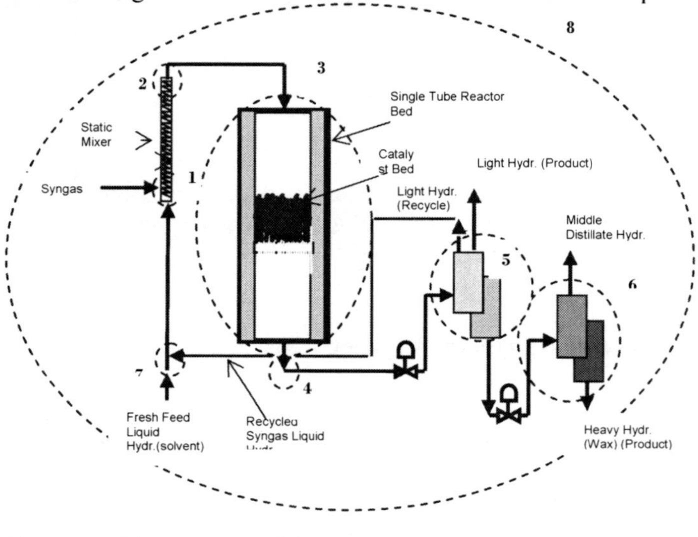

Figure 1: Diagram describing an advanced Fischer-Tropsch synthesis process for operation in near critical and supercritical phase media.

Description of the units in the advanced FTS reactor configuration

1. Inlet Static Mixer	**2.** Outlet Static Mixer Inlet Reactor Bed	**3.** Reactor Bed	**4.** Reactor Outlet
. Separation Process I & Solvent Recovery	**6.** Separation Process II	**7.** Fresh and Recycled Solvent Mixing	**8.** Overall Proces Energy Control Setup

Elsewhere (Elbashir et. al, 2009; Elbashir et. al, 2010; Elbashir and Roberts, 2005). The parametric studies needed for the design includes: thermodynamics and phase behavior of non-ideal reaction mixtures; reaction kinetics of the synthesis process as well as studies on catalyst performance, stability, and characteristics; process control studies; energy integration and process optimization for individual units as well as for the overall process. Highlights of these studies are illustrated in Elbashir et al. (2006), while units and processes under considerations are numbered and identified by dotted circles as shown in Figure 1.

A novel strategy to design this technology is proposed and includes the incorporation of fundamental studies on the phase behavior of the near-critical and supercritical phase FTS, development of reaction pathway, kinetics and modeling for the non-ideal SCF-FTS medium, and advanced process integration and process control studies. The outcome of the synthesis and simulation models will be used to build the reactor bed whereby all experimental investigation will be conducted. Figure 2 illustrates the interrelation between the different phases of this project and evaluation criteria for each outcome of the project at its different stages.

A. *Synthesis and Simulation Models:* The models will be develop in two phases: Phase one (I) of this project is to synthesize a base-case flowsheet of FTS process that includes the key processing reaction and separation units for fixed bed and slurry reactor technologies. This simulation will allow for the determination of process variables needed to model a typical FTS reactor. The developed models will be verified using Qatari Data from the most advanced existing industrial FTS reactor, Shell's Pearl Project (Fixed Bed Reactor) and ChevronSasol's OryxGTL Project (Slurry Bed Reactor).

   Phase two (II) of this project, will include the synthesis of a novel reactor configuration for FTS that maximizes the advantages of the existing technologies while overcoming their limitations using high pressure conditions for near critical and supercritical phase. The reactor configuration is depicted in Figure 1. The second phase of the project will also compare mass and energy flows, performance criteria, and equipment specification of the novel reactor technology to the existing commercial technologies (fixed-bed reactors and slurry reactors).
   At this stage it will be feasible to verify whether the proposed reactor configuration will achieve the expected performance; otherwise, changes to the flowsheet and process condition will be suggested accordingly.

B. *Fundamental Studies on Phase Behavior and Reaction Kinetics:* This phase will be carried simultaneously with the Synthesis Simulation Model to generate necessary data needed to quantify the influence of reaction parameters such as temperature and pressure on phase behavior of the reaction mixtures and on the kinetics of the non-ideal reaction media. Generated data will continuously be fed for the Synthesis and Simulation process to model reactor bed on typical reaction conditions similar to those of the unit that will be built.

C. *Optimization of Simulation Models:* The dynamics of the verified flowsheet of the advanced reactor as well as the control methods will be identified at this stage. This process will be coupled with series of optimization studies targeting desired operating conditions, as well as suggesting modifications to reactor setup. The results generated in this phase will be verified via the experimental investigations of the build reactor as described in the next step.

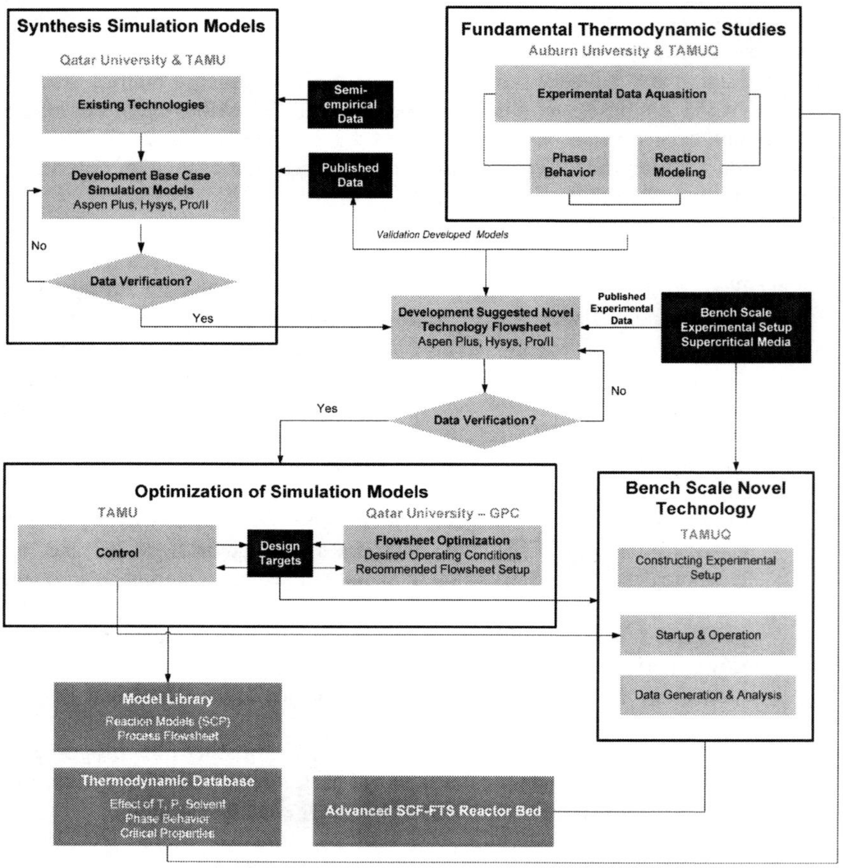

Figure 2: The flowsheet of the different phases of the research and illustration of research approach.

D. *Catalyst Selection for the Process:* Cobalt-based catalysts are known to provide the best compromise between performance and cost for the synthesis of hydrocarbons from syngas (CO/H_2 mixtures). Cobalt catalysts were also found to give the highest yields and longest lifetime producing mainly linear alkanes (paraffins). Since cobalt catalysts are not inhibited by the water-gas-shift (WGS), it gives a high productivity at high syngas conversion(van Berge and Everson, 1997). Optimum catalysts with high cobalt concentration and site density can be prepared by controlled reduction of nitrate precursors introduced via melt or aqueous impregnation methods (Iglesia, 1997). Most of the industrial cobalt-based catalytic systems used in the FTS reaction are supported on either Al_2O_3 or SiO_2. Therefore, cobalt dispersion on the supported catalyst surface (Tauster and Fung, 1978) and the interaction of the Co with the metal oxide support that affect the electronic density as well as the structure of the metal crystallites (Iglesia et al., 1992) are assumed to play significant role on the catalyst performance measured by the hydrocarbon selectivity (C_{5+}) and the catalyst activity. Three cobalt-based catalysts have been selected for this study whereby the activity and selectivity data needed for the model will obtained from the performance of these catalysts on both conventional media and supercritical phase media (Dutta et al., 2004; and Elbashir et al., 2005). The variety in the characteristics (Dutta et al., 2004) of the three different cobalt-based catalytic systems offers our study an opportunity to compare the influence of the catalyst surface characteristics on reaction performance.

E. *Bench Scale Novel Technology:* The construction of the novel FTS reactor unit will be conducted at this stage utilizing the outcomes from the previous phases of this project. The initial set of operating conditions at optimum reactor performance will be used for the start up demonstration. Simulated reactor performance will then be verified experimentally while data generated will be used to further improve model predictability of reaction behavior over wide range of operating conditions.

The aforementioned approach provides several attractive and novel features including:

o *Decomposition and integration of multi-scale models:* Because of the various scales of the important phenomena, it is important to develop the right level of details at each scale. As such, molecular interactions are studied during the thermodynamic and kinetic studies. Meso-scale models are developed for the various process units. Finally, in process control, synthesis, and integration, macroscopic models are developed to connect the various models at the different scales.

o *Integration of theoretical and experimental studies:* The proposed approach provides a systematic framework for linking the modeling, synthesis, control, and optimization work with the experimental work. Insights and guidelines from the theoretical work are used to guide the experimental work and results from the experimental work are incorporated into the theoretical work.

o *Applicability and scale up:* Because of the targeted experimental work, the multi-scale modeling work, and the successive verification of actual data, the developed framework will result in the development of reliable, applicable, and innovative process configurations.

The ultimate goal of this project is to design advanced Fischer-Tropsch reactor technology to operate in both conventional and non-conventional fluid media. The difference between this design approach and others is that it is capable of utilizing the unique properties of near-critical and supercritical-phase reaction media. These media leverage certain advantages of the current commercial technologies (slurry reactor and multitubular reactor) while at the same time overcoming several of their major limitations (Elbashir et. al, 2010). Supercritical solvents have become unique media for chemical reactions because they offer single phase operation with densities that are sufficient to afford substantial dissolution power, while also providing effective diffusivities that are higher than normal liquids and viscosities that are lower than their liquid counterparts (Elbashir et. al, 2009; Elbashir et. al, 2010; Elbashir and Roberts, 2005).

3. Conclusions

The scale-up of novel FTS reactor technologies from lab-scale to pilot plant scale is still a challenge because product yield and selectivity of a large scale reactor requires a deep understanding of phase behavior of reaction mixture, reaction kinetics, and catalytic system and its FTS chemistry. In general, application of non-conventional reaction media in FTS reactor technology has long way to go before commercialization. This paper describes a novel approach in designing and optimizing such complex process that requires multi-disciplinary specialties. The introduction of near-critical and supercritical solvents as a new parameter to influence FTS performance shifts the problem from a classical catalytic reaction engineering process to one combined with a thermophysical puzzle (Elbashir and Roberts, 2005). Better understanding of the phenomena associated with operating under non-conventional FTS reaction conditions, combined with the optimum utilization of those properties in improving the process performance could lead to breakthroughs in advancing SCF-FTS technology.

Acknowledgment

The authors would like to express gratitude to the collaborators in this research work, Dr. Mahmoud El-Halwagi at Texas A&M University, Dr. Christopher Roberts at Auburn University, and Dr. Juergen Hahn, at Texas A&M University for their thoughtful contribution to the work plan. The authors also extended appreciation to scientist and engineers from Sasol, Shell, and ExxonMobil for their thoughtful discussion on this approach. The authors would like to acknowledge the finical support of this project by Qatar National Research Funding under grant (NPRP 08-261-2-082).

References

de Deug, R. M.; Kapteijn, F.; Moulijn, J. A., 2003, Top. Catal, pp. 26, 29-38.
Duvenhage, D. J.; Shingles, T. Catal, 2002, Today, pp. 71, 301-305.
Dutta, P.; Elbashir, N. O.; Manivannan, A.; Seehra, M. S.; Roberts, C. B., 2004, Catal. Letters, pp. 98, 203-210.
Elbashir, N. O.; Bao, B.; and El-Halwagi, M. M., 2009, "An Approach to the Design of Advanced Fischer-Tropsch Reactor for Operation in Near-Critical and Supercritical Phase Media", in H. E. Alfadalla, G. V. Reklaitis, M. M. El-Halwagi Editors "Advances in Gas Processing: Proceedings of the 1st Annual Symposium on Gas Processing". Elsevier, Amsterdam, 1st Ed, pp. 423-433.
Elbashir, N. O.; Bukur, D. B.; Durham, E.; Roberts, C. B., 2010, AIChE J. in press in volume 56 issue 3.
Elbashir, N. O.; Dutta, P.; Manivannan, A.; Seehra, M. S.; Roberts, C. B., 2005, Appl. Catal. A: Gen., pp. 285, 169-180.
Elbashir, N. O.; Roberts, C. B., 2005, Ind. Eng. Chem. Res., pp. 44, 505-521.
Editor, Shell to Build GTl Plant in Qatar. Focus Catal., 2003, issue 12, 6.
Fischer, F.; H., Tropsch, 1926, Brennstoff-Chemie, pp.97-104.
Iglesia E., 1997, Appl. Catal. A: Gen., pp.161, 59.
Iglesia E, Soled S.L., Fiato R.A., J. Catal, 1992, pp. 137, 212.
Jager, B., 2003, American Institute of Chemical Engineering (AIChE) 2003 Spring Meeting, New Orleans.
Lang, X.; Akgeman, A.; Bukur, D. B.,1995, Ind. Eng. Chem. Res., pp.34, 72-77.
Pichler, H., 1952, Twenty-five years of synthesis of gasoline by catalytic conversion of carbon monoxide and hydrogen. New York, N.Y., Academic Press Inc.
Tauster S.J., Fung S.C., 1978, J. Catal., pp. 55, 29.
Van Berge P.J., Everson R.C.,1997, Stud. Surf. Sci. Catal. (Natural Gas Conversion IV), pp. 107 and 207.

Proceedings of the 2nd Annual Gas Processing Symposium
Farid Benyahia and Fadwa T. Eljack (Editors)

Recent Developments in Identification from Step Response

Salim Ahmed

Department of Chemical Engineering, Qatar University, P.O. Box 2713, Doha, Qatar

Abstract

Sustainable operation of process industries depends on the control of the plants and performance of advanced control applications depends on the quality of available process models. For advanced control, transfer function models are widely used and identification from step response is a common choice in process industries to obtain such models. Over the last decade significant developments have taken place in the field of identification from step response using the so called integral equation approach. This work highlights the recent development on this topic and also presents some new results. Common problems encountered in industrial applications of identification namely the presence of disturbances, unsteady initial states, non-ideal nature of the step input are discussed and solutions to these problems are reviewed.

Keywords: step response, identification, integral equation

1. Introduction

Identification from step response is one of the most significant topics in process control. Although there is a concern about its persistency of excitation, the step input might be the most commonly used excitation signal for identification in process industries. This might be due to the simplicity of the test procedure and the advantages the step input offers in the parameter estimation stage.

The early development of system identification centered around identification from step responses that led to the development of a special category of methods, namely the graphical techniques (Oldenbourg and Sartorius, 1948). These methods are commonly used in industrial applications because they involve minimal computation. Due to the limitations of the graphical techniques in terms of their applicability and accuracy (Sundaresan et al., 1978), more computationally involved methods have been developed subsequently.

An important development in identification from step response is the introduction of the integral equation approach (Diamessis, 1965) for simultaneous estimation of the delay proposed by Wang and Zhang (2001). Based on this procedure, over the last few years a number of new methods (Ahmed et al., 2008 and 2007; Liu et al., 2007; Hwang and Lai, 2004) have been proposed to deal with the different practical issues in step response based identification. This article summarizes the development in the integral equation approach based step response identification method. Also, we present new solutions to deal with different practical implementation problems encountered in process industries.

2. Mathematical Formulation

In this section, the mathematical formulations for identification from step response using the integral equation approach are presented. For the purpose of simplicity in the presentation, we discuss the methods using an example of a first order plus time delay model. However, the methods are not limited in application to first order processes.

2.1. Integral Equation Approach

Consider a first order model described by the following differential equation

$$\frac{dy(t)}{dt} + ay(t) = bu(t) + e(t) \tag{1}$$

Here, $y(t)$ and $u(t)$ are the process output and input, respectively. $e(t)$ is the error term, evolving from the measurement noise. The objective of identification is to obtain an estimate of the parameter vector, $[a \;\; b]^T$, from a given set of input and output data $[u(t) \;\; y(t)]$ for $t = t_1, t_2, \cdots t_N$, with N being the length of the data set.

To estimate the model parameters using eqn(1) we need the derivative of the output signal which is not always available. From the measured output, the derivatives are not generally obtained because of the noisy nature of output signals. Integral equation approach is one of the most commonly used method to overcome the problem. In this approach, the differential equation is integrated to express the model equation in terms of the integrals of the signals instead of their derivatives. Integration of eqn(1) once assuming an initial steady state of the output gives

$$y(t) + ay^{[1]}(t) = bu^{[1]}(t) + e^{[1]}(t) \tag{2}$$

For a signal $x(t)$, $x^{[1]}(t)$ is its first order integral, i.e. $x^{[1]}(t) = \int_0^t x(\tau)d\tau$. The estimation equation is then obtained as

$$y(t) = \begin{bmatrix} -y^{[1]}(t) & u^{[1]}(t) \end{bmatrix} \begin{bmatrix} a \\ b \end{bmatrix} + e^{[1]}(t) \tag{3}$$

Or equivalently

$$y(t) = \phi^T(t)\theta + \xi(t) \tag{4}$$

where, $\gamma(t) = y(t)$, $\varphi^T(t) = \begin{bmatrix} -y^{[1]}(t) & u^{[1]}(t) \end{bmatrix}$, $\theta = \begin{bmatrix} a \\ b \end{bmatrix}$, $\xi(t) = e^{[1]}(t)$. Equation (4) can be written for $t = t_1, t_2 \cdots t_N$ and combined to give the estimation equation

$$\Gamma = \Phi\theta + \xi \tag{5}$$

The parameter vector θ is then obtained as the least squares solution

$$\theta = (\Phi^T\Phi)^{-1}\Phi^T\Gamma \tag{6}$$

So far we have considered a general input signal and the methodology is valid for any type of input. However, the step input offers a unique advantage in the estimation of time delay.

2.2 Estimation of time delay

Consider a first order model with delay described by

$$\frac{dy(t)}{dt} + ay(t) = bu(t - \delta) + e(t) \tag{7}$$

where, δ is the time delay. If we follow the above integration approach, we end up with an estimation equation

$$y(t) + ay^{[1]}(t) = bu^{[1]}(t - \delta) + \xi(t) \tag{8}$$

It has been shown in Wang and Zhang (2001) that if the input is a step, the integral of the delayed input term can be decomposed to give an explicit appearance of the delay. For a step input, the integral of the delayed signal can be expressed as

$$u^{[1]}(t - \delta) = ht - h\delta \quad \text{for} \quad t > \delta \tag{9}$$

Equation (8) can be written as

$$y(t) + ay^{[1]}(t) = bht - bh\delta + \xi(t) \tag{10}$$

Then the following estimation equation would give the parameters along with the delay.

$$y(t) = \begin{bmatrix} -y^{[1]}(t) & ht & -h \end{bmatrix} \begin{bmatrix} a \\ b \\ b\delta \end{bmatrix} + \xi(t) \tag{11}$$

2.3 Transient initial conditions

So far we have considered steady initial conditions of the input and the output signals. If the output is at a non-zero initial state $y(0)$, the integrated model equation(7) takes the form

$$y(t) - y(0) + ay^{[1]}(t) = bu^{[1]}(t-\delta) + \xi(t) \tag{12}$$

In the least squares formulation, for a step input we get

$$y(t) = \begin{bmatrix} -y^{[1]}(t) & ht & 1 \end{bmatrix} \begin{bmatrix} a \\ b \\ -hb\delta + y(0) \end{bmatrix} + \xi(t) \tag{13}$$

It has been proposed in Ahmed et al. (2008) that using higher order integration of the model equation an estimation equation can be formulated to give all the unknowns. One more step of integration of eqn(12) yields

$$y^{[1]}(t) - y(0)t + ay^{[2]}(t) = bu^{[2]}(t-\delta) + e^{[2]}(t) \tag{14}$$

For a step input, $u^{[2]}(t-\delta) = \frac{ht^2}{2} - ht\delta + \frac{\delta^2}{2}$ for $t > \delta$. Rearranging the estimation equation we get

$$y^{[1]}(t) = \begin{bmatrix} -y^{[2]}(t) & \frac{ht^2}{2} & t & 1 \end{bmatrix} \begin{bmatrix} a \\ b \\ -hb\delta + y(0) \\ \frac{b\delta^2}{2} \end{bmatrix} + e^{[2]}(t) \tag{15}$$

From the solution of eqn(15) we can get the parameters, the delay and $y(0)$. A step wise procedure has been presented in Ahmed et al. (2008) that uses both eqn(13) and eqn(15) for the solution. In this procedure eqn(13) is solved first and the estimated parameters are then used as known to estimate the last parameter of eqn(15). The step wise procedure gives better estimate of the parameters in terms of bias and variance (Ahmed et al., 2008) .

2.4 Identification using raw data

The methodologies described so far assume that the input and output data are available in their deviation from, i.e. the initial steady state values of the variables are subtracted. However, the initial steady state values are not always correctly known (Ahmed et al, 2007) due to noise. Also the step input might be applied before the process reached at steady state. For a data set with unknown initial steady state (Ahmed et al., 2007) proposed an identification scheme from raw data. Let us consider that $y_r(t)$ denotes the output in its raw form. Then

$$y(t) = y_r(t) - q \quad \text{and} \quad y^{[i]}(t) = y_r^{[i]}(t) - q\frac{t^i}{i!} \tag{16}$$

where, q is the initial unknown steady state value of the output. In terms of the raw data we can write eqn(12) as

$$y_r(t) - q - y(0) + a y_r^{[1]}(t) - aqt = bu^{[1]}(t-\delta) + \xi(t) \tag{17}$$

Or

$$y_r(t) = \begin{bmatrix} -y_r^{[1]}(t) & t & 1 \end{bmatrix} \begin{bmatrix} a \\ bh + aq \\ -hb\delta + y(0) + q \end{bmatrix} + \xi(t) \tag{18}$$

Also eqn(15) can be written as

$$y_r^{[1]}(t) = \begin{bmatrix} -y_r^{[2]}(t) & \frac{t^2}{2} & t & \frac{h}{2} \end{bmatrix} \begin{bmatrix} a \\ bh + aq \\ -hb\delta + y(0) + q \\ b\delta^2 \end{bmatrix} + e^{[2]}(t) \tag{19}$$

We see that from the solution of eqn(19) we cannot extract all the parameters. However, following the sequential integration approach, one more step of integration yields

$$y_r^{[2]}(t) = \begin{bmatrix} -y_r^{[3]}(t) & \frac{t^3}{3!} & \frac{t^2}{2} & \frac{ht}{2} & \frac{h}{3!} \end{bmatrix} \begin{bmatrix} a \\ bh + aq \\ -hb\delta + y(0) + q \\ b\delta^2 \\ b\delta^3 \end{bmatrix} + e^{[3]}(t) \tag{20}$$

Following a step wise solution procedure we can extract all the unknown terms in the above equations.

2.5 Non-ideal step inputs

A step change may not be feasible for some variables or their might have the risk of a process upset. For such a case, variables are often increased gradually during the initial stage or multiple steps are applied. All these result in, what we call, non-ideal step inputs. Certain types of these inputs have been considered under the title of pulse inputs by Hwang and Lail (2004). Piecewise step inputs have been considered in Liu et al. (2007). We will describe the identification methodology for one specific non-ideal step input, namely the saturated ramp because of its wide applications. The mathematical procedure is similar to that in Liu et al. (2007); however, Liu et al. (2007) considered only piecewise step inputs. For a saturated ramp we can express $u(t-\delta)$ as

$$u(t-\delta) = \begin{cases} 0 & \text{for } t \le \delta \\ \frac{h}{L} t & \text{for } \delta < t \le L + \delta \\ \frac{h}{L}[t-\delta] - \frac{h}{L}[t-\delta-L] = h & \text{for } t > L + \delta \end{cases} \tag{21}$$

As shown earlier for a process initially at a transient state, the estimation equation can be expressed as in eqn(12)

$$y(t) - y(0) + ay^{[1]}(t) = bu^{[1]}(t-\delta) + \xi(t) \tag{22}$$

Following eqn(21), the first order integral of the delayed input signal then becomes

$$u^{[1]}(t-\delta) = \begin{cases} 0 & \text{for } t \le \delta \\ \frac{h}{L}\frac{(t-\delta)^2}{2} & \text{for } \delta < t \le L+\delta \\ \frac{h}{L}\frac{(t-\delta)^2}{2} - \frac{h}{L}\frac{(t-\delta-L)^2}{2} = h[t-\delta] - \frac{hL}{2} & \text{for } t > L+\delta \end{cases} \tag{23}$$

The estimation eqn(14) can then be written as

$$y(t) - y(0) + ay^{[1]}(t) - \xi(t) = \begin{cases} b\frac{ht^2}{2L} - b\delta\frac{ht}{L} + b\delta^2\frac{h}{2L} & \text{for } \delta < t \le L+\delta \\ b[ht - \frac{hL}{2}] - b\delta h & \text{for } t > L+\delta \end{cases} \tag{24}$$

Or equivalently

$$\gamma(t) - \xi(t) = \begin{cases} \phi_1^T(t)\theta & \text{for } \delta < t \le L+\delta \\ \phi_2^T(t)\theta & \text{for } t > L+\delta \end{cases} \tag{25}$$

where, $\gamma(t) = y(t)$, $\phi_1^T(t) = \begin{bmatrix} -y^{[1]} & \frac{ht^2}{2L} & -\frac{ht}{L} & \frac{h}{2L} & 1 \end{bmatrix}$,

$\phi_2^T(t) = \begin{bmatrix} -y^{[1]} & ht - \frac{hL}{2} & h & 0 & 1 \end{bmatrix}$. $\theta = \begin{bmatrix} a & b & b\delta & b\delta^2 & y(0) \end{bmatrix}^T$.

Φ_1, Φ_1 and γ can be written for appropriate time intervals to give

$$\Phi_1 = \begin{bmatrix} \phi_1(t_{d+1}) & \cdots & \phi_1(t_{d+l}) \end{bmatrix}^T, \Phi_2 = \begin{bmatrix} \phi_2(t_{l+d+1}) & \cdots & \phi_1(t_N) \end{bmatrix}^T$$

$$\Gamma_1 = \begin{bmatrix} \gamma(t_{d+1}) & \cdots & \gamma(t_{d+l}) \end{bmatrix}^T, \Gamma_2 = \begin{bmatrix} \gamma(t_{l+d+1}) & \cdots & \gamma(t_N) \end{bmatrix}^T.$$

where, $l = \frac{L}{\Delta t}$. The final estimation equation equation is then obtained as

$$\Gamma = \Phi\theta \quad \text{with} \quad \Gamma = \begin{bmatrix} \Gamma_1 \\ \Gamma_2 \end{bmatrix}, \Phi = \begin{bmatrix} \Phi_1 \\ \Phi_2 \end{bmatrix} \tag{26}$$

Note that both Φ_1 and Φ_1 matrices are rank deficient. However, the matrix Φ is full rank.

2.6 Presence of disturbance

In this section, we provide an estimation procedure proposed in [Ahmed et al. (2009) for identification in the presence of one particular type of disturbance. Consider a model equation of a process described in the Laplace domain as

$$Y(s) = G(s)U(s) + W(s)H_w(s) + E(s) \tag{27}$$

where, $G(s)$ is the process transfer function, $W(s)$ is the disturbance transfer function, $H_w(s)$ is the unknown disturbance and $E(s)$ is the noise term. If $G(s)$ and $W(s)$ have the same denominator polynomial, for a first order process transfer function $G(s) = \frac{b}{s+a}e^{-\delta s}$ and $W(s) = \frac{w_1' s + w_0'}{s+a}$ and considering that H_w is of step type with an unknown size h_w, we can write the model equation as

$$sY(s) + aY(s) = be^{-\delta s}\frac{h}{s} + w_1 + \frac{w_0}{s} + E(s) \tag{28}$$

where, $w_1' h_w = w_1$ and $w_0' h_w = w_0$. One step of integration yields

$$Y(s) + a\frac{Y(s)}{s} = be^{-\delta s}\frac{h}{s^2} + \frac{w_1}{s} + \frac{w_0}{s^2} + \frac{E(s)}{s} \tag{29}$$

Taking inverse Laplace transform and rearranging, we have

$$y(t) = \begin{bmatrix} -y^{[1]}(t) & t & 1 \end{bmatrix} \begin{bmatrix} a \\ bh + w_0 \\ -hb\delta + w_1 \end{bmatrix} + e^{[1]}(t) \tag{30}$$

One more step of integration gives

$$y^{[1]}(t) = \begin{bmatrix} -y^{[2]}(t) & \frac{t^2}{2} & t & 1 \end{bmatrix} \begin{bmatrix} a \\ bh + w_0 \\ -hb\delta + w_1 \\ \frac{bh\delta^2}{2} \end{bmatrix} + e^{[2]}(t) \tag{31}$$

To estimate all the parameters we need another step of integration to yield

$$y^{[2]}(t) = \begin{bmatrix} -y^{[3]}(t) & \frac{t^3}{3!} & \frac{t^2}{2} & t & 1 \end{bmatrix} \begin{bmatrix} a \\ bh + w_0 \\ -hb\delta + w_1 \\ \frac{bh\delta^2}{2} \\ \frac{bh\delta^3}{3!} \end{bmatrix} + e^{[3]}(t) \tag{32}$$

2.7Identification of MIMO systems

A multiple input multiple output (MIMO) model can be considered as a set of multiple input single output (MISO) models. In this section, we describe an identification methodology for a MISO model. The method is based on an identification technique presented by Li et al. (2005) for closed loop identification of two input two output (TITO) models. Although based on step change of the set points this method cannot use the advantages of the use of step test. We consider here open loop step test for MIMO processes where all the inputs are changed simultaneously in a step manner. Let us consider a two input single output model described in the Laplace domain as

$$Y(s) = G_1(s)U_1(s) + G_2(s)U_2(s) \quad (33)$$

where, G_i is the transfer function between the output and the $i-th$ input. Two sets of experiments are to be performed as there are two inputs. If we change the inputs simultaneously in a step manner, for two experiments we can write

$$Y^1(s) = G_1\frac{h_1^1}{s} + G_2\frac{h_2^1}{s} \quad (34)$$

$$Y^2(s) = G_1\frac{h_1^2}{s} + G_2\frac{h_2^2}{s} \quad (35)$$

Here, the superscript refers to the number of experiment. Now, doing a simple manipulation (34)× h_{22} -(35)× h_{21} we get

$$Y^1(s)h_{22} - Y^2(s)h_{21} = G_1(s)\frac{h_1^1h_2^2 - h_1^2h_2^1}{s} \quad (36)$$

Expressing $Y^1(s)h_2^2 - Y^2(s)h_2^1 = Y_1(s)$ and $h_1^1h_2^2 - h_1^2h_2^1 = h$, we can get an equivalent SISO expression relating the transfer function between the output and the first input.

$$Y_1(s) = G_1(s)\frac{h}{s} \quad (37)$$

So, we get a fictitious output Y_1 that would have been generated if the processes was excited through only the first input channel with fictitious step input of size h. Now using the procedure described above we can estimate the parameters of $G_1(s)$. Similar expression can be derived for $G_2(s)$ and its parameters can be estimated. Necessary expressions to obtain $Y_i(s)$ and $h(s)$ for processes with more than two inputs and outputs are provided in Ahmed (2006).

2.7Estimation of step response

To take the advantage of the step response method, in this section, we describe a procedure [5] to estimate step response from output data and the mathematical expression of the input where the input is not a step. Let us consider the input output relation of a process in the Laplace domain

$$Y(s) = G(s)U(s) \tag{38}$$

If the input is a unit step, we get the relation as

$$Y_{step}(s) = G(s)\frac{1}{s} \tag{39}$$

Comparing (38) and (39) we get the relation to obtain the step response from output data due to other type of excitation signal

$$Y_{step}(s) = Y(s)\frac{1}{sU(s)} \tag{40}$$

Now from the estimated unit step response we can find the parameters of the transfer function model of the process.

3. Concluding remarks

Recent developments in identification from step response using the integral equation method is presented. In process industries, step tests are often carried out to obtain plant models. However, to get the models, the so called graphical methods are widely used despite their limitations in terms of accuracy and applicability. The recently developed integral equation based algorithms address the implementation issues and are applicable to practical scenario. Nevertheless, these methods are yet to find widespread application. This article summarizes the recent developments on this topic using a simplified mathematical formulation to provide an intuitive understanding of the developed algorithms.

References

Ahmed, S. Parameter and delay estimation of continuous-time models from uniformly and non-uniformly sampled data. PhD thesis, University of Alberta, Edmonton, Canada, 2006.

Ahmed, S. and Huang, B. and Shah, S. L. Identification from step response in the presence of deterministic disturbance. Preprints of the 15th IFAC Symposium on System Identification, :886-891, 2009.

Ahmed, S. and Huang, B. and Shah, S. L. Identification from step response with transient initial conditions. Journal of Process Control, 18:121-130, 2008.

Ahmed, S. and Huang, B. and Shah, S. L. Novel identification method from step response. Control Engineering Practice, 15(5):545-556, 2007.

Ahmed, S. and Huang, B. and Shah, S. L. Process identification from sinusoidal test data by estimating step response. Preprints of the 15th IFAC Symposium on System Identification, :396-401, 2009.

Diamessis, J. E. A new method for determining the parameters of physical systems. Proceedings of the IEEE, February 1965, :205-206, 1965.

Hwang, S. and Lai, S. Use of two-stage least-squares algorithms for identification of continuous systems with time delay based on pulse response. Automatica, 40(9):1561-1568, 2004.

S. Y. Li and W. J. Cai and H. Mei and Q. Xiong. Robust decentralized parameter identification for two-input two-output processes from closed-loop step responses. Control Engineering Practice, 13:519-531, 2005.

M. Liu and Q. G. Wang and B. Huang and C. Hang. Improved identification of continuous-time delay processes from piecewise step test. Journal of Process Control, 17(1):51-57, 2007.

R. C. Oldenbourg and H. Sartorius. The Dynamics of Automatic Control. The American Society of Mechanical Engineers, New York, USA, 1948.

K. R. Sundaresan and C. C. Prasad and P. R. Krishnaswamy. Evaluating parameters from process transients. Industrial & Engineering Chemistry Process Design and Development, 17(3):237-241, 1978.

Wang, Q. G. and Zhang, Y. Robust identification of continuous systems with dead-time from step responses. Automatica, 37(3):377-390, 2001.

Proceedings of the 2nd Annual Gas Processing Symposium
Farid Benyahia and Fadwa T. Eljack (Editors)

Multi-Objective Optimization for Operational Excellence

Hasan Y. Alhammadi

Chemical Engineering Department, University of Bahrain, Kingdom of Bahrain
halhammadi@eng.uob.bh

Abstract

In these days of increasing concern towards energy conservation and environmental protection, process systems engineers are required to integrate their processes to achieve economic, environmental and social objectives, while at the same time keeping their plants readily operable. This process integration presents a dilemma between control and operation on one hand and process design on the other. Thus, it forces the process engineers to consider the plant-wide controllability and operability issues in the early design stages. An integrated multi-objective optimization framework is proposed to identify the trade-offs between various goals in process synthesis, design and operation. Life Cycle Assessment (LCA) is used to quantify and qualify the environmental consequences of any selected flowsheet in the proposed framework. Heat integration and Heat Exchanger Network (HEN) designs are integrated within the framework as process improvement options. Process controllability and operability assessments are carried out to explore the integrated flowsheets. A case study of vinyl chloride monomer (VCM) plant, with varying degrees of heat integration, is used to show the usefulness of the proposed framework.
Keywords: Process Optimization, LCA, Heat Integration, Controllability

1. Introduction

Commonly, process integration leads to tight the designs and force process designers to consider potential control problems and assess the process controllability and operability in the early stages. In general, the success of an integrated design is measured based on its agreement with its ability to be controlled. Therefore, a systematic procedure is required to evaluate the operability and/or controllability issues of the integrated processes. In this paper, a plant-wide control approach is incorporated to design the control strategies for both integrated and non-integrated processes. Additionally, process modeling and simulation are included in the proposed framework as a mean of validating and testing the steady state and dynamic performance of the designed plant and its control structure.

Life Cycle Assessment (LCA) is systematic approach for estimating the environmental impacts of a product, process or activity through its life cycle from the extraction of raw materials to the usage and disposal stages (Consoli et al., 1993). Being an accepted and widely used tool in this area, it was employed in this study to map the environmental impact potential of any given alternative for the selected process in the optimization framework.

During the past decade, process integration techniques are becoming increasingly significant for both improving process economics and reducing environmental impacts (Roster, 1994; Linnhoff, 1994). In the proposed framework, thermal pinch analysis technique is used to identify the utility targets for the Heat Exchanger Network (HEN) designs of the examined process.

On the other hand, process integration leads to tight the designs and force process designers to consider potential control problems and assess the process controllability and operability in the early stages. In general, the success of an integrated design is measured based on its agreement with its ability to be controlled. Therefore, a systematic procedure is required to evaluate the operability and/or controllability issues of the integrated processes. To illustrate the effectiveness of the proposed approach, a case study on the production of vinyl chloride monomer (VCM) process is used.

2. Framework and Methodology

Chemical process design problems are multi-objective in nature; hence several objectives are required to be satisfied, maximized or minimized, simultaneously within a specified range of constraints. The proposed optimization framework presented in this paper formulates the decision maker preferences into mathematical forms and then integrates between these goals into a multi-objective optimizer. Figure 1 is a schematic of the proposed methodology's structure, showing the inter-linking of the software tools used and the flow of data between them.

LCA is employed in the framework to provide as complete picture as possible of the interactions of the examined activity with the environment and identifies opportunities for environmental improvements. In this framework, LCA maps the environmental impact potential of any given alternative to formulate the environmental objectives. The economic objective is formulated based on the operating profit using the mass and energy data transferred from the simulation package.

During the past decade, the pressure of the environmental stakeholders has increased considerably demanding effective usage of natural resources and reducing pollutant emissions. 'Pinch analysis' is a simple thermodynamically based method used for the design of heat and power systems. It groups the heat sources and sinks of process streams into temperature intervals to maximize the heat recovery by using process-to-process heat exchange to minimize the heating and/or cooling 'utility' requirements (Linnhoff et al., 1994). This provides the targets for the Heat Exchange Network (HEN) designs of the examined process. The pinch analysis is performed to determine the minimum utility requirements for any given operating point as a heat recovery target for the HEN designs. These pinch results are used as the economic and environmental models as "best estimate" for the energy integrated version of the process. The HEN design stage is a complex procedure involves combinatorial problems in the pairing between the hot and cold streams to enhance the heat recovery within the process (Gundersen and Naess, 1988).

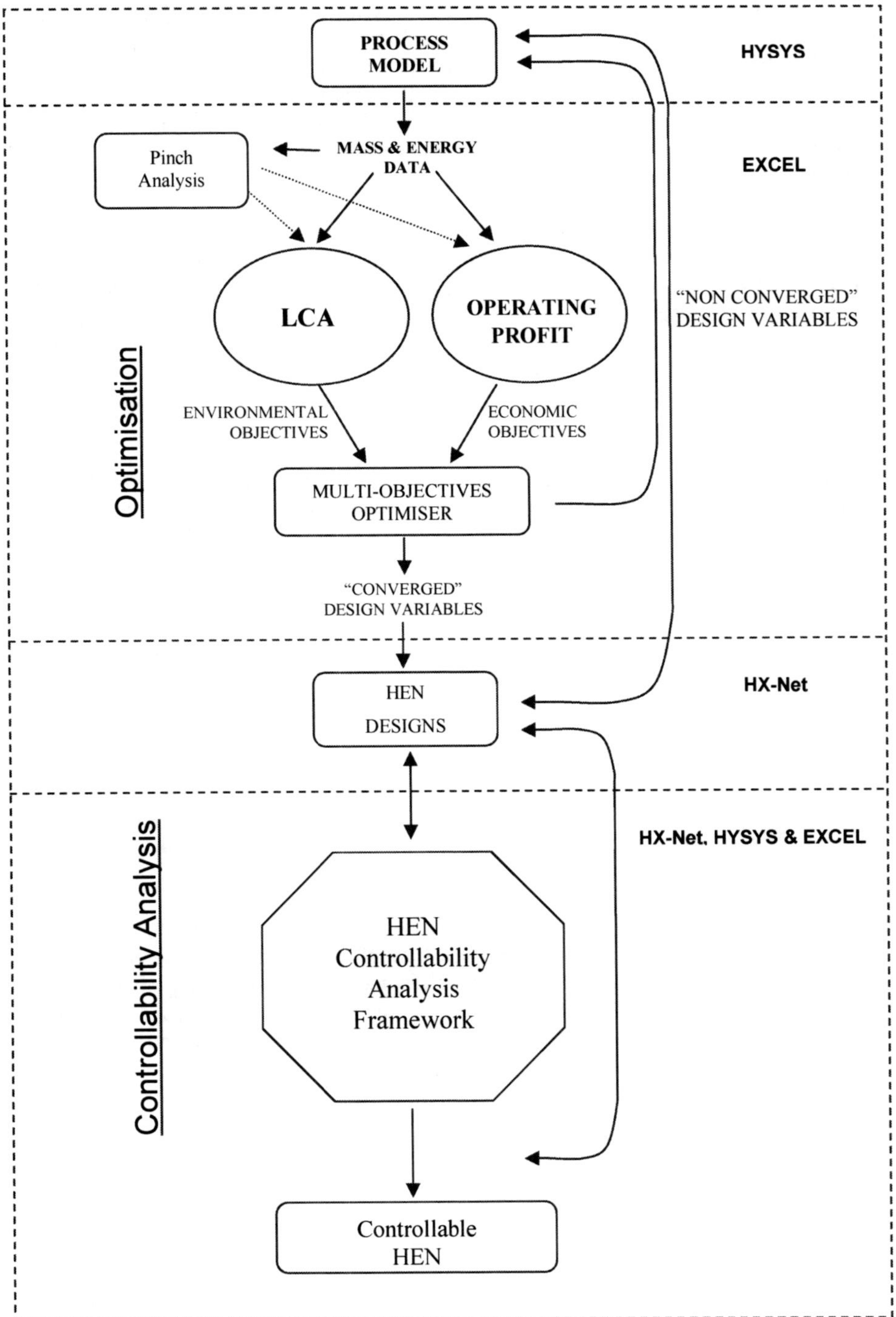

Figure 1: Structure of the proposed framework

The main objective of this study is to optimize the formulated objectives as much as possible within the specified constraint region. However, the multi-objective optimization problem lies in the conflict between the objectives and goals. The decision maker wants to attain more than one objective in selecting the course of action within the specified constraints. In multi-objective optimization problems, it is very rare to get a single solution that simultaneously optimizes all objectives. Though, a Pareto curve is produced to visualise the trade-offs between the objectives through a set of design alternatives and it is defined as a set of 'non-inferior' solutions defining a boundary beyond which none of the objectives can be improved without sacrificing at least one of the other objectives (Miettinen, 1999).

Figure 2 shows the details of the HEN controllability assessment framework shown in Figure 1. This framework is developed where it extends and incorporates the work of Glemmestad and Gunderesen (1998) and Westphalen et al. (2002). This sub-framework is integrated within the overall framework targeting for an optimal design and operation conditions. The aim of this sub-framework is to develop a systematic procedure that helps the design and process engineers in the evaluation and assessment process to select the best controllable and operable designed HEN among a number of different alternatives.

The degree of freedom methodology proposed by Glemmestad and Gundersen (1998) is adapted in this study as the first step of the controllability assessment procedure of the designed HEN. The number of degrees of freedom (NDOF) is expressed as follow:

$$NDOF = R + NU - NT \tag{1}$$

Where R is the rank of the inner HEN, NU is the number of utility streams and NT is the number of targets. No further clarifications will be provided due to the space limitation.

Figure 1 shows a schematic structure of the proposed step-wise procedure to accomplish the performance analysis of a designed HEN and then of the optimized process. The approach starts by developing the HEN designs where the required data to develop a HEN is obtained from the process model and the pinch analysis provides the utility targets of the HEN.

Then the designed HEN is required to be analyzed in term of sub-network and loops identifications where each sub-network goes through the degree of freedom analysis where a zero degree of freedom is the minimum requirement of a controllable HEN. Then the process gain matrix of the HEN that passed the degree of freedom analysis is calculated. Square matrices from the overall gain matrix are developed where the available manipulated variables and the required variables to be controlled are grouped in different combinations.

The relative gain array (RGA) analysis is used as a guide in the pairing process between the variables as it measures the interactions within each square matrix. This is followed by the condition number (CN) analysis to measure the ease of controlling the system. The condition number is commonly used as an index of controllability which is a ratio of the largest singular value to the smallest nonzero singular value of a matrix, i.e. the HEN gain matrix.

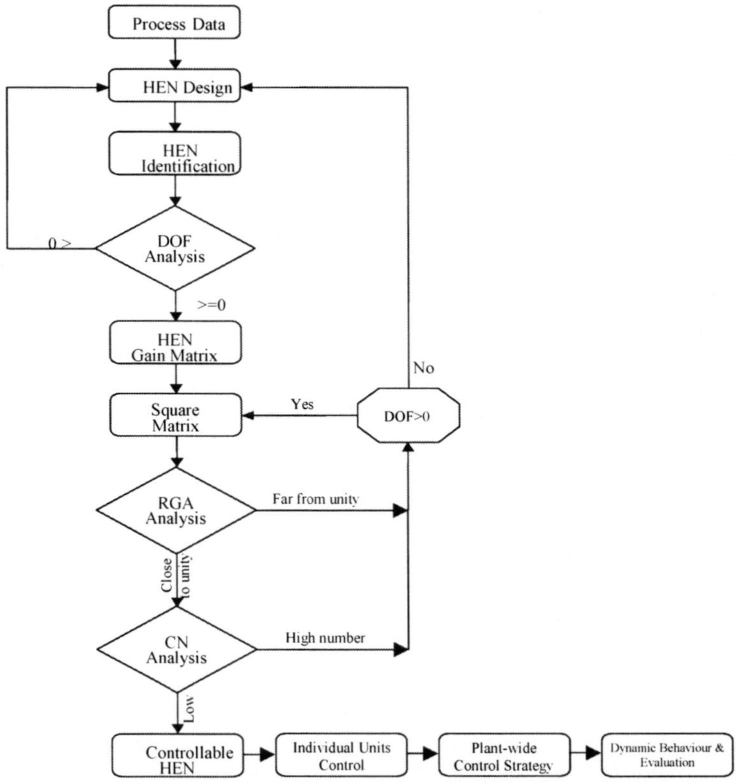

Figure 2: Integrated approach for operability/controllability assessment of integrated processes

Advanced control strategies have received wide acceptance in the process industry in the recent years as an effective means of implementing multivariable constrained control on real processes. For complex units, such as distillation column, a number of integrating variables are required to be controlled where it forms a challenging task for control engineers. The aim of the proposed individual units control sub-framework is to develop and evaluate satisfactory control structures of each individual unit of the entire plant. Each considered unit is modeled in a dynamic mode using a simulation package, e.g. HYSYS, where the available manipulated variables and the controlled variables are identified. HYSYS is linked with Excel which is used as a recorder of each individual process output response for a step change of each process input. These responses are used to identify the relationships between the process inputs and outputs through process identification tools. The transfer function matrix of each evaluated unit is developed and existing analysis tools (RGA, conditions number, etc.) are used to pair the most suitable controlled and manipulated variables so that the minimum interacting loops can be achieved.

The proposed hierarchical procedure for plant-wide control developed by Luyben and co-workers (1998) is utilized, as stepwise guidelines to perform the plant-wide process control strategy. In the proposed framework, the development of the plant-wide control system is performed into two stages. First, the plant-wide control system is developed for the base design where no heat integration is utilized. This stage is performed and evaluated, according to its dynamic performance, as a first step to make sure that the basic designed process is controllable. Then the selected designed HEN, based on the decision maker's preferences, is integrated within the entire plant and the plant-wide control structure is adjusted accordingly.
Process simulation packages provide a great cut down in the required time for process development. Moreover, they provide the opportunity of comparing process alternatives on a consistent basis where a number of different ideas can be analyzed in a short time. In terms of plant-wide control considerations, process simulation of highly integrated processes within a plant directs the process design/control engineers to the available interactions between the processes.

In the proposed framework, HYSYS.PLANT simulation package is used to validate both the steady state and dynamic models even though the switchability from steady state to dynamic mode is not a trivial procedure.

3. Case Study: Vinyl Chloride Monomer (VCM) Plant

Figure 3 is a simplified block diagram of a typical vinyl chloride monomer (VCM) plant (McPherson et al., 1979; Cowfer and Gorensek, 1997). This integrated process produces VCM from ethylene, chlorine, oxygen and a portion of the by-product hydrogen chloride (HCl). The major sections of this plant are as follows: (1) A direct chlorination of the ethylene to produce ethylene dichloride (EDC). (2) An oxy-chlorinator to produce EDC by reacting ethylene with oxygen and HCl. (3) The two crude EDC streams are mixed and purified in a pair of distillation columns (essentially to remove water and unwanted reaction by-products). (4) The pure EDC undergoes (partial) thermal cracking in a pyrolysis furnace to yield VCM and HCl. (5) VCM is separated from the HCl and EDC in another pair of distillation columns. Note that a portion of the HCl is recycled to the oxy-chlorination reactor to make EDC, while any unconverted EDC is recycled (via the purification) to the furnace.

The process design variable selected here for (multi-objective) optimization was the portion of HCl recycled to the oxy-chlorinator. It would be a straightforward extension to the framework to include multiple design variables - however, in this paper only a single variable was considered for ease of demonstration.
The inventory analysis evaluates not only the environmental burdens of the VCM process, but also assesses the burdens of the productions of raw materials (C_2H_4, Cl_2, O_2) and utilities and the treatment of the waste streams. Also, as the environmental potentials in this case all trend in the same direction, the impact potential most sensitive to this design variable (ie GWP) was chosen as an example.

The economic objective chosen was the operating profit which is the difference between the total value of the products and the total cost of the raw materials and utilities.
The ε-constraint method was used to solve the multi-objective optimization problem and obtain the Pareto curve. Here, the economic objective was optimized while the environmental objective was converted into a constraint with a specified upper bound. Each objective function was normalized (over the specified range of the selected process variable) and scaled between 0 and 1 where 0 represents the best value and 1 represents the worst value of the objective. This scaling is required in multi-objective optimization problems to ease the comparison between the formulated objectives and avoid the computational confusion that is due to different scaling and different path of optimization (maximization or minimization).

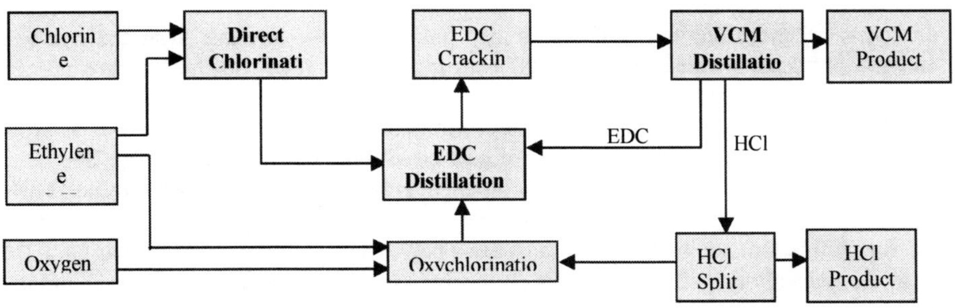

Figure 3: A simplified block diagram of a typical VCM *plant*

Pinch analysis is performed based on the utility streams of the VCM process. This analysis draws the guidelines for the HEN designs. The multi-objective optimization problem was performed for different developed HEN designs that include no heat integration option (Design 1) where the utilities are at their maximum load. In the second case (Design 2), the process was examined for optimal heat integration with the minimum (target) heating and cooling requirements being determined by the pinch analysis technique. In three different cases (Designs 3, 4 & 5), different developed HEN designs for the process were examined.

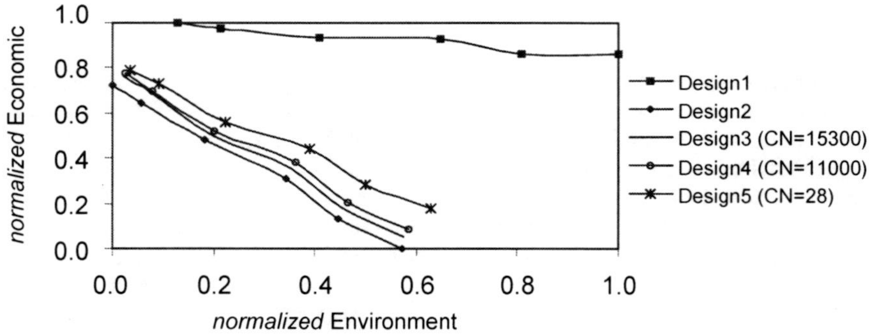

Figure 4: Pareto curves for different designs

4. Results and Discussions

Figure 4 represents the optimization results to the decision makers in a transparent way through the Pareto curves for the entire designs over the whole range of the selected process variable. The curves for designs 1 and 2 provide the lower and upper bounds for all possible levels of heat integration at all operating points. The practical HEN designs shift the Pareto curve of no heat integration condition towards the optimal heat integration curve. The 'optimal heat integration' curve (Design 2) shows the maximum possible reduction achievable for the objectives.

In terms of controllability and operability point of view, moving from the 'no heat integration' level towards the 'optimal heat integration' level results in more process interactions through the HEN that leads to difficult operation and control. Therefore, plant controllability is required to be formulated as one of the trade-offs to be considered in addition to the economic and environmental objectives. The degree of freedom analysis is performed for each independent sub-network of the examined HEN. This step is performed to check the feasibility of the HEN design. The details of the degree of freedom analysis will not be shown here because of the space limitation. After that, the pairing analysis between the controlled variables and the possible manipulated variables is performed through the relative gain array (RGA) analysis of the process gain matrix of the examined HEN design. Finally, the condition number analysis of the process gain matrix is used as a dimensionless measure of the interactions between the control loops paired by the RGA analysis. The condition number analysis provides an excellent measure of controllability for the designed HENs. For the tested HEN designs, the smallest condition number leads to a better controllable process, while regarding the RGA analysis the closer to identity matrix, the less interaction between the control loops leading to better controllability. For the designed HEN, designs 3 and 4 shows the best improvement toward the optimal design based on pinch analysis. However, the condition numbers for designs 3 and 4 are 15300 and 11000 respectively, which are very big and indicate a poor controllability. Design 5 is produced from design 4 where small exchangers are removed or combined and more sub-networks are introduced causing the condition number to be improved reaching a value of 28. This improvement in the controllability is at the price of a reduction in the other objectives.

Plant-wide process control forms the final stage of the process synthesis, design and operation assessments. The proposed plant-wide control procedure for the VCM case study shows the importance of employing the engineering judgments and experiences together with the available systematic analyses. Moreover, a rigorous dynamic model was used to implement and validate the developed plant-wide control structure and to test the overall dynamic performances of the plant. The proposed stepwise plant-wide control procedure and the dynamic validation using non-linear plant model for the optimized plant – design 5 - are not provided here due to the space limitation.

5. Conclusions

In this paper, a methodology has been proposed that incorporates economical, environmental, heat integration and operational considerations within a multi-objective optimization framework. The methodology as it stands enables the design engineers to draw 'boundary' Pareto curves corresponding to the maximum and minimum levels of heat integration for all operating points achievable by the process. It is also possible to use the proposed approach to draw the Pareto curve for any designed HEN between the calculated limits, and thus to quantify the trade-offs between economic and environmental objectives. Improving energy efficiency generally increases plant complexity and which leads to significant impacts on plant operability and/or controllability. The controllability of the designed HEN is analyzed explicitly through the RGA and condition number techniques. Plant-wide control and dynamic evaluation are integrated within the framework as a means of including control and operational considerations in the early stages of design.

References

Consoli, F., et al. (eds.), 1993, Guidelines for Life-Cycle Assessment: A "Code of Practice". SETAC, USA.

Cowfer, J. and Gorensek, M., 1997, Vinyl Chloride: Encyclopaedia of Chemical Technology, 24, 851-882.

Glemmestad, B. and Gunderesen, T., 1998, A Systematic Procedure for Optimal Operations of Heat Exchanger Networks, AIChE Symposium Series, 320(94), Snowbird.

Gundersen, T. and Naess, L., 1988, The Synthesis of Cost Optimal Heat Exchanger Networks, Computers & Chemical Engineering , 12, 503-530.

Linnhoff, B., 1994, Use Pinch Analysis to Knock Down Capital Costs and Emissions, Chemical Engineering Progress, August, 32-57.

Linnhoff, B., Townsend, D., Boland, D., Hewitt, G., Thomas, B., Guy, A. and Marsland, R., 1994, A User Guide on Process Integration for the Efficient Use of Energy, IChemE, Rugby.

Luyben, W., Tyreus, B. and Luyben, M., 1998, Plantwide Process Control, McGraw-Hill.

McPherson, R., Starks, C. and Fryar, G., 1979, Vinyl Chloride Monomer – What You Should Know, Hydrocarbon Processing, March, 75-88.

Miettinen, K., 1999, Nonlinear Multi-objective Optimisation. Kluwer Int. Series.

Rossiter, A. P., 1994, Process Integration and Pollution Prevention. In: Pollution Prevention via Process and Product Modifications, AIChE, 90, 12-22.

Westphalen, D., Young, B., Svrcek, W. and Shethna, H., 2002, 3rd International Symposium on Process Integration, Vancouver, Canada.

Proceedings of the 2nd Annual Gas Processing Symposium
Farid Benyahia and Fadwa T. Eljack (Editors)

On the PρT, PηT and Phase Envelope Behavior Characterization of Qatari Type Natural Gas Mixtures

M. Atilhan[a], S. Aparicio-Martinez[b], K.R. Hall[c], M. El-Halwagi[c], G.A. Iglesias-Silva[d]

[a]*Qatar University, Chemical Engineering Department, Doha, QATAR*
[b]*University of Burgos, Chemistry Department, Burgos, SPAIN*
[c]*Texas A&M University, Artie McFerrin Chemical Engineering Department, TX, USA*
[d]*Instituto Tecnológico de Celaya, Chemical Engineering Department, Celaya, MEXICO*

Abstract

In this work, we intend present the strategy and latest results on our current project in which we investigate the thermophysical properties of natural gas-like mixtures using a base composition similar to that inform the North Field of Qatar. We intend to measure density, phase boundary and viscosity data on several mixtures. Thermophysical properties will be measured in four different locations with six different apparatus located in Qatar, USA, Mexico and Spain. The apparatuses being used in this project are: Two high-pressure vibrating tube densimeters (University of Burgos - Spain, Celaya Institute – Mexico), high-pressure magnetic suspension densimeter and low/high pressure isochoric apparatus for phase envelope measurements (Texas A&M University, USA) and low-pressure magnetic suspension densimeter (Qatar University – Qatar). Using a variety of apparatus provides an internal consistency check on the data. We will establish a database consist of very accurate P-ρ(density)-T and P-η(viscosity)-T for the measured gas mixtures which will be used to describe the Qatari type natural gas mixtures. The results of the project will be used to better predict and enhance the performance of gas-processing facilities. Typical mixtures used in this project have compositions similar to mixture compositions such as 84.99 CH_4, 5.529% C_2H_6, 2.008% C_3H_8, 0.986% C_4H_{10}, 0.316% C_5H_{12}, 0.344% C_{6+} and 3.496% N_2. All the mixtures prepared for this purpose are characterized gravimetrically and compositions are certified via using gas chromatography. Preliminary data for samples called QNG–S1 and QNG–S2 are presented in this work on PρT and PηT measurements. Densities of the samples are measured by vibrating tube densimeter and viscosities of the samples are measured by electromagnetic viscometer. Both the measurements are conducted within 250 – 450 K temperature range. Pressure range of density measurements is 10 – 65 MPa and viscosity measurements are 10 – 65 MPa. Density measurements are within a ±0.1 % and viscosity measurements are within 2% accuracy in the whole ranges.

Keywords: Multi-component mixtures; Natural gas; Isochoric; Density; Isochoric; Viscosity of Natural Gas.

1. Introduction

Natural gas is the most environmentally friendly fossil fuel because it burns more efficiently and produces less greenhouse gas emissions than coal or oil (Natural Gas Issues and Trends, 1998). The emergence of natural gas as a clean-burning fuel as well as an essential feedstock for the petrochemical industry has generated intensified interest in producing it from unconventional sources. Some of these sources provide gas that contains a significant fraction of heavy hydrocarbons. These "heavies" influence the thermophysical properties of the gas, and the extent is generally uncertain. Most equations of state (EOS) that purport to describe natural gas cannot match the phase loops for pressures above that of the cricondentherm (CT), and most also cannot match the high pressure density data (Atilhan, 2007). The experimental and theoretical results of the project will be used to better predict the performance of gas-processing facilities and to enhance their performance. Therefore, we intend to incorporate the project results in process simulation, design, and operation. Consequently, we can optimize the process performance, reduce energy consumption, mitigate negative environmental impact, enhance safety, and increase yield. These results will have a positive impact on the economy, environment, and well being of the State of Qatar.

Based on 2006 figures, with its 911 trillion ft^3 natural gas reserves, Qatar is the worlds 3rd largest natural gas supplier with 14.9% share in the natural gas stake (Patil, 2005; International Energy Outlook Report 2009). Oil and gas sectors have great role in Qatar's national economic structure. Based on 2008 figures, overall 60% of the Qatar's GDP is from oil and gas industry, which used to be 55% in 2004 (International Energy Outlook Report, 2009). Having stressed the importance of natural reserves for Qatari economy, we will measure high accuracy thermodynamic properties of Qatari natural gas like mixtures, implement measured properties to real life problems by using process simulators and establish better understanding of the utilization of the Qatari natural gas resources. We believe that for more efficient natural gas custody transfer, we need accurate equations to predict thermophysical properties. In our past works (Patil et al. 2007), we have shown that current equation of states lack of high accuracy prediction capability for multi-component gas mixtures especially natural gas type mixtures. Despite globally accepted industry standard equations, we have shown that the high accuracy thermophysical property knowledge in natural gas industry is must (Atilhan et al. 2009). For this purpose we believe that this work will meet such needs in Qatar Natural Gas industry. Moreover, with the implementation of our high accuracy data to the process simulators, we will be able to compare the deviations from traditional computational techniques and improved database models for natural gas processing facility process calculations.

2. Experimental

Among the thermodynamic properties, density is the most directly predicted property using EOS (BP Statistical Review, 2008). The measured densities should be approximated by suitable EOS and the measurements should be traceable to the

International System of Units (Kleinrahm and Wagner, 1984). According to Wagner and Kleinrahm (2004), the following considerations are important when deciding upon a density measurement technique:

i) Large pressure and temperature range for wide operations.
ii) Low total uncertainty and high accuracy of the method for the overall range.
iii) Simplicity in design and ease of maintenance and operation.
iv) Little time required for each data point measurement.

Vibrating tubes and vibrating forks are common density measuring techniques. These devices measure the fluid density of interest by determining the oscillation frequency of the vibrating element in the fluid. These instruments provide accurate results quickly. However, frequent calibration is necessary for this apparatus to maintain its accuracy (Zhou, 2005). Moreover, when the density of the fluid is vastly different from air or pure water (frequently used as reference fluids because of their well-known thermophysical properties) the uncertainty of the measurements increases as reported by Padua et al. (1998). In this work, we used vibrating tube densimeter for quick density measurements and densities presented in this paper are collected from Anton Paar DMA 512P densimeter unit. In below figure 1, schematic of the densimeter is shown.

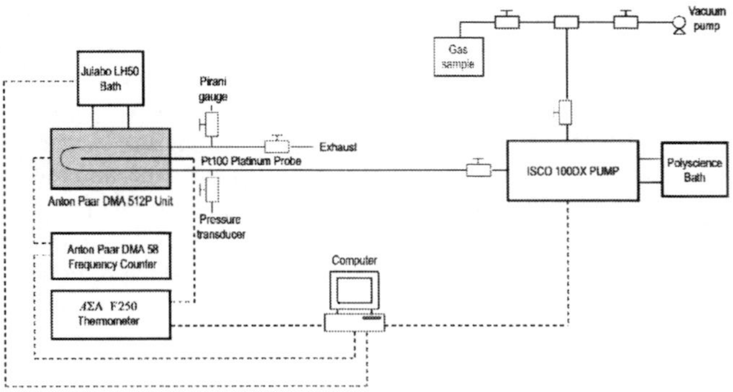

Figure 1: Schematic of vibrating tube densimeter used in this work.

A hydrostatic balance densimeter combined with magnetic suspension coupling mechanism currently is the most accurate densimeter for collecting PρT data over wide ranges of temperature, pressure and density (Zhou et al., 2006). Buoyancy-based magnetic suspension densimeters using Archimedes' Principle were first introduced by Beam and Clarke in 1962 (Majer et al., 2003). In that design, the floating object (sinker) was a magnetic, ferrous material. They used several solenoid coils in the pressure environment to levitate the sinker. The buoyancy force on the sinker was determined from the coil required to lift and suspend the sinker in the density-measuring cell. After Beam and Clarke, Brachthäuser et. al. (1993) developed the modern magnetic suspension densimeter equipped with non-magnetic sinkers, coupling of a permanent magnet and an electromagnet connected to a separated digital balance. The Brachthäuser densimeter was initially a single sinker densimeter. Kleinrahm and

Wagner (2004) designed and developed more complicated two sinker densimeters. To overcome limitations in achievable accuracy, the need for frequent calibration of the apparatus with reference fluids, complexity of operation, limitations on temperature and pressure, they introduced an MSD based upon magnetic levitation of the sinker in the measuring cell. The novelty of the magnetic suspension coupling was that it used non-physical-contact force transmission between the sinker in the pressurized cell and the weighing balance at atmospheric pressure, thus allowing a cell design that covered a very wide temperature and pressure range. In this project, we are using 2 different kinds of MSD, a high pressure and a low pressure MSD. First MSD in Middle East is purchased with the grant of the project and commissioning of the MSD is planned to be completed in early 2010. High pressure MSD is located in Texas A&M University and it is one of a kind 200 MPa device, highest pressure rated MSD ever manufactured by the company. These densimeters are commercially available through Rubotherm®, Bochum, Germany.

Viscosity is also an important property required for simulation of gas production at reservoir conditions to the design and operations of flow lines. Few data are in the literature for natural gas viscosities over wide temperature/pressure ranges and even systematic studies of the effect of heavy fractions on this property are absent (V. Vesovic, 2001; Langlandski et al. 2007). In addition, the development of accurate, theoretical viscosity models is imperative to meet the increasing demand for reliable and accurate data for reservoir operations and gas transportation, but the absence of reliable experimental data hinders this development. Several viscosity measurement techniques appear in the literature but most of them have several disadvantages: low accuracy, small pressure and/or temperatures ranges or long times for each data measurements. Electromagnetic viscometer provides accurate data over wide pressure/temperature ranges in a reasonable time that permits systematic studies of complex mixtures not available with other techniques. In below figure 1, schematic of the viscometer is shown.

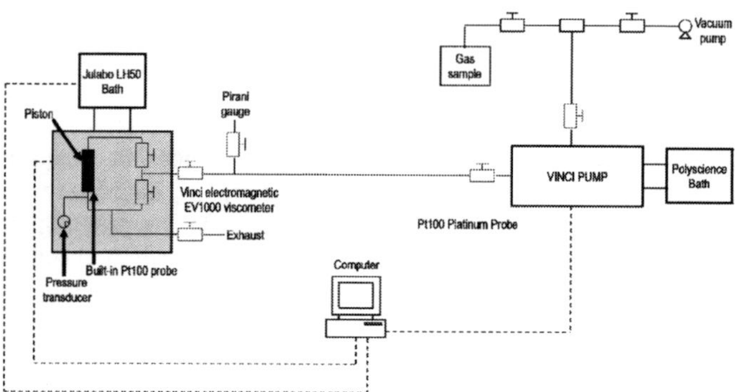

Figure 2: Schematic of viscometer used in this work.

Electromagnetic viscometer works basically with two coils move the piston back and forth magnetically at a constant force perpendicular to the flow direction of the fluid. Meanwhile, proprietary circuitry analyzes the piston's two-way travel time to measure absolute viscosity. A set of six calibrated pistons is used to cover viscosity ranging from 0.01 cP to 10,000 cP. With existing setup, operations within temperature range of up to 190°C and pressure range of up to 15,000 psi are possible.

2.1. Calibration

We measured two synthetic natural as samples and presented in this paper. Prior to main sample measurements, calibration for densimeter and viscometer has been done by using 99.9995% methane as a calibration fluid. In figure 3a and 3b we present the calibration results for vibrating tube densimeter and the temperature effect on the calibration.

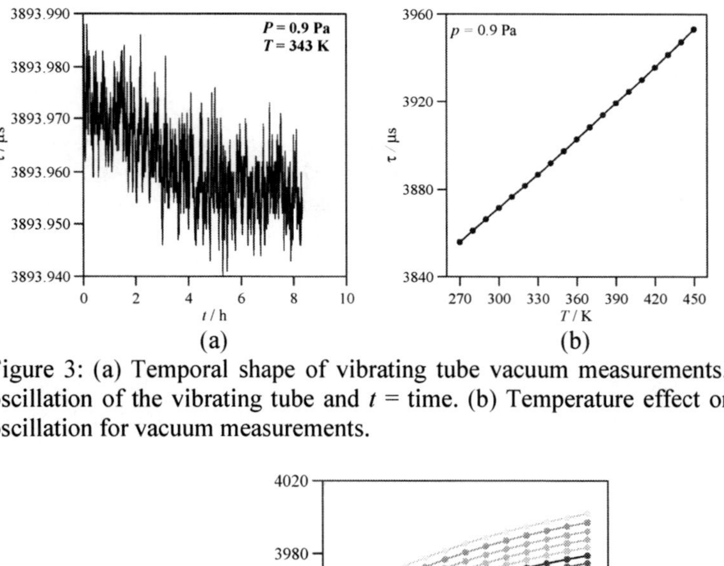

(a) (b)

Figure 3: (a) Temporal shape of vibrating tube vacuum measurements. τ = period of oscillation of the vibrating tube and t = time. (b) Temperature effect on the period of oscillation for vacuum measurements.

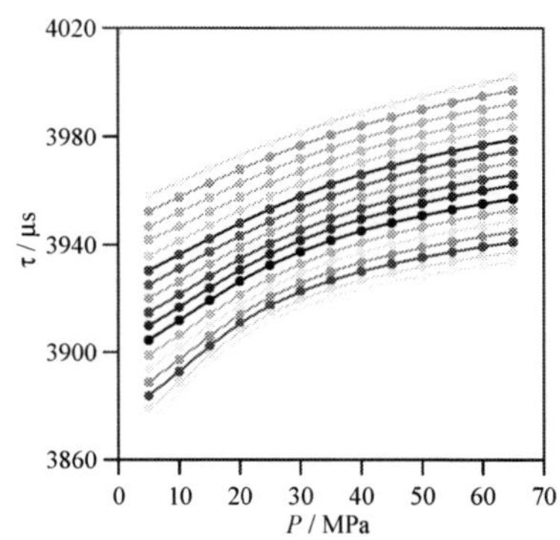

Figure 4: Period of oscillation of the vibrating tube, τ, for methane in the 5 – 65 MPa pressure range and 290 – 450 K temp range (10 K step from bottom, 290 K - 450 K).

Oscillation period of the vibrating tube for methane at varied temperature and pressure is given in figure 4. On the other hand, calibration for viscometer is performed. Dynamic viscosity and its deviation with respect to NIST database for pure methane is given in below figures 5.

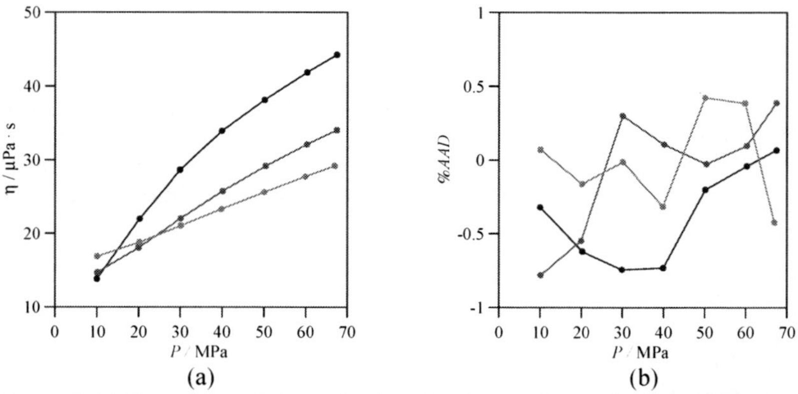

Figure 5: (a) Experimental dynamic viscosity data, η, for methane (99.9995 % purity by GC) obtained with electromagnetic viscometer. T = 270 K (black), T = 350 K (blue), T = 450 K (red). Solid circles = experimental points, (lines) trends for guiding purposes. (b) Percentage deviation between experimental viscosity data for methane (99.9995 % purity by GC) obtained with electromagnetic viscometer and NIST reference data (http://webbook.nist.gov/chemistry/fluid/). T = 270 K (black), T = 350 K (blue) and T = 450 K (red).

2.2. Experimental Results

Two synthetic natural gas samples are ordered from Linde Inc. in Spain and they are prepared gravimetrically in their facilities. The sample compositions are given in below table 1.

Table 1: Mixtures studied in this work.

Component	QNG-S1 [%]	QNG-S2 [%]
CH_4	0.84990	0.90260
C_2H_6	0.05529	0.05828
C_3H_8	0.02008	0.02106
i-C_4H_{10}	0.00401	0.00412
n-C_4H_{10}	0.00585	0.00641
i-C_5H_{12}	0.00169	0.00214
n-C_5H_{12}	0.00147	0.00162
C_{6+}	0.00354	0.00382
CO_2	0.02331	-
N_2	0.03496	-

Experimental density values for sample QNG-S1 and QNG-S2 are given in below figure 6. Densities for QNG-S1 and QNG-S2 are measured between 250K – 450 K and up to 65 MPa.

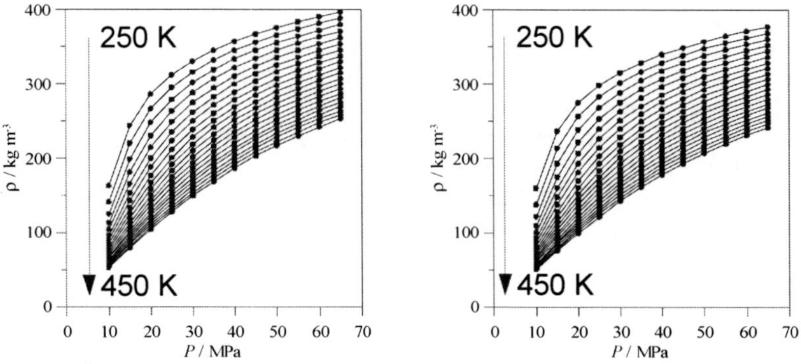

Figure 6: (a) Density, ρ, for QNG-S1 (b) Viscosity, η, for QNG-S1 (Arrows indicate increasing temperature in 10 K steps.)

Experimental viscosity values for sample QNG-S1 and QNG-S2 are given in figure 7. Viscosities for QNG-S1 and QNG-S2 are measured similar to density temperature and pressure range.

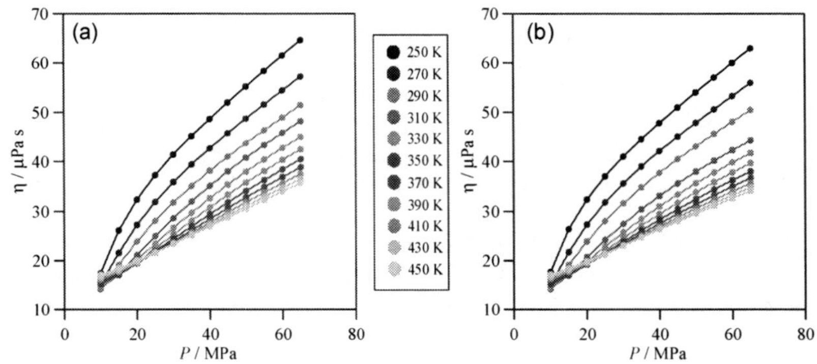

(a) (b)

Figure 7: Dynamic viscosity, η, of QNG–S1 (a) and QNG–S2 (b) mixtures as a function of pressure, P, and temperature, T. In both Panels, symbols are experimental values obtained using the electromagnetic viscometer, and lines are for visual effect. Data appear in 20 K steps for the sake of visibility.

3. Future Work

Our final objective with this project is to develop a Qatari National Database for thermophysical properties on multi-component mixtures. Specifically targeted mixtures in this regard are synthetic natural gas mixtures, which represents typical North Field reservoir gas after sweetening. The presented data on measured density data via vibrating tube densimeter will be used in developing Qatari National Database and the presented results in this paper are the first batch of data in this aim. Followed by these measurements, MSD in Qatar University and in Texas A&M University will be conducted. Moreover, cross check on this vibrating tube density data will be made by Celaya Institute group. On the other hand, isochoric measurements will be conducted on the same samples for phase envelope points at Texas A&M group laboratory.

4. Acknowledgements

This publication was made possible by a NPRP grant from Qatar National Research Fund (a member of The Qatar Foundation). The statement made herein are solely the responsibility of the author(s).

References

BP Statistical Review of World Energy, London, 2008.

K. Brachthäuser, R. Kleinrahm, H. W. Lösch, and W. Wagner, Entwicklung eines neuen Dichtemeßverfahrens und Aufbau einer Hochtemperatur-Hochdruck-Dichtemeßanlage, Fortschr.-Ber. VDI, Reihe 8, No. 371 (VDI-Verlag, Düsseldorf, 1993).

M. Atilhan, S. Ejaz, P. V. Patil, J. Zhou, D. Cristancho, J. C. Holste and K. R. Hall, High Accuracy Phase Behavior and PρT Measurements via Isochoric and Isothermal Methods for Methane Natural Gas-Like Mixtures and an AGA8-DC92 EOS Validity Check, SYMP 17 Conference Proceedings, Boulder, CO, USA 2009

M. Atilhan, J. C. Holste and K. R. Hall, Phase Envelopes and Densities for Natural Gas Samples Including Heavy Components, SYMP 17 Conference Proceedings, Boulder, CO, USA 2009

M. Atilhan, High Accuracy P-r-T Measurements Up To 200 Mpa Between 200 K And 500 K Using A Compact Single Sinker Magnetic Suspension Densitometer For Pure And Natural Gas Like Mixtures, PhD Dissertation, Texas A&M University, College Station, 2007.

M. Atilhan, S. Ejaz, S. Aparicio-Martinez and K.R. Hall, in: Sixteenth Symposium on Thermophysical Properties, Boulder, CO, 2006.

International Energy Outlook Report 2009. DOE/EIA-0484; Energy Information Administration, U.S. Department of Energy: Washington, 2009. Available at http://www.eia.doe.gov/oiaf/ieo/

R. Kleinrahm and W. Wagner, Progress Reports of the VDI Journals, 3 (1984) 92.

L. I. Langelandski, S. Solvang, M. Rousselet, I. N. Metaxa, M. J. Assael, J. Int. J. Thermophys. **2007**, 28, 1120 – 1130.

V. Majer and A.A.H. Padua. in: Goodwin, A.R.H., K.N.Marsh and W.A.Wakeham, ed. (Eds.), Measurement of the Thermodynamic Properties of Single Phases, Experimental Thermodynamics, Amsterdam, 2003, pp.

Natural Gas Issues and Trends. DOE/EIA-0560; Energy Information Administration, U.S. Department of Energy: Washington, 1998. Available at http://www.eia.doe.gov/pub/oil_gas/natural_gas/analysis_publications/natural_gas_1998_issues_trends/pdf/it98.pdf

A.A.H. Padua, J.M.N.A. Fareleira, J.C.G. Calado and W.A. Wakeham, Review of Scientific Instruments, 69 (1998) 2392-2399.

P.V. Patil, Commissioning of a Magnetic Suspension Densitometer for High-accuracy Density Measurements of Natural Gas Mixtures, PhD Dissertation, PhD Disseratation, Texas A&M University, College Station, TX, 2005.

P. V. Patil, S. Ejaz, M. Atilhan, J. Zhou, K. R. Hall, and J. C. Holste, Journal of Chemical Thermodynamics, Volume 39 (2007), Issue 8, 1157-1163.

J. Zhou, P. Patil, S. Ejaz, M. Atilhan, J.C. Holste and K.R. Hall, Journal of Chemical Thermodynamics, 38 (2006) 1489-1494.

J. Zhou, Isochoric Apparatus for PVT and Phase Equilibrium Studies of Natural Gas mixtures, PhD Dissertation, Texas A&M University, College Station, TX, 2005.

V. Vesovic, V. *Int. J. Thermophys.* **2001**, *22*, 415 – 426.

W. Wagner and R. Kleinrahm, Metrologia, 41 (2004) S24-S39.

Proceedings of the 2nd Annual Gas Processing Symposium
Farid Benyahia and Fadwa T. Eljack (Editors)

Studying Influence of Changing Fuel on the Operation and Total Annual Cost of the Total Site Using R-curve

Shervin Karimkashi,[a] Majid Amidpour [a]
[a]*Department of Energy system Engineering, Mechanical Engineering Faculty, K.N. Toosi University of Technology, No. 15, Pardis St. Mollahsadra Ave., Tehran, Iran*

Abstract

In large processing sites the cost of fuel and power can be very significant and better management of the utility system can lead to significant cost savings. The R-curve is an analysis tool that provides guidelines for the cogeneration efficiency for a given site power-to-heat ratio demand of an ideal utility system.

In this paper, considering a total site as the case study, by changing the operating fuel, different actual R-curves have been constructed and then compared. In all of cases, the gas turbine exists and obviously its fuel is natural gas and does not change. In fact in each case, by changing the fuel injected to the boilers of the site, a different actual R-curve is constructed and then another curve of 'R-ratio vs. Total Annual Cost' is sketched to demonstrate the cost of each point of the R-curve. After comparing these curves for different operating fuels, some useful conclusion has been obtained.

Keywords: operating fuel, total annual cost, total site, R-curve, cogeneration.

1. Introduction

Utility systems are an important part of most processing sites. Cogeneration is usually an important feature of such systems, particularly from steam turbines, gas turbines or a combination of both. In these sites the kind of operating fuel, used in boilers, has a big influence on the operation and total annual cost of the site (Varbanov et al., 2004). In fact, choosing a better fuel for the site being studied, will lead to fewer costs.

The R-curve [2,(Kimura and Zhu, 2000)] is an analysis tool that provides guidelines for the cogeneration efficiency for a given site power-to-heat ratio demand of an idealized utility system. Developing the R-curve concept, it is possible to estimate total annual cost of an idealized utility system. This total annual cost is directly dependent on the fuel price and thus the kind of operating fuel.

R-curve was first introduced by kenney (Kenney, 1984), who studied different configurations within utility systems, producing plots of cogeneration efficiency vs. site power-to-heat ratio. This plot was referred to as a 'fuel utilization curve' or R-curve, Fig. 1. The terms 'unfired boiler', 'supplementary fired boiler' and 'fully fired boiler' in

Fig. 1 refer to the different firing modes of a heat recovery steam generator, connected to a gas turbine. This was intended to provide quantitative design guidelines for the equipment most likely to produce maximum fuel efficiency for a utility system. Later, aiming to provide better insights to site operational management and debottlenecking, the R-curve concept was extended by Kimura and Zhu (Kimura and Zhu, 2000). They employed the site cogeneration targeting methodology developed by Mavromatis and Kokossis (1998) to estimate the power generation of an ideal utility system. This allowed Kimura and Zhu (2000) to develop the so called 'ideal R-curve'. By introducing a path efficiency analysis, Kimura and Zhu (2000) attempted to redistribute the steam turbine flows through an existing utility system so that it operated at maximum fuel efficiency (Kimura and Zhu, 2000) and (Mavromatis and Kokossis, 1998). This allowed the construction of the so-called 'actual R-curve', which gives a tighter bound on the utility system fuel efficiency. However, their analysis had a number of limitations. It used simplistic models for steam and gas turbines. Power import and export and steam generation by processes were not taken into account. Also, their R-curve analysis did not take economic term into account and it was only a thermodynamic tool, considering the amount of steam and power potential to be saved.

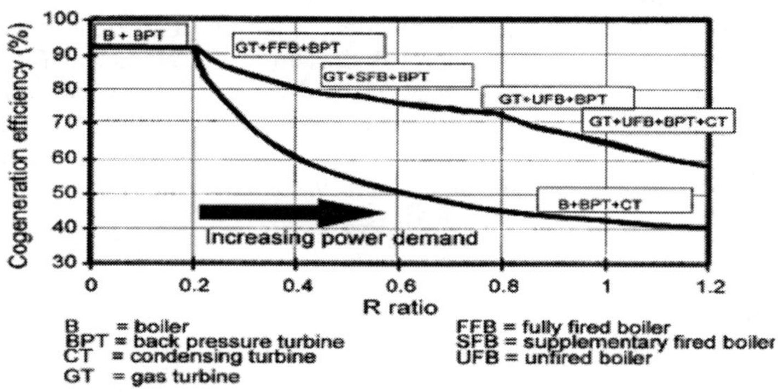

Figure 1: Qualitative sketch of the fuel utilization curve by Kenney (Kenney, 1984)

In this paper, a new curve (R-ratio vs. TAC curve) which is useful for thermoeconomic analysis of the utility system of a total site is first introduced. Then using the R-curve concept and this new analysis tool, it has been possible to compute total annual cost of the utility system of a total site for each of different fuels and then choose the most beneficial one. Also pollution degree of each fuel has been considered as another factor for choosing an operating fuel. Finally the method has been applied to a case study and the results have been discussed.

2. R-curve Concept

In an industrial plant, the fuel energy is converted into heat and power, and part of the energy is lost in the transformation. The cogeneration efficiency (η_{cogen}), which is the fuel utilization efficiency, is defined as a ratio of the useful part of energy and the fuel consumption (eq. 1). A power-to-heat ratio is defined in eq. 2 which specifies the operating condition of a site (Smith, 2005).

$$\eta_{cogen} = \frac{(W+Q_{heat})}{Q_{fuel}} \quad (1)$$

$$R = \frac{W}{Q_{heat}} \quad (2)$$

For a fixed steam heat demand, by varying power generation, we can obtain different *R*-ratios and corresponding cogeneration efficiencies. As a result, we can generate the so-called *R*-curve as shown in Fig. 2. The *R*-curve shows the relation between the maximum cogeneration efficiency (η_{cogen}) and power-to-heat ratio (*R*). It also indicates the optimal configuration for a simple utility system for the required *R*-ratio. The *R*-curve constructed above has two major limitations. First, it is developed based on a simple utility system and cannot deal with a complex system, which includes multiple steam distribution levels and complex steam turbine configuration. Secondarily, it assumes constant isentropic efficiencies for steam turbines, which leads to nonrealistic results. The above two limitations prevent this *R*-curve from practical applications (Kimura and Zhu, 2000).

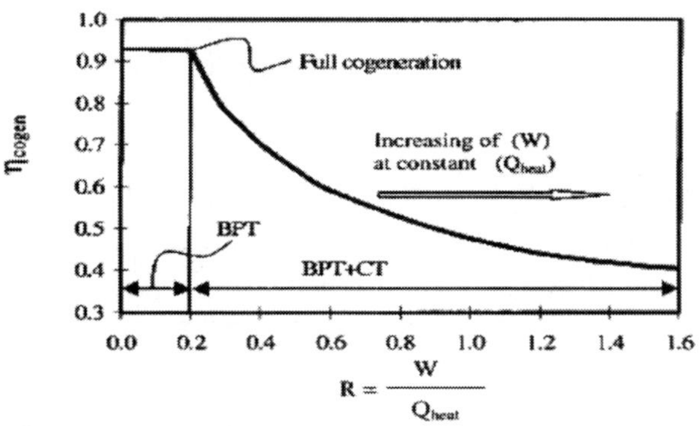

Figure 2: A typical R-curve

By applying some newer methods and using a newer model for steam turbine, it is possible to eliminate the limitations of the R-curve. Then, drawing R-curve for any complex system will be possible with approximately good answers.

3. R-ratio vs. TAC Curve

R-curve is a thermodynamic-based curve, describing cogeneration potential available in a site. To make R-curve more comprehensible, inserting economic term will be useful. In fact economic term makes the site easier to be analyzed, and then it is possible to choose the best point of operation, thermo-economically. To insert economic term, Total Annual Cost (TAC) of the site in each R-ratio should be computed using eq. 3.

$$TAC = Q_{fuel} \times C_F + Q_{fuel,GT} \times C_F + W_{import} \times C_{P,import} + f \times (C_{ST} + C_{GT}) \quad (3)$$

where C_{ST} and C_{GT} are the capital investments for a new steam turbine and a gas turbine, respectively. C_{ST} and C_{GT} include the installation costs and the initial start-up cost. f-factor is an annualizing factor. In the most of references, the prices are given in constant dollars for the current year. It is important to note that working in constant dollars does yield a value of capital charge factor that is lower than expected. All of the costs are brought to their present worth in the initial year and then the annual equivalent of each cost is calculated by multiplying it by an annualizing factor or f-factor. Annualizing factor is calculated using eq. 4 in which *i* is the interest and *L* is the unit life.

$$f = \frac{i}{(1+i)^L} - 1 \quad (4)$$

In this examination, the capital investments are calculated as shown in Table 1 (Kimura and Zhu, 2000).

Table 1: Calculation of Capital Investment

steam turbine	$0.73\,(1 + 0.15X) + 0.0634\,W$ + cost (install) 100% MM$ (W in MW, X is number of extractions)
gas turbine	$0.90 W_{GT,max}$ MM$ (includes installation cost) ($W_{GT,max}$ in MW)

4. Mathematical Model

In order to construct these two curves (R-curve and R-ratio vs. TAC curve), a MATLAB Code has been prepared. Using this MATLAB Code, these two curves may be constructed for any case study. The MATLAB Code needs only some initial input information about the site being studied. The output will be the 'R-curve' and 'R-ratio vs. TAC curve'.

5. Emissions of Different Fuels

Estimation of annual emissions in this paper is based on emission factors. An emission factor is a representative value that attempts to relate the quantity of a pollutant released to the atmosphere with an activity associated with the release of that pollutant. These factors are usually expressed as the weight of pollutant divided by a unit weight, volume, distance, or duration of the activity emitting the pollutant (e.g., kilograms of particulate emitted per megagram of coal burned). In most cases, these factors are simply averages of all available data of acceptable quality, and are generally assumed to be representative of long-term averages for all facilities in the source category (i.e., a population average). Emission of each fuel is computed by product of its emission factor and the fuel consumption (EPA, 1998).

The emission factors that we used were mostly obtained from the EPA's AP-42 (EPA, 1998). We used the updated version of the emission factor database FIRE 6.23, available at EPA's web site. For CO2 estimation, emission factors were not available for all the fuels or boilers; therefore we used Sener emission factors for the same. In Table 2, the emission factors used for fuel oil and natural gas are given for a specified combustion type and boiler configuration, as an example.

Table 2: Emission factors used for power generation

Fuel	*Combustion Type*	*Boiler Configuration*	CO_2[1] $t\ m^{-3}$	NO_x $Kg\ m^{-3}$	SO_2[2] $kg\ m^{-3}$
Fuel Oil	*External*	*Normal*	*3.04*	*5.63*	*18.81*S%*
Natural Gas	*External*	*Normal*	1.92×10^{-6}	2723.73×10^{-6}	9.61×10^{-6}

Source: EPA AP-42 (1998), from FIRE V. 6.23
1. Sener (2003)
1. S% indicates percentage of S in the respective fuel, by weight.

6. Methodology

In this paper, R-curve concept has been developed to study the influence of changing operating fuel, on the operation and also total annual cost of the total site. For this purpose, after considering a utility system of a total site as the case study, different R-curves are constructed for burning each of different operating fuels in the boiler of the total site. It is useful to compare these R-curves to choose the better fuel, but this is not enough. R-curve is a thermodynamic-based curve and does not take the economic term into account. To obtain more useful results for choosing a better fuel available in each case study, another curve of "R-ratio vs. Total Annual Cost (TAC)" has been introduced and employed. Using this new curve it is possible to determine the total annual cost of the total site in each R-ratio. By constructing these curves for each kind of fuel and comparing them, the fuel leading to a lower total annual cost of the total site will be determined.

Then, using these two curves (R-curve and R-ratio vs. TAC curve), it is possible to determine more efficient operating fuel for a total site. But as mentioned in the previous section of the paper, considering the pollution degree of each fuel is also another factor for choosing the most appropriate fuel.

Often there is a contrast between choosing a fuel that is economically verified or a fuel that is not so much economical but it has really less pollutants.

To find a solution for this problem and make the best decision, it is recommended that examine all available fuels in the case being studied and arrange them respecting to their clarity. Then, fuels that are clean enough respecting to the local laws are considered and analyzed using "R-curve" and "R-ratio vs. TAC curve" analysis tools. The most beneficial fuel between these determined fuels will be chosen as the most appropriate one for the case being studied.

This method has some advantages to the previous approaches on process cogeneration:

- This method includes economic considerations other than thermodynamic one.
- A very simple method has been introduced to analyze a total site thermo-economically, i.e. using this method optimal options for energy cost reduction in a utility system can be determined with minimal engineering effort.
- Total annual cost of the total site in any operation point on the R-curve is available.
- It is possible to develop this new method by combination with approaches on application of the R-curve analysis. Then it will be possible to have a more detailed analysis that includes economic approaches too. Examples of these applications are retrofitting, debottlenecking and operational management issues which are available in (Kimura and Zhu, 2000).

7. Case Study

7.1. Introduction

A site as a case study shown in Fig. 3 has four levels of steam mains (VHP, HP, MP, and LP) and two extraction turbines (T1 and T2). T2 consists of three backpressure sub turbines, while T1 has one backpressure sub turbine and one condensing sub turbine. The power deficit is satisfied by importing power or a gas turbine, may be installed.

Condensate loss is returned to boiler by percentages mentioned in the figure, mixing with makeup water.

The following equations are given to calculate the fuel consumption in the GT ($Q_{fuel,GT}$) and the heat content in HP steam generated at the HRSG ($Q_{heat,HRSG(HP)}$), which are added later to the site (Nishio et al., 1980; Chou and Shih, 1987). It is assumed that the installed GT and HRSG are exhausted to the HP level.

$$Q_{fuel,GT} = 2.75 \times W_{GT} + 8.01 \quad (5)$$

$$Q_{heat,HRSG(HP)} = 0.37 \times W_{GT} + 2.55 \quad (6)$$

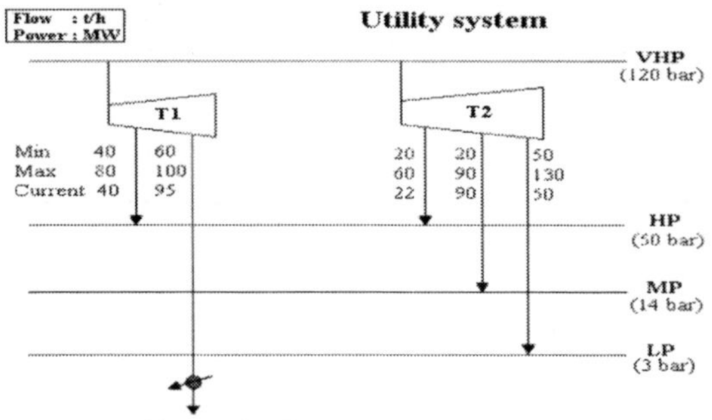

	condensate loss	heat demand (MW)
HP	75%	30
MP	60%	70
LP	80%	85

Figure 3: Schematic figure of the utility system of the case study

Where W_{GT} is generated power from the GT. Then, the GT system can be linked with the rest of the utility system by using the following equations (Colmenares and Seider, 1989; Manninen and Zhu, 1999):

$$Q_{heat(HP)} = \sum_i^P Q_{exhaust(i)}^{HP} + Q_{heat,HRSG(HP)} + \sum_i^P Q_{LD(i)}^{HP} \quad (7)$$

$$W_{demand} = \sum_i W_{(i)} + W_{GT} + W_{import} \quad (8)$$

$$Q_{fuel,total} = \sum_i Q_{fuel(i)} + Q_{fuel,GT} + Q_{fuel,import} \quad (9)$$

Also it should be noted that the shaft power of steam turbines, is calculated using the Willan's line model (Varbanov et al., 2004; Makwana, 1998):

$$W_{(i)} = n \times m - W_{int} \quad (10)$$

All of the information necessary for calculating n and W_{int} as the specified turbine's parameters are presented at (Varbanov et al., 2004).

The shaft power of the gas turbine used in this case study is also calculated with a same method:

$$W_{GT} = 1.2 \times m_f - 0.8 \quad (11)$$

In this case study mechanical efficiency of all turbines is 0.95 and boiler efficiency is 0.92 and the import power price is 0.01 ($/kWh). Unit life of 5 years and interest of 10% has been considered in this case study. It is important to note that with other import power price or fuel price, the results may change significantly but efficiencies of turbines and boiler has very little influence on the results. Also increase or reduction of annualizing factor is not that important because it will affect all of the results (for different fuels) the same. Thus the results are very sensitive to fuel and import power price only.

Some available fuels are considered to be used as the operating fuel in this case study. There is possibility to examine other fuels too. In this case study, sample fuels are natural gas, asphalt and fuel oil with prices given in Table 3.

Table 3: Sample fuel prices of the case study

Kind of Fuel	Price ($/kWh)
Asphalt	0.0079
Fuel Oil	0.0160
Natural Gas	0.0183

7.2. Screening Results

The results are obtained for the case study shown in Fig. 3, for different operating fuels. First it is assumed that the operating fuel is asphalt. Fig. 4 shows the result.

Also Fig. 5 and Fig. 6, show the result for using fuel oil and natural gas as the operating fuel, respectively.

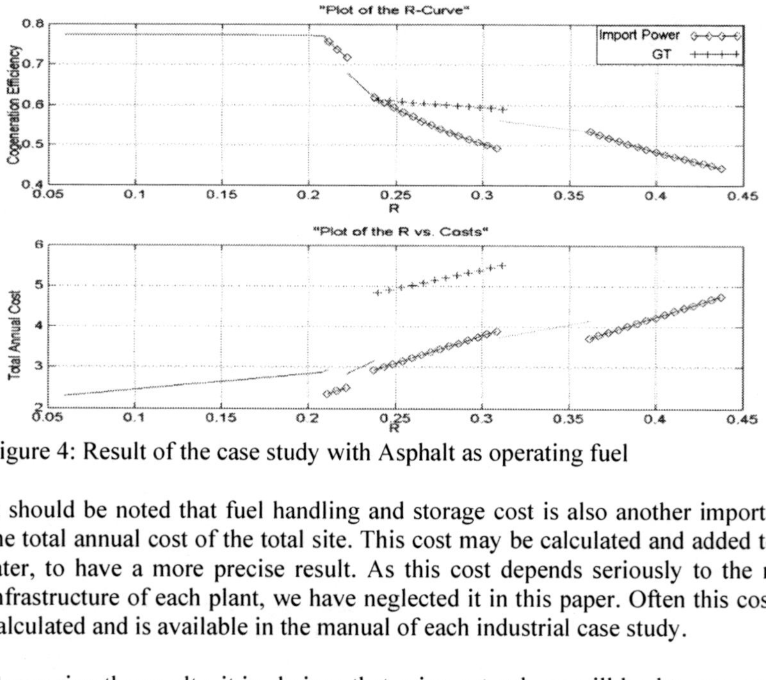

Figure 4: Result of the case study with Asphalt as operating fuel

It should be noted that fuel handling and storage cost is also another important part of the total annual cost of the total site. This cost may be calculated and added to the TAC later, to have a more precise result. As this cost depends seriously to the region and infrastructure of each plant, we have neglected it in this paper. Often this cost has been calculated and is available in the manual of each industrial case study.

Comparing the results, it is obvious that using natural gas will lead to more total annual cost than fuel oil and asphalt in all R-ratios. Also using asphalt, will lead to the least total annual cost between these three fuels. This result is obtained by comparing "R-ratio vs. TAC" curves. For example the total annual cost of installing and operating of gas turbine system, in the case of using natural gas as the operating fuel is much higher than in the case of asphalt as the operating fuel. Also this is true for all other parts of the total site, like condensing turbine, backpressure turbines and imported power.

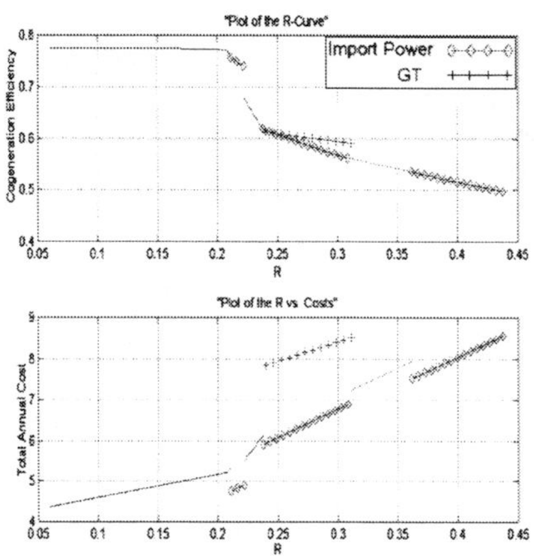

Figure 5: Result for Fuel Oil as operating fuel

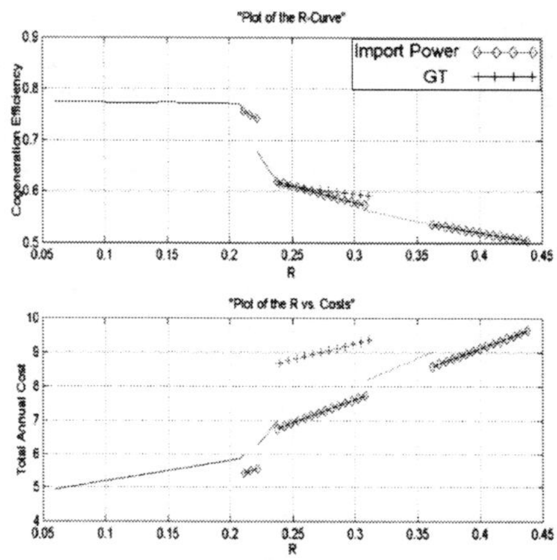

Figure 6: Result for Natural Gas as operating fuel

Comparing R-curves will also results that the R-curve obtained for asphalt as the operating fuel has better performance than two other R-curves. This is true because totally, cogeneration efficiency of the R-curve obtained for asphalt as the operating fuel is higher than cogeneration efficiencies of other two R-curves. Then comparing "R-curves" and also "R-ratio vs. TAC" curves will result that it is more economical to use asphalt as the operating fuel for this case study.

But the other factor for choosing a fuel is its pollution degree. The emission of each fuel is obtained from product of its emission factor and the fuel consumption. Also in each case study, fuel consumptions of all examined operating fuels are equal in the same R-ratios. Then for comparing the emission of these fuels, it is enough to compare only the emission factors of them to determine their relative pollution degree. Comparing emission factors of these three fuels will result that pollution degree of asphalt is very much higher than natural gas and fuel oil. Also it is concluded that natural gas is the cleanest fuel between these three kinds of fuel.

As mentioned, in this case study asphalt has very high pollutants and natural gas leads to a very high TAC for the site. But fuel oil meets both restrictions. It is clean enough to be verified by local laws and also it has a lower TAC respect to natural gas. Then it is better to choose fuel oil between these fuels, as the operating fuel in this case study.

8. Conclusions

The R-curve concept is developed and employed to analyze the effect of changing operating fuel on the performance and cost of the total site. "R-ratio vs. TAC curve" is the new graphical tool that is introduced here by developing R-curve concept.
After constructing "R-curve" and "R-ratio vs. TAC curve" for each operating fuel in a total site, it is possible to compare them and determine the effect of each fuel on the operation and TAC of that total site. Finally considering local emission restrictions, it is possible to choose the most efficient and economical fuel that is also verified by local emission laws for the case being studied.

Overall, the R-curve analysis can provide insights and solutions for choosing the most efficient and economical operating fuel in a total site with little engineering effort.

References

Chou, C. C. and Shih, Y. S., 1987, A Thermodynamic Approach to the Design and Synthesis of Plant Utility Systems, Ind. Eng. Chem. Res., 26, pp. 1100-1108.
Colmenares, T. R. and Seider, W. D., 1989, Synthesis of Utility Systems Integrated with Chemical Processes, Ind. Eng. Chem. Res., 28, pp.84-93.
EPA. 1998. Compilation of air pollutant emission factors, volume 1: Stationary point and area sources, fifth edition, AP-42. US Environmental Protection Agency, Office of Air Quality Planning and Standards. Research Triangle Park, North Carolina.
Kenney, W.F., 1984, Energy Conservation in Process Industries, Academic Press, Orlando, London.
Kimura, H. and Zhu, X.X., 2000, R-Curve Concept and its Application for Industrial Energy Management, Ind. Eng. Chem. Res., 39, pp. 2315–2335.
Makwana, Y., 1998, Energy Retrofit and Debottlenecking of Total Sites, Ph.D. thesis, UMIST, Manchester.
Manninen, J. and Zhu, X. X., 1999, Optimal Gas Turbine Integration to the Process Industries, Ind. Eng. Chem. Res., 38 (11), pp.4317-4330.
Mavromatis, S.P. and Kokossis, A.C., 1998, Conceptual Optimization of Utility Networks for Operational Variations—I. Targets and Level Optimization, Chem. Eng. Sci., 53(8), pp. 1585–1608.
Nishio, M., Itoh, J., Shiroko, K. and Umeda, T., 1980, A Thermodynamic Approach to Steam Power System Design, Ind. Eng. Chem. Process Des. Dev., 19, pp. 306-312.
Smith R., 2005, Chemical Process Design and Integration, John Wiley & Sons.
Varbanov, P.S., Doyle, S. and Smith, R., 2004, Modelling and Optimisation of Utility Systems, Trans IChemE, Part A, Chem. Eng. Res. Des., 82(A5), pp. 561–578.
Varbanov, P.S., Perry, S., Makwana Y., Zhu, X. X. and Smith, R.,2004, Top-level Analysis of Site Utility Systems, Trans IChemE, Part A, Chem. Eng. Res. Des., 82(A6), pp. 784–795.

Proceedings of the 2nd Annual Gas Processing Symposium
Farid Benyahia and Fadwa T. Eljack (Editors)

Chemicals in Gas Processing (CHIGP): An industrial project for the thermodynamics of complex petroleum fluids

Georgios M. Kontogeorgis, Ioannis Tsivintzelis, Michael L. Michelsen, Martin P. Breil and Erling H. Stenby

Department of Chemical and Biochemical Engineering
Technical University of Denmark
Building 229, DK-2800 Kgs. Lyngby, Denmark

Abstract

Since 1999, a thermodynamic project funded by major petroleum, chemical and gas companies has been carried out at the Technical University of Denmark (Department of Chemical and Biochemical Engineering). The project is called CHIGP (Chemicals in Gas Processing).

The purpose of the project is:

i. to provide a rigorous thermodynamic model, the CPA (Cubic-Plus-Association) equation of state together with the parameters required for multiphase multicomponent phase equilibrium calculations of mixtures containing oil & gas, water and chemicals e.g. gas hydrate inhibitors (methanol, glycols) and organic acids
ii. to develop robust algorithms for complex thermodynamic calculations and parameter estimation
iii. to conduct experimental measurements to support the model development when necessary
iv. to disseminate the results of the project to the participating companies via CAPE-OPEN compliant modules and via a user-model specifically developed for the ASPEN process simulator

This article highlights results for the project with emphasis on newest results developments for:

- mixtures with acid gases (CO_2, H_2S), water and chemicals
- glycol-containing systems

Keywords: Cubic-Plus-Association, Phase Equilibria, Complex fluids

1. The CPA (Cubic Plus Association) Equation of State

CPA is a versatile model that can be used to predict thermodynamic properties of pure fluids and mixtures with highly non-ideal behavior, such as mixtures with hydrogen bonding fluids in gas, liquid or supercritical state. It is a combination of the SRK (Soave-Redlich-Kwong) cubic equation of state with the Wertheim's first order thermodynamic perturbation theory [Kontogeorgis et al. 1996]. During the last years the model has been successfully applied for the modeling of various systems. CPA is shown to be highly successful for systems that contain associating compounds (water, alcohols, glycols, acids, amines) and inert non-associating compounds (e.g., hydrocarbons) (Kontogeorgis et al. 2006a, 2006b).

The CPA equation of state for mixtures expressed in terms of pressure P, is:

$$P = \frac{RT}{V_m - b} - \frac{\alpha(T)}{V_m(V_m + b)} - \frac{1}{2}\frac{RT}{V_m}\left(1 + \rho\frac{\partial \ln g}{\partial \rho}\right)\sum_i x_i \sum_{A_i}(1 - X_{A_i}) \quad (1)$$

where, ρ is the molar density ($=1/V_m$), X_{A_i} is the fraction of sites type A on molecule i that do not form hydrogen bonds with other active sites and is related to the association strength $\Delta^{A_iB_j}$ between two sites belonging to two different molecules, e.g. site A on molecule i and site B on molecule j. The parameters of the association term are found in the expression for the association strength:

$$\Delta^{A_iB_j} = g(\rho)\left[\exp\left(\frac{\varepsilon^{A_iB_j}}{RT}\right) - 1\right] b_{ij}\beta^{A_iB_j} \quad (2)$$

with the radial distribution function $g(\rho) = \frac{1}{1 - 1.9n}$ and $n = \frac{1}{4}b\rho$ while

$$b_{ij} = \frac{b_i + b_j}{2}$$

In equation (2), the parameters $\varepsilon^{A_iB_j}$ and $\beta^{A_iB_j}$ or β^{cross} are called the association energy and the association volume, respectively. In total, five pure compound parameters are needed for hydrogen bonding compounds.

No mixing rules are needed in the association term, while typically the van der Waals one fluid mixing rules are used in the SRK part of the model, together with the arithmetic mean combining for the co-volume parameter. For the cross-energy parameter, the geometric mean rule is used for the energy parameter a_{ij}, and one binary interaction parameter k_{ij} is often employed:

$$a_{ij} = \sqrt{a_i a_j}(1 - k_{ij}) \quad (3)$$

For cross-associating mixtures, the following combining rule (CR-1 rule) is used:

$$\varepsilon^{A_iB_j} = \frac{\varepsilon^{A_iB_i} + \varepsilon^{A_jB_j}}{2} \quad \text{and} \quad \beta^{A_iB_j} = \sqrt{\beta^{A_iB_i}\beta^{A_jB_j}} \tag{4}$$

In the case of solvating systems (i.e. in cases where cross associating interactions exist and only one of the components is self-associating) the modified CR-1 rule is used:

$$\varepsilon^{A_iB_j} = \frac{\varepsilon_{associating}}{2} \quad \text{and} \quad \beta^{A_iB_j} = \beta^{cross}$$

where β^{cross} is fitted to the experimental data for binary systems (5)

2. Characteristic examples

Some typical results are presented in this section, while more will be shown in the presentation. All results presented here refer to hydrogen bonding mixtures and especially to solvating systems (i.e. cross associating systems between one self-associating and one non self-associating fluid). Characteristic calculations are presented for mixtures of alcohols, glycols or water with acid gases, while one characteristic example is presented for the liquid-liquid equilibrium of MEG-benzene system. For obtaining satisfactory description of the phase equilibria for such mixtures, CPA (and possibly other similar models from SAFT family) must account explicitly for the solvation between the associating compound (methanol, glycol or water in these cases) and the acid gas or aromatic hydrocarbons. In order to do that with CPA, typically the modified CR-1 combining rule is used (eq. 5) which introduces an additional adjustable parameter, the cross-association volume, thus two parameters must be fitted to experimental data (including the k_{ij} from the SRK term).

Using this approach, CPA results in accurate calculations for the phase behavior of such hydrogen bonding binary mixtures. Furthermore, having the binary parameters from the corresponding binary systems the model satisfactorily predicts the phase behavior of ternary or multicomponent mixtures as shown in Figure 1 for the CO_2 – methanol – ethanol system. One approach to avoid the need for two adjustable parameters which is illustrated here is by using the so-called homomorph approach. According to this approach, the k_{ij} parameter can be obtained from a mixture which is expected to have similar (to the investigated mixture) "physical" interactions, but no cross-association or solvation. For example, for CO_2-methanol, k_{ij} can be obtained from CO_2-ethane, and in the case of MEG-benzene, it can be obtained from MEG-hexane. Then, once k_{ij} is known from the homomorph system, only the cross-association volume β^{cross} needs to be estimated (see eq.5). As figures 2-6 illustrate this approach yields satisfactory results for these systems, thus effectively reducing the number of adjustable parameters when CPA is used for solvating systems.

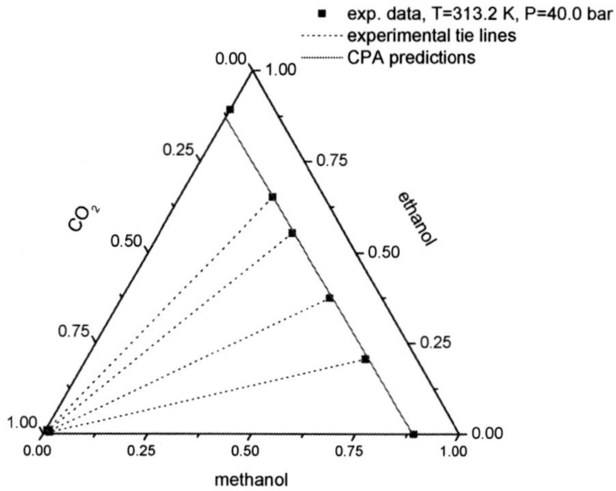

Figure 1: CO2 – methanol – ethanol VLE. Experimental data (points) and CPA predictions (lines).

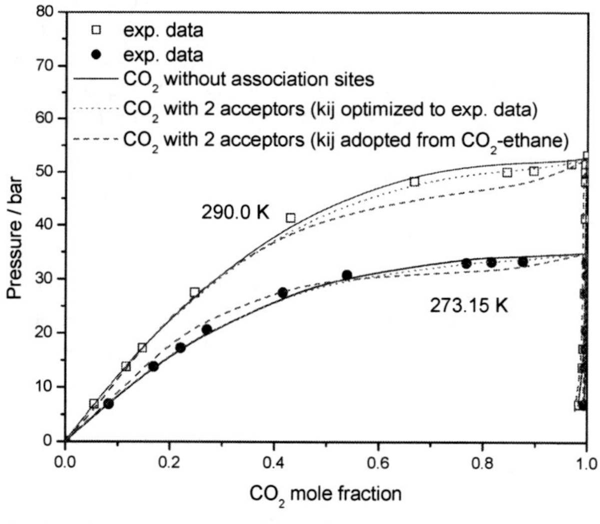

Figure 2: CO2 – methanol VLE. Experimental data (points) and CPA calculations (lines) considering that CO2 has no association sites (solid lines) or that CO2 has two (proton acceptor) association sites (dotted lines).

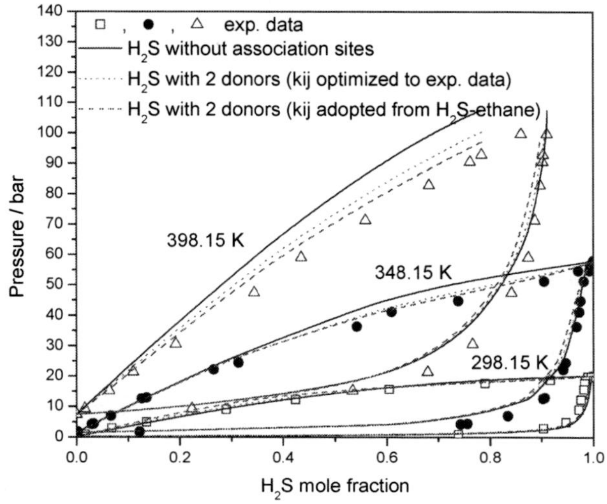

Figure 3: Methanol – H_2S VLE. Experimental data (points) and CPA calculations (lines) considering that H_2S has no association sites (solid lines) or that H_2S has two (proton donor) association sites (dotted lines).

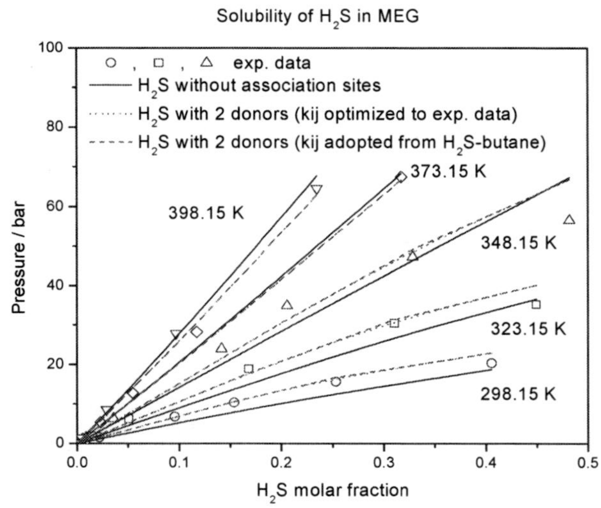

Figure 4: MEG – H_2S VLE. Experimental data (points) and CPA calculations (lines) considering that H_2S has no association sites (solid lines) or that H_2S has two (proton donor) association sites (dotted lines).

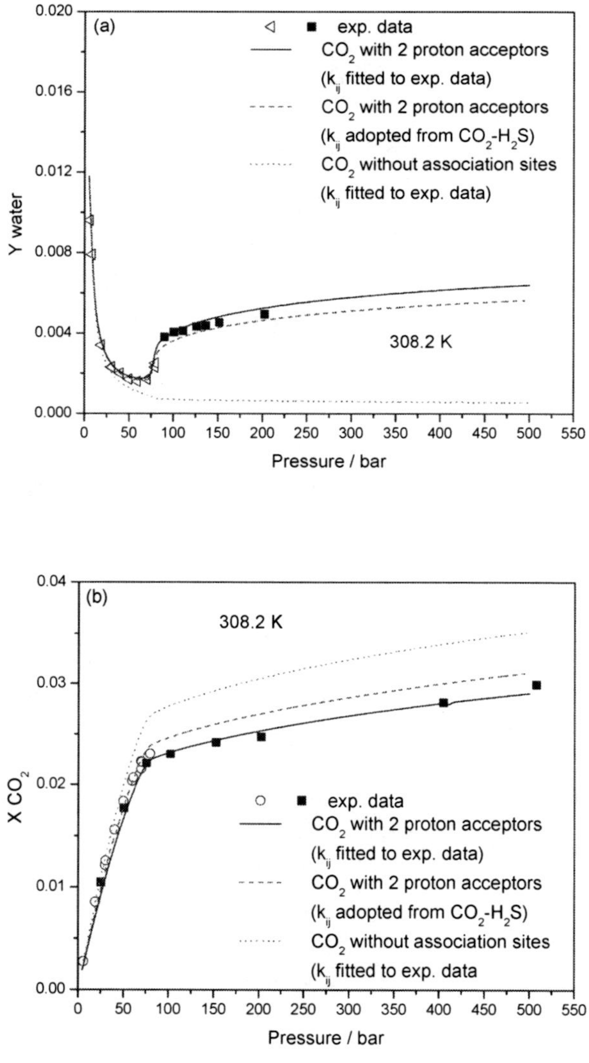

Figure 5: CO_2 – water VLE at 308.2 K, vapor phase (a) and liquid phase (b) compositions. Experimental data (points) and CPA calculations (lines) considering that CO_2 has no association sites (dotted line) or that CO_2 has two (proton acceptor) association sites (solid and dashed lines).

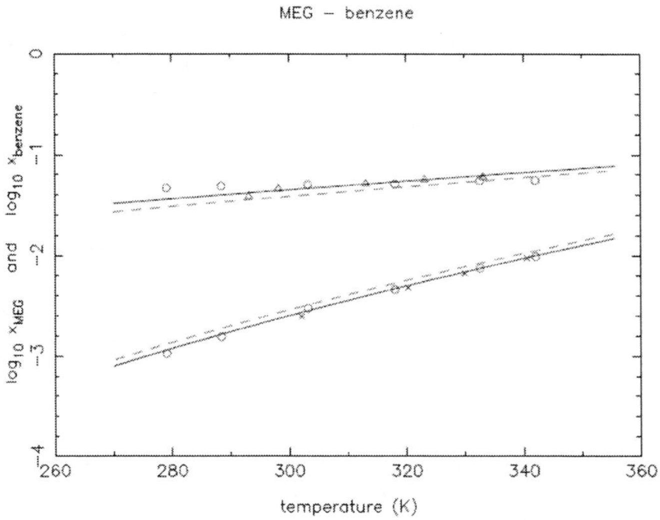

Figure 6: The LLE calculations of MEG-benzene. The full lines are CPA with two optimized parameters. The dashed ones are CPA calculations using the homomorph approach with k_{ij} = 0.059 and β^{cross} = 0.048. The upper curve is solubility of benzene in MEG, and lower curve is solubility of MEG in benzene.

3. Conclusions

Since its appearance in 1996, the Cubic-Plus-Association (CPA) equation of state has been shown to be a versatile and highly successful thermodynamic model for mixtures of interest to the petroleum and gas industries. Most specifically, CPA has been shown to predict successfully multiphase, multicomponent equilibria of mixtures containing gases, hydrocarbons, water and polar compounds used as gas hydrate inhibitors or other chemicals (methanol, glycols, organic acids). CPA combines the classical simple Soave-Redlich-Kwong (SRK) equation with an advanced association term. It provides a balance between accuracy and simplicity, as the model reduces to (the often used in oil & gas industry) SRK in the absence of hydrogen bonding compounds (water, alcohols, acids, etc). Besides simplicity and accuracy, the numerical implementations of the association term ensure that the computation time is not much higher than that of SRK and other simple models.

In this work it is shown that CPA can be used successfully for various solvating mixtures such as acid gases with alcohols and water as well as glycols with aromatic hydrocarbons. These mixtures are characterized by complex cross-interactions even

though one of the molecules is not self-associating. It is also shown how a simple approach can be utilized to reduce the number of adjustable parameters.
Future plans of the project include the extension of CPA to mixtures with electrolytes especially for mixed salt-hydrate inhibitor mixtures, reservoir fluids as well as a systematic investigation of the performance of the model for multicomponent mixtures containing acid gases, water, alcohols/glycols and hydrocarbons.

Acknowledgment

The authors wish to thank the industries participating to the CHIGP consortium (Statoil, BP, TOTAL, Mærsk Oil and GASSCO) for financial support of this work.

References

G.M. Kontogeorgis, M.L. Michelsen, G.K. Folas, S. Derawi, N. von Solms, E.H. Stenby, *Ind. Eng. Chem. Res.* 45, 2006a, 4855-4868.
G.M. Kontogeorgis, M.L. Michelsen, G.K. Folas, S. Derawi, N. von Solms, E.H. Stenby, *Ind. Eng. Chem. Res.* 2006b, 45, 4869-4878.
G.M. Kontogeorgis, E. Voutsas, I. Yakoumis, D.P. Tassios D.P., *Ind. Eng. Chem. Res.*, 35, 1996, 4310-4318.

Proceedings of the 2nd Annual Gas Processing Symposium
Farid Benyahia and Fadwa T. Eljack (Editors)

Hybrid Models for Monitoring & Optimization of Hydrocarbon Separation Equipment

Vladimir Mahalec [a], Asaad Hashim [b], Yoel Sanchez [a]

[a]*McMaster University, 1280 Main St. W., Hamilton, ON L8S 4K1, Canada*
[b]*Saudi Aramco, Dhahran, Saudi Arabia*

Abstract

Hybrid models of separation equipment developed in this work consist of two parts: (i) material and energy balances that determine external and internal material and energy flows in the equipment and (ii) linear PLS models that predict product properties based on the properties of the feed and the internal equipment flows. The models are highly accurate, capable of predicting product properties and internal equipment flows with approx. 0.5% error with respect to the rigorous models of the same equipment. Typical separation equipment (flash, absorber, stripper, two product distillation, and an atmospheric pipestill) have been modelled.

Keywords: hybrid models, separation equipment, PLS

1. Introduction

Separation equipment such as distillation, absorption, and stripping are the fundamental processes found in refineries and petrochemical industries. Accurate prediction of the performance of the separation equipment is essential for optimization of real-time operation, production planning and scheduling.

Over the last two decades rigorous models (tray to tray distillation models, detailed kinetic models or reactors) have been introduced for use in on-line optimization. Models of crude separation units describe crude feedstocks (a complex hydrocarbon mixture) in terms of pseudocomponents, predict product composition in terms of pseudocomponents and then compute distillation curves of the product streams. This enables model use for predicting properties such as 5% or 95% point on a distillation curve and comparison to the corresponding specifications. The resulting models are quite large (e.g. 20,000 to 50,000 equations for an atmospheric pipestill), nonlinear, computationally intensive and requires a specialized expertise to be maintained.

In order to reduce the models size and its complexity, we are proposing to use hybrid models of petroleum separation equipment. These models provide prediction accuracy that is typically within 0.5% of the rigorous tray to tray models, while being much smaller (several hundred equations vs. 20,000 to 50,000) and mostly linear. At the same time, these models compute product quality (distillation curves), mass balances, and energy balances. Hence, these models open a possibility to model an entire refinery by a model that would be substantially smaller than the current RTO practice and yet provide the same accuracy as the rigorous models.

2. Previous Work

Most to the previous work on approximate modeling of distillation equipment has focused on predicting product quality, assuming that the models would be used for on-line inferential predictions.

Mejdell and Skogestad (1991) used all tray temperatures in a distillation tower to estimate product compositions. Hence, the model is suitable for inferential predictions, i.e. monitoring online but cannot predict product quality from operating data (product flows, heat duties).

Safavi, Nooraii, and Romagnoli (1999) developed a hybrid model of a binary distillation column. The hybrid model estimates the separation factor in a distillation column and is constructed in two stages. The first stage is to develop a wave-net model that predict the separation factor and the second is to compute mass and energy balances.

More extensive review of distillation column modeling is presented by Abdullah, Aziz and Ahmad (2007).

3. Hybrid Model Structure

3.1. Selection of Independent Variables for Product Quality Prediction

Crude feedstocks are characterized by their distillation curves. In this work tru boiling point (TBP) curves are used, since other distillation curves can be readily converted to TBP curves. For instance, if crude is fed into a 10 stage absorber tower, then the distillation curves of the feed and the products can be as shown in Fig. 1. If separation in the distillation tower was perfect, then the distillation curves of the products would be sections of the feed distillation curves. Each product distillation curve would start at its front-end cut point and would end at its back-end cut point. However, since the separation is not perfect, there is an overlap between the adjacent product streams. Operating and design variables (internal reflux, number of stages, relative volatility) influence the product quality. Hence, a hybrid model will use:

- Feed cutpoint temperatures
- Relative volatility
- Internal reflux

as independent variables to predict distillation curves of the product streams in a tower with a given number of stages.

In order to use relative volatility we need to define key components for each separation cut. Since petroleum separations are not sharp, the components are distributed between adjacent products. Hence, to represent distribution of components, this study uses relative volatility between the pseudocomponents corresponding to the middle of the product LV% range and the adjacent cutpoint. For separation presented in Fig. 1, there are five pseudocomponents that will be used to develop hybrid model; they correspond to cutpoint 1, cutpoint 2, and 50% points of the three product streams. Then the relative volatilities are calculated according to the following expressions:

$$\alpha_{1,2} = \frac{k_1}{k_2},\ \alpha_{2,3} = \frac{k_2}{k_3},\ \alpha_{3,4} = \frac{k_3}{k_4},\ \alpha_{4,5} = \frac{k_4}{k_5} \quad (1)$$

where k_i is the equilibrium constant of pseudocomponent i.

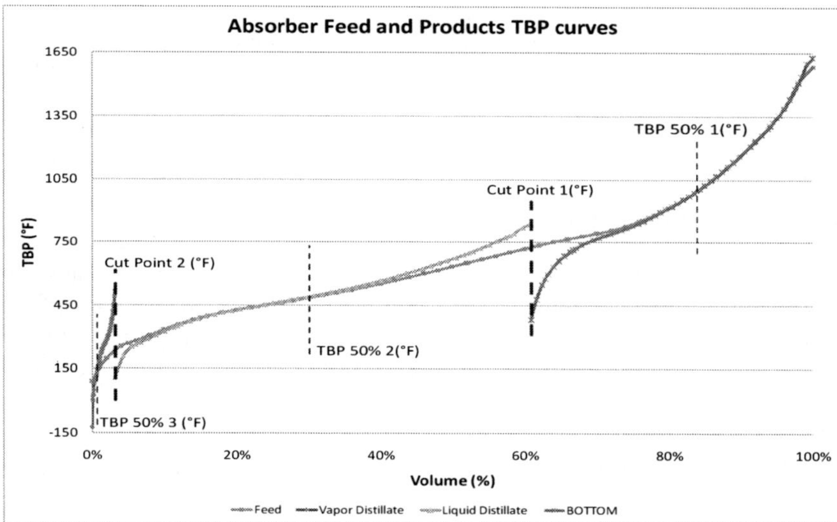

Figure 1: Cutpoints and product 50% points on TBP curve define pseudocomponents

3.2. Product Quality Prediction via Partial Least Squares (PLS) Model

Points on the product distillation curves are predicted via a linear PLS model. PLS is a projection method (S. Wold, 1989) that in the simplest case with two blocks X's and Y's, simultaneously solves two problems, namely:

- to compute a projection that well approximates the matrix X, and
- to compute a projection that well predicts one or several dependant variables collected in matrix Y.

Two data matrices, X and Y, are related to each other by a linear multivariate model. The main advantage of PLS method is its ability to analyze data with many variables in both X and Y, regardless that the variables could be noisy, collinear, and even incomplete. Further explanation of PLS is available in many references.

3.3. Mass and Energy Balances

It is assumed that the reboiler and condenser duty, enthalpy of the streams, as well as product and feed flows, are known. Therefore, the internal vapor and liquid flows are given by (Treybal, 1968):

$$L_i = f(Q_{Cond}, Q_{Reb}, stream\ flowrates, stream\ enthalpies \quad (7)$$

$$V_i = f(Q_{Cond}, Q_{Reb}, stream\ flowrates, stream\ enthalpies \quad (8)$$

Liquid and vapor stream enthalpies are calculated via the following approximation:

$$h = h^0 + Cp_L(T - T^0) \quad (9)$$

$$H = H^0 + Cp_V(T - T^0) \quad (10)$$

where superscript “0” denotes base operating conditions.

4. Distillation Tower Example

Prediction of product quality must be valid for a variety of operating conditions and various feedstock compositions. In order to verify predictions under a variety of feed compositions, two different crudes (Fig. 2) have been used (AspenTech, 2006) and separation of various feed mixtures has been simulated under a variety of operating conditions by a rigorous 20 stage AspenPlus distillation tower model (feed in the middle, partial condenser). Fig. 3 shows dependency of bottoms TBP point on internal reflux; Fig. 4 shows dependency of liquid distillate TBP point on the cutpoint temperature between the distillate and the bottoms, while Fig. 5 shows dependency on

Prediction of product quality has 3 PLS models which are summarized in Table 1. All three models have excellent goodness of fit and predictive capabilities. Clearly, separation between liquid distillate and the bottoms product was the easiest to model, since the overlap is only 50 deg F. On the other hand, vapor distillate and liquid distillate have a very large overlap (more than 300deg F) and the vapor distillate TBP curve is very nonlinear, resulting in a PLS models that has 4 components.

Similar results have been obtained for a flash, absorber, and stripper separation equipment. Invariably, all models have an excellent goodness of fit and predictive capabilities while having a small number of components in the PLS models.

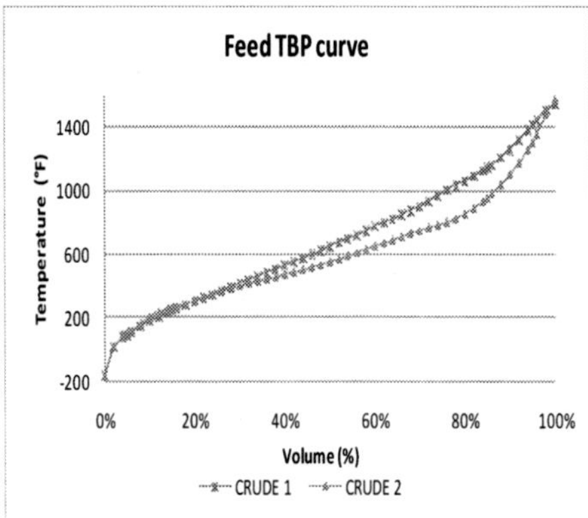

Figure 2: Crude Feed TBP data

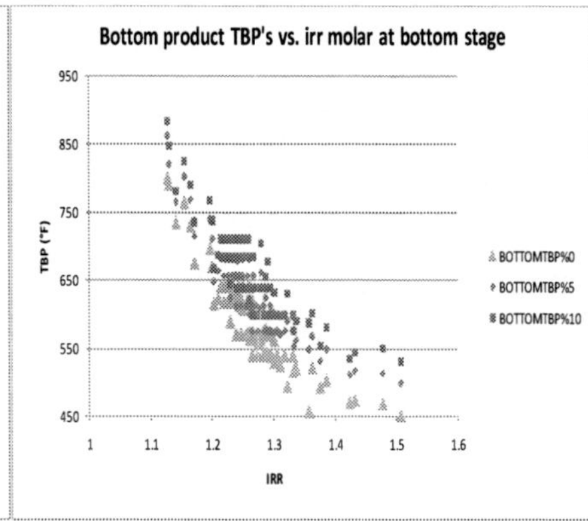

Figure 3: Bottoms TBP vs. Internal Reflux

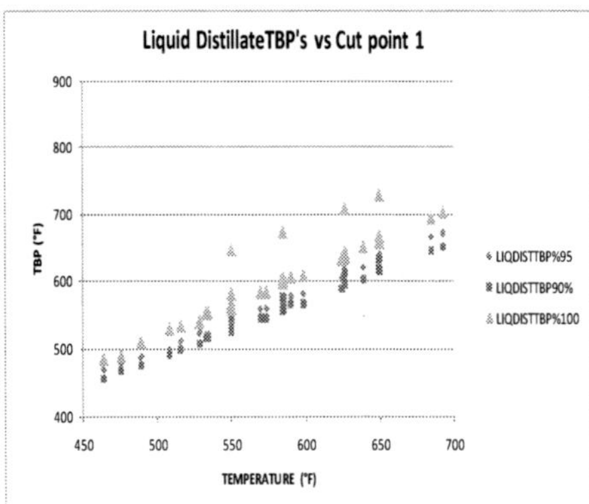

Figure 4: Distillate TBP vs. (D,B) Cutpoint

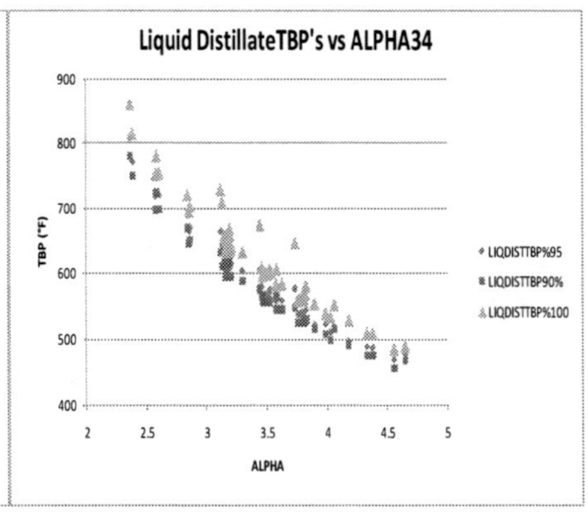

Figure 5: Distillate TBP vs.(D-LV50%, D-LV100%) volatility

Table 1: Summary of Distillation Tower PLS Models

Model	Front/Back of	R^2	Q^2	Number of components
1	Bottoms/Liq.Dist.	0.98	0.97	2
2	Liq.Dist/Vap.Dist	0.96	0.95	4
3	Vap. Dist; Bottoms	0.98	0.99	5

4.1. Model Verification- Product TBP Curves

All hybrid equipment models have been tested for a variety of operating conditions and their results compared to rigorous AspenPlus model results for identical sets of operating conditions. Hybrid model predictions invariably were very close to the results from the rigorous models. Sample product TBP curves from AspenPlus and from hybrid models of the distillation tower are shown in Fig. 6 and 7.

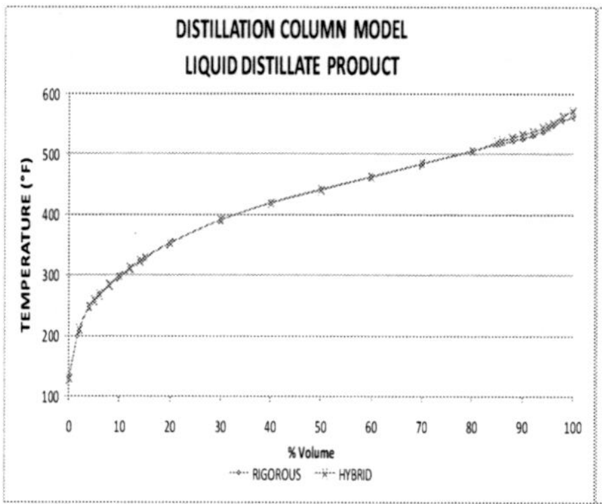

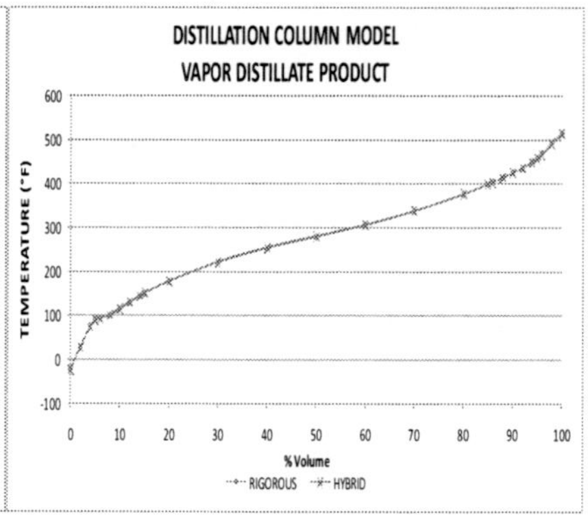

Figure 6: Liquid Distillate TBP Prediction Figure 7: Vapor Distillate TBP Prediction

Particularly interesting is prediction of the 10 stage absorber, since its vapor and liquid distillate have highly nonlinear curves due to their compositions. As shown in Fig. 8 and 9, the hybrid model matches predictions from AspenPlus extremely well.

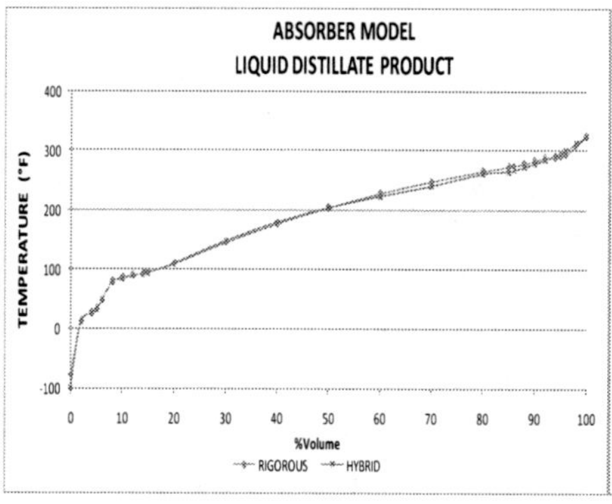

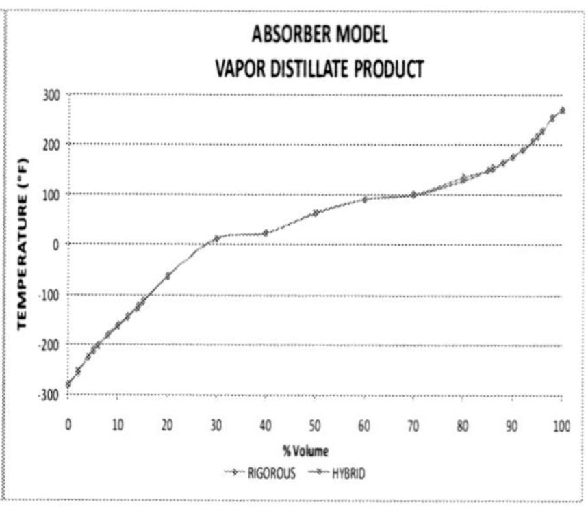

Figure 8: Absorber Liq. Dist.TBP Prediction Figure 9: Absorber Vap. Dist. Prediction

4.2. Model Verification- Optimization of the Complete Hybrid Model

Optimization of a tower performance requires ability to predict performance under the set of operating conditions chosen by the optimization algorithm. As an illustration of the hybrid model accuracy, let us consider the following optimization problem:

minimize: Qreb + Qcond

subject to

Inequality constraints:

- Liquid Distillate TBP 95% ≤ 545 °F
- Bottom TBP 5% ≥ 580 °F

Equality constraints:

- PLS model of Liquid distillate TBP 95% , bottoms TBP 5%
- Material and energy balances
- Internal reflux calculation

Table 2: Optimal Results from Hybrid and from Rigorous Model

Variable	Initial value	Optimized	Rigorous model
Bottom Rate (bbl/D)	55043.1	55323.6	55323.6
Liquid Distillate Rate (bbl/D)	21912	22131.6	21545.1
Vapor Distillate Rate (bbl/D)	5894.4	5394.8	5981.3
Bottom Rate (lb/hr)	758697	762563.9	757457.8
Liquid Distillate Rate (lb/hr)	264507	267152.8	257926.9
Vapor Distillate Rate (lb/hr)	57406	50895.9	66445.3
Qcond (BTU/hr)	119046600	117856134	117856134
Qreb (BTU/hr)	241470507	239055801.9	239594312
Irravg	0.87	0.87	.86
Liquid Distillate TBP95% (°F)	548	544.7	542
Bottom TBP 5% (°F)	584	580.2	579

The optimization problem was solved using "fmincon" function of Matlab, using as free variables bottoms rate, liquid distillate rate, and heat duties in the reboiler and the condenser. The results were then compared with results of solving the same optimization problem with equation oriented AspenPlus (see Table 2). In order to enable closer examination of the results, Table 2 purposely contains more significant figures that it would be used in the industrial practice.

5. Hybrid Model of an Atmospheric Pipestill

Atmospheric pipestill ("Crude Tower") is a very complex separation tower, comprised of a main column, several side strippers and between two and four pumparounds for heat removal. Rigorous pipestill model presented by AspenTech (2006) as a part of an entire crude unit model (see Fig. 10) has been used to simulate the tower performance under a variety conditions and various mixtures of crudes.

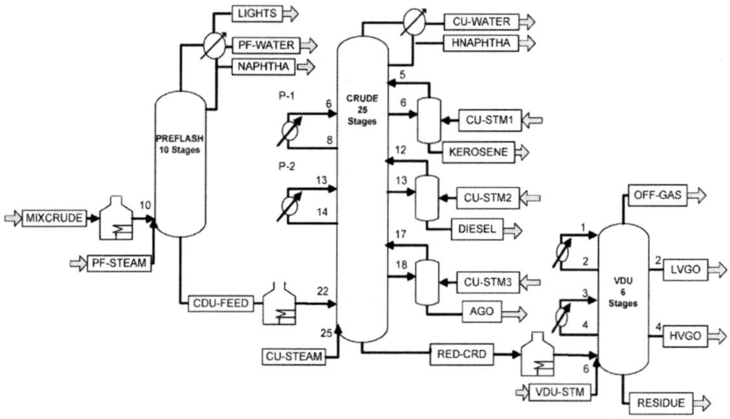

Figure 10: Crude Distillation Unit (AspenTech, 2006)

Hybrid model, analogous to the hybrid models described above, has been developed. The model predicts 0%, 5%, 10%, 20%, 80%, 90%, 95%, 100% points on the product TBP curves. Material and energy balances have been used to compute internal flows and reflux ratios. Several versions of the model have been tested:

- Identify feed TBP curve from known tray temperatures and product data.
- Predict product properties from tray measurements (inferential model)
- Predict product properties from known feed TBP curve and operating data.

The model accuracy is within 0.5% to 1% of the predictions from a rigorous AspenPlus model. Therefore, for practical purposes, the hybrid model matches the rigorous model. The model predicts product TBP curves extremely well (Fig. 10). Moreover, material and energy balance calculations accurately predict the internal reflux profile (Fig.11).

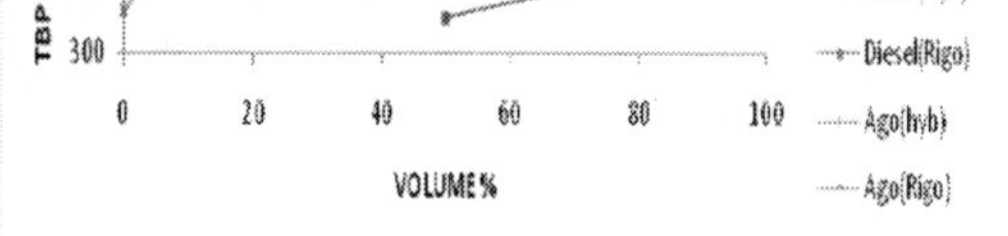

Figure10: Rigorous & Hybrid TBPCurves Overlap

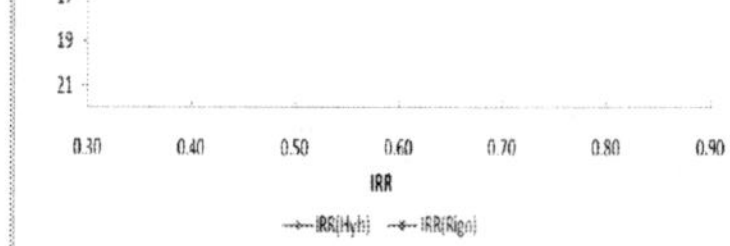

Figure 11: Internal Reflux Prediction

6. Conclusions

Prediction of product quality based on TBP of the feed combined with material and energy balances has enabled us to derive a new class of hybrid models that predict performance of separation equipment with the same accuracy as rigorous models. Hybrid models as proposed in this work have hundreds of mostly linear equations vs. tens of thousands of nonlinear equations contained in rigorous models. Hence, it becomes feasible to model an entire refinery with assurance that the model will be robust and converge reliably. The work opens the possibility for a new class of much simpler optimization applications in plants processing complex hydrocarbon mixtures.

References

Abdullah, Z., N. Aziz and Z Ahmad, "Nonlinear Modeling Application in Distillation Column", Chemical Product and Process Modeling, Vol. 2, iss. 3, Art. 12 (2007).

AspenTech, "Aspen Plus - Getting Started Modeling Petroleum Processes" manual, Aspen Technology, Inc. Cambridge, MA (2006)

Mejdell, T. and S. Skogestad, "Estimation of Distillation Compositions from Multiple Temperature Measurements Using Partial Least Squares Regression", Ind. Eng. Chem. Res. 30, 2543-2555 (1991)

Safavi, A. A., A. Nooraii and J. A. Romagnoli, "A hybrid model formulation for a distillation column and the on-line optimisation study", Journal of Process Control 9, 125-134 (1999)

R. E. Treybal, "Chapter Nine Distillation", in "Mass-Transfer Operations", Second Edition, McGraw –Hill Inc. (1968), pp. 281-406

S. Wold, "Multivariate Data Analysis: Converting Chemical Data Tables to Plots", in "Computers Applications in Chemical Research and Education", Alfred Hüthig Verlag, Eds., Huethig publishing, Heidelberg (1989), pp. 197-216.

Proceedings of the 2nd Annual Gas Processing Symposium
Farid Benyahia and Fadwa T. Eljack (Editors)

Optimal Unloading Procedure for a Mixed Operation of Above-ground and In-ground LNG Storage Tank using Dynamic Simulation

Chul-Jin Lee, Youngsub Lim, Chansaem Park, Sangho Lee, Chonghun Han*
School of Chemical and Biological Engineering, Seoul National University
San 56-1, Shillim-dong, Kwanak-gu, Seoul, Korea

Abstract

The process of unloading liquefied natural gas (LNG) from a carrier ship to a storage tank consists of three steps: recirculation, depressurization and unloading. Because LNG is typically maintained at a cryogenic temperature near -160 °C, a recirculation process is needed to keep the unloading pipeline cool and prevent vaporization of the LNG. The unloading line is then depressurized to a pressure valve that lies between the ship and the storage tank. Finally, the LNG in the carrier ship is transferred to the storage tank. As a rule, there are two different types of LNG storage tanks: above-ground and in-ground tanks. When a single type of tank is used for storage, there are no critical problems encountered between the recirculation and unloading steps. However, for the mixed operation of above-ground and in-ground LNG storage tanks, the depressurization of an unloading pipeline can generate vapor on top of the unloading pipeline of the above-ground tank due to the pressure head. The vapor produced from the above-ground tank can congest depressurization, which can in turn cause excessive BOG(Boil-Off Gas) inflow. In this paper, we suggest a safe and reliable LNG unloading procedure and control strategy for the mixed operation of above-ground and in-ground storage tanks that we designed using dynamic simulations.

Keywords: liquefied natural gas, LNG, LNG storage tank, unloading operation, dynamic simulation

1. Introduction

Natural gas is widely used as a heating fuel and its usage has increased due to its cleanliness. At the source, natural gas is usually transformed into liquefied natural gas (LNG) so as to decrease its volume. It is then transported to the demand region by carrier ship. LNG is then transferred from the carrier ship to an onshore storage tank.

Upon unloading, abrupt temperature increases caused by either heat input or pressure decreases due to the pipeline height can result in vaporization of LNG. This process may be accompanied by a sudden volume expansion of up to 600 times, which could damage the pipeline and storage tank. Therefore, it is important to examine the potential factors that may bring about such a sudden expansion and to design safe and reliable unloading operating procedures.

To accomplish such a goal, we can model the entire unloading process of a given operating procedure using dynamic simulations. In this study, we performed modeling and dynamic simulations for an LNG receiving terminal, which contains two kinds of storage tanks, in order to analyze the causes of sudden vaporization and to suggest a safe and reliable unloading operating procedure.

2. Theoretical Background

2.1. LNG and LNG terminal

LNG is mainly composed of methane and ethane. The composition of LNG varies according to its source, as shown in Table 1. In this research, LNG composed of 90 mol% methane, 8 mol% ethane, and others was used for the modeling and simulations.

Table 1. Examples of LNG composition

Source	Methane	Ethane	Propane	Butane	Nitrogen
Alaska	99.72	0.06	0.0005	0.0005	0.2
Algeria	86.98	9.35	2.33	0.63	0.71
Baltimore Gas	93.32	4.65	0.84	0.18	1.01
New York City	98	1.4	0.4	0.1	0.1
San Diego Gas & Electric	92	6	1	-	1
Qatar	89.86	6.56	2.33	0.94	0.31

Natural gas is liquefied to LNG at a cryogenic temperature near -160 °C at atmospheric pressure, then transported by a carrier for storage in a LNG tank. The LNG receiving terminal refers to the required unit sets used for both LNG storage and for sending natural gas to market (this includes the main pipeline, unloading pipeline, LNG storage tank, compressor, vaporizer, sending pump, and other facilities).

LNG is transferred to a storage tank through a main pipeline and an unloading pipeline from the carrier ship. There are two kinds of storage tanks: above-ground and in-ground. The former stands on the ground, so LNG is transferred via a vertical unloading pipeline which is several tens of meters high over the main pipeline. In contrast, since in-ground tanks are buried, the vertical pipeline is only several meters high (Figure 2). Due to the difference in height between the two types of tanks, a more deliberate operating procedure should be devised.

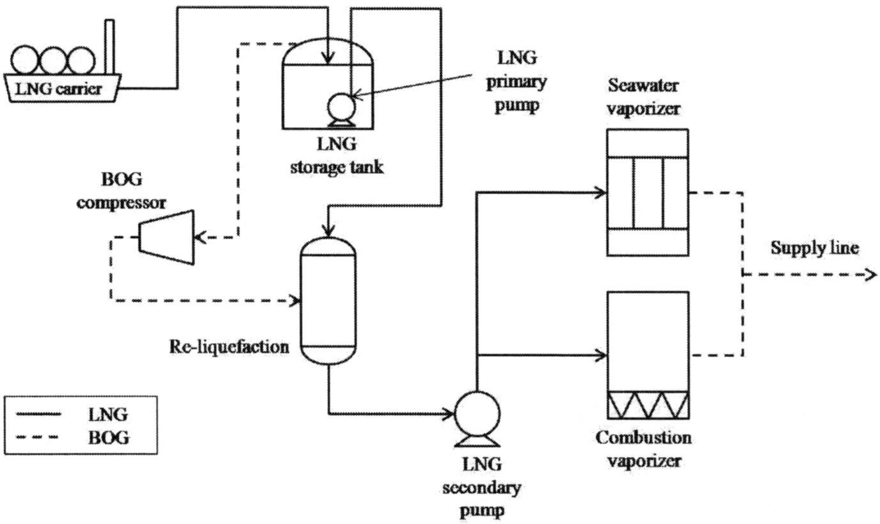

Figure 1: Schematic diagram of the LNG receiving terminal.

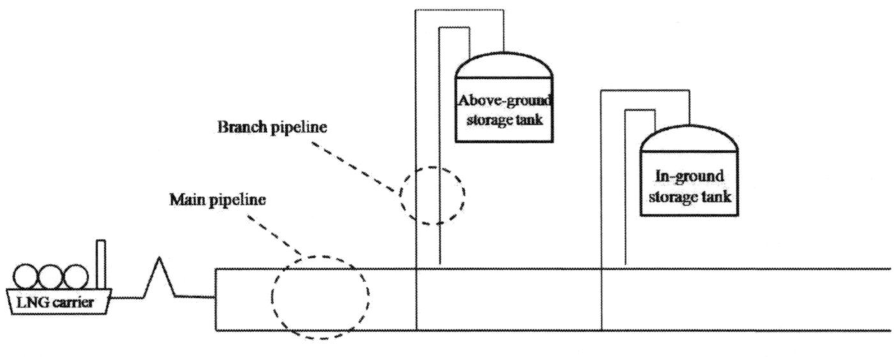

Figure 2: The path that LNG takes from the unloading pipeline to the storage tanks.

2.2. LNG unloading procedure

The LNG unloading process consists of three steps: recirculation, depressurization and unloading. Before unloading LNG, both the main pipeline and the unloading pipeline must be kept cold so as to prevent warming of the pipeline by heat input. To keep the pipeline cold, a small amount of LNG from the storage tank is continuously recirculated through the main and unloading lines; this is called the recirculation stage. In the depressurization step, the pressure in the main pipeline should be lowered to a value near the pressure of the storage tank so that LNG can be transferred from the carrier ship to the storage tank. In the final stage of depressurization, the unloading valve is opened in order to equalize the pressure between the main pipeline and the unloading pipeline. After the pressure is equalized, the unloading process starts by operating a cargo pump on the LNG ship, which transfers LNG from the carrier ship to the storage tank. When the entire unloading procedure is finished, the process returns to the first step of recirculation.

In the pressure equalization step, a large amount BOG can be caused by the pressure difference between the bypass valve in the unloading pipeline when the unloading valve is opened. Therefore, we will focus on a method to minimize the quantity of BOG that may be produced.

2.3. Flashing

LNG has a very low boiling point of around -160 °C at atmospheric pressure. As such, when heat enters from an outside source or the pressure gets too low, LNG vaporizes with a volume expansion up to 600 times larger than its initial volume. Such a sudden volume expansion is called flashing and the produced vapor is known as BOG. In the reliable operation of a LNG receiving terminal, it is crucial to handle the BOG effectively. This requires a more delicate and reliable operating procedure, especially for the mixed operation of above-ground and in-ground storage tanks, since unexpected BOG can be generated due to the height difference between the two kinds of tanks.

3. Modeling and Simulation

We simulated the LNG unloading process for the mixed operation of an above-ground and in-ground storage tank using HYSYS Dynamics. Peng-Robinson Stryjek Vera, (PRSV), which is suitable for the cryogenic hydrocarbon process, was used as a property method.

3.1. Pipeline modeling

We constructed two kinds of models for the main pipeline and the unloading pipeline. As shown in Figure 3, a dynamic model for the main pipeline was used for the unloading process from recirculation to depressurization. A dynamic model for the above-ground and in-ground unloading pipelines, shown in Figure 4, was used to obtain a volumetric quantity of BOG inflow at the pressure equalization step.

The heights of the above-ground and in-ground tanks are assumed to be 50 m and 5 m, respectively, in order to reflect the height difference between the two tanks. This difference creates a pressure head and the pressure at the highest point of the unloading pipeline is reduced compared to the main pipeline by the pressure head. Consequently, a large amount of BOG can be generated on top of the unloading pipeline due to the lowered pressure, which flows to the LNG storage tank.

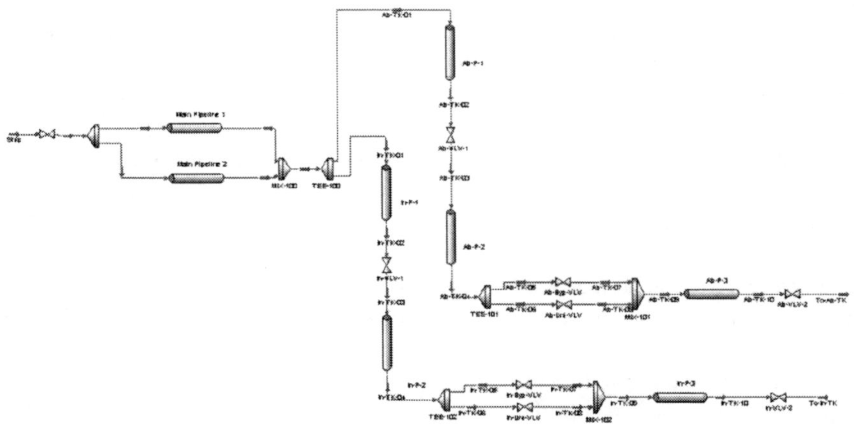

Figure 3: Dynamic model for the main pipeline and the above-ground unloading pipeline.

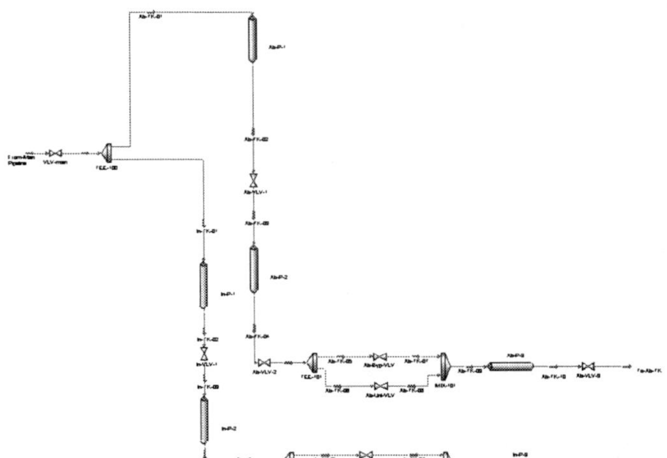

Figure 4: Dynamic model for the above-ground and in-ground unloading pipelines

3.2. Dynamic simulation

We performed a dynamic simulation in order to calculate the quantity of BOG produced when the unloading valve is opened. Recall that this valve is opened in order to equalize the pressure between the main and unloading pipelines in the depressurization step. Because generated BOG inevitably flows to the storage tank, if a large volume of BOG is produced, it can fatally attack the process units. Therefore, we inspected the unloading and operating procedures using a dynamic simulation so as to minimize the amount of BOG produced.

There are two decision variables for the mixed operation of an above-ground and in-ground tank; the unloading priority for the two types of tanks and the final pressure in the depressurization step.

For a receiving terminal composed of a combination of above-ground and in-ground tanks, the depressurization process is stagnated due to vaporization on top of the unloading pipeline of the above-ground tank. At this point, if we unload LNG to the in-ground tank first, excessive BOG flows to the in-ground tank when the unloading valve is opened for pressure equalization. Thus, for a hybrid unloading system, the LNG should be unloaded to the above-ground tank first in order to reduce the generation of BOG and to minimize the damage to the storage tank.

We can also decrease the quantity of BOG by minimizing the pressure difference between the bypass valve on the unloading pipeline. The quantity of BOG can be calculated by the following expression:

$$Q_{gpm} = C_v \sqrt{(\Delta P / G)} \tag{1}$$

Therefore, we can reduce BOG generation by minimizing the pressure difference.

Table 2: BOG inflow at pressure equalization for an above-ground and in-ground TK.

	Unit	Above-ground TK	In-ground TK
LNG density	kg/m^3	450	450
Maximum height	m	30	5
Final P of depressurization	kPa	243	243
ΔP between bypass valve	kPa	3	110
BOG inflow	m^3/h	7,004	62,842

4. Results and discussion

4.1. Simulation results

The simulation results of the unloading priority are summarized in Table 2. When the above-ground tank is selected to be unloaded prior to the in-ground tank, there is only 7,004 m^3/h of BOG inflow. On the contrary, when LNG is transferred to the in-ground tank first, as much as 62,842 m^3/h of BOG suddenly flows to the tank, which can fatally impact the storage tank and the internals.

Simulation results were also generated for the BOG inflow when the pressure difference across the bypass valve was varied. The pressure in the main pipeline decreases until the pressure difference across the bypass valve becomes 5, 3, and 1, respectively. By opening the unloading valve in order to equalize the pressure between the main and unloading pipelines, we can then obtain the amount of BOG inflow.

As shown in Table 3, the quantity of BOG inflow increases proportionally to the pressure difference. Therefore, we should allow the depressurization to progress further so as to induce a minimum in the pressure difference. In addition, we need to keep the unloading pipeline sufficiently cold in order to progress the depressurization up to the target pressure.

Table 3: BOG inflow according to variation in the pressure difference across the bypass valve.

	ΔP between the bypass valve	Final P of depressurization	boG inflow
Unit	kPa	kPa	m^3/h
Case 1	5	245	8,967
Case 2	3	243	7,004
Case 3	1	231	4,100

5. Conclusion

In this study, we investigated a reliable LNG operating procedure for the mixed operation of an above-ground and in-ground storage tank using a dynamic simulation. For unloading priority, the LNG should be unloaded to an above-ground tank prior to the in-ground tank in order to reduce BOG generation. Furthermore, a large amount of BOG can be produced when opening the unloading valve to equalize the pressure between the main pipeline and the unloading pipeline. To minimize BOG generation, we should minimize the pressure difference across the bypass valve on the unloading pipeline by progressing depressurization to a maximum. Using the suggested operating procedure and methodology, it should be possible to handle BOG so as to operate the LNG unloading process in a safer and more reliable manner.

References

Batres, R., Soutter, J., Asprey, S. P., Chung, P., 2002, Operating Procedure Synthesis: Science or Art?, *The Knowledge Engineering Review*, vol. 17, pp 261-294.

Cho, J. H., Kotzot , H., Coyle, D., Durr, C., Cost Effect Vapor Handling in an LNG Receiving Terminal

Hammer, M. , 2004, Dynamic simulation of a Natural Gas liquefaction Plant, Ph.D. diss., Department of Energy and Process Engineering, Norwegian University of Science and Technology.

Jung, M.-J., Cho, J. H., Ryu, W. R., 2003, LNG Terminal Design Feedback from Operator's Practical Improvement, presented at the 22nd World Gas Congress.

Kim. D., Park J. H., Yoon, I., Baek, Y., 2001, Study on the improvement of BOG recondensation process at LNG receiving terminal, *Journal of the Korean Institute of Gas*, Vol.5, No.3.

Lee, S., Hong, S., Lee, J., Park, S., 2000, A Computer Programme Development for Thermal-Hydraulic analysis on LNG Pipeline System, *Journal of the Korean Institute of Gas*, Vol.4, No.2.

Melaaen, E., 1994, Dynamic Simulation of the liquefaction section in Base load LNG plants, Ph.D. diss., Department of Refrigeration Engineering, Norwegian University of Science and Technology.

Seider, W. D., Seader, J. D., Lewin, D. R., 2004, Product & Process Design Principles, John Wiley & Sons.

Zaim, A., 2002, Dynamic optimization of an LNG plant, Case study: GL2Z LNG plant in Arzew, Algeria, Ph.D. diss., RWTH Aachen University.

Proceedings of the 2nd Annual Gas Processing Symposium
Farid Benyahia and Fadwa T. Eljack (Editors)

Inhibition of Gas Hydrate Formation by Low-dosage, Environmentally Benign Inhibitors

Lars Jensen[a], Hans Ramløv[b], Kaj Thomsen[a], Nicolas von Solms[a]

[a]*Technical University of Denmark, Department of Chemical and Biochemical Engineering, Søltoftsplads, 2800 Kgs. Lyngby, Denmark.*

[b]*Roskilde University, Department of Science - Systems and Models, Universitetsvej 1, 4000 Roskilde, Denmark.*

Abstract

Ice-structuring proteins (ISPs) have been considered as an environmentally-friendly alternative to commercial kinetic inhibitors of hydrate formation. ISPs obtained from certain species of pout (a cold-water fish) have been tested in the literature and in our laboratories and the results are promising. However, insect-based ISPs have been shown to be much more active than fish-based ISPs for freezing point depression. We have done initial testing with insect and fish ISPs in our labs for activity in hydrate inhibition and results are promising.

Keywords: gas hydrates, ice-structuring proteins, inhibition, low-dosage hydrate inhibitors

1. Introduction

Gas hydrates are crystalline compounds formed when water and suitably sized gas molecules are combined at high pressure and low temperature. They consist of polyhedral cavities formed from networks of hydrogen-bonded water molecules in which small gas molecules can enter. Depending on the type of gas molecules present gas hydrates form different structures, known as structure I (sI), structure II (sII) and in special cases structure H (sH). The gases that form hydrates are normally small molecules, many of which are encountered in natural gas. Examples are methane, ethane, propane, iso-butane, butane, carbon dioxide, nitrogen and hydrogen sulphide. Gas hydrate formation is a problem that the oil and gas industry is most concerned with. Oil- and gas transmission lines, tie-backs and off-shore process equipment are prone to being blocked by hydrates, causing potential hazards or economic loss (Sloan & Koh, 2008).

Traditionally the formation of hydrates has been prevented by addition of thermodynamic hydrate inhibitors such as methanol and glycol. However the amounts needed to avoid hydrate formation may reach 50 wt% in the water rich phase (Kelland, 1995). Since water production from fields can be quite severe, especially in cases where water injection has been used to enhance the oil recovery, large amounts of hydrate inhibitor are required. For these reasons a particular interest in hydrate formation kinetics has arisen. If the hydrate kinetics are known, and can be controlled, it may then be possible to operate the transmission lines at hydrate formation conditions, while still ensuring that hydrates will not grow sufficiently to cause flow blockage. If the kinetics of gas hydrate formation is very fast, it is possible to slow their formation by the addition of low dosage hydrate inhibitors (LDHI). These are most often water soluble polymers and are effective at concentrations 10-100 times less than conventional thermodynamic inhibitors (Sloan et al., 1996).

Polymeric compounds like polyvinyl caprolactam (PVCap) or polyvinylpyrrolidone (PVP) have shown quite effective to retard to crystallization of gas hydrate under low and moderate degrees of sub cooling (Sloan et al., 1996). Their use has though been limited due to their cost or by environmental restrictions due to their poor biodegradability.

Ice-structuring proteins (ISPs) are compounds that exist naturally in fish or insects and these have likewise shown able to retard the nucleation and growth of gas hydrates (Zeng et al., 2003; Zeng et al., 2006). Fish antifreeze proteins can confer freeze resistance in icy seawater down to its freezing point (-1.9 °C) approximately 1 °C colder than the freezing point of fish plasma (Scotter et al., 2006). The millimolar ISP concentrations (>10–20mg/ml) found in fish blood lower freezing temperatures below the melting point, a phenomenon termed thermal hysteresis (TH), by about 0.5–1.0 °C.

Compared to fish the ISPs found in insects are much more potent. The remarkable specific activities of these proteins have though not been appreciated until they were purified and expressed as recombinant proteins. The difference in TH as a function of concentration between insect and fish ISPs is approximately two orders of magnitude. For example a study has shown that Tenebrio molitor (Tm) ISP depresses the freezing point by 1.4 °C at a concentration of 20 μM, whereas more than a 100-fold higher concentration (2.8 mM) of type III ISP from a fish (ocean pout) is required to produce the same freezing point depression (Scotter et al., 2006).

In this work the effect of TmISP on the growth of methane hydrate has been investigated. The effect of type III ISP from ocean pout on the nucleation of methane hydrate has also been investigated.

2. Experimental Equipment and Procedure

2.1 Experimental Equipment

The growth kinetics of methane hydrate and the effect on adding TmISP in small amounts to the aqueous phase was investigated experimentally at constant pressure and temperature. This was done using stainless steel hydrate equilibrium cell with a fixed volume of 66.5 cm^3 and a maximum working pressure of 150 bar. The complete experimental set-up can be seen in Figure 1. The cell allows for visual observation of hydrate formation through two sapphire windows. A safety valve is attached to the cell which opens when pressure exceeds an adjustable value (40-100 bar). The pressure in the cell is monitored using a digital pressure sensor (EPC, Bronckhorst High-Tech) with a range of 0-100 bar. The pressure censor is connected with a mass flow meter (MFC, Bronckhorst High-Tech El flow) with a flow range of 0-10 ml/min N_2 and a maximum working pressure of 100 bar. The mass flow meter is attached to a gas filter which is attached to a gas reservoir.

The pressure sensor and the mass flow meter work as an electronic backpressure control system which ensures that the pressure in the cell is maintained constant during the experimental run. The pressure sensor receives a digital set point from the flow-bus (High Tech RS232) and when the measured value becomes less than the set point value i.e. pressure drops due to hydrate formation, an analog set point will be send to the mass flow meter. This will cause the mass flow meter to open its valve in accordance with the provided set-point. The backpressure system is PID controlled and parameters have been optimized for the systems of interest to secure highest level of control. Data logging and instrument control is performed using FlowPlot (Bronckhorst) and FlowView (Bronckhorst). The gas supply can by-pass the backpressure control system by closing valve 2 and opening valve 1 and 3. This is done when starting up the experiment in order to quickly obtain the pressure needed in the cell.

The temperature in the cell is controlled by circulating coolant (water-ethanol solution), in a jacket surrounding the cell. The temperature is monitored by using a platinum resistance probe (±0.01 K) placed inside the cell. An extra pressure transducer is attached to the cell (BD Sensors, 0-100 bar) as an extra control. The cell is placed on a stir plate which allows a stirring bar (L = 4 cm) to rotate within the cell. The pressure and temperature in the cell can be recorded continuously on a computer. The experiments were performed using distilled water and methane supplied by AGA with a purity of 99.995.

For investigating the effect of OpISP on methane hydrate nucleation the same set-up was used however here measuring the pressure over time and not needing the PID controlled back pressure set-up to carry out the measurements.

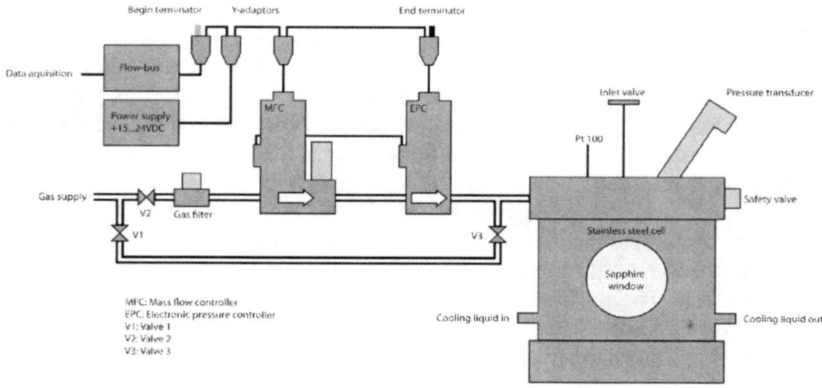

Figure 1: Experimental set-up.

2.2. Experimental Procedure

The cell was cleaned with distilled water and loaded with 10 g of distilled water or distilled water containing small amounts of TmISP. A stirring bar was placed in the cell. The cell lid was screwed on and the cell evacuated using a vacuum pump for approximately 1 hour. The temperature bath was adjusted so the temperature in the cell was 277.15 K. When the temperature in the cell was constant the methane gas was injected to the cell, by by-passing the backpressure control system, until the desired pressure at the chosen temperature was obtained. Then valves 1 and 3 were closed and valve 2 opened. A set-point, corresponding to the start pressure in the cell was given to the EPC. The experiment was the initiated by starting the stirring bar (450 rpm) in the cell and data acquisition.

The data recorded by the MFC is based on N_2 ml/min. This was subsequently converted to ml/min CH_4 using an empirical conversion factor provided by Bronckhorst. The CH_4 volume measured was then converted to moles as there is a tradition in the literature of representing growth kinetics of gas hydrates as moles of gas accumulated in the hydrate as a function of time.

For the nucleation experiments we have used the methodology reported by (Dicharry et al., 2008) since that has shown to increase experimental reproducibility. Hydrate formation was detected as pressure starting to decrease.

3. Results and Discussion

All the experiments were carried out under constant pressure. The maximum deviation between the experimental pressure and the set-point pressure during an experimental run was never more than 1%. The growth kinetics of methane hydrate was investigated at 277.15 K and at three different pressures. The result can be seen in Figure 2. In all three cases a linear growth pattern, in terms of total moles of methane consumed over time, can be observed. At higher pressure, i.e. higher driving force, the growth rate is seen to increase.

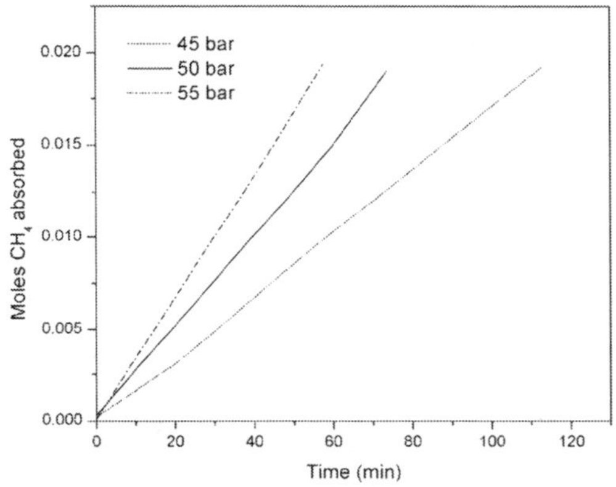

Figure 2: Growth kinetics of methane hydrate at pressures of 55, 50 and 45 bars and a temperature of 277.15 K. The growth kinetics is represented as the total number of moles accumulated in the hydrate as a function of time.

The reason why the curves end at a gas uptake of about 0.02 moles is because a drastic decrease in the growth rate above this value is observed. This is most likely due to a lack in the availability of free water molecules thus a substantial amount of water must have been converted to hydrate. Phase equilibrium calculations reveal that methane hydrate at 277.15 K have an equilibrium pressure of 33.36 bar and the following composition:

$$7.63CH_4 \cdot 46H_2O \tag{1}$$

This corresponds to a total conversion of free water to hydrate of 25 %. However it is likely that a lot of the remaining water is trapped in between hydrate crystals, thus limiting the mass transfer.

In order to asses the growth rate a simple rate expression similar to the one proposed by Skovborg and Rasmussen (1994) however using the solubility, x_{H-L}, of the hydrate former in the liquid phase at the hydrate-liquid interface under H-L_w equilibrium at T_{exp} and P_{eq} instead of x_b which is the mole fraction of gas in the bulk water phase in equilibrium with the hydrate phase at system P and T. The rate of mass transfer and thus the total gas consumption can be described using the following equation:

$$\frac{dn}{dt} = k_L A_{g-l} C_{w0}(x_{L-V} - x_{H-L}) \tag{2}$$

Where k_L is the mass transfer coefficient in the liquid film, A_{g-l} is the gas liquid interfacial area, C_{w0} is the initial concentration of water (mol/m^3), x_{L-V} is the mole fraction of gas in the water phase at system P and T and x_{H-V} and x_{H-L} as described before. The mole fraction of gas in water is found using Henry's law:

$$x_{L-V} = \frac{f(P_{\exp}, T_{\exp})}{H} \tag{3}$$

And x_{H-L} is found as:

$$x_{H-L} = \frac{f(P_{eq}, T_{\exp})}{H} \tag{4}$$

Henry's constant has been calculated using the relation (Carol and Mather, 1997):

$$\ln H_{iw} = 5.1345 + 7837/T - 1.5090 \cdot 10^6/T^2 + 2.060 \cdot 10^7/T^3 \tag{5}$$

The fugacities have been calculated using the SRK equation of state. The equilibrium P and T is found using CSMGem (Ballard & Sloan, 2004). As the gas liquid interfacial area is not known we substitute $k_L A_{g-l}$ with K. dn/dt in (2) is obtained by making a linear regression of the experimental data presented in Figure 2. As seen the linearity of the experimental data is high thus good estimates of the slope is obtained. In table 1 the calculated mass transfer coefficients are provided. The mass transfer coefficients found are quite similar which would also be expected as primarily the driving force in (2) should compensate for the faster growth rate.

Table 1: Calculated mass transfer coefficients for 3 different pressures at 277.15 K.

T	P	System	Mass transfer coefficient (m^3/s) $\cdot 10^8$
277.15	45	No inhibitors	7.96
	50		6.58
	55		5.97
	55	0.1 wt% PVP	3.42
	55	0.004 wt% TmISP	3.78

The growth rate of methane hydrate when adding 0.004 wt% TmISP has been investigated. In Figure 3 the growth curve at 55bar is shown. As seen the TmISP causes the growth rate to decrease. The same is found for PVP however to obtain the same level of inhibition a concentration of 0.1 wt% is required. This suggests that TmISP is a stronger kinetic inhibitor than PVP. The mass transfer coefficients are also provided in Table 1.

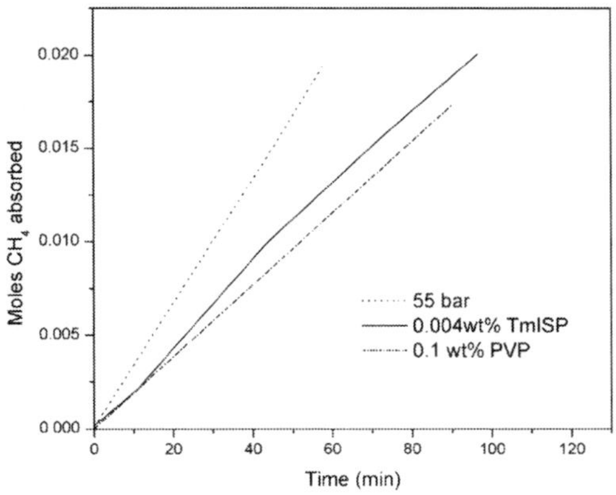

Figure 3: Growth kinetics of methane hydrate in the presence of 0.004wt% TmISP and 0.1wt% PVP at 277.15 K and 55 bar.

The nucleation periods has likewise been investigated to see if ISPs are also capable of inhibiting the induction period of hydrate formation.

Table 2: Results of nucleation experiments for methane hydrate with and without inhibitors. The induction time listed is based on the average of 7 experiments. The number in parenthesis is the standard deviation.

T (K)	System	P (bar)	Nucleation period (h)
277.15	No inhibitor	54	1.15 (0.28)
	0.1 wt% OpISP	60	10.49 (2.41)

As seen the induction period is relatively short at 54 bar. A standard deviation of 0.28 for a hydrate forming system is considered relatively short considering the stochastic nature of hydrate nucleation. This is attributed to the experimental method applied. When adding the OpISP the nucleation period of methane hydrate is prolonged noteworthy even though that the pressure is increased by 6 bar compared to the first experiments with no inhibitor. Relatively the standard deviations in the two experiments are in the same order. The reason that nucleation periods of hydrate formation can be prolonged can be numerous. Previously studies have showed that nucleation of gas hydrates as in the case of ice is heterogeneous and therefore highly affected by the presence of impurities (Bishnoi and Natarajan, 1996). A recent study has revealed that nucleation inhibition using polymers like PVP is mainly due to adsorption of these molecules on impurities like dust particles causing a change in the hydrate-substrate contact angle (Jensen et al., 2008). The same observations have been made for the nucleation of ice when ISP III was added to the water (Du et al., 2003). With this in mind it seems reasonable to assume that OpISP inhibits the nucleation of methane hydrate by a similar mechanism.

While it is not likely that the nucleation prolongation is due to adsorption of ISPs on the nucleus surface, it is however likely that the growth decrease is caused by such an effect. Studies on ice growth inhibition by ISPs have shown that irreversible adsorption of ISP molecules on the ice surfaces are the primary mechanism by which growth is retarded (Partaya et al., 2007). The exact mechanism is however not well defined but one suggestion is that the ISP induces a dense ISP-water layer, which can significantly decrease the mole fraction of the interfacial water and, thus, lower the temperature for a seed ice crystal to grow in a super-cooled ISP solution (Mao & Ba, 2006). Another theory suggests that ISP binding to ice results in facet formation due to the inhibition of growth on the binding plane. Such inhibition is thought to result from the Kelvin effect, whereby the ice surface is pinned by distantly spaced, bound ISP molecules, causing the ice front to grow with local positive curvature, which is less thermodynamically favorable (Knight, 2000). Whether the TmISP is capable of modifying the crystal shape of methane hydrate during growth has not been possible to investigate in this work however the results indicates that the TmISP is somehow affecting the growth mechanism.

3. Future work

In our current work we are planning to increase the potency of these insect ISPs by genetic manipulation as well as bring down the production cost by producing them in bacterial fermentations. Also the ISPs will be tested on sII hydrate formers.

References

Sloan, E.D., Koh, C.A., Clathrate hydrates of natural gases, 3rd edition, CRC Press, 2008.

Kelland, M., (1995). Studies on New Gas Hydrate Inhibitors, SPE 30420, 531-539.

Sloan, E.D., Lederhos, J.P., Long, J.P., Sum, A., Christiansen, R.L. (1996). Effective Kinetic Inhibitors for Natural Gas Hydrates, Chem. Eng. Sci., 51, 1221-1229.

Zeng, H., Ripmeester, J.A., Wilson, L.D., Walker, V.K. (2003). The inhibition of tetrahydrofuran clathrate hydrate formation with antifreeze protein, Canadian Journal of Physics 81 (2003) 17-24.

Zeng, H., Moudrakovski, I.L., Ripmeester, J.A., Walker, V.K. (2006). Effect of Antifreeze Protein on Nucleation, Growth and Memory of Gas Hydrates, AIChE Journal, 52, 3304-3309.

Scotter, A.J., Marshall, C.B., Graham, L.A., Gilbert, J.A., Garnham, C.P., Davies, P.L. (2006). The basis for hyperactivity of antifreeze proteins Cryobiology, 53, 229–239.

Dicharry, C., Duchateau, C., Glénat, P., Hidalgo, M., Peytavy, J.L., Pou, T.E., Proceedings of the 6th International Conference on Gas Hydrates, Vancouver, British Columbia, CANADA, July 6-10, 2008.

Skovborg, P. Rasmussen, P. (1994). A mass transport limited model for the growth of methane and ethane gas hydrates, *Chem. Eng. Sci.*, *49*, 1131-1143.

Carol, J.J., Mather, A.E. (1997). A model for the solubility of light hydrocarbons in water and aqueous solutions of alkanol amines, Chem. Eng. Sci., 52, 545-552.
Ballard, A., Sloan, E.D. (2004). A next generation of hydrate prediction Part III. Gibbs energy minimization formalism, Fluid Phase Equilibria, 218, 15-31.
Bishnoi, P.R., Natarajan, V., 1996. Formation and decomposition of gas hydrates. Fluid Phase Equilibria 117, 168--177.
Jensen, L., Thomsen, K., von Solms, N. (2008). Propane hydrate nucleation: experimental investigation and correlation, Chem. Eng. Sci., 63, 3069-3080.
Du , Ning, Liu,. Xiang Y., Hew, Choy L. (2003). Ice nucleation inhibition, The Journal of Biological Chemistry, 278 (38), 36000-36004.
Pertaya, N., Marshall, C.B., DiPrinzio, C.L., Wilen, L., Thomson, E.S., Wettlaufer, J.S., Davies, P.L., Braslavsky, I. (2007). Fluorescence Microscopy Evidence for Quasi-Permanent Attachment of Antifreeze Proteins to Ice Surfaces, Biophysical Journal (92), 3663–3673
Mao, Y., Ba, Y. (2006). Ice-surface adsorption enhanced colligative effect of antifreeze proteins in ice growth inhibition, The Journal of Chemical Physics, 125.
Knight, C.A. (2000). Structural biology. Adding to the antifreeze agenda, Nature (406), 249–251.

Proceedings of the 2nd Annual Gas Processing Symposium
Farid Benyahia and Fadwa T. Eljack (Editors)

Strategies for Mitigating Impacts on LNG Plant Capacity by Variations in Ambient Temperature

Stanley Huang, Meredith Chapeaux and Bobby Martinez

Chevron Energy Technology Companyy, 1400 Smith Street, Houston, TX 77002, U.S.A.

Abstract

Delivery of LNG product is of the utmost importance for any LNG plant, due to potential penalties for failure to achieve a contractual obligation. Equipment design margins are included to counter the inadvertent impacts of various factors which may be beyond the operators' control. However, overly generous margins can result in inefficient utilization of plant facilities and negatively impact the overall project economics. Achieving a balance between margins and economics remains more of an art than exact science.

Ambient temperature variation is one of the major factors that will affect LNG productions and hence will be the primary focus of this paper. Detailed site temperature data can be acquired from meteorological records. This paper summarizes a new methodology in converting the chronological data into temperature distribution functions suitable for plant designs. Following this, common industrial methods are categorized and their pros and cons are addressed. The choice of an optimal approach is project-specific and general guidelines are provided.

Keywords: liquefied natural gas, LNG, plant capacity, ambient temperature, temperature variation

1. Introduction

Variations of ambient temperature exert significant impacts on LNG production capacity. However, not all LNG plants would face the same level of impacts. This paper provides a comprehensive discussion on the mitigation strategies, from identifying the impact levels to selection of mitigation methods.

The paper is divided into three parts. The first part summarizes climate patterns and a new methodology to covert meteorological data to distribution functions for plant designs. The second part discusses different categories of mitigation methods. Lastly, a general discussion on decision making and an example are provided.

It should be emphasized that each project has its unique features and constraints. Detailed studies beyond what are presented in this paper would require specific cost data. Nonetheless, this paper provides useful guidelines for future LNG projects.

2. Main Features of Climate Patterns

The climate data of this paper were based on the works of Casey (2008). Figure 1 shows the average temperatures versus location latitudes of existing or planned LNG plants. The three climate regions of tropical, desert, and Arctic are marked and LNG plants are grouped accordingly. There is a correlation between average temperature and latitude. The correlation is represented by a band.

It is noted that all LNG plants are located close to the ocean. Therefore, climatic moderation is anticipated. Impacts of local factors are also visible. For example, Peru LNG is relatively chilly due to the Pacific cold current. Conversely, Statoil LNG located on an island offshore Norway is relatively warm, thanks to the Atlantic warm current.

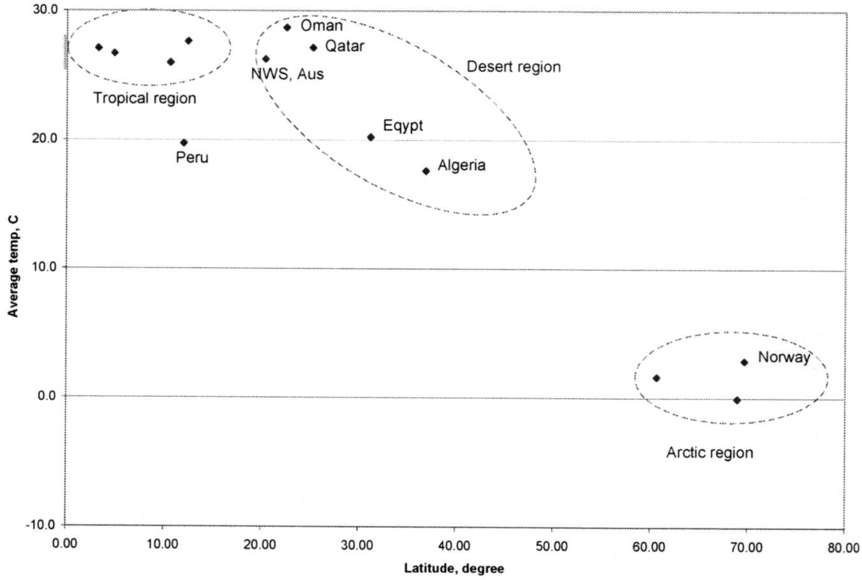

Figure 1: Global LNG plants clustering in three climate regions

Figure 2 plots the temperature swings versus the average temperature for the same data. Again, a trend can be observed. Generally, the higher the average temperature, the narrower the seasonal temperature swings. Hence, plants in the Arctic region require serious considerations of temperature impacts, but not those in tropical regions.

However, the desert climate deviates from the correlation band. As shown in the figure, the temperature swings in the desert area are higher than normal. Unlike tropical regions, where temperature swings are within 5 °C, the temperature swings in desert area can be as high as 18 °C.

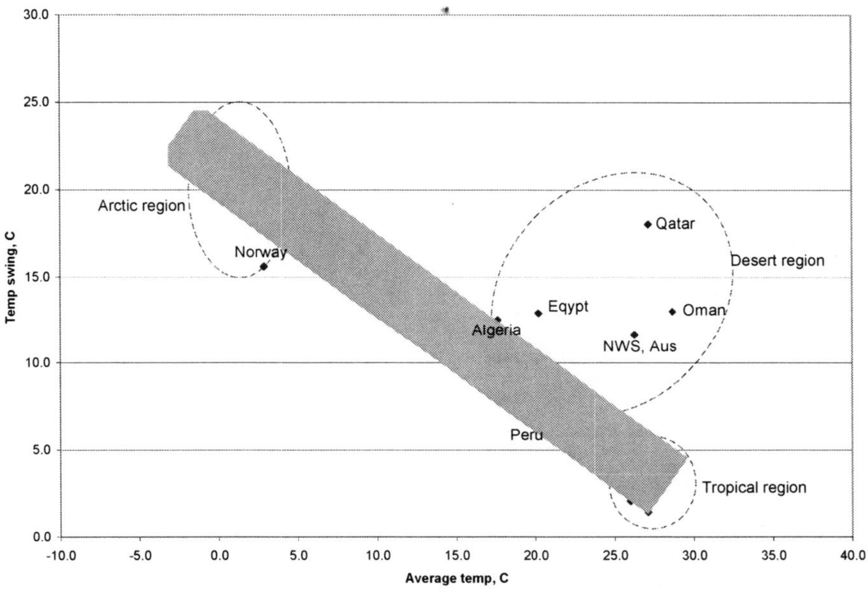

Figure 2: Correlating temperature swing versus average temperature

3. Mathematical Modeling of Temperature Distribution Functions

A new methodology of converting meteorological data to temperature distribution functions was presented by Chapeaux and Huang (2009). Briefly, the new methodology uses twelve normal distribution functions to represent monthly data. The temperature distribution function is then derived from summing the monthly contributions. This method bypasses the tedious inverse mapping operations between time-temperature domains.

Figure 3 shows temperature distributions at three locations: Lagos (tropical), Murmansk (Arctic), and Doha (desert). Similar to the observations in Figure 2, tropical regions have relatively high average temperature and small swings, which are reflected in sharp distributions. Conversely, the Arctic region has relatively low average temperature and wide swings, which are reflected in wide distributions. Desert areas do not follow the pattern, they have high temperatures and wide swings.

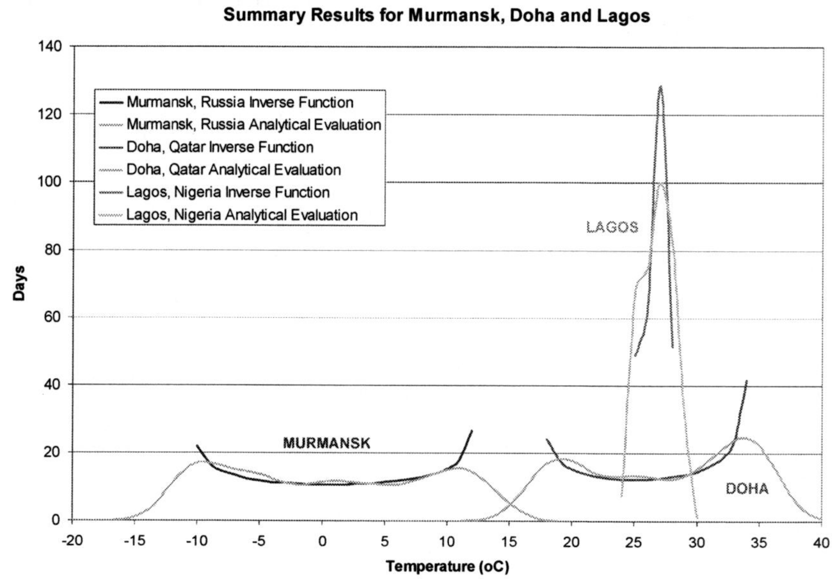

Figure 3: Example temperature distributions at different locations

4. Common Methods to Mitigate Impacts of Temperature Variations

If an LNG project decides to account for the impact of temperature variations, there are many methods to choose from. Common industrial methods are classified into three categories as shown in Table 1. Each category has its pros and cons. Furthermore, not all categories are applicable for all projects. For example, the annual average ambient temperature in the Arctic region is close to 0 °C, the freezing point of water. Many gas turbine vendors require the inlet air temperature to be maintained above freezing to avoid possible clogging. Consequently, the colder temperatures in the winter would not result in power enhancements. Category C has limited applicability in the Arctic region. Only one example will be provided for each category in the following discussions.

Table 1: Three categories of mitigation methods

Cat.	Description	Notes
A	Overall Solution	Focusing on specifications for entire plant without going into equipment domain
B	Loss Mitigation	During high ambient temperatures, trying to maintain normal production rate by altering standard specifications for selected facilities
C	Gain Enhancement	During low ambient temperatures, trying to enhance production rate to make up for lost productions during high-temperature periods; not always feasible due to other constraints

Category A: Overall Solution

Figure 4 shows the exceedance curve of a location in the desert region. Three possible design temperatures (DT) are marked: DT50 for meeting the design capacity 50% of the time, DT80 for 80% and DT95 for 95%. Although possible, it is a rare practice to use DT100 due to the long tail of the exceedance curve.

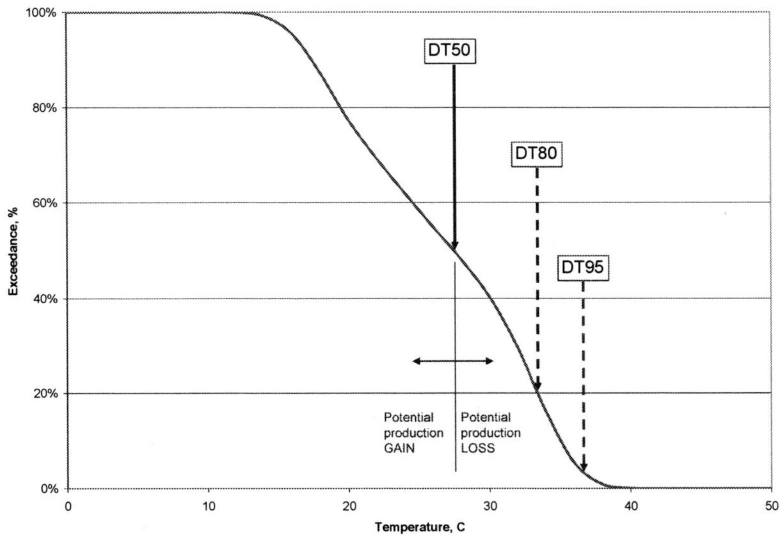

Figure 4: Exceedance curves of an LNG plant in desert region

Example A: Increasing plant design temperature

DT50 will be the basis of comparison because it is close to the average temperature which is commonly used. There are different criteria for choosing a different DT. For example, a project may decide to maintain the annual accumulated productions at the originally designed level. The calculations will need production gains and losses with respect to temperature variations. This aspect will be covered in the next two categories.

Category B: Loss Mitigation

Figure 5 shows the temperature distribution and relative production rate for the same LNG plant depicted in Figure 4. Again, the DT50 is shown as the basis of comparison. Above DT50, the plant capacity would suffer due to the elevated ambient temperature. There are two main reasons for the decreased plant capacity as temperature rises: (1) reduced Gas Turbine (GT) output and (2) reduced condensing temperature approach. Category B focuses on mitigating the production loss. Graphically, the intention is to raise the declining production line back to horizontal.

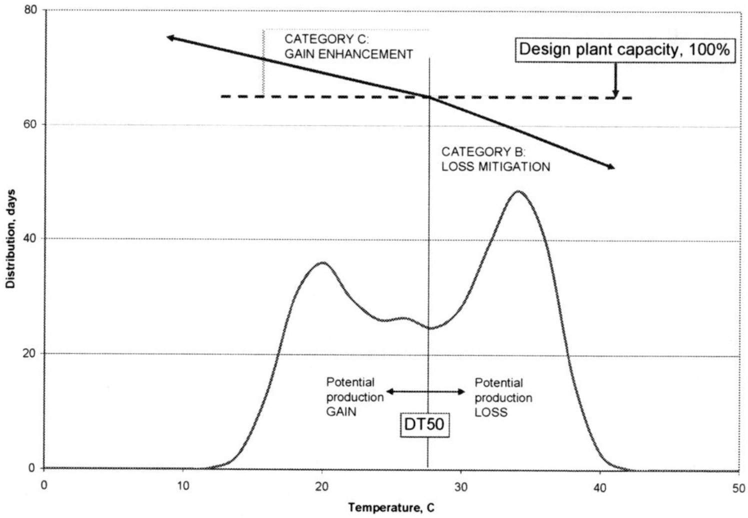

Figure 5: Temperature distribution coupled with LNG relative production rate

Example B: Increasing design margin of helper power

This example provides additional helper power to compensate for the reduced GT output. Also, additional power would help to overcome the heightened vapour pressure resulted from condensing at a higher temperature. Since power generation also uses GT drivers, their power output also decreases as temperature rises. Therefore, this example requires additional PowerGen GT's during the hot seasons.

Category C: Gain Enhancement

Refer to Figure 5 for illustrations. Below DT50, the plant capacity potentially would increase due to the decreased ambient temperature. Category C focuses on measures to enhance the production gain so as to maximize the plant revenue. Graphically, the intention is to raise further the slope of the production line.

Example C: Increasing design capacity of plant, except drivers

Any time the plant instantaneous throughput goes beyond design capacity, the process system hydraulics may become excessive, hence negatively impact plant efficiency. Therefore, the piping, nozzle, and vessel sizing should be re-evaluated hydraulically. Another area of concern is in the heat exchanger applications. This is particularly true for the Main Cryogenic Heat Exchanger (MCHE), for which the vendor information is often kept proprietary.

5. Strategies in Real Applications

A realistic case based on Category A is presented below as an example. Figure 6 shows a case similar to the ones described by Martinez et al. (2007) and Omori et al. (2001). The case has a sensitivity of 1.7% production loss per °C temperature increment. On the other end, there is only 0.6% production gain per °C temperature decrement. The asymmetry in production gain/ loss per degree temperature change is purposeful in order to emphasize the fact that it takes additional investment to realize potential production gains at lower temperatures.

Three DT's are presented: DT50% (28 °C), DT80% (33 °C), and DT95% (36 °C), each representing the percentage of exceedance. As the DT increases, the accumulated production loss decreases. The loss is represented by the area below the reference plant capacity. In other words, the plant has improved resistance against temperature rises.

The production gains are represented by the areas above the reference line. To achieve that additional production rate, equipment enhancements are usually required, for example, additional heat exchanger areas.

Methods in Categories B and C can directly enlarge or shrink the areas above or below the baseline. The inclusion of a specific method by a project demands cost analysis, for example, Life Cycle Cost (LCC). Since no two LNG plants are identical, the optimal solutions for two plants are rarely the same.

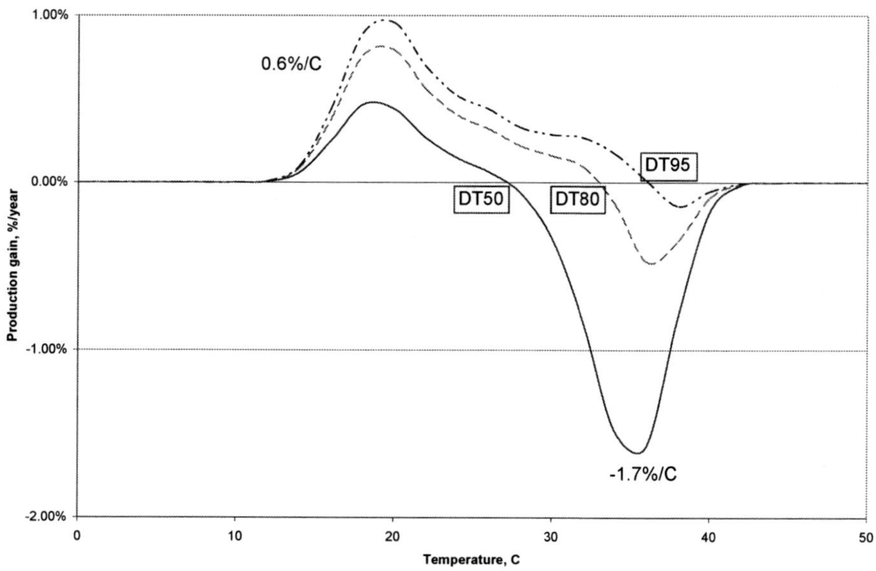

Figure 6: Impact of increasing plant design temperatures

6. Conclusions

The conclusions of this study are:

1. The meteorological data of most existing and planned LNG plants were compiled. Existing plants are located in one of the three regions: tropical, desert, and Arctic. Only those plants in the tropical regions are not anticipated to be significantly impacted by temperature variations.
2. A new method for converting meteorological data to temperature distribution function was summarized. The results show that temperature swings in desert and Arctic regions are fairly significant. Moreover, mid-summer and mid-winter temperatures (extreme values) have higher probability than the averaged values, hence more representative of actual situations.
3. Common industrial practices in managing temperature variations were grouped into three categories: total solution, loss mitigation (during high temperatures), and gain enhancement (during cold temperatures). The pros and cons of each group were explained. The strategies for dealing with real scenarios were described.

References

Casey, J., "World Climate" at website: http://www.climate-charts.com/, from which data of this work were extracted in 2008.

Chapeaux, M., Huang, S., 2009, "Improved Methodology to Handle Meteorological Data for LNG Projects", paper presented at AIChE Spring Meeting, Tampa, FL, April 26-30.

Martinez, B., Huang, S., McMullen, C., Shah, P., 2007, "Meeting Challenges of Large LNG Projects in Arctic Regions", paper presented at GPA 86th National Convention in San Antonio, Texas, March 11-14.

Omori, H., Konishi, H., Ray, S. A., de la Vega, F. F., Durr, C. A., 2001, "A New Tool – Efficient and Accurate for LNG Plant Design and Debottlenecking", paper presented in LNG 13, Seoul, Korea.

Proceedings of the 2nd Annual Gas Processing Symposium
Farid Benyahia and Fadwa T. Eljack (Editors)

Environmental Sustainability: Industry-Academia Perspective on Future Trends

Farid Benyahia and Fadwa T. Eljack

Department of Chemical Engineering, College of Engineering, Qatar University

Abstract

The second annual gas processing symposium technical program was enriched by five discussion forums related to carbon capture and storage, water, energy efficiency, safety and risk assessment, and reliability. The round table discussions and debates were well attended and generated thoughts that inspired this chapter aimed at putting in perspective future trends on environmental sustainability in gas processing driven by collaboration between academics and people from industry. The synergy between academic creativity and industry experience will take the joint efforts to higher levels of achievements in areas critical to sustainable gas operations on a massive scale. This approach is not only desirable, but actually required at a time when emerging economies in the Middle East and Far East consume raw materials, fossil fuels and fresh water at unprecedented rates. The environmental impact of such rapid growth can potentially be disastrous if not accompanied by proper measures to mitigate the effects, improve energy efficiencies, recycle used water and ensure safe and reliable operations. This chapter of the second annual gas processing symposium proceedings attempted to shed light on how industry and academia may work together to address environmental sustainability challenges and establish a perspective for future trends in the areas concerned.

Keywords: carbon capture and storage, water, energy efficiency, process safety and risk assessment, reliability

1. Introduction

Environmental sustainability is becoming a serious challenge due to the global economic growth coupled with a massive surge in energy demand from new economic powers. The nature of the environmental challenge is extremely complicated since it is a composite of scientific misunderstanding, technological advances, public perceptions and political decisions. This unusual combination is not random since it is directly related to the intimate connection between strategic energy policies, public pressure for green house gas emission reduction and an aspiration for a better quality of life.

The second annual gas processing symposium program included a discussion forum that brought academics and people from industry round a table to discuss timely and important topics linked directly or indirectly to environmental sustainability of gas processing operations. This chapter is an attempt to put in perspective future trends driven by joint efforts from academics and people from industry for environmental sustainability in the oil and gas industries. The section of this chapter represent the themes debated in the five discussion forums during the second annual gas processing symposium.

2. Carbon capture and storage: options and challenges

The carbon challenge is greater than ever following public opinion and pressures on governments to reduce green house gas emissions on a global scale. Putting aside CO_2 emissions from transportation, there are two broad sources of CO_2 releases into the atmosphere from industrial activity: Power generation (dilute CO_2 stream) and natural gas processing (concentrated CO_2 stream). Using the same technology as in gas processing (gas sweetening operations), concentrated CO_2 streams may be obtained from fossil fuel power generation plants. The operation aimed at recovering CO_2 from any of the two sources highlighted is commonly called "carbon capture" and makes use of an established technology: amines based solvent absorption followed by solvent regeneration that eventually releases CO_2 in a highly concentrated form. Although effective, this operation is highly energy intensive and has its own significant carbon footprint. The second step in the carbon challenge is the fate of captured CO2. Currently, the vast majority is simply vented into the atmosphere, thus contributing to the green gas house emissions. In that respect, a legitimate question would be what are the options for reducing or eliminating these emissions?

2.1. CO2 underground storage

Amongst the proposed solutions for carbon capture and storage, underground storage has by far been the method widely discussed and preferred by industry. This is mainly due to its potential to storage large amounts of captured CO2 and has been the subject of pilot testing in some parts of the world. It has also been used successfully to boost pressure in some gas reservoirs. Despite this apparent great potential, there are some issues with this method: suitability of geological formation, in particular the cap rock to prevent seepage of CO2 back into the atmosphere. Though this serious drawback may not be quick because transport through porous rocks is rather slow, if it happens, consequences could be disastrous and costly. For this very reason, public opinion is rather sceptical on its long term safety aspect and a great deal of testing and public relations exercise needs to be done in the future. Other variants of underground storage also include pumping captured CO2 in underground aquifers or salt caverns. This option is obviously limited to certain parts of the world and therefore underground storage is not a universal option for globally captured CO2.

2.2. CO2 conversions and mineralization

Conversion of CO2 into valuable products and mineralization of captured CO2 have also been cited as additional options to reduce carbon emissions. Mineralization has a great potential since CO2 will be converted into a rock (bicarbonate or carbonate) and therefore will be more stable when disposed of. However the process chemistry and kinetics are far from being fully mastered and a great deal needs to be done in this area. Some work on mineralization is underway at Qatar University and elsewhere and the outcome will be disseminated in the future. Because this option is reaction based, it will have a capacity to absorb CO2 proportionate to the reaction stoichiometry. It is clearly obvious that even if this method is successful, it will not treat globally captured CO2, but rather a significant quantity. The advantage of this option is its independence from geological issues.

2.3. CO2 biofixation

This option is not exactly new because it has been naturally occurring for millennia with respect to tree growths and vegetation in general. However, there has been suggestions to widen the scope and include artificial growth of micro-algae in the sea under controlled conditions. This option is limited by the slow nature of biofixation and the optimum conditions required unless these are artificially created, adding to the cost.

2.4. The way forward for green house gas emission reduction

It is clear that a great deal remains to be done to make the carbon capture and storage feasible on a global scale. Partnerships between academia and industry will undoubtedly help in finding solutions that are cost effective, effective and safe. For instance the traditional carbon capture technology through amine based solvents is very energy intensive and should be investigated to reduce the energy requirements and thus reduce associated costs. The long term fate of underground stored CO2 also needs to be investigated. There are opportunities of physical and chemical phenomena underground that are yet to be understood and exploited to a safer carbon storage. In the Arabian Gulf region, mineralization offers an attractive and complementary solution to partially offset the issues of carbonate formations that are not naturally suitable for CO2 storage where there is a risk of carbonic acid formation damage to rocks. Until a long term solution is found, a transition solution option could involve using captured CO2 in enhanced oil and gas recovery. This may partially offset carbon storage costs but clearly some of the stored carbon dioxide will be extracted with hydrocarbon in due course.

To summarise the green house gas issue and potential solutions, it can be stated that reduction of emissions at source will be the best starting point. This can be facilitated by energy saving measures and responsible behaviour of citizens. Beyond this "zero cost" solution, the carbon management of the future will really a combination of methods such as underground storage, mineralization and biofixation. The relative importance of each depends will vary around the world depending on circumstances and geology.

3. Water: A challenge increasing in complexity

The demand for fresh water has been continuously increasing and has recently reached alarming quantities, mainly due to the population growth globally and also growth in emerging economies in the Middle-East and Far East. Other parts of the world have also been affected by fresh water shortages or shortfalls. These factors put considerable strain on fresh water resources to the extent where desalination plants started being built in countries where traditionally fresh water was not a major concern.

A great deal of fresh water is used in industrial activities and ends up as wastewater. In some petrochemical processes, water is actually produced as by-product (eg: GTL process). It can also be extracted with oil and referred to as "water cut" or water content of oil. Such wastewater can be treated to a standard where it can be recycled in some applications such as cooling, boiler feedwater or landscape irrigation.

In the Middle-East, specifically in the Arabian Gulf States, most of the fresh water for industrial and municipal uses is produced by desalination. In Qatar, only thermal desalination is currently used to produce fresh water for all applications. The rising production costs of desalination processes and associated environmental issues, including the carbon footprint, is now leading to a review of water usage policies. Such moves will necessitate thoughts about water conservation measures and water reuse as and where appropriate.

3.1. Another source of water: wastewater and "produced water"

Given the huge amounts of wastewater generated in refining and petrochemical processes, plus an equally large amount of water extracted with oil and gas, it makes sense to consider these as potential sources of fresh water after appropriate treatment to bring them up to certain standards required for specific applications.

3.1.1. Industrial wastewater

The nature of impurities present in industrial wastewater is diverse making the treatment process quite complicated and specific, compared to the classic treatment processes encountered in municipal wastewater. Characterization of industrial wastewaters from a variety of hydrocarbon processing indicates the presence of salts, heavy metals, hydrocarbons that include soluble aromatics, chemical additives like corrosion inhibitors, foam inhibitors, emulsion breakers, etc… This "cocktail" of impurities makes it practically impossible to use traditional activated sludge processes, but rather a series of specialist physical, chemical and biological steps. Often, treated industrial wastewater can be either reused in the plant, used in landscape irrigation or simply discharged into the sea. The latter option is gradually being phased as the "zero liquid discharge" regulation in Qatar is being introduced, albeit in phases to give industries the chance to adjust their treatment processes.

3.1.2. Municipal wastewater

Municipal wastewater treatment processes are quite well established and most of the challenges arise as a result of increased load due to increased population, unexpected malfunctions that lead to off-spec water that needs to be re-treated or very heavy storms with rainfalls that overload the sewage plant. Currently treated sewage water can be used for landscape irrigation.

3.2. water treatment horizons: industry- academia partnerships

In order to make the "zero liquid discharge" a reality and reduce the demand for fresh water from costly desalination plants, innovative and breakthrough research work in water treatment needs to be done. Partnerships between industry that is well known for its applied experience and academia that is respected for its innovation would provide a major boost in water treatment research. Such joint research programs may cover work from lab scale to pilot plant scale and possibly up to industrial scale applications involving undergraduate and postgraduate students where suitable.

4. Energy efficiency: so much more can be done

Energy demand has been steadily growing in the past decade, mainly due to emerging economies in the Middle East and Far East. This remarkable trend has had its impact on energy costs that have seen rising sharply. The world economies still heavily rely on fossil fuels for thermal energy despite recent policies around the world to promote renewable energy sources. In the medium term and until nuclear and renewable energy sources reach a level that can offset the reduction in fossil fuel usage, common sense clearly point at energy efficiency gains. Because the cost of energy is variable around the world, process economics are rather distorted and inconsistent, especially for operating costs. The best way to rationalize energy consumption in large hydrocarbon processing industries, energy saving and recovery measures must be taken seriously. These obviously have a long term benefit that is not immediately measurable in financial terms but certainly measurable in environmental terms. When fully integrated in a process, thermal energy recovery from flue gases (boilers, furnaces, gas turbines) can have a very significant impact on operating costs and green house gas emissions and in the long term offset the capital cost invested in such schemes.

There are great opportunities for partnerships between academia and industries in this area. The combination of industrial experience and academic innovation will lead to successful stories on global energy efficiency gains that will positively impact

5. Safety and risk assessment: more important than ever

The economy of scale principle is no longer just a statement but a reality on a massive scale: the recent mega LNG trains and GTL plant in Qatar have demonstrated the viability of such industrial giants. However to ensure safe and reliable operations, a number of measures have to be adopted. First comes the degree of safety integrity.

The Safety Integrity Level (SIL) can be defined as a relative level of risk-reduction provided by a safety function, or to specify a target level of risk reduction. In other words, SIL is a measurement of performance required for a Safety Instrumented Function (SIF). There are several ways to determine SIL for a SIF that include: Risk Matrix, Risk Graph and Layers of Protection Analysis (LOPA). This is a dynamic area of safety research very well suited to collaboration between industry and academia.

The important field of risk assessment and decision analysis is also gathering momentum despite being at the forefront of safety research for over a decade now. This is important for cases of decision making under a great degree of uncertainty. This is another potential area of collaboration between industry and academia on risk assessment in recent state of the art mega scale gas and petrochemical plants where accident data is scarce or inexistent and yet a whole new safety philosophy needs to be fostered. In addition to research collaboration between academia and industry, extensive training to engineers, faculty members and students needs to be put in place.

6. Reliability: achievable or elusive targets?

The term reliability and reliability engineering are often quoted for the ability of a system or component to perform its required functions under specified conditions for a specified period of time. It is usually reported as a probability.

In complex environments such as LNG mega plants, state of the art GTL plants and other petrochemical installations, reliability will have multiple facets:

Reliability of operations, reliability of strategic equipment like compressors, turbo expanders, gas turbines, furnaces, etc...Clearly there is an element of operational excellence, engineering management and total quality management. Indeed Deming's 85/15 rule clearly states that in problem areas 85% of cases are directly related to systems, organizations and only 15% were found to be related to people and individuals' skills.

Therefore, reliability engineering is an important area of collaboration between industry with its vast experience not recorded anywhere in the public domain and academic creativity free from any constraints. It is the combination of both that would lead to a better approach to reliability and total quality management.

7. Conclusions

The second annual gas processing symposium was very successful in bringing together a large number of people from industry and academia in five discussion forums to address, debate and discuss themes related to environmental sustainability: carbon capture and storage, water, energy efficiency, safety and risk assessment and reliability. These key areas were found to be unanimously critical to sustainable gas operations. The solutions to the challenges posed in these critical areas are best sought by teams made up of academics and people from industry due to the synergies. The combination of academic creativity and industrial experience is very potent in making major breakthroughs. A number of potential projects was identified in this chapter and may constitute the first batch of future collaborations between industry and academia to help sustain gas operations on a massive scale whilst caring about the living environment for future generations.

8. Acknowledgements

The authors wish to express their gratitude to the discussion forums session chairs who collected ideas and thoughts throughout the round table discussions and shared them with the audience. Some of these thoughts inspired this chapter.

Index

absorption 31, 321
acid gases 139, 177, 221
activated carbons 187
ADI/ADE algorithm 99
adsorbed natural gas 187
adsorption 187
adsorption isotherm 187
ambient temperature 455
amine 1, 11, 229
amine solvent scrubbing 11
amine wastes 1
amino acid salt 339
Anaerobic digestion 77

Biocide 99
biodegradation 1, 77
bio-energy 69
bio-ethanol 69
biogasification 1

capture 31
carbon capture and storage (CCS) 11, 465
Carbon Dioxide (CO_2) absorption 339
Carbon Dioxide (CO_2) capture 321
Carbon Dioxide (CO_2) Capture and Storage 163
Carbon Dioxide 21, 31, 115, 311, 321, 339
catalyst 145
Cesar 1 31
CFD modeling 349
Chromium 131
climate change 87
cogeneration 409
Complex Fluids 419
compression 177
Computational fluid dynamics 349
consition monitoring 209
Construction 59
controlled variable selection 331
Controllability 389
Cubic-Plus Association 419

dehydration 177
Density 399
Desalinated Water 107
desorption 339
diffusion 299
Domestic 59
Dual Firing 41
dynamic simulation 437

Electricity Demand Modeling 257
Electricity Generation Modeling 257
Electro-catalyst 115
Electrochemical reduction 115
Emissions 51
energy 87
energy efficiency 465
Energy Optimization 163
environmental impacts 69
environmental sustainability 87
Energy analysis 321

Finite Difference 99
Finite Volume Method 99
Fischer-Tropsch synthesis 197, 247, 369
flame length 237
flamelets 237
flare header 359
flares 237
Fugitives 51

gas absorption 229
gas hydrates 445
Gas Processing 121, 163
Gas to liquid process (GTL) 275, 287
Gas to liquid process (GTL) technology 369
gasification 247
greenhouse gas emissions (GHG) 11

Hazardous pollutants 154
Heat Integration 389
heat of adsorption 187
heat/power integration 247
hybrid biomass/coal 247
hybrid models 427
Hydrocarbon Product Distribution 197
Hydrocarbons 115
hydrogen sulfide 229
HySWEET process 221

ice-structuring proteins 445
identification 379
Industrial Usage of Water 107
Infrared 209
inhibition 445
injection 177
integral equation 379
Isochoric 399

Kinetics 21

LCA 389
Leak detection 153
Liquefied natural gas (LNG) 11, 87, 163, 311, 349, 437, 455, 437
liquefaction 311
LNG storage tank 437
low-dosage hydrate 445

magnesia 145
Mass and Energy Indicators 267
membranes 1 39
mercaptan 222
methane 145
3-(methylamino) propylamine 339
micro-aeration 77
modeling 77
molecular dynamics 299
Monitoring 51, 154
monoethanolamine 1, 77
Multi-component mixtures 399
municipal solid waste 69

N,N-Diethylethanolamine 21
natural gas 139, 349, 399
Navier-Stokes Equation 99
nonlinear finite element analysis 359
NOx emissions 41

Operating fuel 409
operating window 311
operation 311
optimal operation 331
Optimization 107, 275, 359

parabolic transport PDE 99
permeability 139
Phase Equilibria 419
PHEVs 257
pilot plant 339
pipe support 359
Piperazine 21
plant capacity 455
Plantwide control 275, 331
PLS 427
polysulfide 131
post combustion capture (PCC) 11
Process Control 275, 331, 369
Process Design 267
process integration 287, 369
Process Optimization 369, 389
process safety 465

Q-Chem 209

radiation 237
R-curve 409
reactor design 369
reforming 145
reliability 465
Reuse 59
risk assessment 465
Road Construction 121

Safety Indices 267
Sarcosine 339
SEA 121
selective removal 229
selectivity 299

Self-Optimizing control 275
separation equipment 427
Shell Thiopave 121
Silicon Polymers 299
Software 267
solubility 299
soot 237
SPE 115
Spills 349
SpyGlass 209
stabilization 131
step response 379
stresses analysis 359
Sulphur 121, 131, 229
supercritical solvents 197
Sustainability 87, 121
Sustainability metrics 267
Switchgears 209

techno-economic analysis 287
temperature variation 455
Thermography 209
thiodiglycol 222
total annual cost 409
total site 409
Treatment 59

unloading operation 437
unstructured and curvlinear grids 99

vapor-liquid equilibrium 31
View Ports 209
Viscosity of Natural Gas 399
VOC 154

Waste management 69
Wastewater 59
water 465
Water management 107

zirconia 145

Lightning Source UK Ltd.
Milton Keynes UK
171562UK00007B/8/P

9 780444 535887